Prometheus Tamed

Library of Economic History

General Editors

Peer Vries (*International Institute of Social History, Amsterdam*)
Jeroen Touwen (*Leiden University*)

VOLUME 13

The titles published in this series are listed at *brill.com/lehi*

Prometheus Tamed

Fire, Security, and Modernities, 1400 to 1900

By

Cornel Zwierlein

BRILL

LEIDEN | BOSTON

Cover illustrations: (*left*) Schwäbisch Hall, the major city fire of 1728 destroyed large part of the city center; (*right*) Engraving of Schwäbisch Hall's *Ground Zero* after the city fire of 1728 by the architect I.P. Meyer (Stadtarchiv Schwäbisch Hall, Grafiksammlung S10/2340).

Library of Congress Cataloging-in-Publication Data

Names: Zwierlein, Cornel, author.
Title: Prometheus tamed : fire, security, and modernities, 1400 to 1900 / Cornel Zwierlein.
Description: Leiden ; Boston : Brill, 2021. | Series: Library of economic history, 1877–3206 ; vol. 13 | Includes index.
Identifiers: LCCN 2020048752 (print) | LCCN 2020048753 (ebook) | ISBN 9789004424128 (hardback) | ISBN 9789004431225 (ebook)
Subjects: LCSH: Fire ecology. | Wildfires–Environmental aspects.
Classification: LCC QH545.F5 Z85 2021 (print) | LCC QH545.F5 (ebook) | DDC 577.2/4–dc23
LC record available at https://lccn.loc.gov/2020048752
LC ebook record available at https://lccn.loc.gov/2020048753

Typeface for the Latin, Greek, and Cyrillic scripts: "Brill". See and download: brill.com/brill-typeface.

ISSN 1877-3206
ISBN 978-90-04-42412-8 (hardback)
ISBN 978-90-04-43122-5 (e-book)

This book is printed on acid-free paper and produced in a sustainable manner.

Printed by Printforce, the Netherlands

Contents

Acknowledgements

A simple English translation of what was a German second book and Habilitation thesis (*Der gezähmte Prometheus. Feuer und Sicherheit zwischen Früher Neuzeit und Moderne*, published with Vandenhoeck & Ruprecht, Göttingen) would not have served the English-speaking reader nor the research community as such. The generous opportunity offered by the editors Peer Vries and Jeroen Touwen of the *Library of Economic History* and the favorable and very helpful reviews of the two anonymous reviewers to publish this new English version confirmed that it was a worthwile enterprise to change, adapt, enlarge, and partially cut the text at some places, and to update it as far as I could concerning the bibiography. The German original was based—next to the archival sources—on some 700 titles in the bibliography. The number of about 280 titles read, added, and integrated in this version that have appeared between 2012 and 2020 might serve just as indicator how quickly the field is developing and how many important contributions are published each year—though I am apologizing already at this point for surely having overseen several that should have found their way into the footnotes. This ratio can also serve as an indicator for the amount of text I added and changed: The whole might hopefully serve as an up-to-date reminder of past, a prism of current, and perhaps a catalyst of future research on the history of (urban fire) disasters, not only in Central Europe, on the history of insurance and risk and the methodological questions linked with it, embedded in the framework of international comparative and global history. Great thanks go to Brill, Wendel Scholma, Arjan van Dijk, and Gera van Bedaf for carefully shepherding this book production.

Figures and Tables

Figures

Tables

CHAPTER 1

Introduction

Since at least the Enlightenment, if not continuously since the time of Aeschylus, Prometheus has symbolically represented the technical rationality of emancipated man.[1] In recent decades, while many historians have used the figure of "Prometheus unbound" as a metaphor for the development of industrialization and mechanization in the eighteenth century in a positive sense (David Landes),[2] they have also employed the mythical fire bringer to invoke the potential negative effects of such developments on the environment (Hans Jonas).[3] While Prometheus appears unbound in these works, Stephen Pyne, the historian of fire, has employed the figure of "Prometheus Bound" as the primary symbol for the European history of firefighting.[4] As such, this bringer

---

1 Cf. for example, the entry in the eighteenth century German standard encyclopedic work on ancient mythology as used by Goethe, "the Hederich": "He [i.e. Prometheus] was a good confidant of Jupiter, because governors definitely need provision. He stole the fire, as rationality had discovered its usefulness; he is bound to a mountain and is eaten into by the eagle, because humans have started to do many things that, in their effect and outcome, have caused them the greatest sorrow and problems: he has invented many arts, because reasoning and provision [*fore-thinking*] is also necessary for this [*Er war aber mit dem Jupiter erst gar vertraut, weil Regenten der Vorsichtigkeit nöthig haben. Er hat das Feuer entführet, weil der Verstand dessen Nutzen erfunden; er ist an den Berg angeschmiedet, und durch den Adler angefressen worden, weil die Menschen durch ihren Verstand sich vieler Dinge unterfangen, welche ihnen hernach die größte Sorge machen:* er hat viele Künste erfunden, weil auch dazu der Verstand und die προμέθεια nöthig sind]" (Benjamin Hederich, *Gründliches mythologisches Lexicon, worinnen so wohl die fabelhafte, als wahrscheinliche und eigentliche Geschichte der alten römischen, griechischen und ägyptischen Götter und Göttinnen* [...] *zusammen getragen* [...] [1770], reprint Darmstadt 1996, col. 2098).

2 David Landes, *The Unbound Prometheus*, 2nd ed., Cambridge 2005; Peter Lyth and Helmuth Trischler (eds.), *Wiring Prometheus: Globalisation, History and Technology*, Aarhus 2004; Thomas P. Hughes, *Rescuing Prometheus*, New York 2000; Timothy V. Kaufman-Osborn, *Creatures of Prometheus: Gender and the Politics of Technology*, Lanham 1997.

3 "Prometheus, definitely unbound, whom science gives unknown opportunities, uninterruptedly boosted by the economy, is crying for an ethic which will prevent him from becoming an evil to man by deliberately bridling his power [*Der endgültig entfesselte Prometheus, dem die Wissenschaft nie gekannte Kräfte und die Wirtschaft den rastlosen Antrieb gibt, ruft nach einer Ethik, die durch freiwillige Zügel seine Macht davor zurückhält, dem Menschen zum Unheil zu werden*]" (Hans Jonas, *Das Prinzip Verantwortung. Versuch einer Ethik für die technologische Zivilisation*, Frankfurt am Main 1979, p. 7; *The imperative of responsibility: in search of an ethics for the technological age*, transl. by Hans Jonas and David Herr, Chicago 1984).

4 Stephen Pyne, *Fire in America: A Cultural History of Wildland and Rural Fire*, Seattle 1997, pp. 126–42, 137f.

 | DOI:10.1163/9789004431225_002

of fire and breaker of norms represents not only the emergence of modernity in its dynamism, but also the newly created dangers and risks of modernity that are not necessarily controllable in traditional ways. When the "taming of Prometheus" is referred to here, it is in the sense of this allegorical Prometheus unbound; however, the mythological image also questions the extent to which Prometheus's unleashing corresponds to the development of new modes of security production associated with containment and restriction. This does not imply a deflection of the consequences of the industrial revolution, but rather the prospective construction of the safety conditions that first created a framework allowing for the possibility of such a development in the first place: in particular, the development of insurance against (natural) disasters, primarily regarding the threat of large city fires in premodern Central Europe.

The taming of the Promethean effects of fire in a narrative covering development from premodern to modern times might immediately suggest a simple teleological path leading from the "unsafe" burning of premodern cities, villages and countrysides to a fire-safe world, however, as current news coverage indicates, the world is still, or again "on fire" today.

As a NASA satellite image from August 2019 of burning forests and plains from the Amazonas to Indonesia shows, the world is very much threatened by large conflagrations—more, perhaps, forests in free nature than larger cities, although fires often also endanger large megacities nearby, like Buenos Aires, or those on the American West Coast (Fig. 1). This is a more and more frequent phenomenon nowadays, commonly explained as one of the effects of climate change, which is not comparable to the premodern situation. Journalists are

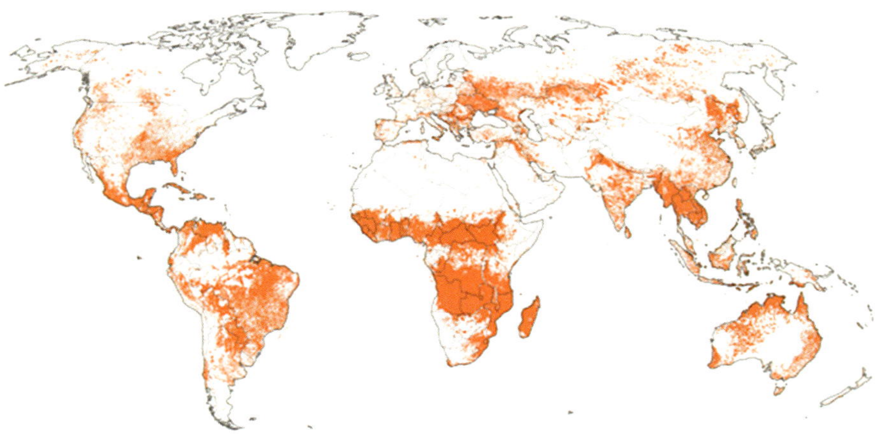

FIGURE 1 NASA Satellite picture, from Pierre-Louis Kendra: The Amazon, Siberia, Indonesia: A World of Fire, published in New York Times August 28, 2019 HTTPS://NYTI.MS/2PKKK8K, ACCESSED JULY 30, 2020

using already the term "fire season" for what seems to have become a phenomenon that occurs now every year at the West Coast.[5] Similarly, the burning roof of Notre Dame cathedral in Paris in 2019—even more than the partial burning of the cathedral of Nantes caused by arson in July 2020—dramatically recalled a situation that was common in medieval and early modern times when large towers and churches were very easily exposed to that threat.[6] The helplessness of the modern firefighters showed the astonished global public how difficult it can still be to *tame* the threat of fire or even to insure for it (taming the financial impact and losses): we are, still, not living in a fire-safe world. Natural fires and those in and of large cities, on the macro- and the microscale have occurred in completely different periods, cultural settings and climatic conditions—and in that way they remind us immediately that this book cannot follow a simple teleological narrative leading from unsafe to safe times. In contrast, I try to investigate the multiple, complex, and eventually even cyclical relationships between timescapes (premodern times and modernit*ies*), the perception and construction of urban space, and the different forms of fire safety provision as they developed in Europe, and were later to be found across the world. The realities of changing fire regimes, and the frequency and conjunctures of fires have likewise to be reconstructed, along with the environmental context of this complex relationship between security regimes in premodern and modern times.

In the broad research on the concept and history of "security," a wealth of specific, "modern" means of security production can be discerned in the areas of politics (internal, external, international security), religion, culture, and economics. This book will investigate the genesis, the transfers, and the dissemination of the insurance principle—particularly in terms of fire insurance—up to the end of the nineteenth century. The principle of premium-based insurance is omnipresent today; currently the most important sectors affecting the everyday lives of individuals would probably be health insurance, accident and liability insurance, and unemployment insurance—thus, primarily private/state hybrids in social insurance sectors. In their present form, these are creations of the nineteenth century that owe their existence to the emergence of the class society and the conflicts and disputes associated with a series of workplace accidents—which meant that social historians of the 1980s were particularly attracted to studying these types of insurance. The German reforms under Bismarck are well known, and parallel developments in France—and later and in different ways in the US—have been the subject of in-depth independent and

5 Cf. Blacki Migliozzi et al.: Record Wildfires on the West Coast Are Capping a Disastrous Decade, New York Times, September 24, 2020, https://nyti.ms/3kFOFCY, accessed October 24, 2020. For a comment cf. Simon Dalby, "Firepower: Geopolitical Cultures in the Anthropocene", *Geopolitics* 23, 3 (2018), pp. 718–742.

6 Cf. below chapter 3, 2.1, specially p. 115ff.

comparative studies.[7] Until around 1900, however, in terms of both genesis and volume, fire insurance constituted the most important sector of private as well as state insurance.

Since about 1994, when the UN officially adopted the concept of "human security"—which in 2003 was elevated to a guiding principle—we once again live in an era in which natural disasters and societal threats are of equal significance to the concept of "security." The subdisciplines of history have responded to this shift, revitalizing the topic of "insurance" in this context: while the history of natural disasters is one of the newest fields to emerge since the 1990s in terms of environmental and early modern history,[8] it has only recently recognized that insurance is the primary institution consistently engaged in

7 Cf. E. Peter Hennock, *The Origin of the Welfare State in England and Germany, 1850–1914*, Cambridge 2007; Florian Tennstedt, "Risikoabsicherung und Solidarität. Bismarck, Lohmann und die Konflikte um die gesetzliche Krankenversicherung in ihrer Entstehungsphase," in Herbert Obinger and Elmar Rieger (eds.), *Wohlfahrtsstaatlichkeit in entwickelten Demokratien. Herausforderungen, Reformen und Perspektiven*, Frankfurt am Main 2009, pp. 65–94; Marcel van der Linden, *Social Security Mutualism: The Comparative History of Mutual Benefit Societies*, Bern 1996; E. Peter Hennock, "Social Policy in the Bismarck Era: A Progress Report," *German History* 21 (2003), pp. 229–38; Ulrich Becker, Hans-Günther Hockerts, and Klaus Tenfelde (eds.), *Sozialstaat Deutschland. Geschichte und Gegenwart*, Bonn 2010; Jan Rüger, "Revisiting the Anglo-German Antagonism," *Journal of Modern History* 83, 3 (2011), pp. 579–617, 582, 598f; Steven King, "Welfare Regimes and Welfare Regions in Britain and Europe, c. 1750s to 1860s," *Journal of Modern European History* 9, 1 (2011), pp. 42–65.

8 The annual conferences of the American Historical Association have often had a whole series of panels on natural disaster history in recent years; while leading monographs will be cited throughout the text, the field of research might best be represented by an (uncomplete) list of collective volumes: Bartolomé Bennassar (ed.), *Les catastrophes naturelles dans l'Europe médiévale et moderne*, Toulouse 1996; René Favier (ed.), *Les pouvoirs publics face aux risques naturels dans l'histoire*, Grenoble 2002; Michael Kempe and Christian Rohr (eds.), *Coping with the Unexpected. Natural Disasters and their Perception*, Strond 2003; Monika Gisler, Katja Hürlimann, and Agnes Nienhaus (eds.), *Naturkatastrophen/Catastrophes naturelles*, Zürich 2003; Gerrit Jasper Schenk and Jens Ivo Engels (eds.), *Historical Disaster Research: Concepts, Methods and Case Studies*, Cologne 2007; René Favier, Claudine Remacle (eds.), *Gestion sociale des risques naturels/Gestione sociale dei rischi naturali*, Valle d'Aosta 2007; René Favier and Anne-Marie Granet-Abisset (eds.), *Histoire et mémoire des risques naturels*, Grenoble 2000; Anne-Marie Mercier-Faivre and Chantal Thomas (eds.), *L'invention de la catastrophe au XVIIIe siècle. Du châtiment divin au désastre naturel*, Geneva 2008; René Favier and Christian Pfister (eds.), *Solidarité et assurance. Les sociétés européennes face aux catastrophes (17e -21e siècles)*, Grenoble 2008; François Walther, *Katastrophen. Eine Kulturgeschichte vom 16. bis ins 21. Jahrhundert*, Stuttgart 2010; Anthony Oliver-Smith, Susanna M. Hoffmann (eds.), *Catastrophe and Culture*, Sante Fe 2002; Gerrit Jasper Schenk (ed.), *Historical Disaster Experience*, Heidelberg 2017; "Natural Disasters, Big Fires and epidemics in the Cities", special issue *Historia urbana* [Editura Academiei Române] 26 (2018).

the serial observation of natural disasters.[9] While there was a long development of contractual and institutional practices in the span of time between the origins of premium-based insurance and the current global assessment of disasters, the recent extreme increase in significant natural and technological disasters once again emphasizes the importance of the early history of insurance, since environmental threats—and the risk of fire in particular—were, of course, its primary focus. State involvement with the insurance principle, in terms of welfare state protections offered to the sick and poor is, in contrast, a downstream development that only intensified in the nineteenth and twentieth centuries.

Fire exists at the fulcrum point between nature and culture and represented, in premodern terms, a threat similar, perhaps, to current anthropogenic technological disasters. In both cases, the starting point, as well as the degree of controllability and predictability of consequences, can no longer be defined along a strict nature/culture boundary line. While large city fires in medieval and early modern cities were mostly caused by the carelessness of human beings, it does not follow that such fires were simply classified as cultural phenomena. Rather, they were treated in all contexts in much the same way as other natural catastrophes, whether in the religious sphere as divine punishment, or within the administrative discourse of technical cameralism. Fire danger was an everyday phenomenon, a quotidian threat whose status and significance is hardly imaginable anymore, disappearing from view in an era beyond fires fueled by wood and coal. A brief examination of the themes of Enlightenment communication, it is clear that the discussion of fire funds and fire extinguishers occupied far more intellectual energy than topics such as freedom, civility, and public opinion, which are typically associated with Enlightenment thought.[10] The fact that premodern cities around the world were exposed to large, widespread fires and that they no longer exist in modern times is part of the self-evident, but largely unconscious knowledge of our history. It remains, after decades or even centuries of professional urban history, still a somehow neglected field of research, at least as regards a systematic analysis, and an

9 As was already the argument in the early German version of this book. Cf. after that the special issue "Uncertain Environments" of *Environment and History* 17, no. 1 (2011) as well as Eleonora Rohland, *Hurricanes in New Orleans, 1718–1965: A History of Adaptation*, PhD thesis, University of Bochum, 2014, pp. 263–314.

10 To verify this, one can compare key words like "conflagration" (*Brand*), "fire" (*Feuer*), etc., in the *Index deutschsprachiger Zeitschriften 1750–1815*, ed. Klaus Schmidt, Hildesheim 1997: the proportion of topics are immediately clear in this older printed version, however cf. now http://www.ub.uni-bielefeld.de/diglib/aufklaerung/.

investigation touching on a wider setting and the cities of a large region or even countries in their collectivity beyond the—nevertheless important!—local view of one given city, as David Garrioch has recently reminded us.[11] In countless city guides and architectural histories, the "last major fire" of a city is mentioned as a definitive event in the history of the city, and there is often even a separate section in the relevant city museum dedicated to this event—be it the fires of London (1666), Hamburg (1842), Istanbul (1660, 1865, 1870, and 1918), Tokyo (1656), Chicago (1871), San Francisco (1906), or even fires in small towns like Heimsheim/Württemberg (1634). The idea that relative safety from major fires is a definitive criterion of "modernity" emerges primarily from the basic assumption, which is later problematized in this book, that there are fixed defining criteria of "modernity." The fact that this particular premodern quotidian threat has disappeared from our world of experience, at least in the so-called "Western" world, leads to the fundamental questions explored in the present work: namely, the relationship between fire danger and security in terms of epochal differences, and how spatial and temporal concepts were affected by the changes in perceptions and expectations regarding this danger. This does not primarily concern the institutional history of fire insurance, but rather the associated changes in the framework of thought and the differences and continuities that can be discerned from the late Middle Ages to the nineteenth century, but which center around the epochal threshold period of 1680–1700. At this point in time, the security one now expects in modern cities was named and conceived of as "doubled": on the one hand, people expected to be (relatively) safe from fire—that fires should not happen; on the other hand, they could expect that insurance against pecuniary losses from fire would be available in the case of a fire. In 1680, Leibniz distinguished between these two forms of security—preventative measures against material loss versus insurance-like forms of value preservation—as "real-assecuration" and "verbal-assecuration" respectively.[12]

The formulation of such a double notion of security as the abstract goal of collective action is not self-evident historically, as confirmed by the older history of the concept.[13] While security discourses around security production

11 David Garrioch, "Towards a fire history of European cities (late Middle Ages to late nineteenth century)," *Urban History* 46, 2 (2019), pp. 202–224.

12 Gottfried Wilhelm Leibniz, *Sämtliche Schriften und Briefe*, Series 4, Politische Schriften, vol. 3 (*1677–1689*), ed. Lotte Knabe and Margot Faak, Berlin 1986, pp. 421–432, 428. For more on this text see chap. 4, 1.2.2 below.

13 Werner Conze, "Sicherheit, Schutz," in *Geschichtliche Grundbegriffe. Historisches Lexikon zur politisch-sozialen Sprache in Deutschland*, vol. 5, eds. Otto Brunner, Werner Conze, Reinhart Koselleck, Stuttgart 1984, pp. 831–62; Franz-Xaver Kaufmann, *Sicherheit als*

and the relationship between "internal" and "external" national, state, and social security has long been a topic in the field of history from the viewpoint of military and (narrowly defined) social/state phenomena, the current emergence of the history of security as a distinct field has opened up the topic and broadened the dimensions of security under investigation. This affects the field of history as well as historiography. Since the concept of security is becoming broader and more multifaceted in terms of political and institutional language, in the fields of political science, and "security studies", sensitivity to the diversity and significance of "security" and security production in the past has also significantly increased.[14]

The main hypothesis of this study concerns the centrality of "security" for settings or constellations of modernity, with the main question investigating how such security is produced, and with what consequences for divergent areas of society. Fire danger and fire safety operate as a strategically chosen example. It is thus useful to employ a very broad notion of "security," one which allows for historicizing the mechanisms of past perceptions of security deficits and security production in a sufficiently differentiated way.

The history of fire safety is necessarily linked to the history of cities: as a result, this study also focuses on the history of German cities, European cities, and, in the last chapter, on globally important port cities. This connection is also a consequence of the work of William Cronon and Martin Melosi, who have successfully integrated urban history into environmental history.[15] The historical analysis of city fires in this context began some time ago in environmental histories that were not limited to or fixated on "green" nature:[16] "Cities,

---

soziologisches und sozialpolitisches Problem. Untersuchung zu einer Wertidee hochdifferenzierter Gesellschaften, 2nd ed., Stuttgart 1973.

14 Cf. Cornel Zwierlein, Rüdiger Graf, and Magnus Ressel (eds.), *The Production of Human Security in Premodern and Contemporary History*, Cologne 2010; Cornel Zwierlein, "Storia della sicurezza in Europa/Storia europa della sicurezza", *De Europa* 3, 1 (2020), pp. 19–35 (DOI: http://dx.doi.org/10.13135/2611-853X/4435).

15 Martin V. Melosi, "The place of the city in environmental history," *Environmental History Review* 17 (1993), pp. 1–23, 2; William Cronon, *Nature's Metropolis: Chicago and the Great West*, New York 1990. Cf. Dominik Collet, "Risk and Resilience", in *Concepts of Urban-Environmental History*, eds. Sebastian Haumann, Martin Knoll, Detlev Mares, Bielefeld 2020, pp. 79–94.

16 On this see Christine Meisner Rosen, *The Limits of Power: Great Fires and the Process of City Growth in America*, Cambridge 1986, pp. 92–176; Ross Miller, *American Apocalypse: The Great Fire and the Myth of Chicago*, Chicago 1990; Karen Sawislak, *Smoldering City: Chicagoans and the Great Fire, 1871–1874*, Chicago 1995; Fionie Koppenleitner, Hole Rößler, and Michael Thimann (eds.), *Urbs incensa. Ästhetische Transformationen der brennenden Stadt in der Frühen Neuzeit*, Munich 2011; Greg Bankoff, Uwe Lübken, and Jordan

too, had their fire regimes. ... Urban conflagration is a remorseless saga of Europe, as endemic to its combustible cities as rats and disease."[17] Johan Goudsblom's remark that there is "*still no complete* overview of urban conflagrations in Europe"[18] suggests that such an endeavor, which a legion of historians could realistically never accomplish, is possible; chapter 3, 1, however, does cover the German and, to some extent, central European "fire ecology" on the basis of the decades-long extensive and continuous urban history research on specific areas—for this region, a study like this is already well prepared by decades of serial urban historical research in comparison to other world regions on the international scale. Beyond a universalized history of fire, there is an abundance of individual studies on city fires in local and regional historiography; however, while such studies are mentioned to some extent in the course of this work, they typically lack the necessary comparative overview and large interrogative framework. Over long periods of European history, singling out cities, especially when small and very small cities are taken into account, has been a reliable method of producing a representative fire ecology for the respective regions and territories; territorial fire funds were first based on networks of cities before any integration of small villages was even considered. Only large fires in major metropolises—which are discussed primarily in the last chapter—are significantly detached from this model.

Traditionally, urban European history generates its own epochal boundaries based on the rhythm of urbanization—modes of city founding (by emperors, bishops, by aristocratic rulers; free cities; refugee cities), and city types (free and imperial cities, residential, commercial, port cities), categorized into eras of the medieval city, early modern city, the era of the metropolis, and the provincialization of smaller cities during the nineteenth century. Usually the perspective of the history of fire (danger) and the history of security are at odds with such an epochal progression and typification, not only due to the fact that for a long time cities of similar size burned at similar rates, regardless of which constitutional type they belonged to. Asking for the relationship between security regimes, danger and threat frameworks, and temporality leads to the consideration of different kinds of epochal structures.

Sand (eds.), *Flammable Cities: Urban Conflagration and the Making of the Modern World*, Madison 2012.

17 For example, Stephen Pyne, *Vestal Fire: An Environmental History, Told through Fire, of Europe and Europe's Encounter with the World*, Seattle 1997, pp. 50–59, 226–28, 287–91 (here 50f.).

18 Johan Goudsblom, *Vuur en beschaving*, Amsterdam 1992, i.e. *Fire and civilization*, Harmondsworth 1992 [qtd. German transl. 1995], pp. 183–95, 184.

If the transition to a state of fire safety is indicative of the entry into "modernity"—as suggested by the first, albeit deceptive, approach to average urban European history—then this would seem to confirm the character of the modern/premodern epochal dichotomy in particularly striking clarity. Additionally, the simplicity of the distinction between "fire safe/not fire safe" would suggest that there was only *one*, albeit universal, notion of modernity in this regard: what good would it do to speak of *multiple modernities*? There may be small differences concerning the various ways that fire-resistant modes of construction could be achieved globally, but on the whole, urban fire resistance seems to be an area where all postcolonial sensibility and ethics fail, when confronted with a historical narrative that inevitably compels a developmental trajectory from a premodern to a modern state. Similarly, for the second form of security—the existence or nonexistence of insurance—the extent of insurance coverage can function as a sort of litmus test for the modernity of a given society. From this somewhat rigid perspective on material safety from huge urban conflagrations, the existence of (always relative and always conditional) fire resistance and a wide distribution of insurance would thus be part of the modernization process in the classical sense.

However, in the current historiographical situation, referring to such developmental stages is often avoided entirely, concentrating instead, for example, on an ethnologically inspired description of social situations and thus avoiding the danger of operationalizing historical changes and conceptualizations that may be based on Eurocentric viewpoints and perceptions. While this is certainly a legitimate approach to history, it provides no argument against a renewed attempt to conceive of epochal changes, thresholds, and important transformations that take the new challenges associated with the relationship between global dimensions and "old" European forms of periodization into account. Effectively, the retreat into a detemporalized historical anthropology risks being reproached as a form of escapism and evasion: the problem is avoided but not faced directly.

Since the 1980s and 1990s, the various theoretical modes of modernization have been subject to an enormous amount of criticism. While social and political science approaches to "modernization" in the 1950s/60s were initially primarily models of active policy recommendations, even and especially concerning developmental policy, in the 1970s such approaches were dismantled from their normative aspects, reshaped as a heuristics of epochal transition and received a great deal of attention in historical sciences research. While on the one hand, such approaches were concerned with identifying goals that would help to politically plan and effect the modernization of a society, on the other hand they served as an instrument of historical analysis. A variety of examples were used to illustrate how a modernization process—both intentionally but also functionally motivated—with its individual aspects (rationalization, social discipline, civilization process, secularization, etc.) had developed around 1800 in a variety of social spheres, leading to a fundamental epochal transition. As early as the 1970s, importers of modernization theory into historiography were aware that the concept often involved a "unilinear logic of development"[19]—i.e., that it implicitly and explicitly emerged from a notion of "Western superiority." World history was split dichotomously into a "phase of 'traditional' [society] or that of 'modernity' " and history was reduced to the transformation from one state into the other.[20] The rapid expansion of research on global historical themes introduced much criticism of the conventional heuristics of modernization in the guise of postcolonial and postmodern critical theory, leading to new interest in other fields, and especially in the increasingly important fields of colonial—and imperial history—albeit with some delay for German-speaking areas.

When at the end of the present work, for example, the establishment of insurance agents outside of European regions, such as in Istanbul, is analyzed, many of the sources (local descriptions of agents, newspaper articles about the fire characteristics of the respective place) are imbued with a certain aspect of "orientalism."[21] The West defines itself as modern and fire-resistant specifically in the act of disparaging non-Western, "nonmodern" states, even though both "the West" and the Other (Asia) are products of a highly problematic process of producing and ascribing identities and attitudes on the other. For almost four decades, postcolonial criticism has gestured toward this kind of *construction* of the completely inferior, non-Western other, as an aspect of Western self-definition.[22] What had previously been discussed as the discourse and practice

19 For a recent contribution still retaking that criticism on notions of modernization rooted, in the end, in discussions between Marxist and Trotzkyist notions of linear or uneven development and proposing a "less deterministic and more relational conception of social change" cf. Fouad Makki, "Reframing development theory: the significance of the idea of uneven and combined development", *Theory and Society* 44, 5 (2015), pp. 471–497, here 492.

20 Cf. Chris Lorenz, *Wozu noch Theorie der Geschichte? Über das ambivalente Verhältnis zwischen Gesellschaftsgeschichte und Modernisierungstheorie*, Stuttgart 2004, p. 118.

21 Edward Said, *Orientalism*, London 2003.

22 Larry Wolff, *Inventing Eastern Europe: The Map of Civilization on the Mind of the Enlightenment*, Stanford 1994; Arturo Escobar, *Encountering Development: The Making and Unmaking of the Third World*, Princeton 1995; Dipesh Chakrabarty, *Provincializing Europe: Postcolonial Thought and Historical Difference*, Princeton 2000. For a current reappraisal of Chakrabarty's criticism of the "epistemic violence" of European colonial concepts of modernity/modernization cf. Saurabh Dube, *Subjects of Modernity: Time-Space,*

of modernization, or even perceived as an incomplete "stage of development" within Europe, has been rediscovered as a discursive product of "orientalism," even within the West: the dichotomy traditional/modern or civilized/uncivilized thus became translatable into a paradigm of orientalism and in a certain sense, the colonizing gaze could also be identified within the Western world (in the "orientalist" view on southern Italy by northerners, for instance).[23]

Rather than completely abandoning the concept of modernity/modernization, it became common to discuss the multiple forms of modernity responsible for shaping the world.[24] The concept of "multiple modernities" continues to suffer from a logical and conceptual indecisiveness: on the one hand, a specific set of modern elements, which can be found everywhere in different forms, is almost always presumed; on the other hand, the plurality and diversity of modernity is emphasized.[25] In terms of the latent but necessary achievement of "modernity" and the goal of being on an equal footing with Europe, the concept of "multiple modernities" changes nothing; rather, it simply divides the world even further into (multiple) modern and premodern societies. In a still more recent debate—which might be termed the post-post-colonial debate—a certain discomfort emerges from an unsuspected place concerning the effect and extent of postcolonial criticism, as speaking about and analyzing inequalities as realities becomes increasingly impossible: the cultural, linguistic, and postcolonial turn should have made it possible to carefully analyze the descriptions of Western travelers making contact with non-Western,

Disciplines, Margins, Manchester 2017 and a short Spanish rendering in *Estudios de Asia y África* 52, 3 (2017), pp. 493–534: the central theoretical problems remain how to address the relationships between timescapes, spatial factors and the effects of social hierarchies and center-periphery relationships in colonial and post-colonial settings. Only some paths leading out of the impasses of perhaps powerfully deconstructing but otherwise "improductive" categories (p. 518) seem promising, such as the admittance of the coexistence of a "plurality of worlds" (p. 521), close to Eisenstadt's approach.

23 Jane Schneider (ed.), *Italy's "Southern Question": Orientalism in One Country*, Oxford 1998.

24 Chakrabarty, *Provincializing Europe*, p. 235f.; Shmuel N. Eisenstadt, "Multiple Modernities," *Daedalus* 129 (2000), pp. 1–29; Sudipta Kaviraj, "Modernity and Politics in India," *Daedalus* 129 (2000), pp. 137–62; Peter Wagner, *Theorizing Modernity: Inescapability and Attainability in Social Theory*, London 2001, pp. 128–33. Some related discussion reviewed in Matthias Koenig, "Modernités multiples et mondialisation," *Archives de sciences sociales des religions* 59, 167 (2014), pp. 127–145.

25 Shmuel Eisenstadt, "The Axial Age: The Emergence of Transcendental Visions and the Rise of Clerics," *Comparative Civilizations and Multiple Modernities*, part 1, Leiden 2003, pp. 197–217. On Karl Jasper's concept of the axial civilization, see Karl Jaspers, *Vom Ursprung und Ziel der Geschichte*, Munich 1949. Cornel Zwierlein, "Frühe Neuzeit, *multiple modernities*, globale Sattelzeit," in *Frühe Neue Zeiten. Zeitwissen zwischen Reformation und Revolution*, ed. Achim Landwehr, Bielefeld 2012, pp. 389–405.

"uncivilized" areas, as *constructions* of inferiority. The verdict against developmental narratives leads to silence regarding real and tangible inequalities in the past and present—a situation, which, for those who aspired (or still aspire) to "modernity" or a higher standard of living, is regularly met with discomfort.[26]

This book thus addresses the problem that both the disrespectful attitude of the "Enlightened" administrators and economists against the "conservative-traditional" stance opposing the introduction of new building materials and insurance institutions in Europe, as well as the equally disrespectful analysis of Western observers regarding non-Western urban construction and preventative measures were part of a discourse of pure self-affirmation. However, the crude and simple reality that some cities burned and were uninsured, while others rarely burned and were insured against losses cannot be denied: there was *certainly* a process in Europe which led to a qualitatively different degree of security production in "modernity," and in "Western/Westernized" regions. Of course, the example of Asian cities in the nineteenth century demonstrates that the equivocation of fundamental fire safety in a city and "modernity," which Europeans themselves took for granted at the time, is absolutely false (Indian cities, for example, were quite fire-safe prior to assuming European building modes). In the process of globalization, the clarity of the developmental narrative stretching from traditional fire control and damage management to modern fire ecologies became obscured by intercontinental comparisons and experiences.[27] The book focuses first on the binome of fire threat/safety first as a part of the historical Enlightenment process. Then it takes into account the evaporation and concussion of (post-)enlightened horizons of expectation during processes of globalization, the book tries therefore to contribute to an ongoing debate on the relationship between the double-bind historical relative as well as systematic-universal concepts, social and institutional practices (e.g. of Enlightenment and its products and values) under the

26 Barbara Weinstein, "Developing Inequality," *American Historical Review* 113 (2008), pp. 1–18. There is much work currently being done on the history of the concepts and practices of development, developmental politics and related issues during colonial times, as precursors to current concepts of modernization and its historiographical theoretical offsshoots, cf. Joseph Hodge, Gerald Hödl, and Martina Kopf (eds.), *Developing Africa. Concepts and practices in twentieth century colonialism*, Manchester 2014, and Cyrus Schayegh, "The Interwar Germination of Development and Modernization Theory and Practice," *Geschichte und Gesellschaft* 41, 4 (2015), pp. 649–684 with recent literature in note 4.

27 For a comparison of the older classical narratives of modernity/modernization, see Danilo Martuccelli, *Sociologies de la modernité. L'itinéraire du XXe siècle*, Paris 1999.

"epistemic pressure" of de-europeanizing global frames of perception:[28] I try to think together the realities, the material and epistemic contexts of houses in very different areas: all houses burnt down in a little Central European city like Hechingen or Seelow in a given year in early modern times on the one hand and the fire threat to houses of Chinese and European inhabitants in a nineteenth century Treaty Port of Shanghai on the other; both are perhaps not part of *one* story but part of one historicizible plurality of the fire threat/safety binome. And as there were common instruments and features such as premium fire insurances at some point in history traceable in both contexts, there are undeniable relationships of historical entanglements between the English, the Central European and the global settings. Within the general methodological discussion about the concept of modernity, Peter Wagner revitalized the concept against a similar backdrop by highlighting its experiential dimension: modernity is not an objective state of the world that is measurable and can be retrieved and held constant; rather, it is a subjective way that people conceptualize their lives, the core content being the "commitment to autonomy". In relation to our subject matter, this would mean the collective as well as individual claim to a certain form of universalized security present in a given society, which first made certain degrees of autonomy possible.[29] This subjective dimension is also stressed when, in contemporary theory, the plurality of perspectives associated with modernity/traditionalism are characterized as the presence of different "horizons,"[30] or as a new copresence of elements of nonsimultaneity in a global, "decelerated" society.[31] This study will not address postmodern deceleration, but will show that the question of simultaneity for different, copresent forms of security claims, security regimes, and subjectivities is also a relevant question for earlier periods. As such, it is not a question of "modernity/ies" but of temporal horizons, their relationship to measures of security production, their copresence, their transformations across time and their spatial diffusion (and partial or complete nondiffusion).

---

28 Cf. a short discussion following Chakrabarty, Felicity A. Nussbaum and Gayatri Chakravorty Spivak by Iwan-Michelangolo D'Aprile, "Critical Global Studies and Planetary History: New Perspectives on the Enlightement," Roldán Concha, Daniel Brauer, Johannes Rohbeck (eds.), *Philosophy of Globalization*, Berlin 2018, pp. 345–356.

29 Peter Wagner, *Modernity as Experience and Interpretation: A New Sociology of Modernity*, Cambridge 2008.

30 Armin Nassehi, *Der soziologische Diskurs der Moderne*, Frankfurt am Main 2006, p. 436.

31 Hartmut Rosa, *Beschleunigung. Die Veränderung der Zeitstrukturen in der Moderne*, Frankfurt am Main 2005, pp. 428–90 and Rudolf Stichweh, "Ungleichzeitigkeit in der Weltgesellschaft. Zur Unterscheidung von Tradition und Moderne," in *Die Weltgesellschaft. Soziologische Analysen*, Frankfurt am Main 2000, pp. 207–19.

In this historiographical and theoretical context, the organization of questions related to a history of fire security and its "regimes" will involve, on the one hand, a reconstruction of the "reality" of fire frequencies, insurance institutions, and the spread of insurance; on the other hand, it will focus on the reciprocal, contradictory, and conflicting perceptions of the processes leading to the proliferation and transformation of security regimes. One could describe such an approach as a historical application and reenactment of the concept of "securitization": just as historians and commentators currently investigate how different sectors (economics, military, politics, the environment, etc.) are treated on different levels (local, regional, national, global) increasingly as objects of *security* politics and governance, one would question how, through an increasingly widespread implementation of the insurance principle, a security regime is constructed in the *longue durée* by including more and more distinct kinds of "accidents" and "hazards" into the field of safety production. As a whole, it would not follow the pattern of an unreflectively narrated process of modernization; rather, by focusing on the object, "security"/"security production," and by continuously switching between a description of "actually" existing and transient processes and developments on the one hand, and the analysis of the contemporary perceptions and demands made on them on the other, it becomes possible to effectively disrupt a simple, teleologically oriented relationship between security and modernity. A main thesis in this work will be that in the development of the "normal secure society" around 1700, the counterfactual normative expectations in relationship to "security" *preceded* the actual process of implementing specific solutions of security production, which essentially corresponds to the idea of "modernity" *as* (a form of) consciousness. Taking this double path—analyzing perceptions and implementation—avoids the simplicity of the older historiography of modernization.

The double perspective on "real development" *and* on the changing level of perception for such a history of security inevitably challenges the historical narrative of the sociology of risk from the point of view of an historian: Historians tend either to 'despise' sociological approaches and terminology then and nowadays, or they tend to 'apply' them unquestioned. However, a historiography sensitive to the problems formulated by sociology might challenge the self-consciousness of this discipline itself (if interested): rather than simply adopting the results and theorems of risk sociology along with their terminology and ways of posing questions, this book engages in a critical dialogue with the historical narrative of the sociology of risk. Although the concepts of a risk society formulated by Ulrich Beck, Anthony Giddens, and many others, are all primarily concerned with particular forms of uncertainties linked to the idea of a "second modernity" or "late modernity," and though they thereby only address our present time as this second modernity, they thus nevertheless also imply—although hardly explicitly—a different historical narrative applicable to "premodern times," as compared to the more common historical narrative of the earlier twentieth-century sociologies of modernization.

The basic argument of the sociology of risk is well known: contemporary societies are characterized by the presence of a constant degree of uncertainty and insecurity—the unintended side-effects of industrialization and the technological transformation of the world. When the sociology of risk first developed in 1986, this argument, in particular, was bolstered by the Chernobyl disaster; subsequently, other global hazards and objects of anxiety have been included—such as the terror attack of 09/11/2001 and the threat of a global climate catastrophe. An expansion of the concept of risks, as well as of security, is implied by that, Beck emphasized—very insightfully in this regard—, the merging and blending of natural and cultural-human phenomena in the society of risk.[32] Anthony Giddens also pursued similar themes.[33]

For the most part, sociologists are not interested in the past, but they still rely on and require a historical narrative—whether as part of a genuine world view, or only as a rhetorical device—that allows them to articulate the difference and peculiarity of current phenomena in contrast to everything that has come before.[34] To give meaning and importance to what is understood and branded as the "risk society," and to establish it as different from all other prior eras in world history, it is important for sociology to develop a schema of historical development that can take the form of a relatively simple teleological three-step sequence. Accordingly, until the beginning of "modernity" (or the early modern period, starting with the Renaissance and Reformation—the German "Neuzeit") there were only *dangers*—natural dangers, in particular—which human beings faced with helplessness, whether they were perceived

32 Ulrich Beck, *Risikogesellschaft. Auf dem Weg in eine andere Moderne*, Frankfurt am Main 1986, p. 9: "das eigentümliche Mischverhältnis von Natur und Gesellschaft. [...] Natur ist [...] von einem Außen- zu einem Innen-, von einem vorgegebenen zu einem hergestellten Phänomen geworden"; Ulrich Beck, *Weltrisikogesellschaft*, Frankfurt am Main 2007. Cf. Peter Itzen and Simone M. Müller, "Risk as a Category of Analysis for a Social History of the Twentieth Century: an Introduction", *Historical Social Research* 41, 1 (2016), pp. 7–29 note that one might question the presentist orientation of risk historiography, but still claim that "Risks are of particular relevance for the social history of the twentieth century." (p. 15).

33 Anthony Giddens, *Modernity and Self-Identity: Self and Society in the Late Modern Age*, Stanford 1991; Giddens, *The Politics of Climate Change*, Cambridge 2009.

34 Important precursors were Frank H. Knight, *Risk, Uncertainty and Profit*, Boston 1921; and later, Adalbert Evers and Helga Nowotny, *Über den Umgang mit Unsicherheit. Die Entdeckung der Gestaltbarkeit von Gesellschaft*, Frankfurt am Main 1987.

and accepted as God-given or not. Providing a secure future would have meant concentrating first and foremost on the divine realm. One could work for security in this life, but it would be realized in the world beyond, not in the earthly realm. In "modernity" dangers were gradually transformed into *risks* through changing perceptions and social developments. These are conceptualized within the "welfare state" of the nineteenth century, which was conceived of, in its completion, as a comprehensive insurance-like institution for all its citizens, as described by Foucault's student François Ewald in 1986 in his book *L'état providence* (The providential state).[35] To summarize, the historical narrative of this generation of thinkers reflected the following three-step structure:
Premodern: God-given dangers, closed future
First/High Modernity: Calculable risks, open future
Second/Late Modernity: Society of uncertain insecurities, extended present.
Without a doubt, this is a highly schematic model, but this model is actually implicitly active and embedded within a great number of sociological and nonsociological articles, texts, and analyses of late modernity. The use of this model had far-reaching consequences because it replaced a long—often equally implicit—pattern of epochs that had been assumed, by sociologists from Durkheim to Luhmann. In the older forms that attempted to describe our present as part of "modernity," it was possible to integrate the often more closely schematized "old European" forms into the subsequent historical narrative in order to categorize the sequence of epochs into old/middle/new era schematics; in sociological terms this could be generalized as a sequence of segmented, stratified, and functionally differentiated societies.

In contrast, the historical narrative of the sociology of risk, while still relatively "blind" to all that came before the so-called "first modernity," invokes and challenges new forms of questioning concerning history writ large, simply by focusing on risk and security as the most central matrices for observing and measuring social communication, and through the aforementioned implicit reconfiguration of epochal sequences in relationship to that so-called new "second modernity."

The transition from premodern to modern times is understood by many analysts of this "second modernity" as a relatively linear process of secularization—according to these models of understanding, the proliferation of the principle of insurance would emerge as a paradigm for a new type of "risk," replacing untamed, unspecified and uncalculated forms of "danger." From this perspective,

---

35 François Ewald, *L'Etat providence*, Paris 1986; François Ewald, *Der Vorsorgestaat*, epilogue Ulrich Beck, Frankfurt am Main 1991.

the expectation would be that those new forms of "rational" encapsulating and planning of the future at the core of the insurance principle would have asserted themselves in a conflictual process of opposition against the earlier idea of an economy of danger oriented around the logic of divine retribution, and a ruling God who punishes and forgives using nature's threats. One would ideally expect that agents of the early capitalistic, modern world of planning—such as merchants, insurers, and insurance brokers—argued with priests and pastors about whether an institution like insurance conflicted with the notion of divine providence. As will be seen, however, this was not or only very marginally the case. Furthermore, it does not make sense to conceptualize the relationship between epistemic forms and fields, like mercantilism, cameralism, and political arithmetic, with religion in an abrupt thetic way that posits first a state of opposition and conflict, and then a secular state (as if religion was superseded and replaced by "rational" economic planning).

For historians, it is important to remember that those phenomena that the sociology of risk claims as definitive and authentic indicators of the second modernity were relative: the unintended side-effects of human rational planning begin with humanity itself, and the dimension and totality of "megarisks" and uncertainties with the potential to destroy the "whole world" depend on the relativity of that what is known as the "world" for a given human at a given time. The schematic sequence of epochs implicitly sets as an ahistorical stable yardstick the current dimensions and possibilities of perceiving "the world" with all the previously achieved material, technological and ideative methods of perception—from microscopes to satellite observation and representations of the "world." Taking that relativity concerning the perception of the self and the perception of the world into account, the whole question of a historical narrative describing the changing concepts of risk and security becomes increasingly complex.

For the second transition between full modernity and the notion of a later, second, or postmodernity, Beck formulated the basic assumptions of the historical narrative more precisely than Ewald had done in his book *L'Etat providence*, where the core case study underlying the major theme of the growth of the welfare state as a comprehensive insurance-like institution for all citizens was the development of worker's compensation insurance against accidents in nineteenth century France.[36] In Beck's reconstruction of Ewald's thesis and in comparison to his own interest in the second modernity, the insurance

36 Ewald, *L'État providence*.

principle assumed a central role as a heuristic instrument for distinguishing between historical epochs:

> The epochal difference that distinguishes the risks of industrial society and the bourgeois social order from the dangers and challenges of the society of risk can thus be clearly articulated: The entry into the society of risk takes place at the moment when the now societally determined and thus produced dangers *subvert or even eliminate the old security systems of existing risk calculations as provided by the welfare state*: nuclear, chemical, ecological, and genetically modified risks are, in contrast to early industrial risks (a) delimited neither by time or space, (b) not attributable to the applicable rules of causation, fault, liability, (c) cannot be compensated, nor are they insurable. ... Anyone who asks for an operational criterion for this transition is presented with *the lack of private insurance coverage,* or even more precisely: with *the uninsurability* of industrial technical and scientific projects. ... Beyond the limits of insurability, the industrial society, transformed into a society of risk through the unintended dangers systematically generated by itself, trembles in the balance.[37]

The criteria formulated here lack sensitivity to the historical relativity of the observed phenomena: again, the horizon of "the world" is relative to each individual's time and place, and thus the characterization of the size or totality is also relative and scalable:[38] a major city fire that destroyed several hundred or several thousand houses was, in premodern times, an uncontrollable megadisaster similar to current threats that could lead to complete annihilation—if not of humanity then of large parts of our "world." Such premodern city fires

37 Ulrich Beck, epilogue, in Ewald, *Vorsorgestaat*, p. 541. For a detailed criticism, see Cornel Zwierlein, "Grenzen der Versicherbarkeit als Epochenindikatoren? Von der europäischen Sattelzeit zur Globalisierung des 19. Jahrhunderts," *Geschichte & Gesellschaft* 38 (2012), pp. 423–52. Wolfgang Bonß, *Vom Risiko. Unsicherheit und Ungewißheit in der Moderne*, Hamburg 1995. Christoph Wehner, *Die Versicherung der Atomgefahr: Risikopolitik, Sicherheitsproduktion und Expertise in der Bundesrepublik Deutschland und den USA 1945–1986*, Göttingen 2017, took up these questions, pointing to the origins of the concept of insurability in insurance mathematics, its transfer to sociology and how we can use it as a heuristical tool in historiography.

38 Cf. for this the observations -and remarks on the radius and outreach of charitable kinship networks in comparison between late medieval, seventeenth, eighteenth and nineteenth century below, ch. 5, 1.1: this is not identical with but might be an indicator for the horizon of perception of a common individual's 'world' and the proportional size of a disaster happening in it.

destroyed the entire native habitat that generations, possibly, had lived and worked in; furthermore, the subject at that point was much less integrated into large and tangible comprehensive relief networks. A resident of the small town of Esslingen, for example, would have been completely devastated in terms of their very existence by the city fire; and, as will be seen, in spite of being subject to a city ruler, or even the emperor, and despite perhaps belonging to a neighborly bond of solidarity, there were narrow boundaries framing the potential assistance on which such a resident could rely. The experience of powerlessness, and therefore also the anticipation, the fearful awareness of the possibility of such major disasters, was to some extent structurally comparable with what the risk sociologists describe as the definitive marker of the second modernity today. Also, on closer inspection, the criteria linked to the concept of "insurability" within the epochal divisions appear inconsistent: for one thing, not being able (being outside of an obligatory framework) to account for risks according to the "rules of causation, fault, liability" mentioned above is a feature that already lies at the basis of insurance in the fourteenth century—this has little to do with late modern development.

Concerning the current relationship between "private" insurance coverage and the state as "insurer of last resort," I will show that the historical development is reversed: in Continental Europe at least, the state-administrative protective mechanisms came first and the proliferation of private insurance followed only in the late eighteenth/nineteenth century. Although the use of these defining criteria for epochal distinctions in the sociology of risk can be easily challenged, the basic underlying assumption is nevertheless helpful and worthy of consideration: namely, that the historical instrument of security production of "insurance" (including its diffusion, its acceptance and the resistance against it, the transformation of the insurance principle from an object and an institution into a different kind of guiding principle, and the investigation of the epistemic substrate of insurance, embedded in the question of insurability posed by administrators and agents in diverse places and in different historical realities) *is* a very specific heuristic key to historicizing the relationship between security, risk, and society within the context of a decisive epochal transformation.

One can understand this as a late fulfillment, in actualized form, of Lucien Febvre's call for a general history of security, spanning the subjective focal areas of care, anxiety, liability, and the search for religious security, and including the associated economic and political developments as well.[39] This was only

39 Lucien Febvre, "Pour l'histoire d'un sentiment: le besoin de sécurité," *Annales* 11 (1956), pp. 244–47. Explicitly addressed here is this historical mentality, and to some degree the religious dimension: Jean Delumeau, *La peur en Occident (XIVe-XVIIIe siècles). Une cité assiégée*, Paris 1978; Jean Delumeau, *Le péché et la peur. La culpabilisation en Occident,*

a brief sketch of an idea, and in the meantime the various historical subdisciplines have provided a wealth of important studies: in the history of economics, for example, these have focused on the development of early forms of "assecuration," especially in the context of maritime trade since the fourteenth century,[40] the early modern forms of maritime transport insurance,[41] and on

---

XIIIe-XVIIIe siècles, Paris 1983; Jean Delumeau, *Rassurer et proteger, Le sentiment de sécurité dans l'Occident d'autrefois*, Paris 1989.

40 On medieval premium-based insurance, see Enrico Bensa, *Il contratto di assicurazione nel Medio Evo*, Genoa 1884; Federigo Melis, *Origini e sviluppi delle assicurazioni in Italia (secoli XIV-XVI)*, vol. 1, *Le fonti*, intro. Bruno Dini, Rome 1975; Karin Nehlsen-Von Stryk, *Die venezianische Seeversicherung im 15. Jahrhundert*, Ebelsbach 1986; Arcadi Garcia i Sanz and Maria-Teresa Ferrer i Mallol, *Assegurances i canvis marítims medievals a Barcelona*, 2 vols., Barcelona 1983; Panayotis Perdikas, "Die Entstehung der Versicherung im Mittelalter. Geschichtliche Grundlagen im Verhältnis zu Seedarlehen, Überseekauf, commenda und Bodmerei," *Zeitschrift für die gesamte Versicherungswissenschaft* 55 (1966), pp. 425–509 and Panayotis Perdikas, "Die Palermo-Verträge und die Entstehung der Versicherung," *Zeitschrift für die gesamte Versicherungswissenschaft* 59 (1970), pp. 151–61; Angelo La Torre, *L'Assicurazione nella storia delle idee. La risposta giuridica al bisogno die sicurezza economica: ieri e oggi*, 2nd. ed., Milan 2000. The forms of Europe-wide guild-based instruments of security, which rarely receive equal attention, are explored once again in a heuristic strongly influenced by Otto von Gierke: Dieter Schewe, *Geschichte der sozialen und privaten Versicherungen im Mittelalter in den Gilden Europas*, Berlin 2000.

41 Giovanni Stefani, *L'assicurazione a Venezia dalle origini alla fine della Serenissima*, Trieste 1956; Alberto Tenenti, *Naufrages, corsaires et assurances maritimes à Venise (1592–1609)*, Paris 1959; Alberto Tenenti, "I tassi assicurativi sulla piazza di Venezia: secc. XVI-XVII," *Studi Veneziani* 10 (1985), pp. 15–55; Alberto Tenenti, Ugo Tucci, *Storia di Venezia. Temi. Il mare*, Rome 1991, pp. 663–86; Alberto Tenenti, *Il prezo del rischio. L'assicurazione mediterranea vista da Ragusa (1563–1591)*, Jouvence 1985; Alberto Tenenti, "Profitti assicurativi mediterranei nella seconda metà del cinquecento," *Atti e memorie della Società Savonese di Storia patria* 32/33 (1996/97), pp. 127–48; M. Basas Fernández, *El seguro marítimo en Burgos (siglo XVI)*, Bilbao 1963; Louis Augustin Boiteux, *La fortune de mer. Le besoin de sécurité et les débuts de l'assurance maritime*, Paris 1968; Antonio-Miguel Bernal, "La Aseguración marítima en la carrera de Indias: Seguros, Crédito y Actividad Financeria (Ss. XVI–XVIII)," *Atti e memorie della Società Savonese di Storia patria* 32/33 (1996/97), pp. 83–125; Frank Spooner, *Risk at Sea: Amsterdam Insurance and Maritime Europe, 1766–1786*, Cambridge 1983; Johan Petrus van Niekerk, *The Development of the Principles of Insurance Law in the Netherlands from 1500 to 1800*, 2 vols., Cape Town 1998; Arnold G. Kiesselbach, *Die wirtschafts- und rechtsgeschichtliche Entwickelung der Seeversicherung in Hamburg*, Hamburg 1901; Eva-Christine Frentz, *Das Hamburgische Admiralitätsgericht (1623–1811). Prozeß und Rechtsprechung*, Frankfurt am Main 1985; Thomas Dreyer, *Die 'Assecuranz- und Haverey-Ordnung' der Freien und Hansestadt Hamburg von 1731*, Frankfurt am Main 1990; Andreas Ebert-Weidenfeller, *Hamburgisches Kaufmannsrecht im 17. und 18. Jahrhundert. Die Rechtsprechung des Rates und des Reichskammergerichts*, Frankfurt am Main 1992; Andrea Addobbati, *Commercio, rischio, guerra: il mercato delle assicurazioni marittime di Livorno (1694–1795)*, Rome 2007; Giovanni Ceccarelli, *Un mercato del rischio. Assicurare*

the new forms appearing at this time,[42] as well as the expansion of the major insurance institutions of the nineteenth century.[43] In its more general historical dimension, however, the principle of insurance, and its implicit general heuristic potential for the fields of history and historiography as such are more or less ignored, as is the case in other related disciplines.[44]

The structure of the present study thus traces the developments and the transfer of the insurance principle: in the context of Mediterranean transport insurance in its northern German guise, this study investigates the process of transfer between southern and northern Europe and between the merchant world and academic legal culture, which together are viewed as a specific spatial and temporal configuration serving as the epistemic center of the instrument of security production (Ch. 2). Following that, the fire ecology of Germany will be reconstructed on the basis of archival sources, providing a detailed reconstruction of the socially influential danger of major city fires through a serial analysis of the available data for German-speaking countries, supplemented by individual case studies as well as internationally comparative

e farsi assicurare nella Firenze rinascimentale, Venice 2012; Guido Rossi, *Insurance in Elizabethan England: The London Code*, Cambridge 2016.

42 The extensive list of references provided in the chapter is not repeated here. In terms of the general economic significance of English fire insurance of the eighteenth century, the work of Robin Pearson is especially important. For an overview, cf. Geoffrey Clark et al. (eds.), *The Appeal of Insurance*, Toronto 2010.

43 Two handbook-like anthologies have recently been published on the global dimensions of corporate history of insurance in the nineteenth century: Robin Pearson and Takau Yoneyama (eds.), *Corporate Forms and Organizational Choice in International Insurance*, Oxford 2015; Peter Borscheid and Niels Viggo Haueter, *World Insurance: The Evolution of a Global Risk Network*, Oxford 2012.

44 For an essay that attempts to show how the history of insurance could become part of general history (rather than remaining a subfield of economic history), see Cornel Zwierlein, "Sicherheit durch Versicherung. Ein frühneuzeitliches Erfolgsmodell," in: Christoph Kampmannn and Ulrich Niggemann (eds.), *Sicherheit in der Frühen Neuzeit. Norm—Praxis—Repräsentation*, Cologne 2013, pp. 381–99; cf. also Richard V. Ericson, Aaron Doyle, Dean Barry, *Insurance as Governance*, Toronto 2003, p. 3: "Insurance also remains in the background in social science. Except for narrow specialisms in law and economics, it has not been subject to extensive analyses. Although the insurance industry is among the most pervasive and powerful institutions in society, the sociology of insurance remains nascent"; Jacques van Gerwen and Marco H.D. van Leeuwen (eds.), *Studies over zekerheidsarrangementen. Risico's, risicobestrijding en verzekeringen in Nederland vanaf de Middeleeuwen*, Amsterdam 1998; Marco H. D. van Leeuwen, *De rijke Republiek: gilden, assuradeurs en armenzorg 1500–1800* (= *Zoeken naar zekerheid—Risico's, preventie, verzekeringen en andere zekerheidsregelingen in Nederland 1500–2000*, eds. Jacques van Gerwen and Marco H. D. van Leeuwen, part 1), The Hague 2000.

data. The development of trends and patterns of perception concerning this threat will be discussed here (Ch. 3). On the basis of the cities of Hamburg and London, which were central for European development and served as "Laboratories of a culture of innovation" for the history of security, I formulate the central thesis of this book concerning the epochal transformation in the early Enlightenment period: I propose the notion of the "normal secure society" as a contrafactual state of expectation in modern times, first conceived of and put into practice in late seventeenth-century contexts (Ch. 4). This is supplemented through a discussion of the expansion of this "normal secure society" following the diffusion of fire insurance, and by questioning potential affinities between the insurance principle and the main religious denominations in Europe during the process of diffusion. For the later, cameralist period, I interrogate the relationship between the insurance principle and the early origins of the concept of sustainability, which were also at play during the period of insurance diffusion (Ch. 5). The concluding chapters of this book explore the development of the next central stage of transfer: while the book begins with the south to north transfer of the principle from the Mediterranean into the northern European realm, it concludes with an overview examining the globalization of fire insurance in the nineteenth century (Ch. 6). Precursors of such diffusion are indeed detectable in the late eighteenth century, but the real rise in the global spread of premium-based fire insurance appears only after the 1830s. Thus the question becomes whether that process can be understood simply as an expansion of the "normal secure society" from Europe outward, or whether this globalization should be conceived of differently, as a merging and intersection of European and different global forms of security regimes.

CHAPTER 2

Spaces, Value, Presentism: Premodern Insurance

In general historical and historical-sociological overviews, "insurance" is primarily characterized as a tool for calculating and developing the future, and as one of the best means of "opening the future" as a time-horizon in the early modern era.[1] According to Geoffrey Clark's somewhat general formulation:

> to the extent, then, that the future could be anticipated and perhaps even manipulated, it belonged to those who could bank on their expectations. Insurance was an invention expressly designed to tame the future, and it therefore provided the credit-driven society of eighteenth-century Britain with a critical tool in the management of risk. ... For merchants and traders, insurance mediated between the hard-headed realities of the present and the speculative futures into which they daily entrusted their fortunes.[2]

Here, Clark alludes to two aspects of a typical element of a "temporally oriented" perception of the world: first, the basic epochal constellation of temporal horizons, and second, the specific sector of life insurance that concerns itself with the temporalized arrangement of a lifespan. However, I believe that interest in the development of probabilistics in the history of science and its focus on the sector of "life" create a false impression of the historical evolution. The main focus of the current project is insurance against natural disasters, in the sense that only really emerged during the seventeenth century. However, in order to conceive correctly of the structure and steps of the development, how the principle of insurance became inscribed within the institutional matrices of the early modern and modern periods well before life insurance in the form of the English societies, it will be necessary to first elaborate on the specific

1 Alberto Cevolini argues strongly for this point of view, which also clarifies and further develops Niklas Luhmann's position; see Cevolini, "Die Einrichtung der Versicherung als soziologisches Problem," *sociologia internationalis* 1 (2010), pp. 65–89; Niklas Luhmann, *Il rischio dell'assicurazione contro i pericoli*, trans. and introd. Alberto Cevolini, Rome 2013.

2 Geoffrey Clark, *Betting on Lives. The Culture of Life Insurance in England, 1695–1775*, Manchester 1999, p. 3.

 | DOI:10.1163/9789004431225_003

conception of insurance in the premodern era—or, in the European context, the Middle Ages. I will show that "insurance" during the Middle Ages and Renaissance was initially strikingly different—and scarcely even related to temporality at all.

1 The Commercial Communication System around 1400

In order to appropriately position the notion of "insurance" beyond the linear development related to the "colonization of the future," it will be important to first review the state of early transport insurance in the context of mercantile communications in the fourteenth through sixteenth centuries. These communications will serve as sources for our explorations of the conceptualization of temporality and spatiality—i.e., the "timescapes"—of the era.[3]

The defining characteristic of the mercantilism of the late Middle Ages and Renaissance was that the ideal type of merchant was one who worked *in situ*—no longer having to travel himself, he sent others out into a realm delimited by the trading centers, or *fondachi*, of the various trade- and port cities comprising the trade network. One particular source provides great insight into this system: the archives of the merchant Francesco Datini, which span approximately from 1370 to 1410 and were assembled and made accessible primarily by the Italian economic historian Federigo Melis. The nearly 600 surviving account books, which are full of information processing and billing accounts, facilitate an understanding not only of the early development of double-entry bookkeeping in a narrower sense, but also allow us to study the particular relationships between different types of account books—notebooks, "mediative" books (*Memoriale*/journals), and the main books—as a complex network and system of notation.[4] In addition to the account books, the other source of data is the Datini letter archive. This archive is unparalleled in terms of quantity: the surviving business correspondence includes 125,549 letters from the years 1363 to 1411 (nearly 80 per cent from 1390–1411), sent from 267 mainly southern European cities and towns

3 For information on earlier medieval message communication, see Cornel Zwierlein, "Gegenwartshorizonte im Mittelalter: Der Nachrichtenbrief vom Pergament- zum Papierzeitalter," *Jahrbuch für Kommunikationsgeschichte* 12 (2010), pp. 1–60.

4 Franz-Josef Arlinghaus, *Zwischen Notiz und Bilanz. Zur Eigendynamik des Schriftgebrauchs in der kaufmännischen Buchführung am Beispiel der Datini/di Berto-Handelsgesellschaft in Avignon (1367–73)*, Frankfurt am Main 2000, p. 115. Mikhail Kuter, Marina Gurskaya, Dmitriy Aleinikov, "The earlier synthetic balance sheet of Datini's Company in Avignon (1410): The combined accounting system", *De Computis. Revista Española de Historia de la Contabilidad* 17, 1 (2020), pp. 7–35, here p. 10.

included within the eight *fondachi* of the company (located in Avignon, Barcelona, Florence, Genoa, Palma de Mallorca, Pisa, Prato, and València).[5]

As such, this collection testifies to the extensive network of the Italian business world operating from the thirteenth century on, which has been previously documented by economic historians on the basis of messenger orders and other sparse and isolated sources.[6] The lion's share of the correspondence in the Datini archive concerns the relationship between the eight *fondachi* themselves: excluding all cities or towns that sent fewer than thirty letters to any of the eight *fondachi*, there are twenty-five places that figure as steady, direct connections, in addition to the *fondachi* themselves. Of these in turn, almost all—except for Bologna, Rome, Venice and Bruges—are important northern cities in the immediate vicinity of *fondachi* sites. In this sense then, the correspondence network comprises a polycentral trade network mainly organized around twelve cities.[7] No other medieval merchant company can claim such a mass level of correspondence being preserved—one that was hardly attainable even for a Pope, King, Emperor, or Duke. Beyond the obvious commercial significance, Melis and his school often attempt to articulate the meaning of this letter exchange as the first stage of a journalism-like messaging system. They go so far as to claim that in "contrast [to] today's trade letter, in which everyday business stands in the foreground, ... in the older trade letter ... information about political, social, and religious events as well as epidemics" was very prevalent and would have fulfilled the function of today's news media.[8] Thereby Melis and his followers seek to emphasize that the city of Prato is home not only to the "archive of a merchant," but "an archive of universal significance and range [*un archivio di portata universale*]."[9]

5 Benjamin Scheller, "(Un-)sichere Häfen: Häfen als Hotspots maritimer Risiken und Risikokommunikation im Mittelmeerraum des 15. Jahrhunderts", *Historische Anthropologie* 26, 1 (2018), 43–65.

6 Richard D. Face, "Techniques of Business in the Trade between the Fairs of Champagne and the South of Europe in the Twelfth and Thirteenth Centuries," *Economic History Review* 10, 2 (1957), pp. 427–38.

7 Federigo Melis, "Intensità e regolarità nella diffusione dell'informazione economica generale nel Mediterraneo e in Occidente alla fine del Medioevo," *in: Mélanges en l'honneur de Fernand Braudel. Bd. 1: Histoire économique du monde méditerranéen 1450–1650*, Toulouse 1973, pp. 389–424/b.

8 Bruno Dini, "Art. Brief, Briefliteratur, Briefsammlungen. F. Handels- und Kreditbrief II. Italien," *LexMA* 2 (1983), p. 681f.

9 "Senza esagerare, mi sembra di poter sostenere che, se si pubblicassero tutti i brani di argomento politico dei carteggi più abbondanti e serrati ... si traccerebbe una profonda e dettagliata storia politica e sociale dei vari popoli, e piuttosto obiettiva, gli osservatori non essendo gente dei luoghi: così si verifica per le dispute in Inghilterra e in Fiandra, per la

Thus far, however, neither Melis nor any of the Datini archive scholars has documented any part of the correspondence to support these far-reaching claims.[10] At this point, beyond Melis's studies (excluding the letters of *Carteggio privato*),[11] approximately 2,450 letters have been made accessible in print; most of the massive collection, however, remains unpublished.[12] In perusing

---

guerra dei Cento Anni e per gli avvenimenti iberici ... tutti avvenimenti narrati da mercanti toscani. Storia, dico, e non mera cronaca, proprio perché è possibile controllare molti dei suoi elementi, che quanto ho ripetutamente constatato autorizza a definire senz'altro esatti" (Federigo Melis, *Documenti per la storia economica dei secoli XIII-XVI*, Florence 1972, pp. 14–27, 17f.; for more detailed information see, Melis, *Aspetti della vita economica medievale. Studi nell'Archivio Datini di Prato*, Siena 1962, pp. 29–42, 33–42. For his lectures, taken from his disciples' notebooks see Melis, *Sulle fonti della storia economica. Appunti raccolti alle lezioni del Prof. Federigo Melis*, ed. von Bruno Dini, Milan 1985, pp. 139–46. Melis himself was indebted to Armando Sapori, "Saggio sulle fonti della storia economica medievale," in Sapori, *Studi di storia economica (secoli XIII, XIV, XV)* [first ed. 1940], Florence 1982, p. 16f. The myth that the Datini letters are full of political and cultural news beyond commercial information is also repeated outside of the Melis school; for example Karl d'Ester, "Fuggerzeitungen," in Walther Heide (ed.), *Handbuch der Zeitungswissenschaft*, vol. 1, Leipzig 1940, col. 1187–98, 1189, or Theodor-Gustav Werner, "Das kaufmännische Nachrichtenwesen im späten Mittelalter und in der frühen Neuzeit und sein Einfluss auf die Entstehung der handschriftlichen Zeitung." *Scripta Mercaturae* 2 (1975), pp. 3–51, 8f.

10 Regarding a particular event, the capture and conquest of Damascus by Tamerlane (Timur, 1336–1405), at the end of the year 1400, Melis compiled nineteen news passages from letters originating between January and July of 1401, which supposedly demonstrated the high quality of the information. However, in reality, they consist of five lines without any great value as news (Melis, *Aspetti*, pp. 30–32; Melis, *Documenti*, p. 168; Federigo Melis, "Movimento di popoli e motivi economici nel Giubileo del 1400," in *Miscellanea Gilles Gérard Meersseman*, ed. Gilles Gérard Meersseman, 2 vols., Padua 1970, vol. 1, pp. 343–67; also included in Federigo Melis, *I trasporti e le comunicazioni nel medioevo, a cura di Luciana Frangioni*, Florence 1984, pp. 237–59).

11 Cesare Guasti, Ser Lapo Mazzei, *Lettere di un notaro a un mercante del secolo XIV*, vol. 2, Florence 1880; see also Christian Bec, *Les marchands écrivains. Affaires et humanisme à Florence 1375–1434*, Paris 1967; Jerôme Hayez, "La gestion d'une relation épistolaire dans les milieux d'affaires toscans à la fin du Moyen Âge," in *La circulation des nouvelles au Moyen Âge*, ed. Société des historiens médiévistes de l'enseignement supérieur, Paris/Rome 1994, pp. 63–83; Iris Origo, *The Merchant of Prato, Francesco di Marco Datini 1335–1410*, New York 1954.

12 Not taking several scattered publications into account, there are five unpublished Florentine *tesi di laurea* containing 1,349 letters. Published collections include, among others, Luciana Frangioni, *Milano fine Trecento. Il carteggio milanese dell'Archivio Datini di Prato*, 2 vols., Florence 1994— which includes bibliographic information concerning the aforementioned scattered publications and the *tesi di laurea* (vol. 1, pp. 345–50); Elena Cecchi Aste (ed.), *Il carteggio di Gaeta nell'Archivio del mercante pratese Francesco di Marco Datini. 1387–1405*, Gaeta 1997; Giampiero Nigro, *Mercanti in Maiorca: il carteggio datiniano dall'Isola (1387–1396)*, Florence 2003; concerning the correspondence in particular, Jerôme Hayez (ed.), "Le carteggio Datini et les correspondances pratiques des XIVe-XVIe siècles",

the 838 fully edited letters of the complete *Carteggio Milano* of the Datini letter archive, one inevitably comes to this conclusion: "the references to the political life of Milan are not very numerous."[13] In quantified and even more general terms: only about 0.06–0.3 per cent of the existing texts do *not* primarily relate to economic conditions affecting the local businesses of the correspondent himself. Accordingly, approximately 99.7 per cent of the contents are made up of goods transport information, price news, calculations, orders for goods, information about the stay and the travel of business partners, etc.[14] In fact, in this sense, this body of texts looks quite similar to the only notable collection of merchant letters concerning the Hanseatic League from the period around 1400— the correspondence of Hildebrand Veckinghusen of Lübeck, who traded mostly in Bruges. Of the 560 edited letters, which in turn reflect the complete northwestern European urban network at the time (Venice being the single and exceptional southern city), only about 15 contain "one or several—mostly short—news accounts, which referred to political events or other circumstances outside the direct trading sphere of merchants."[15] If, more precisely, one ignores whether a letter contains nonbusiness news, but rather estimates the relative proportion of content of business/nonbusiness information, a similar negligible value as the Datini value is reached—a margin of 0.2–0.4 per cent.[16]

Consequently, it would seem that the archival sources demonstrate the opposite of what Melis argues: the mercantile correspondence of the late Middle Ages bears witness to an astonishingly obstinate, autonomous mode of

special issue of *Mélanges de l'École française de Rome. Moyen Âge* 117, 1 (2005); and as a more recent study based on the source material, Ingrid Houssaye Michienzi, *Datini, Majorque et le Maghreb (14e-15e siècles). Réseaux, espaces méditerranéens et stratégies marchandes*, Leiden 2013.

13 Frangioni, *Milano fine Trecento*, vol. 1, p. 19.

14 There are about 4500 characters within approximately 1.5 million characters in Frangioni, *Milano fine trecento*, vol 2; there are only about 1500 characters of appoximately 1.9 million characters in Nigro, *Mercanti in Maiorca*—here, the news nearly always concerns displacements of the king, e.g., pp. 76f., 268, 287, 516, 800.

15 Margot Lindemann, *Nachrichtenübermittlung durch Kaufmannsbriefe. Brief-Zeitungen in der Korrespondenz Hildebrand Veckinchusens (1398–1428)*, Munich 1978, p. 24.

16 Hildebrand Veckinchusen, *Hildebrand Veckinchusen. Briefwechsel eines deutschen Kaufmanns im 15. Jahrhundert*, ed. Wilhelm Stieda, Leipzig 1921. Cf. Lindemann, *Nachrichtenübermittlung*, pp. 46–79. Also here, the very few events referred to are, at least indirectly, almost always linked to commercial interests. There are so few details included in these messages that one could hardly even reconstruct a rudimentary sketch of a particular event from any of them without help from other sources. Cf. the similar judgment by Arnd Reitemeier, *Außenpolitik der Spätmittelalter. Die diplomatischen Beziehungen zwischen dem Reich und England (1377–1422)*, Paderborn 1999, pp. 399–401.

specifically economic communication.[17] The specialized nature of this communication is very much closed—the few social, political, or religious references in the letters appear as a low noise of deviation from the norm. This is not suggestive of a differentiation of the economy as a subsystem of a functionally differentiated society in the sense of the Luhmannian teleological conception of societal development; rather, it refers to the fact that medieval communication circuits served unconnected, paratactically created, specialized purposes, *before* commercial state economic politics took hold to any significant extent, which then—only after skipping a step—resulted in the demand for decoupling. The medieval disconnectedness and specialization of communication is isomorphic, but not homologous, with later modern differentiation.

In his classic essays tracing changing medieval temporal conceptions, Jacques Le Goff sketched the emergence of "merchant time" versus "labor": the mechanized, measurable, predictable time of merchants contrasted with the temporality of the natural environment, characterized by both its seasonal cyclicality and its constant unpredictability.[18] This conception of time would have perhaps been understood in tandem with, though not necessarily in opposition to, "church time"—because "the time in which [the merchant] lives with regard to his work is not the same in which he lives with regard to his religion."[19] Le Goff's essays go on to offer research regarding the special characteristics of "time"; one Le Goffian example that has attracted attention is the effect of displaying hours and minutes on tower clocks.[20] However, the fact that Le Goff formulated these notes on the development of "merchant time" in the context of the exploitation of space has been ignored: as a goal for historical research, he states "that the medieval merchant undertook the conquest

17 Another question concerns whether or not merchants participated in the written communication of the message at all. Their means of transport were certainly used by other senders of messages, as already demonstrated by Renouard's research on the expedition of the Pope's letter: see Yves Renouard, "Comment les papes d'Avignon expédiaient leur courrier," *Revue historique* 180 (1937), pp. 1–29 and Yves Renouard, *Les relations des papes d'Avignon et des compagnies commerciales et bancaires de 1316 à 1378*, Paris 1941.

18 "Pour le marchand, le milieu technologique superpose un temps nouveau, mesurable, c'est-à-dire orienté et prévisible, au temps à la fois éternellement recommencé et perpétuellement imprévisible du milieu naturel" (Jacques Le Goff, "Au Moyen Âge. Temps de l'Église et temps du marchand," *Pour un autre Moyen Âge. Temps, travail et culture en Occident*, ed. Jacques Le Goff, Paris 1978, pp. 46–65, 56).

19 "Le temps dans lequel il agit professionnellement n'est pas celui dans lequel il vit religieusement" (ibid., p. 59).

20 Gerhard Dohrn-van Rossum, *Die Geschichte der Stunde. Uhren und moderne Zeitordnungen*, Munich 1992.

of time *and equally that of space*, and that is what, first and foremost, attracted the attention of historians and art sociologists."[21]

In combining these Le Goffian observations with those concerning the communications system of the merchant letters, it becomes evident how sent and received letters, in the steady rhythm of arriving and departing messengers and later through various relay systems, constructed a systemic framework for the ever more tightly controlled horizon of reality for merchants in terms of the flow of information, goods, and other assets. Without question, this implies a new temporal order. The central question, however, is whether the "conquest of time," in a deeper sense, really implies a decisive shift toward a "colonization of the future," which then would find its most tangible expression within the business of insurance.

2 Insurance as an Accounting Trick between the World of Nature and the World of Values

In his work on fear in the Western world, Jean Delumeau concentrates almost exclusively on the relationship between different religious cultures and processes of social discipline on the spiritual basis of fear in relation to personal salvation; however, he introduces his *opus magnum* with a reflection on fear in the context of the sea for European coastal populations: the *unknowns* of this space—rather than any temporal aspects—are the primary reference objects of this fear.[22]

In fact, in the letters of fourteenth - and fifteenth-century Mediterranean merchants, fear and insecurity express the decisive basic pattern of their world consciousness—elucidated by the concern with which they awaited information about the movements of ships, goods, and assets in the Mediterranean Sea. In the "fight against distance," the two standards of speed and security were often in competition; when in doubt, however, the merchants always opted for transport security and accepted detours and delays.[23] Expressions of correspondents' "melancholia" associated with the care of goods peppers the otherwise sober descriptive style of reports and calculations within the Datini letters. For example, if instead of the six bales sent, only five arrived,

21 Le Goff, "Au Moyen Âge," p. 57.

22 Delumeau, *La peur*, Ch. 1.

23 Bernard Doumerc, " 'Par Dieu écrivez plus souvent!' La lettre d'affaires à Venise à la fin du Moyen Âge," in Société des historiens médiévistes de l'enseignement supérieur (ed.), *La circulation des nouvelles*, pp. 99–109, esp. 106f.

the sender suffered from "melancholy." One merchant was infected with "great melancholy" when he sent forty bales of cloth worth two thousand florins, and there was no news of them. Transporting the goods safely costs eight francs. It was Datini's job to inquire about the whereabouts of the goods in Viglana (Vigliano d'Asti), so that Tieri di Benci in Milan would no longer need to suffer from "melancholy": in this situation, soldiers ("*giente d'arme*") in Turin had prevented the further transport of goods for more than a month.[24]

It was always necessary to insure against damages caused by certain dangers (for example, against "grande rischio [che] achore [*great risk that might occur*]" or, alternatively, to "providere che danno non ve ne seghuisse [*provide that no damage would occur*]").[25] For Datini, traveling to Avignon personally would not have been advisable, because he would have encountered great danger on such a long journey.[26] Upon close analysis of such contexts—concerning maritime travel and transport—the instrument of insurance occasionally also appears in the merchant letters: for example, although a certain Bascano da Pescina is sure that his cloth goods have been sent to Barcelona through the best possible means, he nevertheless thinks that Francesco Datini should personally insure at least half the value of the goods, so that he does not take on too much risk.[27] Such instructions securing shipping insurance for a business partner in another city were common—the route, the ship's strength, its crew and armament, as well as any known hazards were specified. The ban on insuring foreign ships in Florence and elsewhere was often circumvented, even though this risked problems in defending the validity of the policy in the case of a dispute.[28] The

---

24 Francesco Datini wrote upon recieving five *fustani,* that there should have been six: "… dome maraviglia che nula menzione ne fate … òne melanchonia e ancho mi maraviglio chome voy nulla menzione ne fate de la sexena bala … " (Milan, June 26, 1383; Frangioni, *Milano fine trecento*, p. 24, cf. regarding the same case on p. 25); "E vedi s'io dicho vero ch'io avevo meso a chamino da balle 40 i tra più volte e di niuno sapeno novella che valevono più di f. 2000 e grande malichonia n'aveno. … chostono da franchi 8 per falle andare sichure, sì che no ti challe dare malichonia di nulla che, cho lla grazia di Dio, io mi credo andrano a salvamento." (Milan, July 28, 1384; ibid., p. 60f.).

25 Milan, September 30, 1384, Frangioni, *Milano fine trecento*, p. 72.

26 "poi si chore pericholo andare uno sì lungho chamino sì che no mi pare facci per voi" (Milan, March 7, 1386; ibid., p. 101).

27 "E io sono contento che ongni mio fustanio di ghuado e di mezo ghuado che cosà abiate mandiate a Barzallona a' deti per primo buono pasagio ma che prendiatte sechurttà per la mettà di cò che varanno per no chorere a tanto rischio e per prima letera mi ditte che ne pensate fare" (Milan, January 29, 1384; ibid., p. 40). In the part of the Datini correspondence edited by Frangioni, insurance is rarely mentioned, which is probably because Milan was not a major area for maritime transport.

28 See Melis, *Origini e sviluppi*, p. 49 for a detailed letter including a proposed insurance contract.

letters circulating through the communication network constantly exchanged information about current risks: for example, the location of pirate ships (such as those of the notorious Genovese pirate Salagrus di Negro, who in 1399/1400 robbed more than forty ships), bandits along the Flemish coast, the dangers associated with the Hundred Years' War—all hazards that reflected both the necessity of insurance and potential insurability problems.[29]

The letter communication system reported with great sensitivity on news of lost ships or goods, on partial losses, and on all other details. If one "lost sight of" a ship, this too was reported to all interested parties. The constant back-and-forth of letters created a system of constantly pulsating information about *the presence* of business beyond the movement of ships and goods in space and beyond the details of account books: letter communication was the medium of observation that constantly tracked the slow circulation of goods transport. This level of communication belonged to a world that was continuously constructed and stabilized by merchant writing—a world of values, in which real ships, cargo, and individual bales were translated into positive and negative values, costs and assets.

This dynamic becomes even more complex when the role of insurance in merchant billing and account management is addressed. The first distinction concerns insurance as an asset versus a liability: on the credit side, insurance appears in the account books when the merchant himself operates as the insurer—usually only for a part of the total value and in addition to other merchants. Regarding the merchant's own business, in contrast, insurance always appears among the payables—the premium, the cost to issue the policy (which often necessitated a specific form, controlled by specialists), and commissions for the insurance broker. Insurance was thereby always noted as an appendage of shipping costs for a trade item, even in the earliest documents featuring insurance entries:

Messer Lapo e Dosso de' Bardi
deono avere dì xxviii Gennaio anno
milletrecento dicenove, per le spese
fatte d'Avignone a Pisa e in Pisa sopra
undici panni di Cielona iscritti in
questa carta nella faccia di sotto [...]:

Master Lapo and Dosso de' Bardi
require on 28 January 1319 for the costs
[incurred] for 11 rolls of multicolored
tablecloths from Avignon to Pisa and
in Pisa [and which] are here registered
on the front below [...]:

29 Somewhere between Gascony and the North Sea: Melis, *Origini e sviluppi*, p. 58. Benjamin Scheller, "Die Geburt des Risikos", *Historische Zeitschrift* 304, 2 (2017), pp. 305–331.

Per rimanente di vettura da Parigi a Avignone s. v tornesi, e per isciogliere in Avignone e Nizza e rilegare, e caricare a galea s. otto tornesi somma s. xiii tornesi di s. xv d. viii tornesi florini d'oro, vagliono a fiorini .... L. 1 s. iiii	For conveyance by carriage from Paris to Avignon 5 soldi tournois and for unloading in Avignon and Nice, and for recoupling and loading and at Galley 8 soldi, 13 soldi tournois total, each 15 soldi, 8 denari Florentine gold ducats, which equals in ducats .... L. 1 s. iiii
Per la chiaveria d'Aquamorta fior. 1 d'oro vagliono a fiorini........................ L. 1 s. viiii	For the customs in Aquamorta [...] L. 1 s. viiii
Per vettura d'Avignone a Nizza fior. Tre d'oro vagliono a fiorini .... L. iiii s. vii	For the transport from Avignon to Nice [...] L. iiii s. vii
Per navolo e *rischio* da Nizza a Pisa flor. Sedici d'oro vagliono a fiorini ... L. xxiii s. iiii.	For the cargo and the *risk* [*risk assumption*] from Nice to Pisa 16 gold ducats, which together makes [...] L. xxiii s.[30]

In this entry, the cost of insurance is appended to the shipping costs and is specified only for the maritime section between Nice and Pisa, while the land transport from Paris to Avignon and from Avignon to Nice is not insured. Later, these entries become more detailed, and instead of the term *rischio,* the expense is usually listed as *sicurtà*.[31] They appear in full detail only rarely in examples as above or in the *libro mastro* or *maggiore*; more frequently they appear in the *libri di ricordanze/memoriali*, the *quaderni di spese di mercanzie* and the *quaderni di ricevute e mandate di balle* that follow.[32]

Concerning insurance as an asset, various types of *libri di sicurtà* quickly emerged among the merchants (with precursor *sicurtà* payments in other accounts appearing from 1383 on), who recorded various insurance contracts and associated premium revenues.[33] Although no companies focusing solely

30 Bensa, *Il contratto di assicurazione*, p. 185, see also Melis, *Origini e viluppi*, p. 101 and tav. 1.

31 See the account book of the Filippo Borromei society in London in 1437 containing five *sicurtà* for five portions of goods on Catalonian and Florentine galleys: Melis, *Origini e sviluppi*, p. 58.

32 Melis, *Origini e sviluppi*, pp. 77–98.

33 A detailed analysis of the, almost unique *Libro di ricordi di sicurtà* of Salviati of 1524–26, which Melis already addressed, is found in Ceccarelli, *Un mercato del rischio*. A recent transcription and analysis of the premium sums earned by Datini can be found in Mikhail Kuter, Maria Gurskaya, Armina Papakchian, "The Early Practice of Maritime Insurance Accounting in the First Proprietorship of Francesco di Marco Datini in Pisa in 1382–1406", in *Digital Science 2019, Advances in Intelligent Systems and Computing 1114*, eds. *Antipova*

on insurance were formed, these so-called "security" account books do represent a clear specialization in the area of commercial writing—but only there, where it concerned the notation of specific earnings within this business. In this context, there were also occasionally notes on bets, (often without naming the contents) that, but which sometimes even commented on the potentially astounding extravagance of these bets. For example, on 29 January 1503, Alessandro Gondi bet against Neretto Neretti: Neretti gave him a gold ducat, but Gondi agreed to pay him twenty-five gold ducats every year for fifteen years if the Florentine conquest progressed so far in that year that the Florentine empire included both a port on the Tyrrhenian Sea (Livorno) and a port on the Adriatic coast.[34] Also commonly up for wager were the demise of crowned heads[35] and the cancellation of markets as a result of the plague.[36]

In this light, it becomes clear that insurance emerges, above all, as a notion of pure value (see Fig. 2). In contrast to the value of commodities, such as real cloth bales, spices, or jewels, the only "materiality" of insurance lay outside of the account books in the paper policy contracts. Its "invention" or "innovation" can be much more strongly conceptualized by examining the logic of mercantile writing and its associated "effect on the users."[37] The daily thinking and practice of merchants that is reflected in the many thousand records of entries and values on the "credit" and "debit" sides of the account books led to the emergence of an idea: namely, how to prevent undesired effects in the world of values (Fig. 2). Ultimately insurance led to nothing beyond the fact that the credit side of value transcriptions was independent of outside environmental influences. The specific commodity, in all of its material uniqueness and irretrievability, might well be lost, but through insurance its value was retained as an asset. Before quickly (or slowly) coming to the conclusion of a complete change in merchant frames of thought that would lead to the idea that they intended to conquer the future, one should perhaps first consider the duller side of this stream of numbers and entries in the account books. It is not the mindset of the merchants that creates the account books; rather conversely, the world of the account books is a part of the formative forces that bring about

Tatiana and Rocha, Alvaro, Heidelberg 2020, pp. 299–313 with a diagram of the relationship between the account books on p. 311: a reconstruction of the relationship between the entries of four accounts in two books (Archivio di stato di Prato, D357, D377) showing the notation technique.

34 Melis, *Origini e sviluppi*, p. 87, no. 112.

35 According to Melis, *Origini e sviluppi*, doc. 27–29, pp. 214–17: "Life insurance" was taken out on the death of the Lord of Piombino and betting insurance on the death of Pope Nicholas V and the King of Aragón within one year.

36 Melis, *Origini e sviluppi*, p. 218, doc. 31.

37 Arlinghaus, *Zwischen Notiz und Bilanz*, pp. 397, 400.

Insurance

as **liability**
passiva perspective

as **asset**
activa perspective

Ledger / Libro maggiore

‚Dare' / Debit	‚Avere' / Credit
1) Costs of good A a) Price of the good b) Fiscal costs c) Costs of transport → hereafter: **costs of ‚sicurtà / rischio' (premium)** d) Costs of provisions 2) Costs of good B…	2) Price of sale of good A Or (in case of wreckage) **Insurance sum**

Book of securities / libro di sicurtà

Dare / Debit	Avere / Credit
-loss, damage → full or partial insurance sum - in the case of a bet: wager	-Premium income - in the case of a bet: winnings

FIGURE 2 Classification of insurance costs and income in Renaissance merchants' records

their mindset. The constant attempt to at least maintain a balance, if not profit, ought to properly be viewed as the developmental context in which insurance arose. Insurance is, first and foremost, nothing but an accounting trick.

As has already been seen, such accounting tricks were employed for the vast majority of transport, cargo, and shipping insurance—especially maritime transport. Uncertainty was much more strongly associated with the space of "the sea" than with aspects of temporality or the future. As a counterargument, one could certainly claim that the aforementioned betting and betting insurance constituted an aspect of speculation on future events. Insofar as these practices already banned insurance bets on human lives, it is important to note that the exoticism and the cultural-historical interest concerning betting and betting insurance should not obscure the rarity and relative economic irrelevance of this practice. One exception did appear in the form of life insurance policies going beyond playfully speculative bets—those recorded for the lives of slaves. However, since slaves were more or less considered a commodity, such life insurance essentially operated as a derivative of transport insurance.[38] In addition, it is worth highlighting that until the rapid

38 Cf. already in Wilhelm Ebel, "Über Sklavenversicherung und Sklavereiversicherung," *Zeitschrift für die Versicherungswissenschaften* 52 (1963), pp. 207–30; Geoffrey Clark, "Slave Insurance in late Medieval Catalonia," in *Sicherheit in der Frühen Neuzeit. Norm, Praxis,*

growth of English insurance companies in the 1680s, all late medieval/early modern life insurance policies were limited to a few months or a year.[39] Thus, on close inspection, even this type of insurance, which would seem to most easily confirm the future-oriented nature of the principle, offers no clear justification for this idea. Rather, it was the concern regarding uncertain spaces that truly heralded this newly emerging framework of thought in the conceptual world of merchants. Le Goff spoke of a differentiation between "merchant time" and "the time of nature and the environment,"[40] which can be summarized much more generally as a differentiation between the "world of value" as opposed to the "world of materiality." Insurance was an element that perfected the decoupling of these two worlds, because with insurance neither storms, bad weather, pirates, nor fire could cause unpredictable declines in credit values. Merchant concerns associated with the realities of the material world—the uncertainties of the seas, transport distance, etc.—were sublimated to a simple accounting trick in the world of value. Insurance thus simultaneously comprised a unifying and dividing element between the world of nature and the world of value.

3 Premiums of Presentism: Hidden Forces within the History of Law

Premodern insurance can be located in the context of mercantile communication and in the division between the world of values within the account books and the environment outside them. It arose with a primary focus on space—in particular, the long distances associated with the transport of goods. When compared to our current understanding of "insurance," this peculiar characteristic becomes even more clear and increasingly significant viewed in the context of a broader historical dividing line between merchant trade practices and the world of academic law—and thereby also between prereflective practice and the realm of theory and concepts. This intersection between practice and academic discourse is a critical zone, which serves a fruitful point of predilection in the context of a controversial topic in recent research in the history and theory of law: namely, the emergence of a premodern *lex mercatoria* in the sense of a self-determined, if not autonomous, transregional, tracasnsnational trade law, the sources of which are rooted in natural and international law. Interest in

Repräsentation, eds. Christoph Kampmann and Ulrich Niggemann, Cologne 2013, pp. 418–29, 418: "surviving policies are scant and contemporary commentary nearly nonexistent."

39 Clark, *Betting on Lives*, p. 18.

40 See note 18 above.

this discussion feeds mainly from the political concerns of current systematists, who advocate for the (re)creation of such a transnational, universal trade law in the current era of globalization, and who use historical arguments to demonstrate that such a law already existed in the Middle Ages. Historians debunk this argument as a construct, a myth,[41] and a "new medievalism,"[42] and emphasize that the sources of the historical trade law of the European *ius commune* and the *Merchant Law*, so-called by Malynes, were made up of contracts and legislations, rather than universal or universalized trade practices.[43]

In order to clearly and accurately classify the observations discussed below concerning the development of academic law in relation to the insurance contract, it will be necessary to describe the context of this discussion here: it will be shown that something like a natural international trade law *very probably* did arise in the seventeenth century. However, there is neither full consensus concerning this probability among its advocates, nor are the constructivist arguments of its opponents entirely compelling. The aim of the present study is to show that the emergence of this natural international trade law in relation to the insurance contract was due to the confrontation between actual mercantile practices and the theoretical stance of academic discourse.

In addition to the Datini archives and a few other rare early private merchant archives, our understanding of the realities of mercantile practice often relies more on indirect transmission within the context of legal disputes. It was

41 Albrecht Cordes, "Auf der Suche nach der Rechtswirklichkeit der mittelalterlichen *Lex mercatoria*," *Zeitschrift der Savigny-Stiftung für Rechtsgeschichte, Germanistische Abteilung* 118 (2001), pp. 168–84; Charles Donahue, "Medieval and Early Modern *Lex mercatoria*, " *Chicago Journal of International Law* 5, 1 (2004), pp. 21–37; Nicholas H. D. Foster, "Foundation Myth as Legal Formant. The Medieval Law Merchant and the New *Lex Mercatoria*," *forum historiae iuris* (18 Mar 2005, http://www.forhistiur.de/zitat/0503foster.htm); Vito Piergiovanni (ed.), *From Lex Mercatoria to Commercial Law*, Berlin 2005; Stephen E. Sachs, "From St. Ives to Cyberspace: The Modern Distortion of the Medieval 'Law Merchant,' " *American University International Law Review* 21 (2006), pp. 685–812; Emily Kadens, "The Myth of the Customary Law Merchant," *Texas Law Review* 90 (2012), pp. 1153–1206; Emily Kadens, "The Medieval Law Merchant: The Tyranny of a Construct," *Journal of Legal Analysis* 7, 2 (2015), pp. 251–89.

42 Ralf Michaels, "Legal Medievalism in *Lex Mercatoria* Scholarship," *Texas Law Review* 90 (2012), pp. 259–68. For an analysis of such "achronic" argumentation, to which new medievalisms belong, and which since the 1990s increasingly appear in many areas see Cornel Zwierlein: "Return to Premodern Times?—Contemporary Security Studies, the Early Modern Holy Roman Empire, and Coping with Achronies," *German Studies Review* 38, 2 (2015), pp. 373–92.

43 Kadens, "Myth." Guido Rossi opts for speaking of a *consuetudo mercatoria*, even in the plural, instead of a *lex*, to avoid the problems debated in recent literature concerning the existence of a somehow unified transnational *law*. See Rossi, *The London Code*, p. 435f.

these legal disputes about individual problems concerning insurance law that led legal scholars of both Roman and canon law to deal with the *contractus assecurationis*. While there may have been differences from one city or region to another, the rift between actual trade practice and the sphere of academic law tended to be quite large. Of the well-known legists of the fourteenth and fifteenth centuries, only Baldo degli Ubaldi, Paolo de Castro, Paris del Pozzo, and Filippo Decio concerned themselves marginally with this new type of contract; the most important cases were recorded by the otherwise hardly influential Genoese jurist Bartolomeo Bosco. Also significant are the specialized treatises and dissertations on insurance law listed here in appendix 1, beginning with Pedro de Santarém around 1480 (but first printed in 1552) and continuing into the seventeenth and eighteenth centuries—most significantly, the writings of the sons of burghers from Hamburg, Lübeck, and Bremen studying at northern German universities, as well as Dutch students in Leiden, Utrecht, and Groningen. While in the early stages, Iberian and Italian authors schooled in the *mos italicus* set the tone, from the seventeenth century onwards, northern European authors dominated—with them, the influence of *mos gallicus*, the Dutch "Elegante School," and early natural law became increasingly prevalent.

These published treatises respond to a need for orientation, spurred, not least of all, by the judges of the relevant courts, who were tasked with making decisions about insurance law, whether regarding the specific "assecurance" courts, or the other courts of first instance and appeals that handled regional and local matters concerning merchant law.[44] In the northern European treatises and theses of the sixteenth and seventeenth centuries, the older question, which was strongly influenced by canon law, concerning whether insurance was an allowed type of contract at all, was hardly discussed.[45] This is not only because the canonical prohibition of interest or the suspicion of gambling-type activities were no longer problems in Protestant countries—as was common in legal discourse, the relevant discussion was always transdenominational, based on the standard authors on the subject: in this case, Santerna, Straccha, Scaccia, and the late Spanish scholastics, as well as, perhaps, Conrad Summenhart.[46] Rather, at least concerning the northern

44 Concerning Venice, see the Nehlsen-Von Stryk, *Die venezianische Seeversicherung*.

45 On this, see Giovanni Ceccarelli, "Risky Business: Theological and Canonical Thought on Insurance from the Thirteenth to the Seventeenth Century," *Journal of Medieval and Early Modern Studies* 31 (2001), pp. 607–58.

46 Alvaro Vaz [Valascus], *Decisionum consultationum ac rerum iudicatarum in regno lusitaniae libri duo*, Frankfurt am Main 1608 [orig. 1593], cons. XVIII, p. 29f. and cons. LXIII, pp. 107–09 assembles the arguments for the permissability of insurance quite extensively relying on Soto and Navarrus (Martín Azpilcueta). Countering Conrad Summenhart, Luis

European arena, it seems that it was Hugo Grotius's annexation of the insurance contract within general contract law[47] in his *De jure belli ac pacis libri tres* of 1625 that served as the watershed moment which deprived this question of so much relevance that it was henceforth deemed worthy of only a few short paragraphs,[48] or sometimes only a few sentences,[49] or often not even a mention at all.[50]

Molina, *De iustitia et iure tomi sex*, vol. 2, disp. 507 "De contractu assecurationis. Et de quodam alio illi simili, Hispalique consuetudo", Antwerp 1615 (orig. 1602/03), pp. 488–92, 489 quickly dispenses with the question of the permissability of the contract using the argument of usefulness for the *bonum commune*. With the late/neoscholastic Leonardus Lessius, there is no longer any harsh argument against maritime transport insurance. See Leonardus Lessius, *De iustitia et iure ceterisque virtutibus cardinalibus libri* quatuor, Lyon 1653 [orig.1605], lib. 2, cap. 28, dub. IV, p. 292f.

47 The origins of the conception of a general contract law, not constrained by a strict typology of contracts and *actiones* as was the case in Roman law, is a classic topic of private legal history. Older works, such as Klaus-Peter Nanz, *Die Entstehung des allgemeinen Vertragsbegriffs im 16. bis 18. Jahrhundert*, Munich 1985 have been enriched through the focus on the efforts of the Salamanca School prior to Grotius, as seen in Wim Decock, *Theologians and Contract Law: The Moral Transformation of the* Ius commune (*ca. 1500–1650*), Leiden 2013.

48 Johann Adam Schragius, *Disputatio inauguralis juridica de assecurationis contractu [15 Sept.1642]*, Strasbourg 1642, n.1, f. A2r-v; Johann Marquard, *Tractatus Politico-Juridicus De Iure Mercatorum et Commercium singulari*, Frankfurt 1662, lib. II, cap. XIII, n. 3, p. 330; Johann Martin Brandes, *De assecuratione. Resp. Simon Heinrich Sanneman*, Rinteln 1664, p. 3 and 16–22 is still strongly influenced by late scholastic discourse; Heinrich Coccejus, *De assecuratione vulgo Assurantz. Resp. Henricus ab Aschen*, Frankfurt an der Oder 1693, § 10, pp. 7–9; Johann Nicolaus Hertius, "Paroemia XLIII. Wagen gewinnt/wagen verliert," in Johann Nicolaus Hertius, *De paroemiis juris germanicis libri tres*, in Johann Nicolaus Hertius, *Commentationes atque opuscula de selectis et rarioribus ex iurisprudentia universali, publica, feudali et romana, nec non historia Germanica argumentis*, Frankfurt am Main 1716, pp. 455–480, par. IIxf., 461f.; Jacobus van Ghesel, *Dissertatio juridica inauguralis theoretico-practica de assecuratione*, Leiden 1727, § VI, p. 14f. Martinus den Beer, *Dissertatio juridica inauguralis, de assecurationibus [12 Sept. 1729]*, Utrecht 1729, pp. 12–14 dedicates quite a lot of space to the question—however, rather than following the pattern of a serious calling into question of the canonical legal principles, he instead uses this as an opportunity to expand the topic of the *usefulness* of *assecuratio*.

49 Johann Loccenius, "De iure maritimo et navali libri tres [1650]," in *Scriptores De Iure Nautico Et Maritimo*, ed. Johann Gottlieb Heineccius, Halle 1740, p. 980; Heinrich Hahn, *Observata theoretico practica, ad Matthaei Wesenbecii in L. libros Digestorum Commentarios [. …]. Pars Prior*, Helmstadt 1659, p. 1087f.; Reinhold Kuricke, "Ius maritimum Hanseaticum. Olim germanico tantum idiomate editum, nunc vero etiam in Latinam translatum [1667]," in *Scriptores De Iure Nautico Et Maritimo*, ed. Johann Gottlieb Heineccius, Halle 1740, p. 830; Augustin Balthasar, *Dissertatio inauguralis De pecunia assecurationis in locatione conductione praediorum soluta* […]. Resp. Martinus Gustavus Fischer, Greifswald 1739, § II, p. 4f.

50 Franz Stypmann, "Tractatus de Jure Maritimo et Nautico [1652]," in *Scriptores De Iure nautico et maritimo*, ed. Johann Gottlieb Heineccius, Halle 1740; Michael Friedrich Lederer,

It had always been argued that insurance was a highly useful contract for trade and for the community as a whole—considerably more useful even than the bet, which was understood and allowed in Roman law. Martinus den Beer offers an early general formulation: "What is more human, more just, than to disperse public and private threats to the continued health of the republic among those who are willing and in agreement?"[51] Of course, insurance contracts must be conducted in a fair manner, especially concerning the balanced relationship between the premium and the risk. The utilitarian argument invalidated all manner of religious objections. The abstract discussion of whether a person with an insurance contract interfered with God's prerogative to deal blows of fate—which can still be found, at least in fragmentary form, in the governmental insurance discussions of the eighteenth century—is completely absent in legal discourse. Indeed, in the seventeenth and eighteenth centuries the older literature was increasingly "forgotten." The tradition of legal glosses and commentaries from Baldus to the late scholastics disappears, while only the most prominent pioneers of merchant law—Santerna, Straccha, and Scaccia—retain their status as authorities. In addition to them, northern European authors, such as Grotius, Ruland, Loccenius, and others are also cited in this regard.

However, one question stemming from the Roman doctrine of common law continued to persist: namely, the question concerning the contract type and the nature of the contract. This problem appears in the foreground in fifteenth-century legal discourse and retained an important role for a long time. The

Tractatio juridica de jure assecurationum [...]. *Resp. Albertus Fridericus a Syborg*, Wittenberg 1667; Bartholomaeus Leonhardus Svendendörfer, *De Assecuratione. Resp. Georgius Conradus Büttner*, Leipzig 1668; Johannes tho Westen, *De Assecuratione* [*1671*], in Jacob Brandmüller, *Disputationum iuridicarum selectiorum in inclyta Basiliensium Universitate disquisitioni publicae ac solenni expositarum* [...], Basel 1675, pp. 1–46; Johannes Wolfgang Textor, "Disp. XIX De Contractu Assecurationis, habita Heidelb. m. Majo 1674. Resp. Dn. Joh. Ottone Hamburgensi," in Johannes Wolfgang Textor, *Disputationes Academicae, duabus partibus distinctae* [...], Frankfurt am Main 1698; Conrad Vegesack, *Disputatio juridica inauguralis de Assecuratione*, Leiden 1704; Johannes Werlhofius, *Dissertatio de Instrumento assecurationis vulgo polizza* [...]. *Resp.* Martin Lucas Schele [*23 Sept. 1707*] Leipzig 1736; Franciscus Michael Poppe, *Dissertatio inauguralis iuridica sistens partem commentationis de litium assecurationis causa orientium decisione* [...], Göttingen 1752; Rudolph Gülich, *Dissertatio inauguralis iuridica de termino a quo, praescriptionis actionis ex contractu assecurationis secundum ordinationem assecurationis Hamburgensem*, Tit. XVII. Art. I. II, Göttingen 1776.

51 "jam quaeso quid humanius, aequius, quam pericula publica & privata pro salute Reipublicae instituta dividere inter volentes & consentientes?" (den Beer, *Dissertatio Juridica Inauguralis*, p. 13). However, it remains unclear which threats den Beer considered "public," and which "private."

question of the nature of the contract usually comprised the requisite focal point of any dissertation on the topic. The problem concerned how the "new" insurance contract—nonexistent in Roman law—should be classified in the overall system of contract typology. At this point, the dividing line between the world of academic jurisprudence and the world of mercantile legal practice becomes particularly distinct, because this problem was generated only by the intrinsic logic of academic law: the idea that there is no place for the insurance contract in the conventional typology of the contract is a potential obstacle to adjudication only within the Roman legal framework.

Grounding this discussion, deeper effects become apparent: the dispute concerning typological classification with all its consequential problems can be interpreted as a negotiation of different temporal systems—"presentism" versus an early framework of thought oriented toward future-calculations ("futurism" would be too strong a word here). The typology of obligations found in Justinian's *Institutions* leads to this schematic of different obligations, valid for Classical Roman law, as it was transmitted and received with several changes and adaptations in the *Ius commune*:

Lawyers of the fourteenth through eighteenth centuries were always interested in how the insurance contract would most easily be classified, since it did not exist in antiquity; the *foenus nauticum* was a functional equivalent but not a contract per se involving risk as the real and primary object. The classification was important because specific conditions, legal consequences, and other positioning within the legal system were guaranteed only for certain types of contracts. Because of this so-called "closed system of contract types," one could not simply add a new contract to Roman law—at least that was the commonly accepted doctrine (Fig. 3). Things not known in antiquity were problematic. The problem was solved de facto by narrowing the *contractus assecurationis* to another known type of contract in the Roman legal system. Often it was classified as *emptio venditio periculi*—i.e., as risk buying—or as a so-called "unnamed contract," an "innominate contract."

What was essentially an abstract classification problem was very relevant in practice, as an analysis of a northern European dispute will demonstrate: Traditionally, shipping and commodities insurers often wrote policies before receiving the premium, presumably to give the merchants the opportunity to first make a profit.[52] This practice was popular in Hamburg, Germany, and the Netherlands from around 1600 until the end of the 1620s. If, however, bad weather or theft prevented the delivery of goods, the insurer paid out the

52 "[E]ra frequentissima la liquidazione a respiro" (Melis, *Origini e sviluppi*, p. 97).

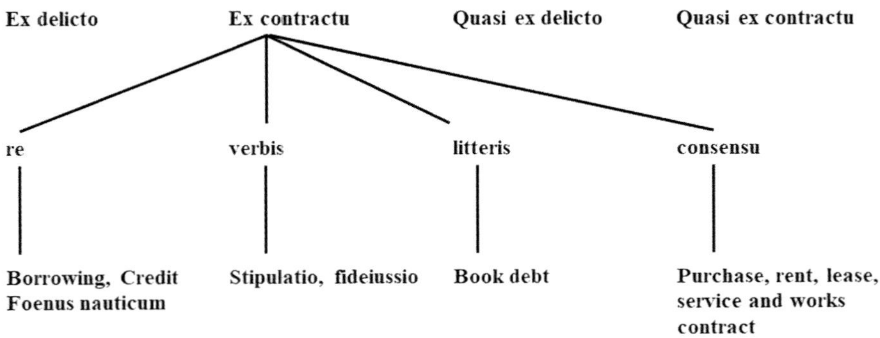

FIGURE 3 Bond types of Roman law (e.g. *Institutions* III, 13ff.)

policy anyway and deducted the amount of the premium from the insurance payment. Of course, by the beginning of the seventeenth century claims began to pile up, especially since the North African Barbary pirates had greatly expanded their activities.[53] This led to a new regulation by the Chamber of Insurance in Amsterdam between 1610 and 1620, making the *prepayment* of premiums compulsory. This regulation, however, was fully disregarded in the Netherlands.[54] Legal disputes arose in Hamburg in 1628/30 because the insurers ended up in dire straits due to high losses, and thus turned to the courts and legal scholars. According to the available information, the following cases represent only the second time in Hamburg—and thus in Germany as a whole—that "insurance" was invoked in legal terms. The first time was in 1590 in a Hamburg High Court (*Obergericht*) and involved policies that were issued in the Netherlands in 1588.[55] One of those cases made it to the Imperial Chamber Court (*Reichskammergericht*) through an appeal in 1606 and gained status as a legal precedent. At that time, it was still disputed whether there were judges at all in Hamburg who were competent in insurance matters, or whether—as supposedly was otherwise the custom in northern Europe—responsibility lay solely with the Chamber of Insurance in Antwerp.[56] However, in the 1620s

53 Only now the Barbary corsairs were changing to war ships competitive with the Europeans.

54 Van Niekerk, *The Development*, pp. 740ff.

55 This is not the place to sketch the diffusion of the premium insurance principle and practice from fourteenth-century Italy to the rest of Europe. A current overview is Adrian B. Leonard (ed.), *Marine Insurance. Origins and Institutions, 1300–1850*, New York 2016.

56 "... sondern es solten die beclagte durffen vast vermessentlich vernehmen lassen, das in dieser Stadt Hamburg niemand sei vorhanden, welcher diesen angezogenen Contractum

insurance claims were no longer unknown to Hamburg judges, and thus, when one of the parties appealed to the Imperial Chamber Court located in Speyer, they hoped for success and the annihilation of Hamburg judgements by the judges of a court and a city in a region far away from a major sea port city.

> In addition, it must be remembered that in this city for a good time now, for 50, 60 or more years, a legally admissible type of contract has been in use among merchants and traders, where one of the contracting parts assumed the risk and danger of ships that were sailing on the sea to far-away places and of their load, for which he [one party] asked that he [the other contracting part] promised and guaranteed to him 1, 2 or more percent [of the insured value] (which is called the *praemium*). That type of contract is now well-known among other nations, such as Italy, England, France, the Netherlands, and is called *contractus assecurationis* by the legal experts.[57]

Such cases, marking a turning point in development (for Hamburg around 1620, for other countries and cities earlier or later) usually resulted in a general dispute about whether the practice of merchants ought to be assigned its own sphere of law, or where there was a *single* legal system based on Roman law. The representatives of the plaintiffs, typically in agreement with the city council, were of the opinion that in matters concerning these new contracts, academic lawyers were at a clear disadvantage.[58]

It was from this context that the first specific treatise on insurance within the northern European realm emerged, entitled *Erledigung einer schweren/in*

---

assecurationis recht verstehen konne [- but this would not be true: the contract type and practice was now well known in Hamburg]" (HStA Hamburg RKG D 19, *acta prioris*, prod. Speyer, 2 November 1590); Kiesselbach, *Die wirtschafts- und rechtsgeschichtliche Entwickelung*, p. 17.

57 HStA Hamburg RKG H 127 Teil 1, Nr. 13b: Der Stadt Hamburg extrajudicial Schreiben an des Hochlöbl. Keiserl. Cammergerichts Praesidenten vnd Assessoren, 10.10. October 1631.

58 "Wan sonst ex Jure civili dieser contractus antiquis juris consultis incognitus zu decidiren, da solten sich noch wohl hundert consulenten vber zu todte disputiren, vnd ist nirgends in C. oder ff. [= in the Codex or in the Digests of the *corpus iuris civilis*] deßhalber eine jeglige decision zu finden, sondern muß per disputationes pro e contra die haubtfrage ratione praemiorum ventilirt werden, so können auch die juris consulti eben so wohl in solcher Kauffmansfrage, vnd zwar viel ehr, als verstendige vnd des gebrauchs jn Hamburg erfahrene Handelsleuthe, fehlen" (HStA Hamburg RKG H 127, part 2, no. 40: "Duplica ad Replica den 18. Augusti 1636 von den Appellanten vbergeben, Pro interesse Buirgermeister [sic] vndt Raht der Stadt Hamburg in Sachen Helbergk contra Budier"—prod[uciert = brought forward, processed] 17 April 1637).

Assecuration Sachen/vorgestalter fragen/Ob nemlich/wann einer auff ein Schiff versichern lest/aber kein praemium bezahlet/vnd hernacher zur See schaden erfolget/als dann denselben er als Assecuratus, von den Assecuratoribus mit rechte zu fordern vermüge? (Response to a difficult question that has arisen in the matter of insurances, whether, if someone is insuring a ship, but has not paid the premium [in advance], and if later a damage occurs on the sea, if then he might be entitled to ask as *assecuratus* [reimbursement of that damage] from the *assecuratoribus*?). This treatise, published in 1630 in Hamburg, was written by the Aachen-born Rutger Ruland, who had long-practiced in Hamburg. In the treatise, Ruland ultimately presents a strict Roman-law interpretation in response to the customary view of maritime trade: since the notion of insurance was unknown in the Roman law, it could only be treated as a *contractus innominatus*. Such a contract, however, required a transfer of something in order to be complete, thus involving, in this case, the transfer of the coins used for the premium payment. The nearly two-hundred-page treatise essentially revolves around this classification discussion, which also involves the question of whether insurance operates as an actual contract or not and whether payment claims of the insured were valid. This discussion affected not only the justification for the existence of the Hamburg insurance market, but also the entire northern European market.

The majority of tradesmen and other insurers defended the perspective of autonomous merchant customary law on what concerned their business against the perspective of Roman academic law on it.[59] Ruland saw himself as a lawyer in the cross fire of criticism: "many have brought forward that it would

59 On the theory and practice of the appeals in Hamburg trade disputes see Ebert-Weidenfeller, *Hamburgisches Kaufmannsrecht*, pp. 16–70; Anja Amend-Traut, *Wechselverbindlichkeiten vor dem Reichskammergericht*, Cologne 2009 treats the question of appealability of commercial issues before the Imperial Chamber for the parallel case of the *cambio*, likewise received from the Italian early capitalist communication. The question of whether the chamber would accept such cases or not was of greatest importance throughout the early modern period; in insurance law, the question of premium life insurance was even more problematic than transport and later fire insurance, cf. Cornel Zwierlein, "Tod und Leben. Das Reichskammergericht und die Lebensversicherung im 18. Jahrhundert," in *Kulturen des Risikos im Mittelalter und in der Frühen Neuzeit*, ed. Benjamin Scheller, Berlin 2019, 147–173. The particular case that gave occasion to Ruland's treatise and inquiry was rendered as such: A ship, "Sancta Maria," belonging to "A," which was supposed to sail from Lisbon to Bahia in Brazil sank, and 500 ducats (= 1/16 of the ship's value) were insured by (probable Hamburg) insurer "B" at a 12% premium; however, the 60 ducats premium was not yet paid by "A" to "B" before the ship sank (Rutger Ruland, *Erledigung einer schweren, in Assecuration Sachen vorgestalter Fragen*, Hamburg 1630, p. 38f.).

not be allowed for academic lawyers to pronounce their judgement in matters of insurance, because they would have no knowledge of that." Insurance matters were considered mercantile "special cases," subject to customary laws (the "*stilus mercatoriae*"), fairness according to mercantile law, and the "conscience" of the merchants. Ruland notes almost angrily:

> Because many merchants—even of those who want to be coplaintiffs in this case—cry *stilus stilus, aequitas aequitas, conscientia conscientia*; or they say "it is a matter of conscience" and therefore dare to enforce and direct everything they can think of; the learned jurists will never appreciate and allow this but will confront them with the truth, that first, a fairness born just in one's own mind [*aequitas crebrina*] can never be admitted ... and second, that the conscience must likewise be brought into conformance not with anybody's particular wishes, but with the laws.[60]

Stilus mercatoriae was thus pitted against the *leges* of Roman law. First Ruland reviews the common classifications: "It was obvious that the insurance contract was not used in this form by ancient legal experts, but that it has been newly invented; therefore, it cannot be classified among the nominate contracts of Roman law."[61] While Santerna suggested that the insurance contract ought to be classified as a purchase contract according to Roman law, Benvenuto Straccha—"the patron of all merchants, as the elder merchants who still knew him from Ancona used to call him at the stock exchange"[62]—argued that it should be classified as a rental contract, and most scholars classified the insurance contract as an innominate contract, thus requiring comparison with other types of contracts for its regulation.[63] In addition to using several southern European authors, Ruland also attempts to verify his stance through reference to a German author, Anton Hering—one of the first authors north of the Alps to address the problem at all.[64]

In the end Ruland accessed a specialized form of proof to support his viewpoint: he wrote to over twenty lawyers in Germany—as well as the Chambers

60 Ruland, *Erledigung*, p. 95.
61 Ruland, *Erledigung*, p. 52.
62 Ruland, *Erledigung*, p. 48.
63 "Auß welchen allen dann scheinet zu schliessen seyn/daß ob gleich dieser contract etwas, von der Natur der contractuum, so keinen Nahmen haben/participieret, doch nach Art der contracten, welchen ein gewisser Name zugeeignet/sich reguliren muß: Traditur enim, quod quando contractus innominatus quid participat de contractu nominato, ejus illum naturam induere" (Ruland, *Erledigung*, p. 50f.).
64 Ruland, *Erledigung*, p. 57.

of Insurance in Amsterdam, Spain, Portugal, and Italy—and printed their (naturally, according to him, positive) responses in addition to other short opinions from other lawyers in the appendix of his treatise. Therefore, by 1630 he had assembled a huge number of professional legal precedents and common law empirical investigations "from the people" (or "from the nations" if we take the later notion of a *laws of nations* seriously). According to general consensus within legal history, comparative law in the context of commercial law begins only in the eighteenth century.[65] However, although Ruland certainly lacked the appropriate methods and systems, in practice he was conducting comparative law. Although Ruland does not use the term *ius gentium*, he definitely wanted to give the impression that he had inductively established a universally applicable law. In the absence of clear legislative competence in trade matters, he proposes a proto-Grotian international trade law in opposition to the alleged *stilus mercatoriae*: Ruland was relying on forms of empirical European-wide investigation regarding customary legal practices and local standards. This was a practice close to the theory of generating natural international law as formulated by Grotius in the prologue of *De jure belli ac pacis* (1625). Ruland's approach clearly distinguishes how such conflicts were handled at the beginning of the seventeenth century from the approach used in the English disputes of the Middle Ages, in which judiciary bodies were involved with the application of special laws regulating merchants (the *lex mercatoria* of the twelfth and thirteenth centuries) and the use of trade experts in the courts. While cases alluding to the beginnings of a systematic and historical legal discussion have been much discussed recently,[66] at the beginning of the seventeenth century this discussion did not concern only the local customs of Hamburg or the legitimacy of local trade law in contrast to the respective local or national legal standards; rather, what was at stake was the empirical inquiry into a general, new type of standard commercial contract, valid throughout Europe—different in all port and trade cities in some practical details, but also integrated by and into compliant mercantile communication networks, and dependent on its inclusion within the universal Roman legal systematics of the *ius commune*.

How might this point of confrontation around 1630 be interpreted? Looking at the "temporal regimes" element that is intertwined through these different perspectives, it becomes clear that embedded within the opposition "merchants vs. academic lawyers," there is also a conjunction between "presentism" and the "primacy of the future." As has been shown, insurance was,

65 Karl Otto Scherner, "Allgemeine Rechtsgrundsätze und Rechtsvergleichung im europäischen Handelsrecht des 17. und 18. Jahrhunderts," *Ius Commune* 7 (1978), pp. 118–35.

66 See Ch. 2, notes 41 and 42 above.

for merchants, simply an accounting trick, and therefore the prepayment of the premium was not a major concern. It was fairly unproblematic in common mercantile practice to settle up the premium after the merchant's ship had successfully returned and the ensuing profit had been generated. The academically trained lawyer, in contrast, demanded the payment—and thus the anticipatory and calculated element of insurance—in advance and only issued the policy after the premium was paid. Lawyers had little to do with what today's sociologists of risk and time consider to be the decisive criterion for the emergence of a "modern concept of risk"—namely, the probabilistic calculus, "calculation," and the notion of making the future available for the needs of present planning and acting. The insistence on the correct execution of the contractual arrangements of a *contractus innominatus* clearly demanded a legal fixation on the future, however—which, while inherent in every contract, is particularly relevant to the insurance contract. Paradoxically the element of insurance as a "taming and mastering of the future" was *not born in the merchants' communication networks*, even though the insurance contract originates there. Rather, the necessity of abstract future-orientation was first invoked by scholars in the field of academic legal discourse, who took note of this new form of contract and the related institutions developing within the commercial sphere. It was the presentism of the commercial world of value that was on trial in lawsuits. With Ruland's treatise, the transfer of insurance matters from the commercial world to the academic legal world was realized for the first time in the northern European arena, followed by a large number of theses and treatises during the later seventeenth and eighteenth century.

4 International Trade Law and Insurance as an Achievement of the *Moderni*

While until, and still around, 1628–30 the entire discussion originating in legal disputes revolved around the problematics concerning the classification of contract types, the sublimation of the insurance contract within general natural and international commercial law that Grotius first addressed, for a long time, hardly even penetrated the consciousness of commercial law experts. It began to assert itself in that realm only slowly and gradually. The *assecuratio* was not specifically defined in any chapter on specific contracts in Grotius's *De jure belli ac pacis libri tres*; rather, it was relegated to the introductory parts of the work, where he developed his doctrine of general contract law. For Grotius, it served as an example of a contract type that was hardly known at the time but is very common today. It was unproblematic to conclude such a contract in terms of natural law because natural law does not acknowledge the precise differences between named and unnamed contracts as recognized within civil law ("at ius naturae ignorat haec discrimina"). Grotius wanted to revert to the very basic notion of contracts as being based on the principle that there is one party of a contract that gives something, and the other that returns something: either something is given for something else, or money is given for money (exchange), or something is given for money, or the use of something is given for something else, or something is given for the use of something else, or the use of something is given for the use of something else, or the use of something is given for money.[67] Through the basic concept of natural law, the classification problem disappears—accompanied, of course, with an almost obvious solution to the problem of premium payments described above by Ruland. The problem posed by the fact that the classification as *contractus innominatus* demands the immediate transfer of a good (the premium as the *res* by which an innominate contract was to be concluded) disappears within this natural law perspective. Practical provisions for the insurance contract itself are, however, rare in Grotius's work.[68]

Grotius's disregard for Roman legal classification was initially greeted with a surprisingly faint echo in the specialized literature of insurance law. Many seventeenth- and eighteenth-century authors continue to raise the issue (presented by Santerna and Straccha) of whether insurance ought to be considered a purchase, a rental, a guarantee, a *stipulatio*, a maritime advance, an innominate contract, etc., and ponder the similarities and differences among named contract types, even though they also quote Grotius in other parts of their writing, and very frequently adopt Grotius's definition of insurance word for word.[69] According to the prevailing doctrine, however, the insurance contract was still an innominate contract in the sense of Roman law, occasionally formally referred to as *stipulatio, contractus, conventio* to be on the safe side.

67 Hugo Grotius, *De iure belli ac pacis libri tres* [*orig. 1625*], eds. B.J.A. De Kanter-van Hettinga Tromp, Robert Feenstra, and C.E. Persenaire, Aalen 1993, lib. II cap. xii § iii, pp. 341–343.

68 Grotius specifies only the rule that the insured object cannot be already known to have arrived safely or to have been lost at time of the commencement of the contract; and the price had to be proportional to the danger: Grotius, *De iure belli ac pacis libri tres*, lib. II cap. XII § xxiii, p. 354.

69 As, for instance, Schragius, *Disputatio inauguralis*, f. A3v-B2r (discussion of the contract types), C1r (reference to *De jure belli ac pacis*); Lederer, *Tractatio juridica*, § 5–10, f. A2v-A4r: of course he made reference to Grotius, but did not import the natural law framework; very traditional: Brandes, *De assecuratione*, pp. 4–6; Svendendörfer, *De Assecuratione*, f. A3r-B1r; tho Westen/Brandmüller, *De Assecuratione*, pp. 15–17; Textor, *Disputationes Academicae*, pp. 295–300; Coccejus, *De assecuratione*, pp. 5–7; Vegesack, *Disputatio juridica*, pp. 7–10.

Therefore, according to the perspective of legal scholars, the "colonization of the future"—the demand of advance payment on the contract—was valid, otherwise the contract itself would not be enforceable.[70]

However, one author, Franz Stypmann of Greifswald (1612–50), quickly adopted and further developed the new interpretation of insurance laid out by Grotius. Stypmann, the son of an old Greifswald merchant family, met Grotius in 1637 in Paris while he was serving as *Hofmeister* (steward/tutor) to a Holstein nobleman on a *Grand Tour*. Henceforth he was a keen student of the famous Dutch founder of natural law; according to his obituary, he was rarely seen without Grotius's major work, *De jure belli ac pacis,* in hand.[71] This bore

70 "quod sit auersio periculi stipulatio vel cautio idonea de praestanda indemnitate rei vel sortis alienae circa casus fortuitos, pro certo periculi pretio." (Loccenius, "De iure maritimo," p. 980); "Assecuratio hic significat conventionem avertendi periculi, seu qua quis praemio sibi dato suscipit periculum rerum alio transvehendarum." (Hahn, *Observata theoretico,* p. 1087); "quod sit mercium, quae alio deferendae, sinistro casu, sive periculo tam terrestris quam maritimi itineris amittuntur, restituendarum, vel damni inde emergentis per conventionem facta cautio." (Marquard, *Tractatus Politico-Juridicus,* p. 331); "assecuratio ... est contractus, quo praemium datur, ut quis suscipiat periculum mercium alio transvehendarum. Vel etiam quo quis incertum periculum rerum, alio transvehendarum in se suscipit, ut praemium detur." (Georg Adam Struve, *Syntagma Juris Civilis* [*bzw. Jurisprudentiae*] ... [*orig. 1655–63*], Jena 1687, p. 919); "quod sit contractus innominatus, quo rerum quarundam periculum pro pretio, aut alio quocunque interesse suscipitur." (Lederer, *Tractatio juridica,* f. A2v); "quod sit conuentio, qua quis periculum traiiciendarum, vel alio terra marique vehendarum mercium, vel etiam nauis ipsiusmet, et indemnitatis eo nomine praestationem in se, pro certo periculi pretio, siue praemio recipit." (Kuricke "Ius maritimum Hanseaticum," p. 829); "Assecuratio est Contractus innominatus, quo periculum rerum alio transvehendarum, alterius securitatis gratia, pro praemio suscipitur." (Svendendörfer, *De Assecuratione,* f. A3r); "Assecuratio est contractus innominatus, quo indemnitas certo interveniente pretio praestatur." (tho Westen/Brandmüller, *De Assecuratione,* p. 15); "Quod sit sponsio mutua in vim contractus innominati ex consuetudine recepta, qua quis mercium vel rerum alienarum suscipit periculum pro eventuali participatione commodi, si scil. salvae sint." (Textor, *Disputationes Academicae,* p. 300); "Conventio de periculo mercium alio transvehendarum suscipiendo" (Coccejus, *De assecuratione,* p. 7); "quod sit contractus, quo quis accepta certa mercede in se suscipit & praestat pericula, quae subiturae sunt merces in alia loca potissimum mari transportanda, ita ut si has interire contigerit, Assecurator earundem pretium domino restituere teneatur." (Vegesack, *Disputatio juridica,* p. 7 [according to Pufendorf]); "... vulgo vocant contractum, quo mercatores sibi rerum periculo exponendarum indemnitatem circa casus fortuitos pro certo periculi pretio stipulantur" (Balthasar, Dissertatio inavgvralis, p. 3); "assecuratio est Contractus, quo quis periculum alterius, citra respectum ad aliam conuentionem, pro certo praemio in se suscipit." (Poppe, *Dissertatio inauguralis,* p. 2).

71 "... Cujus [i.e., Grotius's] incomparabiles de Belli Pacisque jure libros nunquam fere deponere e manibus, & tanquam abstrusissimae prudentiae & divini humanique juris sacrarium solebat venerari. Quem ille, ut fuit comitate insignis, postquam sibi virtute

fruit when he was appointed as professor at the university in his hometown.[72] His significant work for our purposes, *De jure maritimo & nautico,* was only published posthumously in 1652 by Anton Gunther Fritzius.[73] Written in the style of humanist Dutch jurisprudence, later known as the "Elegante School,"[74] the question of the contract type of *assecuratio* systematically penetrates the text—in particular, Stypmann tackles Ruland's most detailed treatise at that time.[75] One after another, he rationally rejects each argument which would classify insurance as a type of contract within Roman law and offers a legitimate foundation for the insurance contract within international natural law, as a new and defined (nominate) contract based on actual contemporary usage.[76] He confidently introduced this as a "new view of the author, that [the insurance contract] is a nominate contract."

> If, in fact, most agreements of international law have received their designation through a process of tacit agreement, why could insurance not also be accepted by many or most peoples? Or, more specifically, if the Romans could name [i.e., had the power to define] the *stipulatio* [as] a [new] nominate contract, who forbids other peoples that same privilege as concerns the insurance contract?[77]

---

sua & eruditione conciliaverat, cottidie fere accessit, & eo tempore, quod Amicis largiri consueverat, de gravissimis rebus disserentem audivit." (Johannes Pomereschius, "Oratio in Mortem [...] Francisci Stypmanni [...]," in *Memoriae jurisconsultorum nostri seculi clarissimorum renovatae decas prima*, ed. Henning Witte, Frankfurt am Main 1676, pp. 256–276, 266). On Stypmann cf. also Herbert Langer, "Handel und Bildung in pommerschen Seestädten im 16. bis 17. Jahrhundert," in *Stadt und Literatur im deutschen Sprachraum der Frühen Neuzeit*, eds. Klaus Garber et al., vol. 1, Tübingen 1998, pp. 729–736, 735; Maciej Ptaszyński, *Beruf und Berufung. Die evangelische Geistlichkeit und die Konfessionsbildung in den Herzogtümern Pommern, 1560–1618*, Göttingen 2017, pp. 424–426.

72 Kjell A Modéer, "Aktenversendung und Greifswalder Juristenfakultät," in *Justitia in Pommern*, ed. Dirk Alvermann and Jürgen Regge, Münster 2004, p. 78.

73 Cf. briefly, Karl Otto Scherner, "Die Wissenschaft des Handelsrechts," in *Handbuch der Quellen und Literatur der neueren europäischen Privatrechtsgeschichte*, vol. 2.1: *Wissenschaft (1500–1800)*, ed. Helmut Coing, Munich 1977, p. 854f.

74 Cf. Govaert C.J.J. van den Bergh, *Die holländische elegante Schule. Ein Beitrag zur Geschichte von Humanismus und Rechtswissenschaft in den Niederlanden 1500–1800*, Frankfurt am Main 2002.

75 Stypmann, "Tractatus de Jure," pp. 438–452.

76 This usage of natural law reasoning is scarcely recognized in the literature (cf. Scherner, "Allgemeine Rechtsgrundsätze"; La Torre, *L'Assicurazione*, p. 177 n 405; only Gian Savino Pene Vidari briefly discusses Stypmann in his "Sulla classificazione del contratto d'assicurazione nell'età del diritto comune," in *Rivista di storia del diritto italiano* 71 (1998), pp. 113–37, 125–27).

77 "Si enim juris gentium pleraeque conuentiones nomen acceperunt tacito consensu, cur non idem potuit adsecuratio a multis vel pluribus gentibus adipisci? Vel si a solis romanis

If even the Byzantine Emperor Leo VI could distinguish the emphyteusis from a rental contract and a purchase, why would it not be possible in 1563 (second edict: 1570) for the Spanish King Philip II to distinguish insurance from rent and other nominate contracts after countless small disputes?[78] The question now posed was the question of primacy between Classical Ancient law and the law of modernity modern times. Stypmann continues—along the lines of Donellus and Grotius— that it is not only the name that creates the *actio* (i.e., the possibility of making a legal claim at court as with the *praetor* in Ancient Rome), but rather the underlying, legally binding *causa*. He therefore urges against formalistically examining the old classification of contracts and the magic of naming and typology; rather, he suggests looking substantively at the habits and necessities of the practice of law. This applies to international contract types generally, since they are not bound by any explicit law, but by the tacit practice which accommodates human needs. The fact that the origins of this practice are unknown is irrelevant; if the memory of the origins of a legal tradition has been extinguished, but the tradition is nevertheless maintained by all or at least by a great number of countries, then it should be deemed accepted law. Once again Stypmann boldly questions why neither the people nor states have the same rights once enjoyed by the Romans and the Roman emperor, who constantly introduced new nominate contracts (*invenerunt*).[79] There is no proof that the Romans ever sought to limit the number of nominate contracts for all time and there is also no justification for their ability to do so, even if they had wanted to. The name of the respective *actio*—the basis of the claim—could therefore also be freely and newly defined. There is no reason one could not have an *actio assecuratoria*—applying the contemporary wording in use among merchants—or, if one wanted to stay true to the classical Latin, an *actio aversi periculi*.[80] Stypmann thus introduced a new nominate contract of commercial law.[81]

---

stipulatio potuit nominata conuentio declarari, quis vetabit alios populos, vt adsecurationi id tribuerent?" (Stypmann, "Tractatus de Jure," p. 448 n. 162–164).

78 Ibid., p. 448 n. 165.

79 Ibid., p. 449. "Nullibi enim dicitur, quod Romani soli hanc potestatem habuerint suo tempore, contractum nominatum faciendi. Quod illis licuit, et hodie aliis gentibus in suis rebuspublicis licet" (Ibid., p. 450).

80 Ibid., p. 469.

81 "Adsecuratio [est contractus nominatus de iure gentium], conuentio de mercibus tuto aliunde transferendis pro certo praemio seu est auersio periculi: Vel contractus indemnitatis pro certo pretio." (Stypmann, "Tractatus de Jure", p. 453); "quod sit contractus Iuris gentium nominatus nudo consensu constans de periculi rerum et personarum siue mari, siue terra transuehendarum auersione pro certo ac definito pretio." (Werlhofius, *Dissertatio de Instrumento*, p. 9).

While in his discussion of general contract law Grotius postulated that, especially on an abstract level, the strict distinction between innominate and nominate contracts was insignificant and argued for a natural law understanding concerning different obligation-based contract types, Stypmann insists on the nominate contract as a "type." Through a theoretical legal approach which was, for this time period in Europe, exceptional in its application of historical relativity and criticism, Stypmann proclaimed that commercial law practice itself was authorized to introduce new nominate contracts based on natural international law. What otherwise had emerged, on the merchants' side, as a vindication of a free and special sphere of commercial law pertaining to their intrinsic needs and rights (based on allegiance to the customary law of the *lex mercatoria*[82]) is developed here as a rationalistic continuation and variation of the Grotian theory of a natural international trade law.

A specifically modern horizon of time and history—a latent competition between *antiqui* and *moderni*—is also quite obviously imbedded within this legal sphere. Insurance was an *inventio* of recent times that could be shifted self-consciously from the obscuring shadows of Roman law into the light of a particularly modern legal order. Thereby, the catch-all of the "innominate contract" lost its validity, and not just concerning insurance contracts. In this way, Stypmann accomplished something beyond the ken of even the nineteenth-century Pandectists—namely, he integrated the insurance contract as a standalone type of contract in a nonetheless Roman-based framework.[83]

Although Stypmann's treatise, which was published in Greifswald in 1652 and reprinted in Stralsund in 1661, enjoyed a hearty reception within insurance literature, few seemed to recognize the entire scope of his radical natural law approach. Occasionally, the categorization of insurance contracts within the Roman classification schema would be abandoned and insurance contracts would simply be described (without reference to Stypmann) as a new type, but there were no reflections on the wider consequences for the entire system of law.[84] However, Stypmann attracted quite a few followers, who incorporated

82 See above Ch. 2, notes 41 to 43.

83 Louis Pahlow, "Ein 'Geschöpf der ganzen civilisirten handeltreibenden Welt.' Der Versicherungsvertrag im Deutschland des 19. Jahrhunderts," *Zeitschrift für neuere Rechtsgeschichte* 29 (2007), pp. 18–46, here 23–25.

84 As in Wolfgang Adam Lauterbach, *Q.F.E.F.S. Theseis de Jure in curia mercatorum usitato. Resp. Lucas Langemann.* [*12. Calend. Octobr.*], Tübingen 1655, no. 91, 17: "Assecuratio non est contractus innominatus [...] non locatio [...] non fidejussio [...] non emptio [...] sed novus, & jure civili incognitus contractus, quamvis simile aliquid apud veteres inveniatur." Similarly, but in a more harmonizing way, Coccejus, *De assecuratione*, p. 6f.: "Vera itaque sententia est, assecurationem non esse certam & definitam contractus speciem,

his natural international law argumentation in their dissertations—for example, Johann Friedrich Rhetius of Frankfurt/Oder in 1666;[85] Matthäus Meier of Bremen, who in 1666 wrote a particularly detailed dissertation (supervised by Jakob Brandmüller) which was, however, not published until 1739;[86] Johannes Werlhofius of Bremen, whose 1707 dissertation, supervised by the Hamburger Martin Lucas Schele, was finally published in 1736;[87] Johannes Nikolaus Hert of Giessen in 1716;[88] and without naming Stypmann explicitly, Jakob van Ghesel of Leiden's 1727 dissertation.[89] Meier appropriates Stypmann's reasoning most fully and completely, arguing that whether one considered insurance an "invention" (*inventio*) of international law or a creation of new civil law (through the ordinance of Philip II), the insurance contract constituted a nominate contract in its own right.[90] Nevertheless, it is striking that many of the texts that finally adopted Stypmann's absolute definition of the insurance contract as a part of a general natural international contract were indeed written soon after the initial editions of Stypmann's work (1652/1661), but were actually only published and made accessible seventy or eighty years later around 1730 (a new edition of Stypmann's treatise was issued in 1740). The public acceptability of

sed in se ac sua natura conventionem seu pactum in genere. Uti vero pacta terminos nudae conventionis varie transeunt, ita & assecuratio; varie enim in aliud genus transit, & varie contrahitur, verbi gratia per emtionem aleae, si ita convenit, ut assecurator existente periculo precium certum mercium perditarum solvat; emit enim periculum seu aleam, quod utique potest [...] Vel per contractum innominatum, cum quid datur assecuratori, ut periculum suscipiat. Vel per donationem cum gratis suscipit; Vel per stipulationem si accesserit stipulatio, & in eam deductum sit pactum: omnibusque his negotiis fidejussio accedere potest." Similarly, den Beer, *Dissertatio Juridica Inavgvalis*, pp. 10–12; Poppe, *Dissertatio inavgvralis*, p. 1f.

85 The first being Johannes Friedrich Rhetius, *Q.D.F.E.J. Dissertatio Juridica De Versura* [...]. *Resp. Walter Neuhaus.* [*24.9.1666.*], Frankfurt an der Oder 1666, f. H3r-H4v.

86 Matthaeus Meier, *De assecurationibus mercatorum tractatus denuo editus* [*1668*], Bremen 1739. It is not known who presided over the *disputatio*; however there is a poem of dedication from Brandmüller (p. 126f.). In the dissertation, defended a bit later by tho Westen of Hamburg and advised by Brandmüller, the *communis opinio* of the insurance as innominate contract prevails (tho Westen/Brandmüller, *De Assecuratione*, p. 15); while he briefly reiterates Stypmann's argument from natural international law, he eventually concludes: "id solum monstramus, cui juris civilis negotio Assecuratio sit magis vicina" (p. 17).

87 Werlhofius, *Dissertatio de Instrvmento*, § VI, p. 9.

88 Hertius, *Commentationes*, no. XXVI, p. 472: However, the influence of Grotius is evident here only in what concerns the quite general position that the difference between innominate and nominate contracts would be insignificant in the realm of natural law.

89 Ghesel, *Dissertatio juridica*, § VII, p. 15—the origins in natural law, are however, not made explicit here.

90 Meier, *De assecurationibvs*, pp. 96–108.

this view therefore appeared at a point in time at which the question of the contract genre was no longer pressing and insurance law literature increasingly limited itself to empirical descriptive recording of local regulations and legal customs.

However, until the eighteenth century, the vast majority of authors presented the classification of the insurance contract as an innominate contract.[91] The boundaries of academic legal discourse were still closely aligned with the typological system of Roman law; insurance therefore always had to be camouflaged, translated into another type of contract. That which is new can only be integrated into an old system through dissimulation. Even when one recognized the usefulness of insurance, it remained a foreign body in the eyes of the authoritative order and left a stale aftertaste on the basis of its atypicality and misfit form. It was considered to be a crude trade practice of port cities on the periphery of the continent, where one had to make due with such makeshift practices.[92] With regard to the *lex mercatoria* discussion, a temporal and spatial evolutionary niche—occurring briefly in 1650/1730—allowed for the development of a true international trade law in a conceptually elaborated form within a prenational timeframe. Since that unexpected development occurs *after* the Grotian innovation and *prior to* the nationalization of all legal systems, it creates the necessity to discuss the new creation of such a sphere of law in a postnational context. In this context, the primary question is to what extent the exchange between academic jurisprudence and commercial practice resulted in a transformation of the insurance principle in terms of the negotiation of temporal regimes as addressed here. The future-orientation associated with the later universal and transferable principle of "insurance" emerged in this context.

91 Loccenius, "De iure maritimo," p. 980; Hahn, *Observata theoretico*, p. 1088; Marquard, *Tractatus Politico-Juridicus*, p. 332; Struve, *Syntagma Juris Civilis*, p. 918; Brandes, *De assecuratione*, pp. 4–6; Lederer, *Tractatio juridica*, §§ 5–10, f. A2v-A4r; Svendendörfer, *De Assecuratione*, f. A3r-B1r; tho Westen/Brandmüller, *De Assecuratione*, pp. 15–17; Textor, *Disputationes Academicae*, pp. 298–300; Samuel Stryk, *Praelectiones Viadrinae de cautelis contractuum necessariis* […], Frankfurt an der Oder 1684, p. 483f.; Vegesack, *Disputatio juridica*, pp. 8–10; Kuricke, "Ius maritimum Hanseaticum," p. 830, and Johann Adolph Cronius, *De jure assecurationum, vom Assecurantz-Recht*, Rostock 1725, p. 6f. deliberately abstain from any determination, as does Poppe, *Dissertatio inauguralis*, p. 1f., who, however, praises Stypmann und Coccejus though both were rejecting the Roman typology as strictly binding.

92 For a more general theoretical outline, see Cornel Zwierlein, "Pluralisierung und Autorität. Tentative Überlegungen zur Herkunft des Ansatzes und zum Vergleich mit gängigen Großerzählungen," in *Pluralisierungen. Konzepte zur Erfassung der Frühen Neuzeit*, eds. Jan-Dirk Müller, Wulf Oesterreicher and Friedrich Vollhardt, Berlin 2010, pp. 3–30, 24f.

The persistence of the classification as innominate contract among the majority of lawyers in this early modern context indicates the persistence, and perhaps even growth, of the abstract future-orientation of insurance in the framework of thought dominating the academic world. The new classification as a natural law contract in its own right initially seems to obscure the matter, because it does not immediately solve the conflict of anticipatory action vs. presentism. In fact, the fundamental significance of this international law solution may be unrelated to this issue. However, it can be understood as a decisive indicator and reflex on the level of legal discourse, in the sense that "insurance" becomes generalized as a new principle in the academic world. A few years before Stypmann's treatise, insurance against "slavery", establishing the payment of ransoms in case someone was taken captive by pirates, was established for the first time in the same area of northern Europe—in both cooperative and premium-based forms.[93] A few years after his treatise, the establishment of public fire insurance funds (and in England, of fire insurance companies) and the institutionalization of life insurance (in England) decisively mark a new era of insurance history.

Stypmann's emphasis on the insurance contract as a legal institution *sui generis*, exposed with all the pride of a part-baroque, part-early Enlightenment academic style of reasoning, was a semantic parallel and mirror to those developments on the level of institutional innovations. "Insurance" was thus liberated from the world of account books and transformed into an "achievement of modernity, of the *moderni*." This consciousness of "insurance" as a new principle of great general societal significance is first suggested in the juridical treatises of the seventeenth century and, for the first time, goes beyond the notion that it would be a "*novus contractus*" that could already be encountered in the writings of glossators and the late scholasticists. Insurance was not mentioned by the Renaissance precursors of the *Querelle des anciens et des modernes*:[94] from the perspective of the Renaissance philosophers insurance was merely an aspect of menial everyday business dealings; furthermore, around 1600, it only

93 Cf. Magnus Ressel and Cornel Zwierlein, "The Ransoming of North European Captives from Northern Africa. A Comparison of Dutch, Hanseatic and English Institutionalization of Redemption from 1610 to 1645," in *Seeraub im Mittelmeerraum*, eds. Nikolas Jaspert and Sebastian Kolditz, Munich 2013, pp. 377–406.

94 General literature on the classic *Querelle* rarely takes into account technical inventions and global comparisons. Cf. for the controversy of the 1680s (Perrault, Fontenelle et al.) Anne-Marie Lecoq (ed.), *La Querelle des Anciens et des Modernes, XVIIe-XVIIIe siècles*, Paris 2001; Christelle Bahier-Porte and Claudine Pouloin (eds.), *Écrire et penser en Moderne (1687–1750)*, Paris 2015; Larry Norman, *The Shock of the Ancient. Literature and History in Early Modern France*, Chicago 2011; Levent Yilmaz, *Le Temps moderne. Variations sur*

concerned the specialized area of maritime transport.[95] After the preparatory work by legal scholars like Stypmann, the topic of insurance increasingly attracted the attention of the academic world. However, it was not until the end of the eighteenth century that it became historicized and recorded as a "technical discovery" of modernity. For example, the historian of technology Johann Beckmann proudly records insurance on the "credit side" of modernity—in competitive comparison with antiquity—along with other discoveries such as the New World, gunpowder, paper, the printing press, porcelain, silk, alchemical distillation, Arabic numbers, and mathematics.[96] While in juridical

---

les Anciens et les contemporains, Paris 2004; for global intersections cf. Cornel Zwierlein, "Orient contra China: Eusèbe Renaudot's Vision of World History (c. 1700)," *Journal of the History of Ideas* 81, 1 (2020), pp. 23–44.

95 One early contribution to what might be classified as an early history of technology was Guido Panciroli, *Discorso delle cose che haueuano gli antichi, & hoggidi non sono & di quelle, che dopo si sono ritrouate* that circulated as a manuscript prior to 1580 (e.g. Archivio Apostolico Vaticano, Misc. Arm. II, 90, f. 227r-289r); in the expanded Protestant edition: Guido Panciroli, *Rerum memorabilium iam olim deperditarum et contra recens atque ingeniose inventarum libri II*, ed. Heinrich Salmuth, 2 vols., Amberg 1599–1602 [further editions published in 1607f., 1612, 1622, 1646, 1660]) there is no mention of insurance. On Panciroli cf. Vera Keller, *Knowledge and the Public Interest, 1575–1725*, Cambridge 2015, passim. No mention of the insurance also in other early works on the history of inventions, cf. S. Lancilotti and Alessandro Tassoni, "Pensieri diversi [1620], lib. x: Ingegni antichi e moderni," in *Prose politiche e morali*, by S. Lancilotti and Alessandro Tassoni, ed. Giorgio Rossi, vol. 2, Bari 1930, pp. 298–340 (while this is an incomplete edition there is also no information about insurance in the original). August Buck, "Aus der Vorgeschichte der Querelle des Anciens et des Modernes in Mittelalter und Renaissance," *Bibliothèque d'Humanisme et Renaissance* 20 (1958), pp. 527–41; Marc Fumaroli and Anne-Marie Lecoq, *La querelle des Anciens et des Modernes: XVIIe–XVIIIe siècles*, Paris 2001; Charles Trinkaus, "Antiquitas versus Modernitas: An Italian Humanist Polemic and its Resonance," *Journal of the History of Ideas* 48 (1987), pp. 11–21; Simone Mazauric, *Fontenelle et l'invention de l'histoire des sciences à l'aube des Lumières*, Paris 2007, pp. 177–224; Brian P. Copenhaver, "The Historiography of Discovery in the Renaissance: The sources and Composition of Polidore Vergil's De inventoribus rerum, I-III", *Journal of the Warburg and Courtauld Institutes* 41 (1978), pp. 192–214; Marcus Popplow, *Neu, nützlich und erfindungsreich: die Idealisierung von Technik in der frühen Neuzeit*, Münster 1998.

96 Johann Beckmann, "Assecuranz," in *Beyträge zur Geschichte der Erfindungen*, ed. Johann Beckman, vol. 1.2, Leipzig 1785/86, pp. 204–22. Concerning Beckmann, cf. Günter Bayerl, "Die Anfänge der Technikgeschichte bei Johann Beckmann und Johann Heinrich Moritz von Poppe," in *Die technikhistorische Forschung in Deutschland von 1800 bis zur Gegenwart*, ed. Wolfgang König, Kassel 2007, pp. 13–32; Ulrich Troitzsch, "Übersetzung, Bearbeitung, Plagiat, kritische Übertragung. Johann Beckmanns Beyträge zur Geschichte der Erfindungen in englischen und japanischen Ausgaben," in *Technik, Arbeit und Umwelt in der Geschichte. Günter Bayerl zum 60. Geburtstag*, ed. Torsten Meyer, Münster 2006, pp. 95–109; Gerhard Banse and Hans-Peter Müller (eds.), *Johann Beckmann und die Folgen. Erfindungen—Versuch der historischen, theoretischen und empirischen Annäherung an*

treatises of the seventeenth and eighteenth centuries, increasingly passages appear cataloging the etymology and the ancient precursors of "insurance," this modern awareness regarding a generalized principle of insurance in the reflections of thinkers who were interested in the origins of "insurance" but were not specialists in the areas of trade, accounting, or cameralistic production of security is not clearly expressed until the age of Enlightenment.[97]

5 Summary

This brief analysis of the early history of insurance, which was initially limited to the maritime transport sector, has demonstrated that premium-based insurance initially emerged as an instrumental "accounting trick" in the context of double-bookkeeping in order to stabilize balance sheets. The specific threat in question was, for a long time, an almost exclusively spatial one: the realm of the sea. The notion of insurance was concerned with securing value in organized spatial contexts. The "future-orientation" of insurance was, in this context, not substantially more significant or different than it is in any legal transaction: after all, every contract (purchase, rental, transport, etc.) contains an element of futurity for both partners. Of course, with the new concept of *risico*—as well as through singling out and differentiating *periculum* as a specific object of contracts, rather than in the meaning of distribution of risk (e.g. of buyers, sellers, and transporters of a good) within the context of indemnity law—the seeds of later developments were sown. The spatial dimension associated with early insurance is hardly mentioned at all in the insurance treatises, commentaries and *conclusa* of the fourteenth through sixteenth centuries because it was self-evident. Literature tracing and defining the impact of insurance history with a view toward the *longue durée* usually also fails to acknowledge the importance of the spatial dimension as a defining characteristic of early insurance, which results in the erasure of the significant qualitative distinction between insurance around 1500 and insurance around 1700/1800. The understanding of maritime insurance merely as "precursor and experimental phase"[98] suggests that it corresponds to a single developmental process within one and the same

einen vielschichtigen Begriff, Münster 2007; Günter Bayerl and Jürgen Beckmann (eds.), *Johann Beckmann (1739–1811). Beiträge zu Leben, Werk und Wirkung des Begründers der Allgemeinen Technologie*, Münster 1999.

97 Johann A. H. Reimarus, "Ueber das Alter der Assecuranzen," *Handlungsbibliothek* 3 (1797), pp. 119–22.

98 Bonß, *Vom Risiko*, p. 179.

instrument of the "colonization of the future," which merely progresses further along the same trajectory. In contrast, according to the conceptualization presented here, insurance around 1500 and insurance around 1700/1800 are heuristically much more fruitfully conceived of as quite distinct practices in regard to their underlying temporal and spatial horizons. Despite the recurrences of the basic elements and vocabulary (policy, premium, concept of risk), which remain structurally similar, these modes of insurance are more isomorphic than homologous phenomena. Between them lies the crucial engine of development and change: the confrontation between actual trade practices and the distinct observation and internal logic of legal academic discourse with its drift toward natural and rational systems. Thus around 1650, the principle of a premium-based insurance, recognized and recorded as a universalizable and transferable entity, appears for the first time in history.

CHAPTER 3

The Danger between Nature and Culture: The Quotidian Threat of Urban Fires in the Premodern Era

The first chapter described the development of premium-based insurance in the Mediterranean and its spread into the northern European realms, and concentrated especially on its counterintuitive, spatially-oriented character. Despite undisputed major continuities in terms of contract modalities and related concepts, there was a change in quantitative terms around 1680–1700 —a much wider diffusion of the principle of premium-based insurance beyond the narrow circles of merchants and port cities. Two further qualitative changes also took place—first regarding the various ways in which insurance was institutionalized, and second, in the sense that the dominant characteristic of insurance shifted from the dimension of space to the dimension of time. Since one of the most important elements of the shift around 1700 was concerned with the thematic object areas associated with premium-based insurance—namely, with the turn toward "fire" and "life"—the current chapter will seek to clarify how the reality and perception of the threat of fire, as well as the means of fire prevention and firefighting, were represented and transformed in the premodern era. In the analysis of this transformation—which should primarily be understood as a true epochal threshold in the history of security, occurring around 1700—I focus, in particular, on major fires, because it is only major fires that can be sufficiently evaluated diachronically and in terms of a large area, such as the German-speaking world. Fire insurance, however, was not created as a response to these major disasters, which in the premodern era constituted something similar to the currently uninsurable "Acts of God". Nevertheless, the great conflagration, the utter annihilation of medieval and Renaissance towns, was the sword of Damocles that hovered over premodern consciousness. The threat of such fires stimulated impulses of fear and anxiety, making fire into the most common and also the greatest fear for the people of this time. It is therefore false to specify the London fire of 1666 as the sole origin of the development of British fire insurance—the reasons for change should be fundamentally located in the cumulative effects of major disasters on the people living in this period. The statistics for major fires will demonstrate that, in light of the continuity of this problem over centuries and the great number of incidents, a simple model that equates "insurance innovation with learning from disasters" hardly addresses the issue. Although not all fires were major and extensive, fires were also less easily controlled than they are now; thus, a simple house fire more frequently resulted in the loss of many houses.

In Part 1 of this chapter, I will reconstruct these dimensions of the danger of fire. In Part 2, I will present an inquiry into how the perception of major city fires changed. I will only briefly discuss the treatment of the perception of fires according to the theology of punishment, because this topic is already prominent in existing research. I will, however, discuss the development of this topic in the seventeenth and eighteenth centuries, which demonstrates that while the theology of punishment did not disappear during the century of the Enlightenment—even after the Lisbon earthquake of 1755—the arguments became increasingly fanned and differentiated in order to correspond to this type of theological interpretation on the one hand, and to developments in philosophy and natural sciences on the other.

The topic of fire also became subject to various forms of visualization, which can be neither reduced to the pure internal development of "l'art pour l'art," (given that premodern art was, to a large extent, already craftmanship, and this holds even more true for works of art produced at the local level of small towns and cities for purposes of commemoration); nor can the characteristics and changes in the forms of visualization be reduced to a simple "reflection" of changes in the real world of developments in the realm of urban planning, building and fire protection. The question is, what change to, and intensification of, emotional motivating forces took place during the centuries that were arising from, and stimulated by the representation and communication of urban fire events? The perception and visual representation of fire danger is therefore understood as a "soft" but relatively significant causal factor driving changes at the institutional level regarding the security provisions put into place both before and after disasters—and as such it is included here on the same categorical level as the developments of such material and useful items such as fire hoses and ladders. This leads, in Part 3 to a brief outline of the notion of pre- and post-fire provisions that were *not* in the form of insurance: from the municipal fire prevention administration (fire *policey*)[1] of the late Middle Ages to the state fire *policey* of the Enlightenment. This also involves the question of whether, and in what sense, something was "learned" from fire disasters. Among other things, the conceptualization and structure of city reconstruction will be briefly addressed at this point. In particular, I will investigate how technical innovations in fire extinguishers, specifically hoses and related instruments were communicated and implemented from an epistemological-historical perspective.

1 On the premodern concept of "policey," cf. below, Ch. 3, 3.

1 The "Reality" of the Danger: Fire Cycles, Fire Frequencies

1.1 *The "Fire Gap"*

How great was the actual average risk of a major city fire and what trends and developments appear between the Middle Ages and the nineteenth century? This question not only concerns the development of insurance demand, but also addresses the general perspective of cultural as well as environmental history regarding the question of the significance of fires in subsequent eras. The systematic assessment of the problem of city fires is still lacking for many regions and periods: "Despite their importance there have been few attempts by historians to measure trends in the number and cost of fires."[2]

In an essay published in 1989, urban historians Lionel Frost and Eric L. Jones plotted the number of houses destroyed in major fires in twenty large North American cities[3] during the course of the nineteenth century in relation to the growth of the urban populations. They defined a major fire as one that destroyed over fifty houses. While damages from urban fires significantly increased, city population totals rose exponentially. They called the discrepancy between these two values the "fire gap." This information from across the US was mapped according to geographical location, which suggested a particular trend: the West was initially marked by extremely frequent fires between 1830 and 1860; thereafter, however, the increase in urban populations was substantially more dramatic than in the East.[4] The "fire gap" quantifiably summarizes the phenomenon at the core of this book: it seems to clearly mark the transition to modernity (Fig. 4).

Urban population growth and fire damage is, of course, linked only graphically—there is no causal connection between these two measures. As the main source for the determination of fire damage, the authors used the *Times Index* and other sources of local history, although they concede that their list (in particular the number of destroyed houses per fire) is by no means absolute due to various issues associated with source preservation and accessibility.[5] The fundamental conclusion, that from the 1840s in the East and the end of the 1850s in the West, the danger of large fires, or at least the danger of major, widespread fires, massively declined, is in any case definitively

2 Robin Pearson, *Insuring the Industrial Revolution: Fire Insurance in Great Britain, 1700–1850*, Aldershot 2004, pp. 32–33.

3 Baltimore, Boston, Buffalo, Chicago, Cincinnati, Cleveland, Denver, Detroit, Kansas City, Los Angeles, Minneapolis-St. Paul, New Orleans, New York, Omaha, Philadelphia, Pittsburgh, Portland/Ore., St. Louis, San Francisco, Seattle.

4 Lionel E. Frost and Eric L. Jones, "The Fire Gap and the Greater Durability of Nineteenth Century Cities," *Planning Perspectives* 4 (1989), pp. 333–47, 338–40.

5 Ibid., p. 346 n 15.

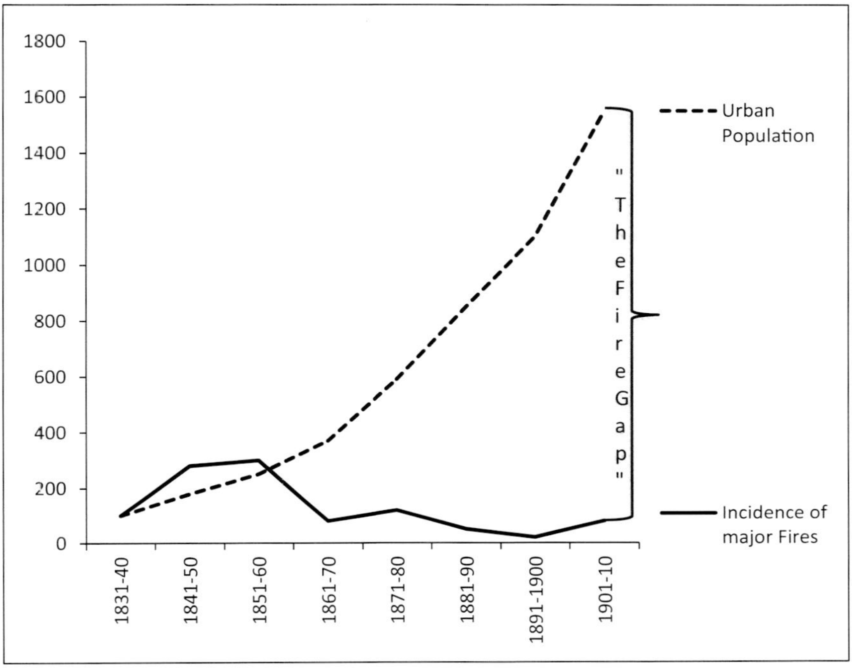

FIGURE 4 The *fire gap* for the entire USA, 1831–1910, based on Frost/Jones (population growth vs. the number of houses destroyed)

clear—although the best-known fires, including the Chicago fire of 1871, the Boston fire of 1872, and the Baltimore fire of 1904, all happen to be exceptions. The correlation with population increases serves as nothing other than a visualization of the paradox—that this decrease in the overall risk of fire occurs in a period in which explosive population growth would suggest a higher risk of fire. Particularly at the threshold point at the beginning of the nineteenth century, one would assume that the rapid growth of cities would indeed correlate with an increased fire risk. Nevertheless, while there is very little actual informational content in Frost and Jones's "fire gap" graphs, as far as I know, they do represent the first attempt to take stock of the seriality and severity of urban fire risk in international urban history. There have only been a few attempts to systematically address the phenomenon of urban fires. A pioneer among them was the 1984 *Gazetteer of English Urban Fire Disasters* by Eric L. Jones, Stephen Porter, and Michael Turner, which identifies exactly 518 "definite" and 100 "probable" fires not caused by war that destroyed at least ten houses for the period between 1500 and 1900.[6] The sources used in this study include

6 I offer Jones/Porter's attempt to formulate a definition of a "major fire" here, because it clearly demonstrates how difficult it is to do so: the "lower-bound definitions of a fire

"charity briefs" (i.e., begging briefs, licences and privileges, that were issued for the collection of money),[7] leaflets, newspapers, fire insurance files (from the eighteenth-century on), liturgical books, diaries, and all other conceivable forms of local historiography (chronicles, travel reports, etc.). It is not entirely clear how Porter, Jones, and Turner were aggregating their data from primary sources; nevertheless, it provides a wealth of general information.[8] Robin Pearson conducted an audit of the *Gazetteer* by examining every fire reported in three monthlies, focusing specifically on the years 1730–79. In addition to the 126 fires identified by Jones, Porter, and Turner for this period, Pearson added a further 814; through extrapolations and calculations, he estimated that the 7,600 houses destroyed during these fifty years represented less than a quarter of the entire extent of destruction.[9] It was thereby confirmed that from 1750/60 on, the fire hazard and the rate of destruction generally declined, while the urban population, especially in England, increased considerably—thus suggestive of an earlier fire gap.

For Finland, Eino Jutikkala attempted a project comparable to the *Gazetteer* on the basis of reports and indications of major fires gathered from urban history in secondary literature.[10] Of course, it is important to remember that the urbanization of Finland occurred later. Jutikkala finds only seventy-six major fires in twenty-two towns and cities from the sixteenth century until about 1860. The greatest number of fires occurred between 1800 and 1860, while the eighteenth century, in contrast to both the seventeenth and the

---

disaster ... are at least ten houses destroyed; a significant value of loss in current £ s.d.; or a descriptive phrase implying destruction of a high order, such as 'great fire,' 'disastrous fire,' or 'half the town burned.' These phrases may also include figures such as 'eighteen rents lost,' or reports of a large number of buildings burned in which houses were clearly included although the number is not specified" (Eric L. Jones, Stephen Porter, and Michael Turner, *A Gazetteer of English Urban Fire Disasters, 1500–1900*, Norwich 1984, p. 12).

7 Concerning this type of source see Mark Harris, "'Inky Blots and Rotten Parchment Bonds': London, Charity Briefs and the Guildhall Library," *Historical Research* 66 (1993), pp. 98–110; for a more recent case study of charity briefs after the Great Fire of London see Jacob F. Field, "Charitable Giving and its Distribution to Londoners after the Great Fire, 1666–1676," *Urban History* 38, 1 (2011), pp. 3–23. The goal of total donations of £ 16.487 is, in any case, astonishingly average in light of this major event: national donations via such letters were collected for English seafarers captured by the North African Barbary pirates in the following years:1670–74—£21,530; 1691–96—£8,309; 1700/1—£16,591. See LMA COL CHD PR 07/001-003.

8 Jones, Porter, and Turner, *A Gazeteer*, pp. 7–12.

9 Pearson, *Insuring the Industrial Revolution*, pp. 32–38.

10 Eino Jutikkala, "Great Fires of Finnish Cities," in Martin Körner (ed.), *Stadtzerstörung und Wiederaufbau*, vol 1, *Zerstörung durch Erdbeben, Feuer und Wasser*, Bern 1999, pp. 239–54.

nineteenth centuries, was blessed with a relative paucity of fires. Jutikkala also provided an average distribution of fires from month to month, with May through September, and especially May and June, being the most fire-prone months—perhaps due to the fact that houses were covered with snow during the winter months in Finland. In central Europe, in contrast, the fire danger was at least as high during the dry winter months as it was in the summer. For Sweden, Sven Lilja's evaluation of medieval and early modern urban-historical research found 238 fires in 122 cities between the thirteenth and eighteenth centuries and concluded, in a somewhat revisionist manner, that the importance of fires was lower than commonly assumed—although it is important to remember that Sweden in this period had a very low level of urbanization.[11] Additionally, it seems that after the Great Northern War (1700–21) in Sweden, the number of fires massively declined over the course of the eighteenth century.

There are no attempts to create such lists of major city fires for most other countries and regions of the globe, although this gap in research is sometimes noted.[12] For England, many monographs exist about the one (certainly important) Great Fire of London 1666, but "there has been no attempt to synthesize local

11 Sven Lilja, "Wooden Towns on Fire. Fire Destruction and Human Reconstruction of Swedish Towns Prior to 1800," in Körner, *Stadtzerstörung*, pp. 255–75; cf. Sven Lilja, *Städernas folkmängd och tillväxt. Sverige (med Finland) ca 1570-tal till 1810-tal (Historisk tätortsstatistik 2.)*, Stockholm 1996—he used as sources for the fire data *Medeltidsstaden. The Medieval Town, Town reports*, published by Riksantikvarieämbetet; also used: Nils Ahlberg, *Stadsplaneförändringar i Sverige 1550–1875*, 2 vols., Uppsala 1995, on which the following is based: Nils Ahlberg, *Stadsgrundningar och planförändringar: svensk stadsplanering 1521–1721*, Uppsala 2005.

12 For the Netherlands, for example: "Een overzicht van alle branden van enige omvang uit deze periode is niet voorhanden" (Marco H.D. van Leeuwen (ed.), *De rijke Republiek. Gilden, assuradeurs en armenzorg 1500–1800*, The Hague 2000, p. 304); for Czech and Slovakia: "Fire statistics for Bohemian towns and villages can be established only after the appearance of all volumes of the city encyclopedia" that Karel Kuča, *Města a městečka v Čechách, na Moravě a ve Slezsku [Cities and small towns in Bohemia, Moravia and Silesia]*, vol. 1–8, Prague 1996–2011 was then only preparing (Josef Hrdlička, "Beten oder löschen? Feuer in den böhmischen Städten der Frühen Neuzeit," in Lorenzetti, Luigi and Giannò, Vanessa (eds.), *Al fuoco! Usi, rischi e rappresentazioni dell'incendio dal Medioevo al XX secolo*. Dibattiti & documenti 13/Studi di storia alpina/Studies on Alpine History 6. (Milan and Lugano: 2010), 91–108, 93n4. Kuča's volumes include a paragraph for each city on catastrophes, with a separate discussion of big city fires (Požáry), as in the German *Städtebücher*. A statistical analysis has not yet been undertaken. For Romania "Istoriografia românească înregistrează foarte puține preocupări pe tema catastrofelor urbane provocate de foc" (Dan Dumitru Iacob, "Incendiul din Iaşi, din 25 Iunie 1833", *Historia Urbana* 26 (2018), pp. 67–96, here p. 69 with reference to some works treating individual fires and some works on fire-fighting).

studies, or to reconcile fires, and the experience of fires, with the growing global and interdisciplinary field of historical disaster studies."[13] In addition to scattered statements regarding selected cities, there is some information on czarist fire statistics from nineteenth-century Russia—which, however, is not comparable to the above examples[14]—as well as some very unreliable individual assessments of small sub-regions or assessments of fires named in an arbitrary selection of early modern encyclopedias.[15] There are some sporadic notes on two hundred fires during the Ch'ing dynasty (1644 and 1673–1912) in China and concerning a few random, large fires in Asian and Islamic cities,[16] but they are not particularly useful for purposes of comparison.[17] The following section is therefore an attempt to construct and analyze a fire ecology for Germany on the basis of Keyser's urban history compendia (i.e., the *Städtebücher*, discussed below). These and other statistical sources from the macro- to the meso- and micro-level of urban networks and particular cities will be evaluated in order to explicate the realities of fire danger at the time. As a result, the contours emerge of a German/Austrian fire ecology and we begin to see the frequencies and distributions of major urban fires. The purpose of this chapter is to create a tangible sense of the dimensions and qualities of fire danger before moving to a discussion of the perceptions and measures of security production that developed in response to that danger.

---

13 John Morgan, "The representation and experience of English urban fire disasters, c. 1580–1640", *Historical Research* 89, 244 (2016), pp. 268–293, p. 269 who adds in n. 8 "Such studies exist for other national and international contexts, but not England."

14 Cathy Frierson, *All Russia is burning! A Cultural History of Fire and Arson in Late Imperial Russia*, Seattle 2002, pp. 68–71.

15 Jacky Theurot, " 'Au feu!' Les bourgs et les ville du comté de Bourgogne face à l'incendie (XIVe et XVe siècles)," in François Lassus and François Vion-Delphin (eds.), *Les hommes et le feu de l'Antiquité à nos jours. Du feu mythique et bienfaiteur au feu dévastateur*, Besançon 2007, pp. 233–251, 236: regional summary of fire for the late Middle Ages; Christine Lamarre, "Les incendies de villes dans les géographies du XVIIIe siècle," in Lassus/Vion-Delphin, *Les hommes*, pp. 283–92 (A list of 35 fires in France according to encyclopedia entries). Similarly arbitrary is the selection of fires found in Maxime Petit, *Les grands incendies*, Paris 1882, pp. 273–92, which combines 82 Parisian fires since the sixth century with a universal-historical collection of fires from around the world; François Jarrige and Bénedicte Reynaud, "Les usines en feu. L'industrialisation au risque des incendies dans le textile (France 1830–1870)," *Le mouvement social* 249 (2014), pp. 141–62, 143, 146: Featuring fires from 1804–70 in five French départements in the textile industry.

16 Cf. Frost and Jones, *The Fire Gap*, p. 336.

17 Some literature on fire "statistics" concerning individual cities in China is included in selected bibliographical references for chapter 6, 4 below; however, these primarily concern the nineteenth century.

1.2 *8,200 Fires in Germany and Austria*

The *Städtebücher* is a multivolume encyclopedic register of all German and Austrian communities once or still possessing the rights and status of a city. These volumes, the publication of which began in 1939, are the result of a large and multipronged archival research project, and they facilitate a uniquely accurate and precise summary of large urban fires for the German and central European region.[18] It is often forgotten that the *Städtebücher* project was a product of National Socialist *Volksgeschichte* (ethnic history). It was founded by the Danzig archivist and professor Erich Keyser (1893–1968) at the seventh International Congress of History in Warsaw in August 1933.[19] The particular element that made the *Städtebücher* unique was the attention devoted to "small and smallest cities."[20] Keyser organized a host of contributors from the Institutes for Regional History Publication (*Konferenz der landesgeschichtlichen Publikationsinstitute*), the Historical Commissions of the States (*Historischen Kommissionen der Länder*), the German Assembly of Cities (*Deutscher Gemeindetag*), and experts

---

18 Erich Keyser (ed.), *Deutsches Städtebuch. Handbuch städtischer Geschichte*, 5 vols., Stuttgart 1939–1974: vol. 1, *Nordostdeutschland* (1939); vol. 2, *Mitteldeutschland* (1942); vol. 3, *Nordwest-Deutschland*, vol. 3.1, *Niedersachsen/Bremen. Niedersächsisches Städtebuch* (1952); vol. 3.2, *Westfalen. Westfälisches Städtebuch* (1954); vol 3.3, *Landschaftsverband Rheinland. Rheinisches Städtebuch* (1956); vol. 4, *Südwest-Deutschland* vol. 4.1, *Land Hessen. Hessisches Städtebuch* (1957); vol. 4.2.1, *Badisches Städtebuch* (1959); vol. 4.2.2, *Württembergisches Städtebuch* (1961); vol. 4.3, *Land Rheinland-Pfalz und Saarland. Städtebuch Rheinland-Pfalz und Saarland* (1964); vol. 5, *Bayern. Bayerisches Städtebuch*, vol. 5.1, *Unter-, Mittel- und Oberfranken*, (1971); vol 5.2, *Ober-, Niederbayern, Oberpfalz und Schwaben* (1974). Hereafter *Deutsches Städtebuch. Handbuch städtischer Geschichte* will be cited as "StB: volume number."

A revised version of the *Handbuch* has been published by the Institute for Comparative Urban History (*Institut für vergleichende Städtegeschichte*) at the University of Münster, edited by Heinz Stoob und Peter Johanek: vol. 1, *Schlesisches Städtebuch*, ed. Waldemar Grosch (1995); vol. 2, *Städtebuch Brandenburg und Berlin*, eds. Evamaria Engel, Lieselotte Enders, Gerd Heinrich and Winfried Schich, with Harald Engler (2000); vol. 3.2, *Städtebuch Hinterpommern*, eds. Peter Johanek and Franz-Joseph Post, with Thomas Tippach and Roland Lesniak, 2003. For Austria, cf. Alfred Hoffmann (ed.), *Österreichisches Städtebuch*, Vienna 1968–1999: vol. 1, *Die Städte Oberösterreichs*, ed. Herbert Knittler (1968); vol. 2, *Die Städte des Burgenlandes*, ed. Renõ Deák, 2nd revised ed. (1996); vol. 3, *Die Städte Vorarlbergs*, ed. Franz Baltzarek and Johannes Pradel (1973); vol. 4, *Die Städte Niederösterreichs*, 3 subvols., ed. Friederike Goldmann (1976–1988); vol. 5, *Die Städte Tirols*, 5.1, *Bundesland Tirol*, 5.2 *Südtirol*, ed. Franz-Heinz Hye (1980/2001); vol. 6, *Die Städte der Steiermark*, 2 subvols., ed. Friederike Goldmann (1990/1995); vol. 7, *Die Stadt Wien*, ed. Peter Csendes and Ferdinand Opll (1999) still to be published are: vol. 6.1, *Graz*, vol. 6.2, *Die Städte der Steiermark: A-H*; vol. 8, *Kärnten*.

19 Erich Keyser, "Neue deutsche Städteforschung," in StB:1, 1–10, 3. This introductory essay is primarily a reprint of one of Keyser's essays, "Neue Forschungen zur Geschichte der deutschen Städte," *Blätter für deutsche Landesgeschichte* 83 (1937), pp. 46–53.

20 Keyser, "Neue deutsche Städteforschung," p. 1.

in race research at the Reich Ministry of the Interior. City archivists and local and regional historians provided Keyser with data for a short "biographical" entry for each town or city according to a predetermined rubric. The schematic also included obviously ideologically biased points—for example, point 16, which recorded the city's Jewish population. Point 5 deals with major disasters in the history of the city—in particular, fires and floods—which allows for the database-like organization of this information. This is not the place for a detailed discussion concerning the very problematic roles played by Keyser and many contributors to Nazi ethnic history and to the framework of *Ostforschung* (an outdated *Weimar Republic* and National-socialist term for Eastern European research) in general.[21] For present purposes, we can accept that not all elements of the data and the schematic are steeped in ideology and that the *Städtebücher* do retain value as an aggregation of source analysis. The definition of "city" underlying the whole enterprise is disputable, as is every definition of the city, but also entirely pragmatic and feasible:

> The city book contains the representation of the history of all cities that belonged to the German Empire or to the area comprising the free city of Danzig on January 1, 1936. Under the term "city" we understand here all communities which were, in the past, considered as cities according to historical legal, business, or settlement parameters and which, before the law of January 1, 1936, were officially described as cities, according to city constitution or official designation. It therefore also includes those places that, in the course of time, have lost their status as cities.[22]

21 Willi Oberkrome, *Volksgeschichte. Methodische Innovation und völkische Ideologisierung in der deutschen Geschichtswissenschaft* (Göttingen: 1993), pp. 32–35, 56–72; Ingo Haar, *Historiker im Nationalsozialismus. Die deutsche Geschichtswissenschaft und der Volkstumskampf im Osten*, Göttingen 2000, pp. 135–39; Kai Arne Linnemann, *Das Erbe der Ostforschung. Zur Rolle Göttingens in der Geschichtswissenschaft der Nachkriegszeit*, Paderborn 2002; concerning Keyser in the context of National Socialist *Ostforschung* (East Studies), see Jörg Hackmann, "Der Kampf um die Weichsel. Die deutsche Ostforschung in Danzig 1918–1945," *Zapiski Historyczne* 58 (1993), pp. 37–57. On populationist and racial governmentality at the roots of the *Ostforschung*, see Ingo Haar, "Ostforschung im Nationalsozialismus. Die Genesis der Endlösung aus dem Geiste der Wissenschaften," in Rainer Mackensen (ed.), *Bevölkerungslehre und Bevölkerungspolitik im "Dritten Reich,"* Opladen 2004, pp. 219–40; in particular concerning Keyser, see Alexander Pinwinkler, "Volk, Bevölkerung, Rasse and Raum: Erich Keyser's Ambiguous Concept of a German History of Population, ca. 1918–1955," in Ingo Haar and Michael Fahlbusch (ed.), *German Scholars and Ethnic Cleansing*, New York 2005, pp. 86–99.

22 StB:1, p. 4, for the concept of the city in more recent literature see Peter Johanek and Franz-Joseph Post (ed.), *Vielerlei Städte. Der Stadtbegriff* (Cologne: 2004).

Of course this also includes originally "German" towns and cities, which are no longer located in Germany and may not even exist as distinct cities at all anymore. The attribution of cities in the *Städtebücher* to a certain region corresponds to twentieth-century political borders. This was a deliberate choice necessary for the purpose of organizing work and fixing spatial boundaries at a given point in time. In most cases relevant to this study, such regional attributions simply served as approximate indicators of geographical location and city and fire distributions (e.g., northwestern, northeastern, southwestern, etc.).

The then unusual nature of the ethnic-historical approach consisted primarily in the inclusion of not only the traditional large imperial cities of Cologne, Augsburg, Nuremberg, Frankfurt, Regensburg, etc., but a multitude of 3,500–4,000 smaller cities, towns, and other communities that together shaped the everyday urban landscape between the Middle Ages and early modern eras. Since they concern the history of the entire "*Volk*" in structural terms, these "average cities" were accorded a new dignity in relation to the exceptional major cities. This innovation was described as a key contribution to the overall historiographic turn to ethnic history in the 1930s.[23] Strikingly, the planned inclusion of small-, micro-, and dwarf municipalities into newer urban historical analysis thus has a hidden root in that early ethnic history. While since the mid-1990s the inclusion of smaller municipalities has been once again recognized as important by several researchers in the international community, they tend to think of this as an emulation of the English model of the "small city" approach.[24] Therefore, just as continuities

23 StB:1, p. 2. Emphasis in original.

24 See the programmatic publications: Peter Clark, *Small Towns in Early Modern Europe*, Cambridge 1995; Peter Clark, "Small towns 1700–1840," in Peter Clark (ed.), *The Cambridge Urban History of Britain*, vol. 2, *1540–1840*, Cambridge 2000, pp. 733–74; Holger Th. Gräf (ed.), *Kleine Städte im neuzeitlichen Europa*, Berlin 1997; Peter Borsay and Lindsay Proudfoot (eds.), *Provincial towns in early modern England and Ireland: change, convergence, and divergence*, Oxford 2002; Peter Johanek, "Landesherrliche Städte—kleine Städte. Umrisse eines europäischen Phänomens," in Jürgen Treffeisen (ed.), *Landesherrliche Städte in Südwestdeutschland*, Sigmaringen 1994, pp. 9–25; Klaus Vetter, *Zwischen Dorf und Stadt. Die Mediatstädte des kurmärkischen Kreises Lebus*, Weimar 1996; Katrin Keller, *Kleinstädte in Kursachsen. Wandlungen einer Städtelandschaft zwischen Dreissigjährigem Krieg und Industrialisierung*, Vienna 2001; Christoph Bernhardt (ed.), special issue "Kleine Städte," *Informationen zur modernen Stadtgeschichte* 2 (1999); Helmut Flachenecker and Rolf Kiessling (eds.), *Städtelandschaften in Altbayern, Franken und Schwaben. Studien zum Phänomen der Kleinstädte während des Spätmittelalters und der Frühen Neuzeit*, Munich 1999; Thomas Rudert and Hartmut Zückert (eds.), *Gemeindeleben. Dörfer und kleine Städte im östlichen Deutschland (16.-18. Jahrhundert)*, Cologne 2001; Holger Th. Gräf and Katrin Keller (eds.), *Städtelandschaft—Réseau urbain—Urban network. Städte im regionalen*

between "ethnic history" and the older structural and social history of the 1950s have been highlighted, it might be wise to recall similar continuities in urban history.[25] However, it is a simple truth that small towns dominated the landscape, particularly in central Europe, for a very long period of time.[26] In this respect, continuing the German *Städtebücher* project and embarking on respective Austrian and Bohemian analogues after World War II, was a logical development.

The entries of the *Städtebücher* regarding city fires conceal some uncertainties in their scanty descriptions, which have nothing to do with their problematic ideological framing: very often the year alone is given as information ("fires in 1517, 1679, 1843"), though perhaps the original archival records contained more information. Occasionally, statements concerning the severity of the fires, the number of destroyed houses, or the state of reconstruction are added to the entries. If such specifics regarding the severity of the fires are given, the descriptions, which are often based on narrative sources, are often very vague, citing, for example, the destruction of "almost the whole town," "half of the city," or "two-thirds". But these forms of description often have direct roots in the source material, as has been confirmed by other attempts to create fire lists.[27] The *Städtebücher* formulated no clear criteria as to which fires were big and significant enough to be recorded under point 5 of the rubric. Because a legion of contributors scattered throughout German archives for decades (at least before World War II) provided this information, a uniform result and homogenous classification cannot be assumed. It therefore seems highly possible that there were very different criteria at work here.[28] In order to achieve greater precision and accuracy, or at least to be able to qualitatively estimate the probable degree of aberration and difference, in the course of building the

Kontext in Spätmittelalter und Früher Neuzeit, Cologne 2004; further literature with Olaf März, *Fließende Übergänge zwischen Stadt und Land. Ein sozial- und wirtschaftsräumlicher Vergleich in Braunschweig-Wolfenbüttel im 18. Jahrhundert*, Hannover 2014, pp. 13–18.

25 Cf. Winfried Schulze, *Deutsche Geschichtswissenschaft nach 1945*, Munich 1989, pp. 281–301; Oberkrome, *Volksgeschichte*.

26 Heinz Schilling, *Die Stadt in der Frühen Neuzeit*, 2nd ed., Munich 2004, p. 8; Ulrich Rosseaux, *Städte in der Frühen Neuzeit*, Darmstadt 2006, p. 5.

27 Jones, Porter, and Turner, *Gazetteer*, p. 12; Frost and Jones, *The Fire Gap*, p. 346 n 15; Lilja, *Wooden Towns*, pp. 273–75; Jutikkala, *Great Fires*, p. 241.

28 On this point, cf. also Thomas Spohn, "Der Stadtbrand als Chance. Wiederaufbauplanungen für westfälische Städte vom 17. bis ins 19. Jahrhundert," in *Feuer! Stadtbrand in Westfalen*, Münster 1990/1991, pp. 49–72: "The references in Erich Keyser, Westfälisches Städtebuch, Stuttgart 1954, concerning the extent of the destruction are incomplete and imprecise" (note 1).

database I collected the research of local historians and other accessible urban research—systematically for specific years and unsystematically for the entire dataset.[29]

Combining all the information in the *Städtebücher* with the further information derived from archival and local history data results in 8,200 fires in 1,964 towns and cities—not all of the 2,315 towns and cities represented in the first edition of the *Städtebücher*,[30] nor all of the 371 in the new editions since 1995, or the 259 Austrian towns and cities, include an entry on fire. A simple distribution of the frequency of fires per year for the total dataset from the years 1000 to 1939[31] is presented in the image below (Figure 6). It is evident that the number of fires becomes important only from the 1370s on and that the absolute number of fires dramatically decreases only beginning around 1870. If one were to superimpose (as Frost and Jones do) the rapid urban population growth beginning at the end of the eighteenth century onto this graphic, the striking concentration of fire frequency between around 1820 and 1865 would lose significance in relation to the much larger urban population numbers, so that the beginning of the spread of the "German" or central European fire gap would appear to be located around 1700, increasing steadily until 1750 (Fig. 5).[32]

29 Cf. Spohn, "Der Stadtbrand als Chance", note 1: "The present study is thus based on an (in any case, incomplete) examination of the regional historical literature as well as on materials that were collected by the author and others at the Westphalian office for monument conservation".

30 StB:5.2, p. 8.

31 The time frame stopping in 1939 suggests that the destruction of cities through the bombing of World War II took on new dimensions in an epochal sense that are not comparable with earlier findings.

32 The data for the city fires comes from my database just described; the graph for the growth of German urban population (in 10.000) in the 64 largest cities is drawn according to the numbers well known as gathered by J. de Vries, *European Urbanization 1500–1800*, London 1984, Appendix I, 281–282, 287 and rendered by Schilling, *Die Stadt*, pp. 11–12. The graph claims not to represent precise data (as did and could not the original graph by Jones/Porter). It does not suggest that there is a mathematical relationship between both values. Likewise, in Frost/Jones' original fire gap graphs, the graphs and data of the urban population were likewise just superimposed over the graphs of the number of burnt houses per year, but using the same scale on the left side does not mean that both numbers are parts of a mathematical function: the 'fire gap' pictures are therefore just a form to illustrate the fact that more and more people of a country live in cities, that cities grow immensely, but that, apparently the danger of very large conflagrations decreases at the same time at least in those cultural contexts and on average.

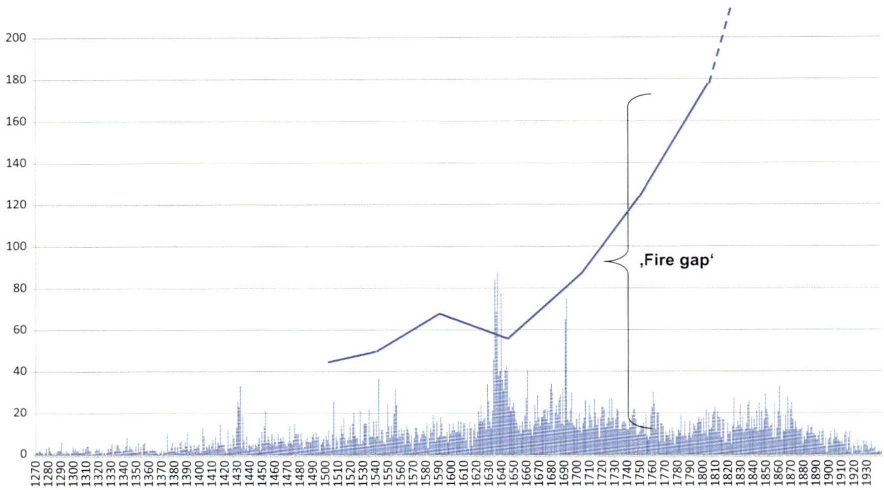

FIGURE 5 The 'Central-European fire gap': Big fires in Germany 1270–1939 and growth of the urban population in Germany (in the largest 64 cities, by 10,000 people)

The frequent peaks in the graph are predominantly war induced. Single years with values of thirty fires or more from the Hussite Wars (1428–30), the Landshut War of Succession (1504), the Second Margrave War (1553), the Thirty Years' War (1626–48)—including the record years of 1632 (84 fires), 1634 (88 fires), and 1637 (78 fires)—the First Northern War (1657–58), the Swedish-Brandenburg War (1676–78), the Nine Years' War (War of the Palatine Succession) (1688/89 and 1693), and the beginning of the Seven Years' War (1758) are all easily recognizable as caused by war-related activities. In the eighteenth century, however, fires were relatively scarce even during wartime (Revolutionary and Napoleonic wars), which is a clear indicator that war tactics had changed from infrastructure destruction to decisive battle plans. The exceptionally high fire frequencies in 1540, 1666, 1684, 1686, 1847, 1858, and 1868, however, are not explainable by wars—a topic which will be further addressed below. It will be useful to construct a regionally differentiated analysis, especially when accounting for the war-induced peak values, for reasons that go beyond simply addressing fire counts for the whole dataset.[33] In this case, it

33 Due to the shifting of national and administrative borders after 1945 and changes in the relevant state/country borders on which the new series of *Städtebücher* draws, there are overlaps and omissions (several cities appear twice, others are no longer defined as "German"), which should be recognized, but are not of such great importance that they would lead to real inaccuracies for the issue of fire characteristics in a given physical/geographic region.

FIGURE 6 Number of large city fires per year in all regions covered by the *Städtebücher*, 1000–1939

is apparent that merging the data to create a fire statistic for *all of Germany* has relatively little meaning, because the respective averages and peak values are still strongly specific and different for each region. A comparison of fire frequency distributions demonstrates that it makes sense to group certain regions[34]—namely Baden, Rhineland, Rhineland-Palatinate and Saarland (Figure 7); Württemberg (Figure 8), Bavaria (Figure 9), Hesse, Westphalia and Lower Saxony—excluding the coastal regions of Lower Saxony (Figure 10); Saxony, Saxony-Anhalt, and Thuringia (Figure 11); Schleswig-Holstein, Mecklenburg, Pomerania, East Pomerania, and (with some exceptions because, particularly in the late Middle Ages, a very unique fire characteristic is still present) East Prussia as a Baltic state (Figure 12); Brandenburg and Berlin (Figure 13); Silesia (Figure 14), Austria (Figure 15). The selected starting year for these statistics reflected the degree of urbanization of the specific regions. For the Southwest, even for very early years, a substantial number of fires were documented, simply because there were already many early urban communities in this region. The datasets for the North and East have respectively later starting points.

Further, more sophisticated consideration of the urban fire phenomenon associated with individual wars is made possible by such regional distribution analysis: in East Prussia, the so-called "Hunger War," the Polish invasion of the Prince-Bishopric of Warmia in 1414, and the Knights' War of 1519–21 can all be considered war events that caused solitary series of city fires. For Baltic Schleswig and Holstein, Mecklenburg, and Pomerania, the Thirty Years' War—except for 1630, the year that Sweden entered the war—appears to be much less important than the years 1656–59, 1676/77, 1688, 1696, and 1719–36, correlating with the Northern Wars, the Swedish-Brandenburg War, and the War of Polish Succession; in later years, only the years 1759 (of the Seven Years' War) and 1807 (of the Napoleonic Wars) distinguish themselves. While the Hussite Wars had no effect in the Northeast, Brandenburg (1432), and especially Saxony, Saxony-Anhalt, Thuringia (1429/30), and Bavaria (1430) were hit hard. In the latter regions, the Saxon Brother War culminated in 1450 with a devastating onslaught of city fires; the Hildesheim Abbey Feud of 1519 was also destructive in this sense. In Saxony, Saxony-Anhalt, Thuringia and Hesse, as well as Brandenburg, the Thirty Years' War years of 1621, 1626, 1631/32, 1636/37, 1641/42 comprise the pinnacles of destruction for these regions. Silesia was only affected

34 According to the modern distribution of the *Städtebücher*, which of course do not adequately represent the historical territories during the time of the Holy Roman Empire. However, for the purpose of fire statistics, such an approximate regional categorization is sufficient.

in the years 1627, 1633/34 and 1642, and Bavaria was only affected in 1631–35 and—less intensively—in 1639–41. The western regions of Baden, Rhineland, Rhineland-Palatinate, and Saarland, were repeatedly hit by fire during the years 1631–47; however, truly destructive peaks representing at least ten major fires per year were not recorded. 1678, 1681, 1684, 1707, 1709, 1711, 1717, and 1719 are very fiery years in Saxony, Saxony-Anhalt, and Thuringia (ten or more fires per year)—the fires were, however, obviously not related to war in these years—and thereafter it is only the year of the Napoleonic military expedition in 1806 that stands out. The Nine Years' War caused a year of blazing cities in Baden, Rhineland, and the Palatinate in 1688, but did not have this effect until 1693 in Württemberg. The Seven Years' War, viewed in terms of the fires it triggered, is actually of minor significance in Brandenburg (1757/58) and Pomerania (1759). The Bavarian region is influenced by some only regionally important fire-filled wars—the Cities' War (1388), the First Margrave War (1449, 1553), as well as the Landshut War of Succession (1504), and the War of the Austrian Succession (1743). Other wars that were only influential on a local level included the Feud of Soest (1447) and the Cologne War (1583).

In 1,511 cases, the number of houses destroyed by each respective fire was directly given or otherwise deducible from information either in the *Städtebücher* or from other local historical sources.[35] In total, there were 145,881 destroyed houses with an average of about 96 houses per fire, with fires ranging from very small to large, and including a few very large fires in Hamburg (1684: 1,714 destroyed houses; 1842: 4,200), Altendresden (1,900), and Aachen (4,600). Basing the size categorization for fires on the number of destroyed houses, the overall characteristics thus differ significantly from the English analysis by Jones, Porter and Turner (Table 1):[36]

While for England it would seem that the number of fires with a small number of destroyed houses (10–49) clearly outweighs the following categories (the ratios for the first six categories of 10–299 houses are 100: 27: 19.6: 9: 1.5: 1.5), for central Europe there is much less spread in the ratios (100: 64: 36: 20: 9: 5). This means that—assuming the English numbers are reliable—in the case of central Europe, the frequency of major fires is much higher. From dividing the

35 In cases where the percentage of destruction was given and where information about the number of houses in the city at the time of the fire was available, I have calculated that number by dissolving the ratio. Since the proportions are given in vague forms such as "half the city/one third of the city burnt," no emphasis is laid on the real accuracy of the absolute numbers, but the degree of precision should be enough for evaluating the general proportions.

36 The latter dates are from Jones, Porter, and Turner, *Gazetteer*, p. 5.

TABLE 1 Number of fires/number of destroyed houses.
Left: Numbers for central Europe. Right: Numbers for England[a]

Houses Destroyed	Number of Fires
10–49	546
50–99	347
100–149	198
150–199	108
200–249	53
250–299	28
300–349	35
350–399	11
400–449	6
450–499	0
500–549	0
550–599	0
600–649	2
650–699	0
700–749	2

Houses Destroyed	Number of Fires
10–49	66
50–99	18
100–149	13
150–199	6
200–249	1
250–299	1
300–349	1
350–399	0
400–449	0
450–499	1
500–549	0
550–599	0
600–649	2

[a] The latter dates are from Jones, Porter, and Turner, *Gazetteer*, p. 5.

relevant number of destroyed houses by the total number of fires (and accepting the inherent imprecision of this),[37] we can conclude that the most likely number of houses that burned down in a typical fire is ten for central Europe.

The fact that the highest probability is ten houses, rather than a lower number, reflects the fact that the data sources primarily concern major fires.[38] However, at the core of these findings the fact remains that the premodern

37 Thus, by replacing the data points of an imagined graph of the number of cases of a given number of houses burnt down with the mean of the surrounding ten points. In this case, each of the disproportionately common 100- and 10-step levels can be explained on the basis of the fact that they seem to only arise in the *Städtebücher* and other information because the values are always fixed at round numbers (e.g., "10 houses, 100 houses, 200 houses destroyed"). A graphical representation of the resulting curve is not given, but would essentially comprise a high peak around the value 10 and a rapidly logarithmically falling curve for all higher values.

38 The countless small and isolated fires involving less than ten houses are typically not included in the *Städtebücher* at all. The value thus corresponds to a certain degree of

danger of city fires consistently focused on major conflagrations rather than on isolatable single fires. The data in this regard are the same for both wartime and peacetime fires—interestingly, in comparison to the ratios in the database as a whole, war-induced fires are significantly underrepresented by a factor of two. This likely reflects the fact that the entries detailing the numbers of houses were usually based on the administrative efforts of the magistrate or other official authorities—of course, the functioning of the responsible municipal management, especially in times of war, was often completely ineffectual. In any case, no statistically significant differences in the destructive effects of normal versus war-related fires can be ascertained.

Since the distribution is strongly asymmetrical, the most probable number of houses destroyed per fire is not equal to the average number of houses destroyed, which, as previously mentioned, is 96. This average value of 96 therefore indicates very little, and the variance—i.e., the width of distribution—is very high. Even if one removes the eleven exceptionally major fires which burned more than 700 houses, the variance value is still 112 houses. This value is also a measure for the uncertainty with which the size of each particular fire could have been predicted—a measure of uncertainty that may have been of interest to fire insurers, but was not accessible to them in early modern times. Today, an insurer might ask the question about what was the most probable size of a fire, and one can answer it retrospectively of a fire: while approximately 40 per cent of all fires destroyed between 3 and 48 houses, about 70 per cent of all fires in the area destroyed between 3 and 100 houses. This demonstrates that a remarkably high proportion of fires were relatively large (i.e., affecting between 49–100 houses).

In the attempt to model the relationship between the number of destroyed houses per fire with a proximity function, another value from the data can be derived, from which the normal size of a "city" can be estimated at between 280 and 380 houses.[39] This illustrates the extent to which the problem concerns small and tiny towns. By including this information, the numbers given above concerning the probable size of fires becomes that much more remarkable—in relation to these tiny communities, the given numbers indicate almost complete destruction of the town in many of these cases. Because the 1,511 documented cases concern, for the most part, smaller damages and the given number of cases represents approximately 18 per cent of the total

self-referentiality, since the more than one thousand contributers to the *Städtebücher* and the further primary and secondary sources integrated here have implicitly defined a major, noteworthy fire as one that destroyed at least ten houses.

39 The approximation formula for the number of fires would be f times x burned houses: $f(x) = A^*x/(1+x^2/Delta^2) * exp(-x/G)$, with the total value of A = 2.9 fires, the range Delta = 10–14 houses as the most probable size of fire.

number, these findings may be taken as representative for the total scope of city fires.

For all regions and for the total dataset it becomes clear that the fire gap for the German-speaking area begins to expand around 1700. One of the most important military developments of this time was the trend away from the wars of siege and devastation characterizing the late Middle Ages and the seventeenth century toward combat wars and greater military professionalism and discipline. While this demanded great resources from cities for quartering soldiers, among other things, the trend no longer aimed at the complete destruction of towns and cities (see section 4 below).

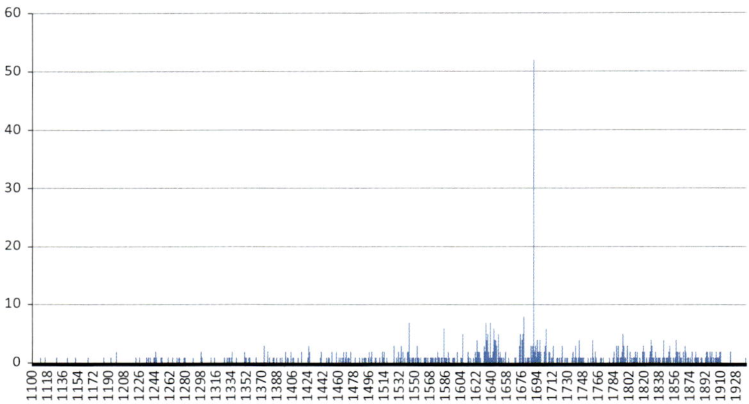

FIGURE 7 Baden, Rhineland, Rhineland-Palatinate, Saarland 1100–1939

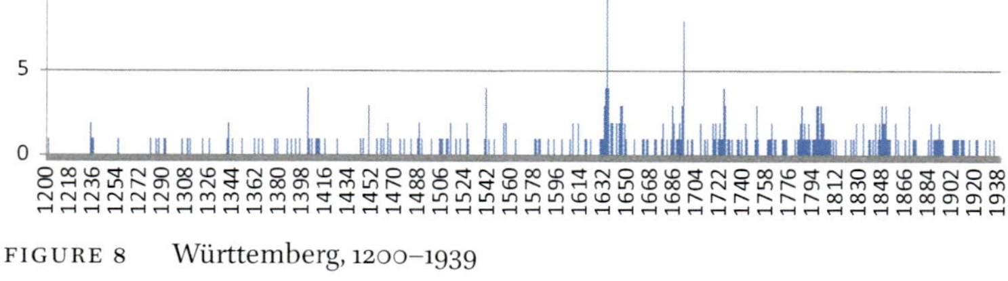

FIGURE 8 Württemberg, 1200–1939

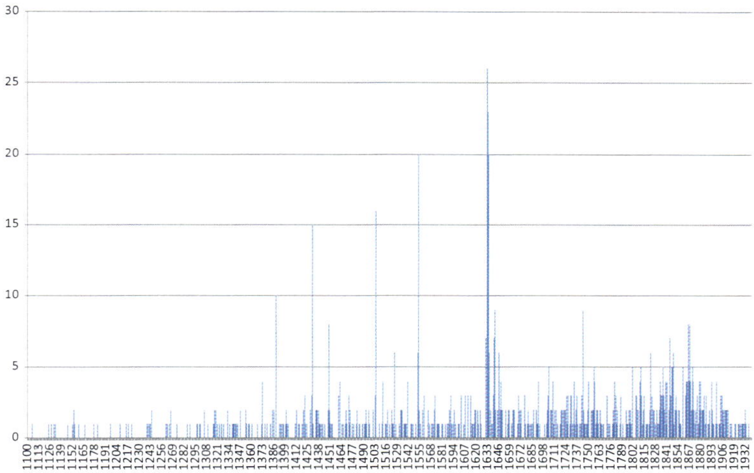

FIGURE 9 Bavaria 1100–1939

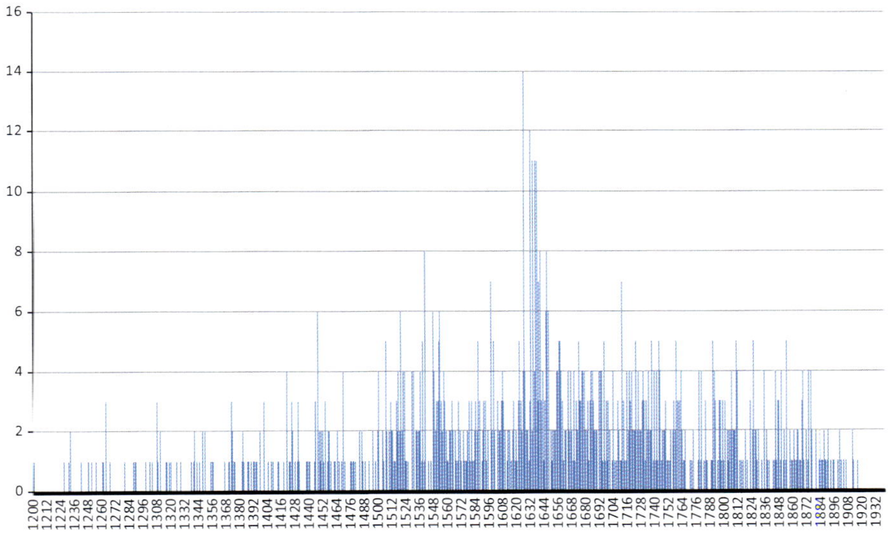

FIGURE 10 Hesse, Lower Saxony, Wesphalia, 1200–1939

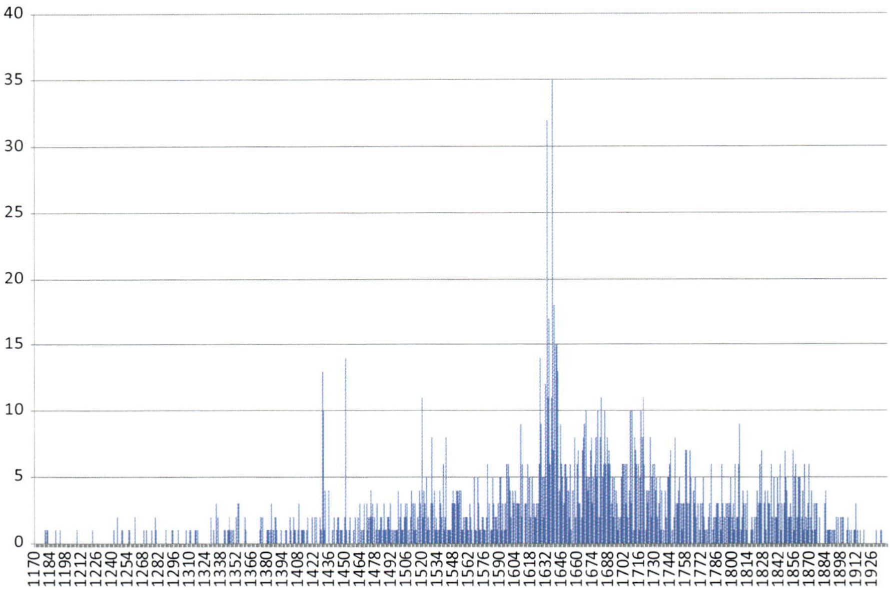

FIGURE 11 Saxony, Saxony-Anhalt, Thuringia, 1170–1939

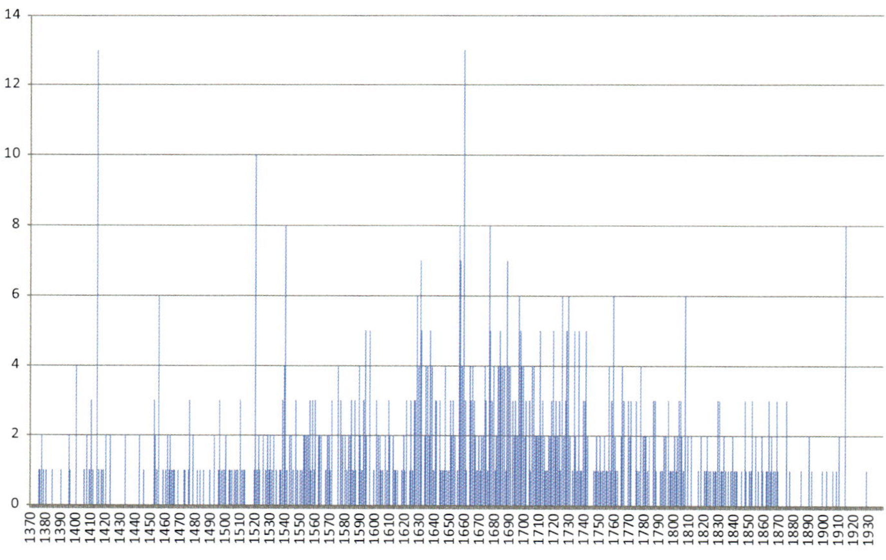

FIGURE 12 Schleswig-Holstein, Mecklenburg, Western and Eastern Pomerania, East Prussia, 1370–1939

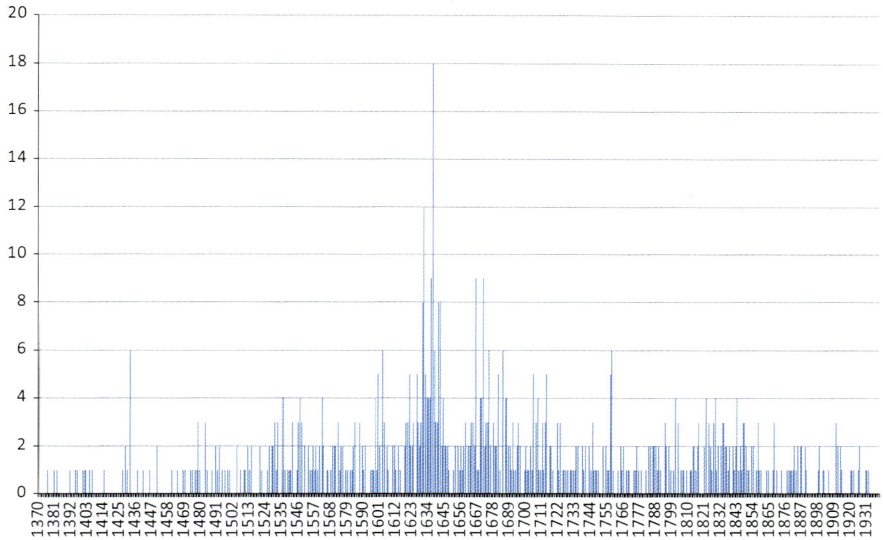

FIGURE 13 Brandenburg and Berlin 1370–1939

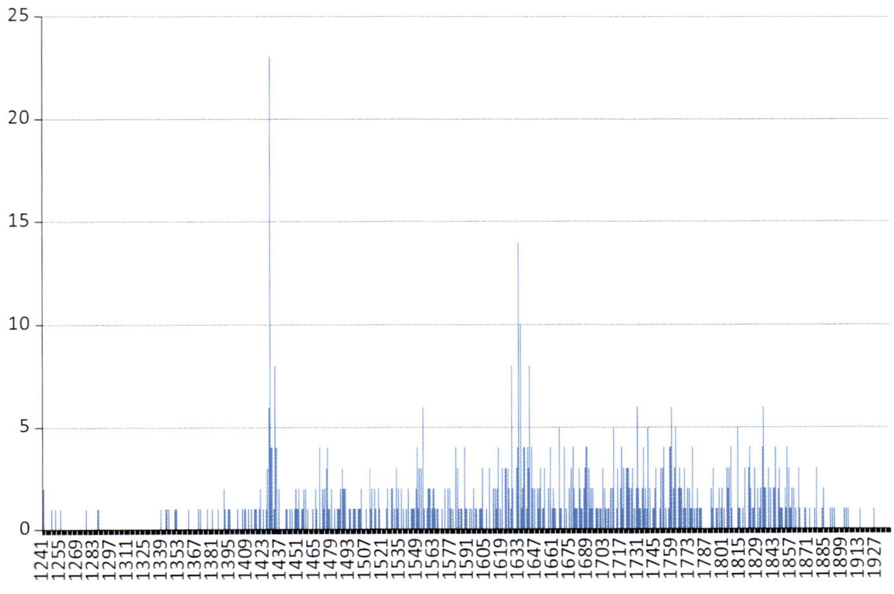

FIGURE 14 Silesia, 1240–1939

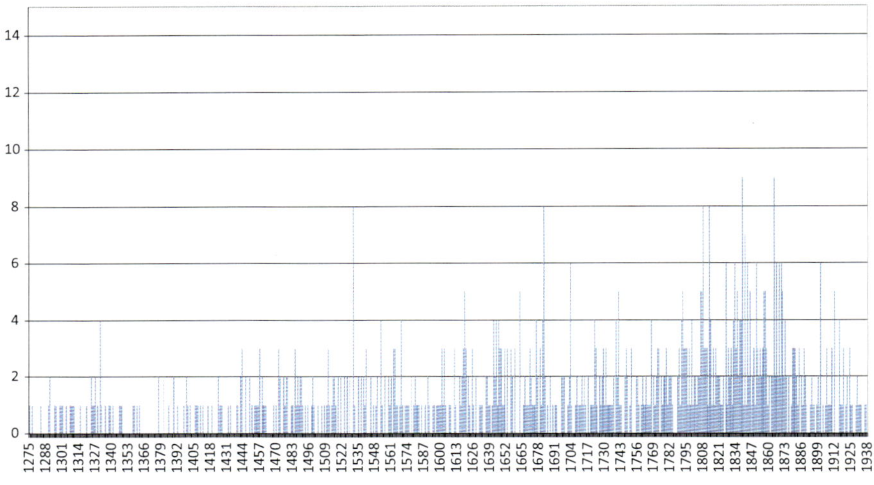

FIGURE 15 Austrian *Städtebücher* 1275–1939

1.3 *Trends in Fire Frequencies According to Fire Insurance Statistics*

Only a grainy image of Germany's urban fire ecology emerges through the analysis of the *Städtebücher*. To develop a more detailed image and reveal the finer details in the picture of annual fire damage in a region, there is hardly a more productive premodern source than the records of fire insurance institutions themselves. However, this does greatly limit the targeted time period and regions, since the establishment of such insurance and the relevant archival transmission is a prerequisite. In this context, the Middle Ages become inaccessible. Of course, it should be noted that there are a few substantive local analyses for earlier time periods. Gerhard Fouquet demonstrates that only 63 fire events were responded to in Basel between 1445 and 1549, and that a single major fire in April 1495 destroyed 40 houses—for a Swiss town of some importance with a substantial portion of the town built in stone, these are hardly dramatic findings, but it shows that for some rare cities we might still find similar data for medieval times.[40] The statistics assembled by Peter Borscheid and Anette Drees reveal twenty-four territorial or municipal fire insurance organizations with values recorded before 1800,[41] the most important being those

40 Gerhard Fouquet, *Bauen für die Stadt. Finanzen, Organisation und Arbeit in kommunalen Baubetrieben des Spätmittelalters. Eine vergleichende Studie vornehmlich zwischen den Städten Basel und Marburg*, Cologne 1999, pp. 414–20. Thomas Spohn, "Zur Einleitung," in Thomas Spohn (ed.), *Bauen nach Vorschrift? Obrigkeitliche Einflussnahme auf das Bauen und Wohnen in Nordwestdeutschland, 14.-20. Jh.*, Münster 2002, pp. 1–68, judges this as the single experiment with quantitative calculation thus far.

41 Borscheid and Drees, *Versicherungsstatistik*.

of Baden-Durlach/Baden, Braunschweig, Hamburg, Hesse-Darmstadt, Hesse-Kassel, Calenberg-Grubenhagen, Lippe, East Prussia (i.e., Königsberg and that for the so-called "small towns" [Kleinstädtische]), Saxony, and Württemberg. Of course, not all cameralistic fire insurance funds existing in Germany before 1800 are represented within this source,[42] and the data is somewhat incomplete even for those addressed by Borscheid/Drees. Nevertheless, the data does offer a more detailed picture of the fire trends in the individual regions associated with each institution, not only concerning the evolution of insurance amounts—i.e., the usually continuous growth in the value of insured building stock—but also by noting the actual yearly amount of fire damages paid. Of course, due to the state of record preservation complicated by centuries of intense territorial and administrative diversification within Germany, historical economic valuations of buildings and documentation related to urban building trends are scarce, while in Great Britain, data regarding building cycles is more easily accessible. Therefore, Pearson could compare the economic development of insurance with building trends and thereby relatively plausibly deduce when a construction boom also created an increase in insurance numbers—simply because there was an increase in the building stock that could be insured—or when there was intrinsic growth in the insurance business that was unrelated to such external factors.[43]

The interrelation between damage payments and the value of, for the most part, relatively continuously rising insurance amounts primarily demonstrates only that the rise of damage payments on average is, of course, based only on the proportions given here (generally, a higher number of insured buildings also leads to increased damages). It is important to pay attention to the peaks in certain years with unusually high damage payments. Do they indicate a very large city fire or a multitude of smaller fires? Do they correlate with simultaneous peaks in German fire ecology?[44] This can be determined by comparing regional fire insurance data with the database of the *Städtebücher*. A major city fire always causes a substantial peak in the data—it is potentially but not necessarily related to a general climatic situation. Therefore, such years should be excluded when evaluating a climate correlation, unless other data confirms that the spike should be interpreted as a result of climate influence. If fire

42 A good discussion of the small scale of many territorial institutions, and also of the limited nature of surviving sources concerning those small territories, which make a true historical statistical accounting impossible, is offered by Spies, *Feuerversicherung*, pp. 253–272. Cf. appendix 2 below for the major portion of fire insurance foundings in the German territories until the early nineteenth century.

43 Pearson, *Insuring the Industrial Revolution*, pp. 38–46.

44 The concept of "fire ecology" is used here in reference to Stephen Pyne's concept of "fire ecology"; it could be defined as the broad overview and analysis of the fire cycles and fire threats inherent to a given regional city environment.

damages in a region are widely distributed, without a major city fire identified as the primary cause, then the probability of a climatic induction is very high. Continuously dry conditions, unusually hot and dry months, scant rainfall before a fire, and strong winds at the moment the fire begins have always been conditions that significantly increase the effect of fires.

In particular, this comparative analysis results in the following conclusions: in Baden, the years 1832 and 1849 seem to stand out, but there are no unusual events for those years recorded in the Baden *Städtebuch*; the peak values in these years could have been caused by a large number of fires scattered throughout the region. In Bavaria, in contrast, the peak in the fire insurance statistics does correlate with the data derived from other city records: the year 1822, with a significant number of six large fires, including Marktredwitz (30 destroyed houses and 36 barns),[45] Deggendorf (92 houses), and Sulzbach-Rosenberg (281 houses); the year 1834 (three major city fires in Wunsiedel [205 houses], Weilheim, and, most significantly Bad Reichenhall, in which three-quarters of the city was destroyed), 1848 (four major town fires: Eibelstadt, Kemnath [103 houses], Pleystein [55 houses], Selach, Windischeschenbach [60 houses]) and 1850 (two town fires in Bischofsheim [137 houses] and Grafenau). The Braunschweig statistics (Figure 16) show high peaks in 1780 (Goslar city fire),[46] 1794 (fire in Altenau, upper Harz), 1823 (Büddenstedt fire: 34 farms, 73,049 Reichstaler (Rtl.) in damages), 1825 (Seesen city fire: 60 houses, 142 outbuildings, 104,187 Rtl. in damages); the peaks in 1816, 1834, and 1836 are, however, not explainable by any major singular city fires in the Duchy of Braunschweig. The Calenberg-Grubenhagen Fire Insurance Fund, which later belonged to the Electorate of Braunschweig-Lüneburg, then later to the Kingdom of Hanover, exhibits swings that are comparable with the *Städtebücher* data: 1824/25 (Elze fire, 1824), 1826/27 (Einbeck fire, 1826), 1834/35 (Dransfeld and Rethem fires), 1844/45 (Clausthal-Zellerfeld fire). The peaks of 1796/97 and 1831/32, however, appear somewhat inscrutable. The Hesse (Hesse-Darmstadt) Chamber of Fire Insurance lists peaks in 1797, 1799, and 1803 (explainable by the impact of the revolutionary wars in rural areas), 1834, 1842–44, and 1848–53, which cannot be clearly explained via coincidental events in the *Städtebücher* database; the

45 Oskar Gebhardt, *Abriss der Geschichte und Topographie von Markt-Redwitz* [...], Markt-Redwitz 1906, p. 69f.

46 The more or less free imperial city of Goslar applied for inclusion in the Braunschweig insurance fund in 1768, was admitted in 1771 and was expelled again after the destructive fire of 1780, the damages of which were, however, still thoroughly compensated by the fund, see Braunschweigische Landesbrandversicherungsanstalt (ed.), *Festschrift 225 Jahre LBVA. Braunschweigische Landesbrandversicherungsanstalt 1754–1979*, Braunschweig 1979, p. 16f.

peaks in 1810 (Beerfelden fire), and 1825 (Offenbach and Michelstadt fires), however, can be explained in this way. In the case of the Nassau Fire Insurance Fund, the years 1819, 1825, 1827, 1843, and 1848–50 are peak years. In this case, there is nothing more than a fire in Haiger (1827) that relates to these numbers. For the Saxon State Insurance Funds (*Sächsische Landesbrandversicherungs-anstalt*), the Wars of Liberation are reflected in 1814/15, and 1832 was a year of plentiful city fires (Hainichen, Waldheim, and Schellenberg/Augustusburg), while 1843 bears the consequences of the Oschatz and Sayda fires of 1842. The Waldeck-Pyrmont Insurance Fund records a completely typical development of over almost 60 years without major incidences; the large peaks in 1843/44, 1847, and 1849 are, however, hardly explainable through city fires (only the Sachsenberg fire of 1844 is relevant here). The major fires in Göppingen (1782), Wangen (1793), Tuttlingen (1803), and Balingen (1809) are easily discernable in the statistics of the Württemberg Fire Fund (Figure 17). The peaks after 1844 are, however, more difficult to explain as the result of major city fires (although there were smaller fires in Ebingen, Grötzingen, Trossingen, Oberriexingen, Dornhan, Schömberg, Veringenstadt, and Tettnang during this period). The East Frisian Rural Fire Insurance Fund (*Die ostfriesische Feuerversicherungs-kasse des platten Landes*; Figure 18) had the advantage of covering only rural areas absent any larger cities. This unusual feature makes it clear that the significant spikes in 1815, 1825, and 1834 are an indication of unusual climatic events, since they concern regionally scattered conditions that are not distorted by major solitary city fires.

For almost all statistical groups, the years 1824/25 and 1834, in addition to the 1840s and 1850s, feature disproportionately high peaks: the peaks in the latter years suggest that the effects of protoindustrial development—as analyzed below (in Ch. 3, 1.6) using Hamburg as an example—are not isolated effects, but that the fire risk definitively rose in these years. In relative terms, this absolute increase was not dramatic—taking into account the population increase and the fire hazards related to the protoindustrial means of production. In absolute terms, however, it was significant.

The analysis of fire ecology characteristics on this mesolevel, therefore, illustrates times of low and high fire frequencies on a local-regional level. One might take the peaks of such periods with high fire frequency therefore also to represent the strong "stress factor" that this great threat was representing for the urban (and rural) population in premodern societies—a stress factor often neglected in manuals and syntheses devoted to the early modern era. This stress factor can be detailed, when the sources allow, even more precisely at the city or regional microlevels (cf. Ch. 3, 1.6). Two causal factors for outbreaks of city fires were clearly war and climate.

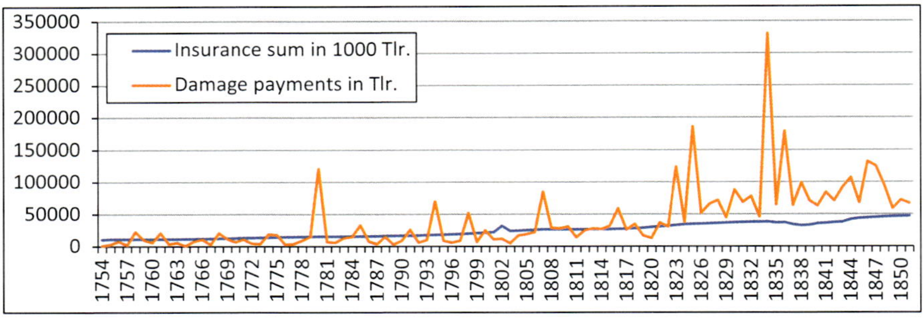

FIGURE 16 Braunschweig Landesbrandversicherungsanstalt (Braunschweig Territorial Fire Insurance) 1754–1850

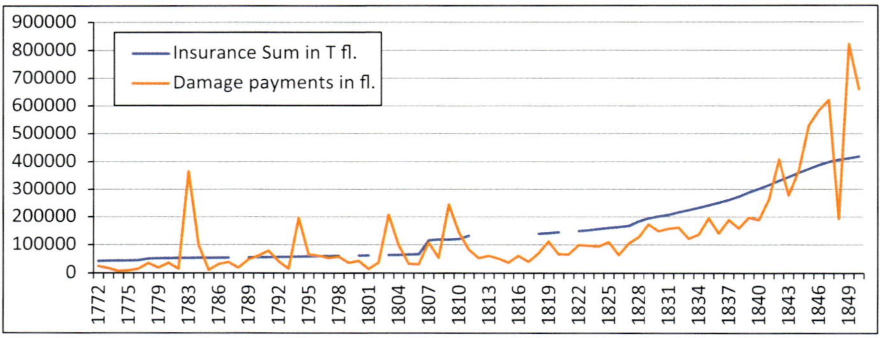

FIGURE 17 Württemberg Gebäudebrandversicherung (Württemberg Building Insurance) 1772–1850

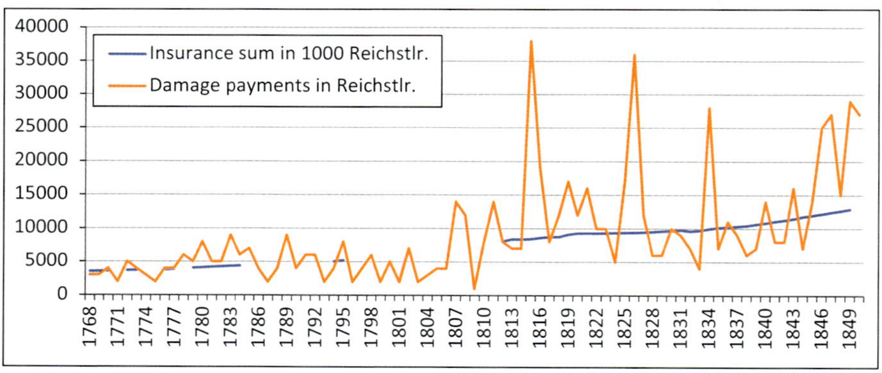

FIGURE 18 Ostfriesische Feuerversicherungskasse des platten Landes (East Frisian Rural Fire Insurance Fund) 1768–1850

1.4 *War and Fire Trends*

In German city fire statistics taken as a whole, decisive war years are always clearly visible; the Hussite Wars, the Thirty Years' War and the Nine Years' War comprise the highlights of the entire list. The 250 years between 1430 and 1700 form the high-point of fire danger in the German-speaking realm, at precisely the time of the much-discussed "military revolution." This suggests that this critical war-fire boom could also be correlated with other European regions.

At first glance, it is not surprising to find that 1634, a single year during the Thirty Years' War, was the most important fire year in Germany as a whole, according to the available fire data, since it is this year, in particular, that was hit by a random cumulative effect, even though there are many other years that seem more important in regional historical memory. This can be easily ascertained through a comparison of statistics for Germany as a whole with specific regions. In Silesia, fires were more prevalent in 1633; in Bavaria they were more prevalent in 1632 and 1633, after Sweden's entry into the war; in southwestern Germany, 1633; in Brandenburg, 1631 and 1637; in Saxony, Saxony-Anhalt, and Thuringia, 1632, 1637, and 1639; in Hesse, Westphalia, and Lower Saxony, 1626 and 1632; in northern Germany, 1630. Only Württemberg reflects this singular peak during the Thirty Years' War: 1634, with 20 city fires. The high peak value of nearly 90 city fires in 1634, never before and never again reached in the nearly 1,000 recorded years, is therefore a cumulative value. This value signifies that while 1634 was not the regional peak year for the destruction and pillaging of the Thirty Years' War (other than in Württemberg), there were never again so many major destructive fires in so many areas during the same year.[47]

---

47 In Baden: Elzach, Hornberg, Schönau; Bayern: Aichach, Bad Tölz, Ebersberg, Eichstätt, Erding, Freising, Furth i. Wald, Fürth, Göggingen, Kronach, Landshut, Lichtenberg, Mainburg, Pleystein, Roding, Schnaittenbach, Schönsee, Stadtsteinach, Velden, Wassertrüdingen. In Brandenburg: Gartz/Oder, Kyritz, Liebenwalde, Lieberose; Hessen: Büdingen, Gelnhausen, Grüningen, Reinheim, Runkel, Windecken. In Farther Pomerania: Bahn, Pyritz; Mecklenburg: Teterow. In Lower Austria: Melk. In Lower Saxony: Clausthal, Dransfeld. In East Prussia: Kreuzburg. In Rhineland: Düsseldorf. In Rheinland-Pfalz: Bingen. In Saxony: Bautzen, Dippoldiswalde, Kirchberg/Zwickau, Sayda, Wilsdruff, Zschopau. In Saxony-Anhalt: Klötze, Sömmerda, Suhl, Wolmirstedt. In Silesia: Glatz, Greiffenberg, Hirschberg, Lauban, Leobschütz, Liegnitz, Muskau, Neustädtel, Ohlau, Patschkau. In Thuringia: Creuzburg/Werra, Kaltennordheim, Rastenberg, Themar. In Westphalia: Medebach, Olpe. In Württemberg: Aalen, Calw, Friedrichshafen, Giengen/Br., Grötzingen, Heimsheim, Heubach, Laichingen, Lorch, Metzingen, Neuffen, Plochingen, Schorndorf, Steinheim/Murr, Stuttgart-Feuerbach, Stuttgart-Zuffenhausen, Trossingen, Tuttlingen, Waiblingen.

A few of the recorded city fires for 1634 were, however, not war induced—the Olpe fire of 1634, caused by the carelessness of blacksmiths, completely incinerated the town, with the exception of the Felmicke area;[48] the fire of Liebenwalde (the so-called "Dancers'" or "Betzen" fire of April 16, 1634);[49] the small fire in Buxtehude during the imperial occupation;[50] the Clausthal-Zellerfeld fire of September 20, 1634 (destroyed 162 houses, as well the city hall, archives, and the market church);[51] and the Düsseldorf fire of 1634, caused by lightning striking the gunpowder tower. In contrast, the fires in Bavaria set by Sweden in the spring and summer of 1634 were clearly war induced: Matthias Merian's *Theatrum Europaeum* documents over 700 burned towns and cities in this region.[52] This number may be highly exaggerated, but without a doubt, the number of communities affected by fire in 1634 was much larger than the number of 90 known fires recorded in the *Städtebücher* according to given archival sources: one must keep in mind that only communities with town charters are included in the data, while the large number of villages is excluded by definition—though the size of the communities was often similar. Local histories note a few details about the 1634 fires in Aichach (June 24);[53] Pleystein,[54] a fire, which like the Bautzen fire, was caused by setting the suburbs on fire by the Austrian Colonel Goltz;[55]

48 Asta Schröder, "Die Stadt Olpe," in Josef Wermert (ed.), *Olpe. Geschichte von Stadt und Land.* vol. 1, *Von den Anfängen bis zum Ende des Ersten Weltkrieges*, Olpe 2002, pp. 219–62, 229.

49 Jörn Lehmann, *Die Stadt Liebenwalde. Ein Streifzug durch die Geschichte der Ortsteile Freienhagen, Hammer, Kreuzbruch, Liebenthal, Liebenwalde und Neuholland*, Karwe 2004, p. 103.

50 Michael Mattern, "Die Stadtbefestigung Buxtehudes im 17. Jahrhundert und ihr Bild in den schwedischen Festungsplänen," *Heimatliches Buxtehude* 5 (1997), pp. 115–29.

51 Heinrich Morich, "Die früheren Brandkatastrophen in Clausthal und Zellerfeld," in *450 Jahre Clausthal-Zellerfeld 1532–1982*, Clausthal-Zellerfeld 1982, pp. 74–76, 75.

52 Johann Nepomuk Sepp, *Die Kriegsthaten der Isarwinkler*, Munich 1874, p. 59f. Henricus Oraeus, *Theatri Europaei Oder Historischer Chronicken Beschreibung Aller Vornembsten vnd Denckwürdigsten Geschichten, so sich hin vnd wieder in Europa, sonderlich im Reich Teutscher Nation von Ao. 1633 biß in Ao. 1638 begeben vnd zugetragen*, part 3, first published by Matthäus Merian, Frankfurt am Main 1670.

53 Konrad Danhauser, *Topographische Geschichte der Stadt Aichach und ihrer Umgebung*, Munich 1858, p. 12f.

54 Colonel Corpus entered the city with 300 cavalrymen and set the city on fire, destroying most of it, including the city hall, hospital, and the city hall archives: Siegfried Poblotzki, *Geschichte der Herrschaft, der Stadt und der Pfarrei Pleystein*, Pleystein 1980, p. 140f.

55 Karl Limmer, *Entwurf einer urkundlich-pragmatischen Geschichte der Lausitzen*, Ronneburg 1839, p. 235; cf. Rudolf Wolkan, "Studien zur Reformationsgeschichte Nordböhmens," *Jahrbuch der Gesellschaft für die Geschichte des Protestantismus in Österreich* 4, 2 (1883), pp. 67–95, 80f.

as well as a wealth of fires set by the swarming imperial army after the battle of Nördlingen,[56] i.e., those in Giengen/Brenz (May 9), Aalen (July 9),[57] Waiblingen (September 18),[58] Heimsheim, Neuffen, Schorndorf,[59] and also Fürth/Bavaria.[60]

The spread of the fire gap after about 1700 also strongly relates to the fact that the number of war-induced fires apparently decreased consistently due to changes in military techniques. Military history no longer characterizes the eighteenth century as an era of outright peace, but rather emphasizes that even during the Enlightenment, war and the military played a significant role. Jeremy Black has repeatedly pointed out that in the discussion about the "military revolution," the developments of the eighteenth century have been severely neglected.[61] The wars of the eighteenth century cannot be viewed as simply more harmless and less cruel than the wars of the seventeenth century; even if it were possible to definitively establish a specific decrease in frequency in view of the *longue durée*, this would be offset by the increase in intensity related to the ever-increasing size of armies.[62] Nonetheless, the fire statistics seem to clearly prove, rather re-revisionistically, that the character of wars had changed. Whether it is possible to ascertain the establishment of standing armies with successful social disciplining of the military corps is anyone's guess.[63]

---

56 Gerd Zillhardt, *Der Dreißigjährige Krieg in zeitgenössischer Darstellung. Hans Heberles "Zeytregister" (1618–1672). Aufzeichnungen aus dem Ulmer Territorium*, Ulm 1975, p. 151.

57 Bernhard Hildebrand, 'Aalen im Jahre 1634. Der große Stadtbrand und seine Ursachen," *Aalener Jahrbuch* (1982), pp. 67–81; Bernhard Hildebrand "Die Ereignisse um 1634. Im allgemeinen und unter besonderer Berücksichtigung der Schlacht bei Nördlingen und des Schicksals der Stadt Aalen," *Nordschwaben. Zeitschrift für Landschaft, Geschichte, Kultur und Zeitgeschehen* 12, 3 (1984), pp. 123–31, 130f.: the city was supposedly set on fire by the brothers Johann and Jesaias Nietner, who set the gun powder stores on fire.

58 Gottfried Lammert, *Geschichte der Seuchen, Hungers- und Kriegsnoth zur Zeit des Dreissigjährigen Krieges*, Wiesbaden 1890, p. 179.

59 Gustaf Roeder, *Württemberg. Vom Neckar zur Donau. Landschaft, Geschichte, Kultur, Kunst*, Nuremberg 1972, pp. 191, 314, 438.

60 Georg Tobias Christoph Fronmüller, *Chronik der Stadt Fürth*, 2nd ed., Fürth 1887, p. 96.

61 Jeremy Black, *A Military Revolution? Military Change and European Society, 1550–1800*, Basingstoke 1991; Jeremy Black, *European Warfare, 1660–1815*, London 1994; Jeremy Black, "A Military Revolution? A 1660–1792 Perspective," in Clifford J. Rogers (ed.), *The Military Revolution Debate, Readings on the Military Transformation of Early Modern Europe*, Boulder 1995, pp. 95–114.

62 Johannes Burkhardt, "Die Friedlosigkeit der frühen Neuzeit. Grundlegung einer Theorie der Bellizität Europas," *Zeitschrift für historische Forschung* 24, 4 (1997), pp. 509–74.

63 Cf. Ralf Pröve, *Stehendes Heer und städtische Gesellschaft im 18. Jahrhundert. Göttingen und seine Militärbevölkerung 1713–1756*, Munich 1995; Jürgen Luh, *Kriegskunst in Europa 1650–1800*, Cologne 2004.

However, the professionalization of the army, the new recruitment and quartering practices, the logistical innovations concerning the organization of army supplies (*étapes*), and the martial-law ban on outright pillaging and raids outside of the logics of a formal "contribution system" corresponding to a military state created new conditions.

While the old theory of the increasing militarization of society is supported in this respect, since the military was a consistent presence and a part of society in everyday urban life,[64] it is conversely true that the separation between the military and society was much more precisely defined and adhered to.[65] For the state of the war itself, this necessitated a shift toward fortress sieges and decisive battles in the context of relatively regulated occupation politics, in contrast with the uncontrolled practice of mercenary warfare, the burning of cities as an act of revenge during the Thirty Years' War, and the rationalistic, thoroughly organized scorched-earth warfare practices employed, for example, in the Nine Years' War by Louis XIV.[66] These are, however, only trends, for which one can always find counterexamples. In pointing out this tendency, one should also mention that the catastrophic town-destroying military logic of the seventeenth century was, in terms of fire history and aggregate serial town destruction, only exceeded by the bombing campaign of the Second World War. Looking at these exceptional periods standing together in an unholy alliance, it becomes clear why, in the collective memory of cities and towns (through guidebooks, chronicles, monuments, etc.) the destruction of the Thirty Years' War, (especially concerning the last major city fire), and the destruction of the Second World War are often mentioned in the same breath. This doubling of two wide-ranging series of urban destruction is certainly worthy of a collective

---

64 Concerning this topic see Peter H. Wilson, "Social Militarization in Eighteenth-Century Germany," *German History* 18 (2000), pp. 1–39.

65 For a summary of the research see Daniel Hohrath, "Spätbarocke Kriegspraxis und aufgeklärte Kriegswissenschaften. Neue Forschungen und Perspektiven zu Krieg und Militär im 'Zeitalter der Aufklärung,'" *Aufklärung* 12 (2000), pp. 5–47, where it is suggested that, regarding the wars of the eighteenth century in particular, "very little has been published in recent times" (14). Cf. also Marian Füssel, *Der Siebenjährige Krieg. Ein Weltkrieg im 18. Jahrhundert*, Munich 2010.

66 Horst Carl, *Okkupation und Regionalismus. Die preußischen Westprovinzen im Siebenjährigen Krieg*, Mainz 1993, pp. 10–17; Colin Jones, "The Military Revolution and the Professionalisation of the French Army under the Ancien Régime," in Rogers, *The Military Revolution Debate*, pp. 149–67, 164; Roland Vetter, *"Die ganze Stadt ist abgebrannt." Heidelbergs zweite Zerstörung im Pfälzischen Erbfolgekrieg 1693*, Karlsruhe 2009; Emilie Dosquet, "Die Verwüstung der Pfalz als (Medien-)Ereignis. Von der rheinländischen Kriegshandlung zum europäischen Skandal," in *Krieg und Kriegserfahrung im Westen des Reiches 1568–1714*, ed. Andreas Rutz, Göttingen 2017, pp. 333–370.

memory study and, to a certain extent, might be considered a truly unique property of the German collective disaster memory on a European and global level.[67]

Stephen Pyne, the American historian who dared to devote almost his entire œuvre to fire and has founded an almost distinct field of fire history, has focused almost exclusively on the history of rural fire management regimes. In his account of the history of fire in Europe, city fires appear only cursorily,[68] and there is no discussion of the relationship between fire and war. Similarly, Johan Goudsblom's history of civilization essay devotes not a single word to the development of warfare in relation to fires.[69] This is all the more surprising because of the fact that premodern warfare was very obviously the main producer of major city fire disasters. It seems that the unconscious influence of the well-known Romantic nature/culture distinction is at play here, so that fire disasters, inflicted by people on people, drop out of the perspective of fire ecology—in direct opposition to the methodological caveat that was long ago articulated by researchers such as Martin Melosi and William Cronon. But just as environmental history in general has increasingly included urban spaces within its purview, when it comes to the material flow and metabolic relationships between the town and its surroundings (Melosi), from a macroperspective, even this secular conjuncture of the relationship between war and the city fire as a relevant topic of environmental history must be grafted into fire history. The partial taming of war in this regard from 1700 until the firebombing campaign of World War II is a historically unlikely phenomenon that is central to an epochal threshold in fire history—one which will be explicated as a central point of investigation further below.

---

67 The topic of how the National Socialist image of history often referred to the "Thirty Years' War" as a tragically fated chain of events reflecting its own, contemporary notion of "total war" is certainly relevant in this context, cf. Laurenz Müller, *Diktatur und Revolution. Reformation und Bauernkrieg in der Geschichtsschreibung des "Dritten Reiches" und der DDR*, Stuttgart 2004; Mathias Lehmann, *Der Dreißigjährige Krieg im Musiktheater während der NS-Zeit* [...], Hamburg 2004. This also includes the metaphors and images that were resurrected when, after the British bomber attack on Hamburg in 1942, "Operation Gomorrah" reinvoked the Biblical references also present in the early modern sermons on divine retribution. For recent contributions to this field see Fabian Lampart et al. (eds.), *Der Zweite Dreißigjährige Krieg. Deutungskämpfe in der Literatur der Moderne*, Baden-Baden 2019.

68 Stephan Pyne, *Vestal Fire: An Environmental History, Told through Fire, of Europe and Europe's Encounter with the World*, Seattle 1997, pp. 50–59.

69 Johann Goudsblom, *Die Entdeckung des Feuers*, Frankfurt am Main 1995, pp. 175–183.

1.5 *Climate and Fire Trends*

Because of the current relevance of the topics of climate change and global warming, climate history has, not surprisingly, increasingly shifted from a mostly neglected, marginal problem of general historical interest, to an absolutely central problem in the postwar era. Inquiries into whether the reality and nature of correlations, or even causal relationships, between the historical climate changes of the last 1,000 years and social developments can be determined, or at least assumed with good reason, are becoming increasingly important.

The basic phases and elements of the climatic evolution of this time period are now well-known, and have been subject to a globally focused and highly specific analysis during the last fifteen years in particular. After Hubert Lamb's 1965 study, which elevated the Medieval Warm Period (Medieval Climate Optimum)—occurring from about 900–1000 AD and then again from 1200–1300 AD—to a leading topic in recent climate-historical interest, the critical inquiries of Malcolm K. Hughes and Henry F. Diaz in 1994 found that this climatic anomaly applied not only to Europe, but that its validity (with phase shifts, in part) could be demonstrated globally.[70] Likewise, the existence of the second large climatic anomaly of the millennium, the Little Ice Age, is indisputable, and though it was originally discovered as a period of glacier growth on the basis of existing Swiss glaciers, its global reach has also been documented.[71] However, historical research still has a long way to go in

70 Hubert H. Lamb, "The Early Medieval Warm Epoch and Its Sequel," *Palaeogeography, Palaeoclimatology, Palaeoecology* 1 (1965), pp. 13–37; Malcolm K. Hughes and Henry F. Diaz, "Was There a 'Medieval Warm Period,' and If So, Where and When?," *Climatic Change* 26 (1994), pp. 109–42; also cf. the contributions from Zhang De'er, Jean M. Grove/ Roy Switsur, Ricardo Villalba, Kenneth Lee Petersen in the same volume, pp. 143–69, 183–97, 243–69; Scott Stine, "On the Medieval Climatic Anomaly," *Current Anthropology* 41, 4 (2000), pp. 627f.; Keith R. Briffa, "Annual Climate Variability in the Holocene. Interpreting the Message of Ancient Trees," *Quaternary Science Reviews* 19 (2000), pp. 87–105; B.G. Hunt, "The Medieval Warm Period, the Little Ice Age and Simulated Climatic Variability," *Climate Dynamics* 27 (2006), pp. 677–94; Samuli Helama, Jouko Meriläinen, and Heikki Tuomenvirta, "Multicentennial Megadrought in Northern Europe Coincided with a Global El Niño-Southern Oscillation Drought Pattern during the Medieval Climate Anomaly," *Geology* 37, 2 (2009), pp. 175–78; Michael G. Loso, "Summer Temperatures during the Medieval Warm Period and Little Ice Age Inferred from Varved Proglacial Lake Sediments in Southern Alaska," *Journal of Palaeolimnology* 41 (2009), pp. 117–28. The standard reconstruction of Michael Mann et al. (first 1998, then 2008) for the climate of the last millenium has been remodeled by Robert Tardif et al., "Last Millennium Reanalysis with an expanded proxy database and seasonal proxy modeling", *Climate of the Past* 15 (2019), pp. 1251–1273.

71 Cf. Jean M. Grove, *The Little Ice Age*, London 1988; for a virtual reader of 'classic papers' addressing the LIA cf. Frank M Chambers, "The 'Little Ice Age'", *The Holocene* 26, 3 (2016), pp. 335–337. Raphael Neukom et al., "No evidence for globally coherent warm and cold

terms of the application of these climatological findings, and it is indeed possible that there is an ultimately unbridgeable gap preventing the creation of absolutely valid narratives of causal history in this field. The idea that Gothic cathedrals with their large windows—in contrast to Romanesque buildings—could be interpreted as an architectural-stylistic adaptation to the new warm medieval climate seems to be a rather weak association.[72] The extent to which the spread of the plague was enabled by the dawning of the Little Ice Age, and whether (or how precisely) the European witch hunts, certain observations concerning suicide statistics, or specific trends in the discourse of melancholy can be best understood in terms of a causal relationship with climatic phenomena, must be further examined, though causal narratives of this kind are already proliferating in large numbers, emerging from the most reputable research institutes and the best academic presses.[73]

However, at least in some cases, it should certainly be possible to draw reliable and statistically plausible correlations between fire trends and annual climatic fluctuations, but this requires sufficient data and statistical evidence. The clearest case in the applicable data sample is the year 1540, which is known in climate history generally as one of the largest anomalies of heat and dryness in the past 500 years: "Extreme warmth and unusual dryness" prevailed. "For several weeks on end, it hardly rained at all; according to source statements, in 26 weeks there were 'sprinkles' on only five days. The heat of the summer of 1540 is the best-documented event of the sixteenth century and can be viewed

---

periods over the preindustrial Common Era", *Nature* 571 (2019), pp. 550–554 argue that, after having shown the global character of both anomalies within the last 1000 years, one needs to stress that those climate epochs were not at all synchronic and coherent for all regions and the continent.

72 "In contrast, the new Gothic style features light and lightness. With its large windows it lets the sunlight of the Medieval Warm Period into the monumental cathedral that were being constructed" (Wolfgang Behringer, *Kulturgeschichte des Klimas. Von der Eiszeit bis zur globalen Erwärmung*, Munich 2007, p. 109; similarly, Rüdiger Glaser, *Klimageschichte Mitteleuropas. 1200 Jahre Wetter, Klima, Katastrophen*, Darmstadt 2008, p. 59f.)

73 As a point of entry, see Wolfgang Behringer (ed.), *Kulturelle Konsequenzen der "Kleinen Eiszeit"*, Göttingen 2005; Gillen d'Arcy Wood, *Tambora. The eruption that changed the World*, Princeton 2014; Sam White, *The Climate of Rebellion in the Early Modern Ottoman Empire*, Cambridge 2011. For a somewhat critical reponse to Behringer's theories see Emmanuel Le Roy Ladurie, *Histoire humaine et comparée du climat*, vol. 1, *Canicules et glaciers, XIIIe-XVIIIe siècles*, Paris 2006, pp. 280–91; for some reminders concerning the learnings from (post-)Popperian discussions about the general problem of historical explanative narratives Cornel Zwierlein, "Klimageschichte und Kulturgeschichte der Frühen Neuzeit. Zum Problem des interdisziplinären Dialogs," in Philip Ajouri, Katja Mellmann, and Christoph Rauen (eds.), *Empirie in der Literaturwissenschaft*, Münster 2013, pp. 331–51.

as one of the major events of the millennium."[74] In fact, the *Städtebücher* clearly identify 1540 as the year with the highest number of town fires (33) without a major war incident in 1540. From the Middle Ages until the nineteenth century this value was only exceeded by the most active phase of the Thirty Years' War (1631–41) and in 1658 and 1688/89: 1540, therefore, is not only memorable as the year of the "millennium wine," but also as one of the most catastrophic years in terms of fire danger in the urban history of central Europe. Individual fires are recorded for Stettin Altdamm, Arnswalde, Biesenthal, Dippoldiswalde, Greußen, Husum, Marktredwitz, Neustettin, Nordhausen, Nossen, (July 22),[75] Pausa, Stettin-Pölitz, Sonthofen, Stargard, Triptis, Velburg, Wangen, Wehlau, Weiden, Wolkenstein, Bingen, Erkelenz, Freudenberg, Hadamar, Herbstein (September 8),[76] Jever, Peckelsheim, Rhoden, Vörden, Tübingen, Leutkirch, Einbeck and Großenhain. One can presume that the real numbers are actually slightly higher, especially if the unknown number of countless village fires were added. Fires were also documented in Memel (Lithuania) and Regensberg (Switzerland), and the landmark of the town of Greiz, the *Oberschloss*, also burned in 1540.[77] However, 1540 is also known in research as the "*Mordbrenner*" year (the "murder-burner" year, i.e., the arsonists' year): as the Protestant/Catholic confrontation neared a climax, Duke Heinrich the Younger of Braunschweig-Wolfenbüttel, in particular, was accused of employing arsonists to burn several Protestant cities, more or less as acts of guerilla warfare. So which interpretation of fire frequency is therefore correct?

The Einbeck town fire on or about July 26, 1540, which destroyed the whole town and (according to different statements) cost the lives of between 100

74 Glaser, *Klimageschichte Mitteleuropas*, p. 108; Christian Pfister, *Wetternachhersage. 500 Jahre Klimavariationen und Naturkatastrophen (1496–1995)*, Bern 1999, pp. 191–93; Oliver Wetter et al., "The year-long unprecedented European heat and drought of 1540—a worst case," *Climatic change* 125 (2014), pp. 349–63; Oliver Wetter and Christian Pfister, "An underestimated record breaking event—why summer 1540 was likely warmer than 2003," *Climate of the Past* 9 (2013), pp. 41–56; Christian Pfister, "The 'Black Swan' of 1540: Aspects of a European Megadrought," in Claus Leggewie and Franz Mauelshagen (eds.), *Climate and Culture, Climate Change and Cultural Transition in Europe*, Leiden 2018, pp. 156–194.

75 Alfred Berger, *Ein Streifzug durch die Nossener Geschichte zur 1000-Jahrfeier der Landschaft, 750-Jahrfeier des Ortes und 100-Jahrfeier der Einführung der Städteordnung. Gedenkschrift zum Heimat- und Schulfest Pfingsten 1936*, Nossen 1936, p. 38: determined to have started spontaneously after a long, major period of heat and drought.

76 Rudolf Bald and A. Dollberger, *700 Jahre Stadtrecht Herbstein. Geschichte und Entwicklung einer Stadt*, Lauterbach/Hessen 1962, p. 35: no mention of arson.

77 Manfred Gläser (ed.), *Lübecker Kolloquium zur Stadtarchäologie im Hanseraum*, vol. 1, *Stand, Aufgaben und Perspektiven*, Lübeck 1997, p. 309; Alvin Jaeggli-Leepin, *Burgen und Schlösser in der Schweiz. Nach alten Vorlagen*, Frankfurt am Main 1962, p. 127.

and 500 people was supposedly one of these arson events—subsequent reconstruction created today's unique cityscape of a completely preserved late Gothic half-timbered town. On July 28, an intoxicated shepherd from Hohenbüchen was arrested, and confessed—admittedly under torture—that the patrician Heinrich Dieck paid him for the arson.[78] Dieck himself, administrator (*Vogt*) of Claus von Mandelsloh of Hohenbüchen in the principality of Braunschweig-Wolfenbüttel, was arrested and tortured. In turn, he testified in the interrogation log of September 4, 1540, that Heinrich the Younger commissioned him; he had been hired, in exchange for money, by Klaus von Oberg and Heinrich's head of mercenaries, Christoph von Wrisberg. Whether these statements corresponded to the truth is uncertain—in any case, Dieck was executed.[79] The fires in Pausa/Vogtland (July 27), Triptis in Thuringia (July 26), Nordhausen (August 11),[80] Hadamar (March 14),[81] and Erfurt are all counted among those fires caused by arson. Arson plans were foiled in Kassel, Göttingen,[82] Braunschweig, Halberstadt, and Magdeburg.[83] This information, however, relies on statements acquired through torture of the accused or actual offenders. In the case of the fire of Sulza (today Bad Sulza), there was initially uncertainty about the causes, but when hay burnings also occurred in Jena, the elector acted:

> After the tiny Sulza, located in our administration district of Roßlau, burned down completely during the last lent (March 2) and it was impossible to understand and recognize if the fire started somehow by accident or if it was due to arson, and as in the following week of Judica (April 3–9)

78 Otto A. Ellisen, "Einbeck im 16. Jahrhundert," *Zeitschrift des Harz-Vereins* 27 (1894), pp. 549–60.

79 Andreas Heege, *Einbeck 1540. Brandstiftung! Der Einbecker Stadbrand vom 26. Juli 1540. Archäologischer Befund und politische Hintergründe*, Einbeck 2005, p. 34.

80 It was also believed that the fires were started through arson in the neighboring city of Mühlhausen: "Wyr haben got erbarms den grossenn verterblichenn brandtschadenn kumer und elende So E.L. auch dernn selbigenn Burgerre und Inwonerre am vergangem mittewochenn durch bosse leutte Zcugefugt, mit bekemmertem und betrubtem gemotthe angehort, und ist uns das got weys hertzlich leidt, unnd wustenn wyr E. L. sampt E. L. Burgerrn, unsers vermogens behulfflichenn zcu sein, wullen wyr uns, das zcuthon hirmit erbettenn habenn." (StadtA Mühlhausen 10/W 1–7 Bd. 16, Bl. 10 v/r, thanks to Peter Kuhlbrodt for communicating the digital reproduction).

81 Karl Joseph Stahl, *Hadamar Stadt und Schloß. Eine Heimatgeschichte anläßlich der 650-Jahrfeier der Stadtrechteverleihung an die Stadt Hadamar 1974*, Hadamar 1974, p. 62f.

82 Bruno Crome, *Kulturgeschichtliche Miniaturen vom Ausgang des Mittelalters aus einer alten Chronik,* Göttingen 1921, p. 63.

83 Heege, *Einbeck 1540*, p. 34.

> in Jena, three fires broke out, two of which were arsons set in barns; on the basis of many suspected crimes, the city council has arrested and imprisoned four individuals who were just passing through, three of them in the district of Leuchtenberg and the fourth in the district court of Hans Puster.[84]

Here too, torture was applied with permission from court authorities in Leipzig and the prisoners testified; the main offender was supposedly a certain Wenzel Voigt. The prisoners were executed on July 19 on the road from Jena to Leipzig.[85] The electorate of Saxony had sent the interrogation files to Regensburg to his ambassadors; they were supposed to be forwarded to Landgrave Philipp. However, there the Protestant princes had already submitted a petition against the *Mordbrenner* arsonists to Duke Heinrich on May 13, which also appeared in print. The court testimonies (*Urgichten*) of the *Mordbrenner*[86] were attached to it. Essentially, these documents reinforced the idea that the 1540 fires concerned anti-Protestant arson events, an idea that was intensified somewhat through polemics such as those by Luther in his leaflet written against Duke Heinrich, "Wider Hans Worst":[87]

> Now—so that I may come to an end—this, I myself think, is the reason Harry [Heintze] the devil undertook to write such evil, slanderous, and deceitful books: he knows that throughout the world he has a most infamous name and that he stinks like devilish filth flung into Germany. Perhaps he would like not to stink so much himself, but would like to make other princes stink as well, so that it is not his stench alone which fills everyone's nostrils, particularly since this year the cry of arson was raised against him. He cannot drown these cries with soft words. Hence he must shriek and scream in his struggle to escape, cursing, slandering, lying, raving, and raging, just to see whether that would help. But it does not, Harry [Heintz], you rave in vain, even if you could thunder and storm like God himself. That mass of innocent blood shed at Einbeck and

84 Kurfürst Johann Friedrich to Fürst Wolfgang v. Anhalt and the others present at the Reichstag in Regensburg in 1541, Zwickau, 28 May 1541 (StA Weimar E Nr. 101, Bl. 232, WA Briefe IX, 345 n 7).

85 Adrian Beier, *Mag. Adrian Beiers Jenaische Annalen (1523–1599)*, ed. Herbert Koch, Jena 1928, p. 22.

86 *Supplication/an Keiserliche Maiestat/Der Mortbrenner halben/Auff dem Reychstag zu Regenspurg/ Kayserlicher Mayestat vberantwortet*, s.l. 1541 (VD16 E 4649, BSB Munich 4 Res 232, Beibd. 2), f. AiijV-EijR.

87 Martin Luther, "Wider Hans Worst [1541]," WA 51, 461–571, at 551.

> elsewhere where your arson cries so strongly to heaven that (if God wills) it will soon cry you and your associates down into the abyss of hell; and it will not cease till then.[88]

Heinrich protested in front of the emperor against the accusations of the Schmalkaldic petition in a 1541 reply.[89] In it, he points out that since his name was consistently suggested as the potential instigator during torture, the prisoners would be inclined to eventually testify to that, and also draws attention to other inconsistencies. He also brings up the point that according to the court testimonies in the petition, the target of arson is always only Protestant towns, and that therefore his own Catholic lands and the Mainz diocese should have therefore been spared; but in fact: "It is notorious and undeniable that there were likewise fires in my own country, in the episcopacies of Mainz, Magdeburg and Halberstadt."[90] Also, the accused mercenary leaders Christof von Wrißberg and Christof von Oberg defended themselves against the accusations lodged in the published court testimonies.[91] The chronicles and other sources concerning the Einbeck fire beautifully illustrate the religiously polarized framing of reality at play: on the one hand, a Hildesheim monk documented fiery clouds above Einbeck, which incinerated the city via punishing lightning strikes from God while explicitly sparing the gunpowder tower, so that this could not later be deduced as the primary cause, since the cause of the fires clearly lay with the guilty residents of Einbeck; on the other hand, the Lutheran Hanover Mayor Berckhusen also recounted the Heinrich Dieck arson story: in this rendering, it was now the St. Alexander church belonging to Duke Heinrich that remained untouched by fire, thereby making it clear that the arsonist had spared his own property.[92]

---

88 Martin Luther: "Against Hanswurst, 1541," trans. Eric W. Gritsch, in *Luther's Works*, ed. Helmut T. Lehmann, Church and Ministry III 41, ed. Eric W. Gritsch, Philadelphia 1966, pp. 179–256, at 242.

89 *Hertzog Heinrichs von Braunschweig Antwort/Der Keiserlichen Maiestat/auff die vbergebene Supplication/der Mordbrenner halben/gegeben*, s.l., 1541 (BSB Munich 4 DHg. 255, Beibd. 9).

90 Ibid., f. aivV.

91 *Gedächtniß Christoffs von Wrißberg, des Kriegs Obristen, aus fast lauter alten ungedruckten Urkunden zusammen gezogen, Und zur Verbesserung verschiedener Geschicht-schreiber des 16ten Jahrhunderts, insonderheit des Schmalckaldischen Krieges, nunmehro zum Druck befördert*, ed. Johann Justus Losius, Hildesheim 1742, supplement 1, 3–8—cf. Heege, *Einbeck 1540*, pp. 111–20.

92 Helge Steenweg, "Einbeck im Zeitalter der reformatorischen Bewegung," in *Geschichte der Stadt Einbeck*, ed. Horst Hülse on behalf of the Einbecker Geschichtsverein e.V., vol. 1, *Von den Anfängen bis zum Ende des 18. Jahrhunderts*, Einbeck 1991, pp. 135–54, 135f. In a

It is therefore difficult to determine retrospectively whether one of the most notorious *Mordbrenner* series of the early modern period truly should have been characterized as such, or whether the religiously framed presuppositions and the always incipient logic of the search for a scapegoat *made* this series of fires *into Mordbrenner* fires. Similar spirals of blame are associated with almost all major fires. Questions of whether antipapal polemics in the context of the hated Frenchifying royal court influenced the trail and execution of the Frenchman Robert Hubert following the 1666 fire of London, or whether the Istanbul fire of 1660 was interpreted as a sign of necessary purification and used as an opportunity for the expulsion of Jews and Greeks and for the Islamization of additional sections of the city,[93] often employ a similar logic and despite or, in fact, because of the large number of sources produced in the course of such discourses of blame, the historian usually cannot definitively answer questions regarding the "actual" cause. Retrospectively, however, in the context of a climate anomaly that is well established in climate history, the observer does have access ex post to a further pattern of explanation. In the context of other city fires in 1540, which did not take place at the northwestern confrontational border between Catholic and Lutheran towns and territories, allegations of *Mordbrenner* were unheard of. In Großenhain, the first victim was the Orthodox Maria-Magdalene convent; in Erkelenz, a Catholic "bulwark"

statement by the council of Einbeck itself to the Göttingen council there is still primarily ignorance concerning process and cause; only unsurprising common sources of fire, such as wind and flying sparks, were reported.

93 Marc David Baer, "The Great Fire of 1660 and the Islamization of Christian and Jewish Space in Istanbul," *International Journal of Middle East Studies* 36, 2 (2004), pp. 159–81; Stephen Porter, *The Great Fire of London*, Phoenix Mill 1996; Adrian Tinniswood, *By Permission of Heaven. The story of the Great Fire of London*, London 2004, pp. 163–89. Concerning the complicated London case one should recall that while Hubert declared himself to be Catholic and pleaded guilty, most witnesses identified him as a Huguenot. However, his testimony was, in the end, decisive for the case and was the primary factor that led to his execution and to the creation of the myth of "papist" arson, later memorialized on the plaque erected in Pudding Lane in 1681: "Here by Permission of Heaven Hell broke loose upon this protestant city from the malicious hearts of barbarous Papists by the hand of their agent Hubert [...]" (ibid., Fig. 30 und LMA London, Journals of the Court of Common Council 49, f. 224r). Similarly the inscription on the Fire Memorial, in honor of the "Furor papisticus," which was supposedly responsible for the "dreadful Burning of this Protestant City" with the goal of "carrying on their Horried Plott, for extirpating the Protestant Religion, and old English Liberty and Introducing Popery and Slavery" (ibid.). Regarding the monument, see Matthew F. Walker, "The Limits of collaboration: Robert Hooke, Christopher Wren and the Designing of the Monument to the Great Fire of London," *Notes & Records of the Royal Society* 65 (2011), pp. 121–43; Field, *London, Londoners*, 45–48.

belonging to the bishopric of Lüttich (Liège) was also completely destroyed in the midst of Lutheran-Calvinist reformation movements; Bingen was also an important town in the region controlled by the Archbishop of Mainz.[94]

All in all, against the background of the fire statistics for Germany as a whole in the climatically exceptional year of 1540, it appears very likely that the whole group of *Mordbrenner* stories, particularly concerning the Einbeck fire, belongs more to the area of scapegoating. Even if these fires were not all directly caused by the heat and dryness, statistically the mere number of fires is clearly attributable to the "megadrought", a climactic condition that would at least significantly intensify fire. This all suggests that 1540 was a year not of *Mordbrenner* fires, but rather one of climatically induced fires—a suggestion which also corresponds with the historical record of fires in other regions of Europe.[95]

In second-place after 1540 in the approximate millennium of the *Städtebücher* data is 1666, in which a disproportionally high number of fires not easily attributable to a war event also occurred (29 are known). The 1650s and 1660s also endured a surprisingly high number of major fires worldwide: the largest fire that ever struck the exceptionally fire-prone city of Constantinople/Istanbul, occurred in 1660. The largest fire ever to hit Edo/Tokyo dates to 1657.[96] The London fire of 1666 figures into general cultural memory as the "mother of all fires." The prominent fires noted in more recent German history

---

94 Wilfried Setzler, Benigna Schönhagen, Hans-Otto Binder, *Kleine Tübinger Stadtgeschichte*, Tübingen 2006, p. 70; Jürgen Sydow, *Geschichte der Stadt Tübingen*, vol. 1, *Von den Anfängen bis zum Übergang an Württemberg 1342*, Tübingen 1974, p. 74; *Von der Ackerbürgerstadt zum Oberzentrum. 90 Jahre Haus- und Grundbesitzerverein Weiden und Umgebung*, Weiden 1998, pp. 76–78; Heinrich Stöcker, *Krieg, Pest und Brand im Großenhainer Land*, Großenhain 1998, pp. 26–30; Hans-Josef Broich, "Erkelenz. Geschichtliches Portrait einer Stadt," *Der Niederrhein* 75 (2008), pp. 114–17, 116; Friedrich Hermann Hoffmann, *Die Kunstdenkmäler von Oberpfalz und Regensburg*, Bezirksamt Parsberg 4, Munich 1906, p. 218; Hellmut Rogowski, *Verfassung und Verwaltung der Herrschaft und Stadt Jever von den Anfängen bis zum Jahre 1807*, Göttingen 1967, p. 26; Walther Zimmermann, Hugo Berger, Friedrich von Klocke, Johannes Bauermann (eds.), *Handbuch der historischen Stätten Deutschlands*, vol. 3 *Nordrhein-Westfalen*, Stuttgart 1963, p. 182 (Erkelenz), 206 (Freudenberg); Jacob Keuscher, *Die Feuerbrände der Stadt Bingen in den Jahren 1403, 1490, 1540, 1689 und 1850. Ein Beitrag zur Geschichte dieser Stadt*, Dannstadt 1853.

95 Cf. Paul Delsalle, " 'L'orvale de feu' dans les villes et les campagnes comtoises (1493–1634)," in Vion-Delphin and Lassus (eds.), *Les hommes et le feu*, pp. 253–73, 269.

96 Noel Nouet, *The Shogun's City. A History of Tokyo*, Sandgate 1990, pp. 47, 105f.; Kaoru Ugawa, *The Great Fire of Edo (Tokyo) in 1657*, in Körner, *Stadtzerstörung*, vol. 1, pp. 213–238; Jordan Sand and Steven Wills, "Governance, Arson, and Firefighting in Edo, 1600–1868," in Greg Bankoff, Uwe Lübken, and Jordan Sand (eds.), *Flammable Cities. Urban Conflagration and the Making of the Modern World*, Madison 2012, pp. 44–62.

of Passau in 1662 and Annaberg in 1664 also fall within this time frame.[97] The two decades between 1650 and 1669 boast approximately 250 fires in central Europe, excluding those caused by the First Nordic War. Thus, taking war-induced fires out of the equation, these two decades—located in a position in climate history immediately preceding the "Maunder Minimum," which coincided with the coolest point of the "Little Ice Age"—comprise the two most fire-dense decades in the period. From the 1720s onwards, however, such values were no longer reached, and the decades of the Thirty Years' War, as well as the period 1670–1719, were so heavily influenced by the war that a valid comparison seems quite impossible (the values are actually only marginally higher and an exact comparison is not possible with the current state of precision in the database, furthermore, since not every fire involves the question of war-induction, one often reaches the limits of research). The year 1666 also shows up as an anomalous year for historical climatologists—from the end of April until September, unusual drought and heat dominated the climate in central Europe and also in Switzerland.[98] All non-war-related fires of the year 1666 occurred exclusively in northern Germany—from Westphalia, to Brandenburg, Saxony-Anhalt, and East Prussia: Beelitz, Dahme,[99] Frankfurt an der Oder, Kyritz,[100] Lenzen, Luckau,[101] Neustadt/Dosse, Rathenow, and Teupitz in Brandenburg; Deutsch Neukirch (Altstett), Deutsch Wartenberg, Löwen, Bischofstal (Ujrest) in Silesia; Ellrich, Gommern, Könnern, Querfurt, Schafstädt, Stendal and Werben in Saxony-Anhalt; Falkenburg in East Pomerania; Mehlsack in East Prussia; Neukalen in Mecklenburg;[102] Altena, Freudenberg and Sendenhorst in Westphalia. In the locations with existing fire data, the

97 Allemeyer, *Fewersnoth und Flammenschwert*.

98 Glaser, *Klimageschichte Mitteleuropas*, p. 160; Pfister, *Wetternachhersage*, p. 189; also cf. the nine (of forty-eight) source documents relating to 1666 in content or temporality from ClimDat: http://mitgliHg.lycos.de/mili04/1640_1699.html; Kathleen Pribyl, "A survey of the impact of summer droughts in southern and eastern England, 1200–1700", *Climate of the Past* 16 (2020), pp. 1027–1041, here 1036.

99 Werner Reinhold, *Chronik der Stadt Dahme und der Umgegend*, Dahme 1845, p. 48. Many town residents were in a church, where a wedding was taking place: "Since there was a great drought at that time and the houses were primarily covered with straw or thatch, the fire grew so quickly that within one hour the city was not only fully engulfed by flames, but, to a large extent, already reduced to ashes."

100 Not mentioned by Hans Gressel, *Die Stadt Kyritz. Entwicklung, Verfassung und Wirtschaft bis zur Städteordnung 1808/09*, Berlin 1939.

101 Rudolf Lehmann, *Geschichte der Niederlausitz*, Berlin 1963, p. 280.

102 Friedrich Schliemann, "Zur Geschichte der Stadt Neukalen," *Archiv für Landeskunde in den Großherzogthümern Mecklenburg und Revüe der Landwirthschaft* 12 (1862), pp. 511–57, 641–58, here 641.

fires consistently occur in the window from late June to late September, when climate-historical drought and heat were exceptionally intense. Thus, the data would suggest that 1666 can also be characterized as a climatically induced fire year. This also means that the London fire, in some respects, can be categorized as "one of many" in northern Europe.

1.6 *The Fire Ecology of Hamburg*

As has been shown, the establishment of fire-trend statistics across an entire land or region is plagued with a variety of uncertainties. For one thing, the data density is too low, for another, factors like war induction vs. climate induction are difficult to precisely isolate within the current state of research.[103] In the end, the view of major fire trends offers a kind of a collective fire biography for Germany, which provides significant insights into the intensity of major fires and how they were controlled. Regarding the question of the quotidian threat and also the question of the danger that people began to insure themselves against, small fires ranging from a single house fire up to several blocks of destroyed houses were more important than the threat of an overwhelming, devastating fire resulting in widespread destruction, against which protective measures were always in place anyway. These major fires contoured the horizon of anxiety and concern, and the need for security stimulated and offered starting points for processes such as "learning from disasters", although we also frequently encounter moments and periods of disaster amnesia in history, where lessons eventually learned became forgotten. If those long term developments represent something like a "danger profile" (as they would not yet speak of a "risk profile" in those premodern times) from the point of view of the human perception of a supranational, national or regional population, for individual cities, however, such a "danger profile" can more easily be shown at the level of a series of house fires. Unfortunately, such a detailed fire history—and therefore the sketch of such a "danger profile"—is not possible for many cities due to the lack of the necessary sources and fire data. Only after fire insurance had been established (during and after the eighteenth century), did such data start to be gathered for many cities. Some cities, however, do provide exceptional data for earlier periods, such as Hamburg, for Germany, which offers access to three different types of data.

103 Even today there is a lot of uncertainty regarding that question: now as before, reinsurers engaging in causal analysis, who attempt to account for the sheer constant growth of damage values caused by natural distasters, cannot give any definitive answer regarding the proportion of causality attributable to "climate change" vs. the simply increasing volume of exposed values.

- the number of destroyed houses per fire since the sixteenth century, regardless of whether they were insured or not.
- the number of firefighting incidents (*Feuerausläufe*)—i.e., the actual instances when firefighting equipment and personnel were mobilized—since 1773.
- the number of claims payments on the part of the General Fire Fund (*General-Feuer-Casse*) since 1683.

The fire and firefighting incident numbers are precisely known thanks to the assistant fire chief, Johann Ehlert Bieber, who recorded them from 1773 until the great fire of 1842. There is no other city in Germany, and only few in Europe, that can lay claim to a similar longly and precise data series—in spite of the discrepancies and uncertainties that will be discussed. Through this material, a true fire physiognomy of the city can be constructed.

Bieber belonged to a Hamburg-based dynasty of fire and insurance specialists. His father, Georg Ehlert Bieber (1761–1845) was head fire chief for decades, alderman (*Oberalter*) for the St. Nicholas parish district, and founded the private Fire Insurance Association in 1795, which went bankrupt after the fire of 1842. Johann Ehlert Bieber specifies that he can vouch for the absolute correctness of the data only for the years from 1772 on, "because the sources [concerning the previous time period] [were] not very reliably recorded." A spot check based on the ledgers of the General Fire Fund shows that, at least for the second half of the seventeenth and for the eighteenth century, the data are indeed almost entirely accurate. Only the smallest fires, causing damages of less than about 1,000–2,000 marks courant, are not listed.[104] Bieber started a tradition of recording verifiable fire statistics at the end of the seventeenth century that was also continued by Fire Chief Ammon.[105] Bieber's list records the number of houses burned per fire and also, from the year 1772 on, the number of "Ausläufe" per year—i.e., all incidents involving the mobilization of firefighting resources—even those concerning minor fires.

In analyzing Bieber's data,[106] the frequency of fires during the eighteenth century is remarkably low (Fig. 19). The relative fire scarcity characterizing the century of the Enlightenment (which, as discussed above, was also the case for the entirety of Germany as well as for England, in terms of major fires) is framed

104 Cf. HStA Best. 331-1/1, I-1, Bd. 2, f. 79–81, 83, 85, 87, 88, 94, 95, the entry on fires for the year 1767 have been checked.

105 Cf. HStA Best. 331-1/1, II-4, Bd. 1, f. 1–77: the fire damages from 1684 to 1710/12 are documented here with few gaps. In HStA Best. 111-1, Cl. VII. Fe Nr. 3 Vol. 5: Ammon's fire list for 1717–50.

106 Ibid.

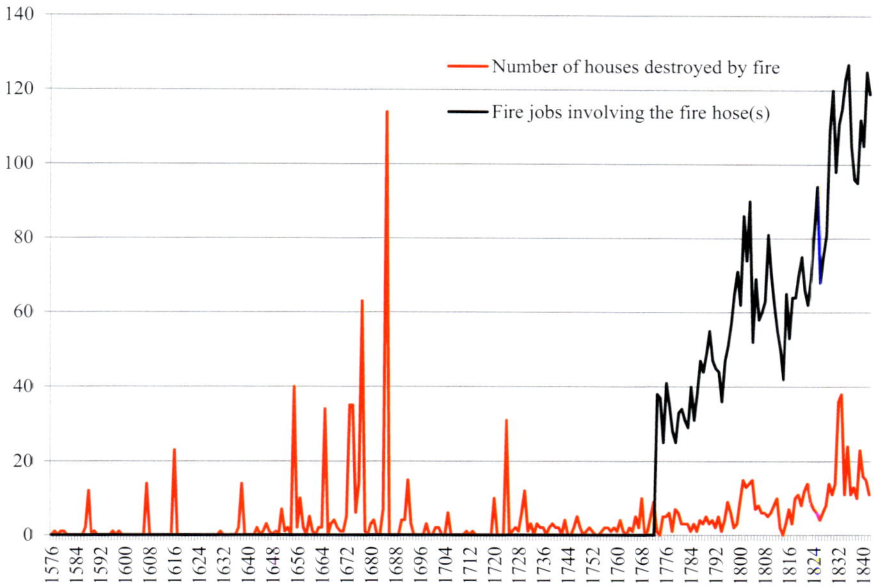

FIGURE 19 Number of destroyed houses per fire (red) and firefighting incidents recorded (black) according to descriptions by Johann Ehlert Bieber, *Verzeichniss aller Feuersbrünste in Hamburg und der Umgegend, von 1281 bis 1830*, Hamburg 1830 (StAHamburg Bibl. A 456/18: continued hand-written until 1841)

by the fire-dense periods of the late seventeenth and early nineteenth centuries. Even based on this observation alone, it is clear that the discursive overmodulation of the enormity of the Michel fire of 1750 (as will be discussed in the section of this study concerning perception and reality [Ch. 3, 2.1]), as well as the impetus for the reform of fire *policey* management during the eighteenth century and the Enlightenment dynamics of lightning-rod proponents (see Ch. 3, 3.1.2), are factors that cannot be reduced to the simple logic of "major disasters and great risks lead to reform reactions." Rather, there seems to be an intrinsic logic at work here, which perhaps, in view of the *longue durée*, actually originated in the crisis-like context of the late seventeenth century. In the case of Hamburg, the element of war-induced fires is of no importance for the relevant time period.

However, it is important to avoid falling prey to the false impression that can be easily generated by the graphical visualization: the high peaks during the late seventeenth century—in particular the peak associated with the great fire of 1684—must not obscure the fact that the sum total of destroyed houses between 1650 and 1700 (435) is less than the total in the period 1800 to 1841 (470); in the nineteenth century, the average fire risk increases steadily and annually until 1842 without the occurrence of a major disaster. It is also important to correlate these values with population numbers. In 1650 there were approximately 60,000 inhabitants in Hamburg, and around 1750, nearly 75,000. The frequent large fires thus can hardly be viewed in a causal relationship to the population increase, which was relatively limited. In contrast, the relatively rapid population increase from 75,000 in 1750 to 100,000 in 1787 and 130,000 in 1800 did not substantially affect the number of houses destroyed by fire. The mean value for the period between 1780 and 1805 was 5.61 houses destroyed by fire per year (with four fire-prone years between 1800–03); during the time of French occupation between 1806–14, when the population dropped dramatically to 55,000 inhabitants, the average number of houses burned remained relatively steady at 5.75; in 1815–27 the number rose to 8.08; and in 1827–41 it rose fairly dramatically to 17.6. The population in 1815–26, directly following the French period, had just reached 122,000 and in 1834–40 it rose slightly to 130,000–136,000—i.e., approximating the population of 1800 before the occupation. The sudden increase in the average number of houses burned after the French period, but particularly after 1827, is thus in no way proportional to population growth. At the time of the major fire of 1842, Hamburg was not a quickly growing city like, for example, Chicago of 1871, where the population almost tripled from about of 109,000 to about 300,000 in the ten years before the fire.[107] On the contrary, after the recovery from the population collapse during the French occupation of 1806–14, population growth in Hamburg was rather stagnant, and it was only after the fire of 1842 that a particularly dynamic growth phase began again. The rapid rise of numbers reflecting fires and firefighting incidences obviously reflects an actual increase in the fire danger for Hamburg (and its early perception and accounting as risk from the middle of the eighteenth century).

For purposes of comparison, the particularities of the fire fund records (*Generalfeuerkasse*) should also be addressed. These numbers were recorded in different formats, and the deviations between different currencies (e.g., bancomarks, Lübisch marks, marks courant) are not always easy to identify in the archival material itself, nor are they always disclosed in the literature. Therefore the number values that Borscheid/Drees give in the historical insurance statistics of Germany for the Hamburg-based fund are in many cases uncertain. In addition to the sources given by Borscheid/Drees and the 1976 overview created by Oakley and Spohnholtz, there are also figures in other archives

---

107 Cf. Rosen, *The Limits of Power*, p. 9.

that differ substantially from the Borscheid/Drees information.[108] Because of this unclear data situation, the following are based on a new reconstruction from these archival sources (Fig. 20).[109]

Concerning the earliest records from 1684 to 1713, the figure below illustrates discrepancies between the Borscheid/Drees data and the numbers recorded in 1794/95 by the long-tenured head fire chief and founder of the first private fire insurance company in Hamburg, Georg Ehlert Bieber:

While according to Georg Ehlert Bieber's figures there were several years recording no damages—which is quite likely—and other years (see especially 1767, 1786, 1794/95) recording relatively high numbers, the Borscheid/Drees numbers are consistently below the 50,000-mark line, but suggest a continuity of damages and payments. These numbers illustrate the difference between damage events and the completion of payment, which was often delayed or extended over many years. The Borscheid/Drees statistics thus merely serve

108 StA Hamburg Best. 333-1/1, Nr. IV a2, vol. 1, f. 58–78 contains, for example an exact listing of all that damages that incurred in the years 1684–1713; admittedly the values in Borscheid/Drees (*Versicherungsstatistik*, p. 195f.) reflect annual damage payments and not the damage values actually incurred, but even the total amount for 1684–1713 differs by almost 400,000 (Bankomark?) (= ca. 30%). Concerning the "Bilance der General-Feuer-Cassa auf ultimo December 1809" in HStA Hamburg Cl. VII, Lit. Fe Nr. 3 Vol. 13, nr. 1 the 1808 Borscheid/Drees numbers in the column "Zahl der Policen in T Bco. M" (5218) and the column "Versicherungssumme in Bco. M"(78.856.477) are in complete correspondence; however, neither the amounts (Borscheid/Drees: 35.247, Balance: 39.684 plus 2.116 for invoicing taxes and increase of insurance amounts), nor the rubric "Schadenszahlungen (damage payments)" (Borscheid/Drees: 26.703, Balance: 13.048) correspond with each other. Furthermore, all values on the balance sheet are clearly given in marks courant, while Borscheid/Drees uses bancomarks. The other balance sheets given in the same volumes for the years 1817–1822 also differ significantly from the Borscheid/Dress numbers in terms of contribution and damage amounts. For all balance sheets between 1823–1859 (with the exception of the years 1849/50) that are presented in their entireties in HStA Hamburg Cl. VII Lit. Fe Nr. 3 vol. 16, it is possible to precisely determine which values have been included in the Borscheid/Drees statistics: while the column "damage payments" represents actual fire damage amounts paid in a given year (excluding other expenditures such as taxation costs of the estates), regardless of whether they were incurred in the same year or earlier, this situation is reversed for the column "Beiträge (contributions)," where the target of the fire fund allowance due that year and the supplementary payments for fire-extinguishing efforts were deducted, irrespective of if and how much of these amounts had been accrued in the current year. Here, the two values were combined, and appear in the balance sheets within an undifferentiated rubric. Additionally, the numbers are presented here not in marks courant (the currency in which the fire fund balance sheets are always given), but in bancomarks instead (in 1859 the exchange rate was approximately 100 bancomarks to 125 marks courant).

109 Source: value of damages: StA Hamburg Best. 333-1/1, Nr. IV a2, vol. 1, f. 58–78; damage payments: Borscheid/Drees.

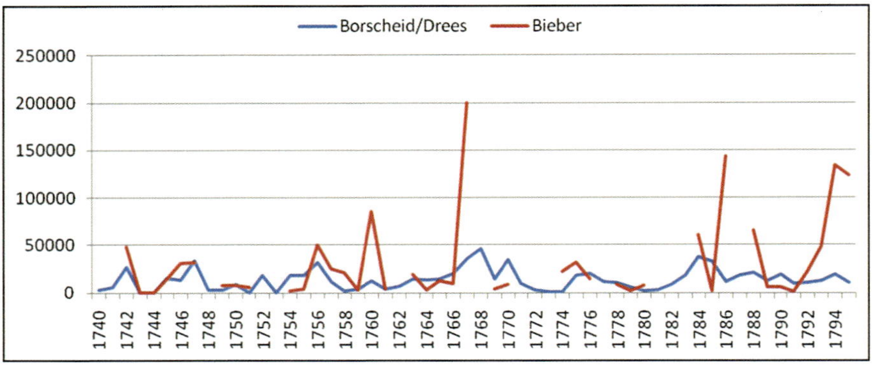

FIGURE 20 Comparison of the Borscheid/Drees damage payment amounts (blue line) and Bieber's damage amounts (red line) 1740–94
SOURCE: damage payment amounts: Borscheid/ Drees, *Versicherungsstatistik*; damage amounts: Georg Ehlert Bieber, *Prüfung der Bemerkungen eines Ungenannten über die Vorläufigen Puncte einer Association Hamburgischer Einwohner, zur gemeinsamen Versicherung ihrer Mobilen-Güter*, S.l., i.e. Hamburg 1795, pp. 1–4—the numbers given there differ in individual cases from the previous version in Georg Ehlert Bieber, *Plan zur Errichtung einer für Hamburg möglichst vortheilhaften Versicherungs-Compagnie gegen Feuers-Gefahr entworfen und seinen Mitbürgern gewidmet*, S.l., Hamburg 1794, p. 3f

to demonstrate the completion of the payment processes and the cash flow in the fire fund, but cannot serve as a measure of the actual damage level associated with fires over the course of a year. Looking at the Bieber data, however, one can easily see how four extremely expensive years (1767, 1786, 1794/95) are hardly even noticeable in the statistics above regarding the numbers of destroyed houses and firefighting incidences: among the fires of 1767 there were two sugar factories owned by Nicolaus Ernst Gädgens and Filster[110] as well as the house (and warehouse?) of the Huguenot merchant Chaunel that burned; in 1786 the estate of Baetke and Rodatz was hit by fire; and in 1794/95, the sugar factory of Thode and Biesterfeld and the Köster estate burned. Sugar factories and stores were typically very fire-prone early industrial businesses and also represented high value exposure for the rich merchants of the trade city—but, in the end, they are simply recorded as "1 house" or "1 firefighting incident" in other forms of accounting. But all in all, the numbers are roughly congruent

110 Description of fire causes in HStA Hamburg Cl. VII, Lit. Fe Nr. 2, vol. 19, nr. 5, p. 46.

with the number of destroyed houses according to Johann Ehlert Bieber (see above): the distribution of the early modern fire characteristics of Hamburg in the fire-prone late seventeenth century, a fire-scarce eighteenth century, and a fire-heavy early nineteenth century is demonstrated by these numbers.

On the basis of various data series, it is clear that contemporaneous administrative specialists already realized these characteristics that have become evident ex-post to historians. In particular, they noticed the increase in the risk of fire beginning at the end of the eighteenth century. In 1787 Senator Ankelmann assembled the following overview of fire damages:

> 1691–1709 195314 MC [marks courant]
> 1701–1710 248351 MC
> 1711–1720 68889 MC
> 1721–1730 114268 MC
> 1731–1740 98155 MC
> 1741–1750 91454 MC
> Thus over 60 years, 816431 MC, or annually 13607 MC.
> However, from 1766–1786, 319630 MC, or annually 15981 MC.[111]

Although these values differ slightly from the above sources, it is interesting, that such observational data suggests that the administration recognized these trends well before George Ehlert Bieber's 1794 calculation of average damage values—which, however, were for the primary purpose of founding his own private insurance company. It is tempting to compare this admittedly very simple form of calculation with the probabilistics already common in the field of life insurance. Ankelmann's calculations—which served as an argument for increasing contributions—incorrectly suggest that the increase in the values took place in the period 1766–86, in particular due to the two exceptional peaks of 1767 and 1786. The pressing perception of the increase in fire danger that also resulted in fire *policey* efforts and discussions within the Patriotic Society was, in any case, obvious in Hamburg in the decades after about 1800: "there is no mistaking, that in the preceding months, the fires everywhere have been dramatically increasing again, and that they have been almost always started by flammable factories" is reported, for example, in 1826;[112] in 1829 it is recalled that "in recent times there are often large fires that ignite in the fire-prone

111 HStA Hamburg Cl. VII. Lit. Fe Nr. 3, vol. 9, nr. 9.

112 HStA Hamburg Cl. VII. Lit. Fe Nr. 2 vol. 36, nr. 1a (January 28, 1826).

factories";[113] in 1827–31 there are constant investigations of this problem; and also in 1835 it is questioned "whether there are general measures that could be taken, in order to prevent the frequent fires?"[114] The ruling elite of the city constantly considered remedies, but apparently without success. In actuality, the effects of early industrialization are clearly demonstrated here, as becomes apparent when the various fire inventories can be qualitatively evaluated. Among the approximately 575 destroyed buildings that Johann Ehlert Bieber recorded for 1772-41, there were 48 sugar factories, 17 distilleries/breweries, 7 bakeries, 5 windmills, 4 calico factories, 2 paper mills, 2 candle factories, and also a dyeworks, a printing press, a wallpaper factory, a tobacco factory, a charcoal factory, a "bone-blacking" [i.e., coal] factory, a cotton mill, a leather factory, a machine factory, and part of a steam mill. Also among the 66 buildings damaged by fire that the Ehlert Association paid for in its early years of existence between 1795–1802, there were 14 sugar factories, 4 calico mills, 1 leather store, 2 carpentries, 1 carriage factory, 1 wallpaper factory, 1 pottery store, 1 distillery, 8 department stores, 2 flour dealers, 1 mill, 1 wood trader, 4 pubs, 1 wine cellar, and 2 goods rooms:[115] this list comprises places of trade and early industrial practices that were extremely risky to insure in terms of both fire risk and damage amounts. The number of burned sugar factories decreased somewhat in the late 1820s, which coincides with the decline of this industry at this point (or reflects its earlier boom due to the Continental Blockade).[116]

A structural reason for the demonstrated increase in the Hamburg fire risk from about 1800 on is a significantly disproportionate amount of specifically fire-prone trades—characterized, above all, by the Hamburg sugar-refining industry,[117] a still very unregulated transitional industry, often located in normal residential buildings, in a quickly evolving landscape of factories and manufacturing industries. Although on the infamous night of May 4–5, 1842, a fire broke out not in such a sugar factory, but in the home of a tobacco merchant,

113 Ibid., Nr. 3.

114 Ibid., Fe Nr. 2 vol. 40.

115 HStA Hamburg Cl. VII Lit. Fe Nr. 2 vol. 28a, nr. 27.

116 Astrid Petersson, *Zuckersiedergewerbe und Zuckerhandel in Hamburg im Zeitraum von 1814 bis 1834. Entwicklung und Struktur zweier wichtiger Hamburger Wirtschaftszweige des vorindustriellen Zeitalters*, Stuttgart 1998, pp. 58, 229. 1807 was the highpoint with ca. 428 businesses, in 1834 the number had declined to 194 businesses; see ibid., pp. 225–33.

117 A 1797 recommendation of the Hamburg Patriotic Society pointed out the "relatively large waste of fire ... in the fire-hungry kettles of the sugar refineries" (*Verhandlungen und Schriften der Hamburgischen Gesellschaft zur Beförderung der Künste und nützlichen Gewerbe*, 7 vols., Hamburg 1792–1807, vol. 6, p. 389, qtd. in Petersson, *Zuckersiedergewerbe*, p. 54).

the spread of the fire was aided in particular by a "reservoir filled with spirits and rum … (of Ross & Vidal)"; with its character as a remote trading center, the fire easily spread across the city through the especially flammable storage buildings.[118] In the initial decades of the nineteenth century, Hamburg found itself in a transitional phase, in which pre- and early industrial trades shaped the image of the city; as they reached a critical mass, it created a high fire risk on a structural level for the city.

Thus, on a purely statistical level and according to an increasingly dominant perception, the great fire of 1842 was a disaster with a message, so to speak—a disaster which corresponded with the image of increased risk characterizing this time period and context in particular. Although in many senses it continues to recall the terrifying premodern pattern of the quotidian threat of major fires, it must simultaneously be categorized much differently—as a symbol of the risk pattern of early industrialization.

These Hamburg data are definitely not broadly representative, because Hamburg, as the largest port and maritime trade city, was a typological exception. The high number of early industrialization-era sugar factories was hardly found in other German cities in the hinterlands. In fact, there are few cities at all with which appropriate comparisons could be made at this point in time. While the old imperial cities, for the most part, lived through early modernity without major fires—probably due to a strong position acquired by investing in stone buildings at the beginning of the late Middle Ages—they also seem to have failed to develop the requisite meticulousness of quantifying introspection. The accounting practice of coding reality in numbers spread earlier in the institutions of the cameralistic administrative state and in centers of technical-economic Enlightenment like Hamburg. For Cologne, for example, it is true that in order to satisfy the firefighters' demand for a gratuity for each extinguished fire, an annual count of the actual fires was performed (cf. Ch. 3, 3.1.1). However, nothing of this practice[119] is preserved in any systematic form and it also does not appear that such statistics were collected over many years. Aachen, the only major imperial city that was completely destroyed by fire in the early modern period—in 1656, by the largest German city fire before Hamburg's—completed reconstruction in the traditional way, without any

118 F.R. Schrader, "Aus den Tagen des großen Brandes," ed. Theodor Schrader, *Mittheilungen des Vereins für Hamburgische Geschichte* 15 (1892), pp. 378–83, 379; Ortwin Pelc, "Hamburg 1842. Eine Katastrophe verändert die Stadt," in Claudia Horbas and Ortwin Pelc (eds.), *Es brannte an allen Ecken zugleich. Hamburg 1842. Ausstellungskatalog des Museums für Hamburgische Geschichte*, Heyde 2002, pp. 9–15, 9.

119 Or "survived," after the collapse of the city archive.

major changes in the city structure and with few indications pointing toward greater reforms in fire *policey*.[120] A fire-statistical assessment was not undertaken by the city—or at least there is no evidence of one. The fact that Aachen then, in the early nineteenth century, served the location of one of the earliest and largest German private fire insurance companies was due not to a previously existing administrative culture—as, for example, was the case in Hamburg with the practical knowledge of the principle of insurance and a tradition of recording and observing its own security regime in the "code of numbers"—rather, it emerged from a completely new sector, gaining access at the economic level of modern industrial capitalism.

1.7 *Summary*

The sheer representation of fire statistics, the demonstration of the influences of war and climate, and a single more-detailed urban fire ecology or "fire biography" of a city like Hamburg should be accompanied with an illustration of the threat encountered in everyday life, as distinguished from the danger in the premodern era. In some respects, one might confidently make the judgment that there is nothing surprising about the distribution of fire, about the war- and climate-induced fires—that they rather confirm, with a bit of accuracy, what one would expect. The relatively early decline in the high risk of fire during the eighteenth century (the early spread of the "fire gap") is more surprising and may help to explain why, especially in that century, city fires begin to increasingly appear as the unspoken exception to normality, although true fire security in towns and cities was still far from comprehensively achieved.

Further detailed research for other single cities like Hamburg, which could punctuate, change, or complement this broad view of everyday fire danger

120 The relevant group of sources regarding the reconstruction (Stadtarchiv Aachen, RA II Allg. Akten 577: correspondence between near and distant cities, dukes, princes, and potentates, but also an interrogation protocol of May 17, 1656 concerning the causes of the fire), the city council protocols that begin after the fire (Nr. 1: 1656–1660), as well as the ordinances and the series concerning the city's officers (Stadtarchiv Aachen, RA II Edikte II 1 und RA II Beamtenprotokolle vol. 1 (1656–1660) bear evidence of quite traditional resilience measures taken by the city administration—no truly innovative impetus can be detected; cf. Claudia Kunitz, *Aachen nach dem Stadtbrand. Wege und Maßnahmen zum Wiederaufbau 1656–1661*, MA thesis, RWTH Aachen University 1994; Thomas R. Kraus, "Der Aachener Stadtbrand vom 2. Mai 1656 und seine Folgen," *Zeitschrift des Aachener Geschichtsvereins* 109 (2007), pp. 35–99. There are no fire *policey* files for the period after the fire that would offer insight into whether a statistical inquiry produced by the city administration: probably such analytical proto-state bureaucracy did not exist.

would be useful, because a large number of prima facie unexplainable peak fire years could still be compared to regional findings within climate history, or other local explanations could potentially be found. While the seventeenth century can be characterized as a century of wars and epidemics, it is also clearly an exceptional period in terms of fire crises. Research has yet to offer more detailed and comparable fire histories beyond the existing pioneering but rudimentary lists—provided, up to this point, by historiography. A short comparison with the English Porter/Jones list demonstrates that very different numbers can be expected in different countries and that thus the fire history of Europe may be far more multifaceted than is currently imagined. For the following chapters, the analysis given here can serve the purpose of representing the "reality of the danger." From that starting point, the question must be pursued of how this reality of a threat was perceived and what developments were catalyzed by it in the means and mechanisms of security production.

2 The Perception of Danger

2.1 *The Theology of Divine Punishment and the Fire Events of the Seventeenth and Eighteenth Centuries: Security Losses, Security Gains*

In the late Middle Ages and early modern period, fire was almost always interpreted as a disaster sent by God to punish the community for sins and immoral behavior, the particularities of which were typically specified after the fact. This interpretation is found, first and foremost, in the sermons presented by local clergy after fire events. Popular sermons featured acts of divine destruction in Biblical cities such as Nineveh, Sodom, and Gomorrah, and it was emphasized that the punishment of Christian cities was also within the purview of divine providence. The basic elements and rhetorical patterns of fire sermons were always similar and the sins subject to divine punishment by fire were typically defined by preachers as immoral behavior—in particular, sporadic church attendance. The fires thus also served as moments that ministers could exploit to gather the lambs of the congregation together through incentives of fear, concern, and the search for consolation. In addition to chastising tirades, the sermons usually also included consoling remarks and calls for devout, Christian behavior in the future. As such, fires were often used by preachers for purposes of periodic reformation and as an opportunity for the re-Christianization of their congregations. For example, as David Underdown argues, the burning of the city of Dorchester in 1613 resulted in a multiyear religious reform of the city collective that transformed Dorchester into the most "Puritan" city in England.[121]

The highpoint of fire sermons occurred between the second half of the sixteenth century through and during the entire seventeenth century. Preachers reliably interpreted events as punishment for a long, unrepentant period of carelessness:

> All that said, Oh city of Giengen, Oh dear fatherland, you have—unfortunately—a very transient, forgettable memory [*labilem memoriam*] [of prior fire conflagrations in other cities]. ... Because we find here a *Magna securitas*, a great carelessness [*Sicherheit*], as one lives here safe and secure throughout the day. Some think that because today there is peace and no hardship, they might always live in safety and security. But it is said: Well, God will not and cannot leave this great safety and security unpunished; he will soon unite us anew in his wrath.[122]

The Giengen evangelist Simon Böckh (about whom there are no in-depth studies beyond local histories) can be recognized as representative of a certain type of orthodox Lutheran preacher: his anti-astrological treatise, published in 1681, positions him as a true anti-Philippist perhaps even opposed to ideas of natural theology.[123] However, there are two noteworthy points in the above passage. First, we can identify a reflection on the relationship between memory (*memoria*) and disaster, which was apparent in the tradition of the medieval and rinascimental *ars memoriae*. Here it is expressed as a duty of believers to remember past punishments for the present and for the future; more recent disaster history, in contrast, questions the relationship between collective memory and post-disaster resilience in terms of a sociohistorical analysis concerning its functions for identity and community building.[124] This passage

121 David Underdown, *Fire from Heaven. Life in an English Town in the Seventeenth Century*, New Haven 1992.

122 Simon Böckh, [...] *Eine Christliche Erinnerungs- und Gedächtnus-Predigt deß höchstkläglichesten Abbrands/der/deß Heil. Röm. Reichs-Stadt Giengen*, Ulm 1655, p. 28.

123 Simon Böckh, *Cometarum praesagitio*, Ulm 1681. On orthodox Lutheran preachers in Böckh's immediate regional environment, see Norbert Haag, *Predigt und Gesellschaft. Die lutherische Orthodoxie in Ulm 1640–1740*, Mainz 1992; Sabine Holtz, *Theologie und Alltag. Lehre und Leben in den Predigten der Tübinger Theologen 1550–1750*, Tübingen 1993.

124 On *ars memoriae* see Lina Bolzoni, *La stanza della memoira. Modelli letterari e iconografici nell'età della stampa*, Torino 1995; on more recent disaster history, see Peter Gray and Oliver Kendrick (eds.), *The Memory of Catastrophe*, Manchester 2004, and for natural disaster in particular, Rohland, *Hurricanes in New Orleans*, p. 50f and passim.

illustrates the fundamental goals the clergy initially had in their roles as disaster preachers: to stress God's intention to castigate.

The second noteworthy point is that the passage contains a rare but interesting use of the word "Sicherheit" (i.e., security/safety) in semantic (rather than morphological) transference from the Latin *se-curitas* in its original meaning: the concept in Classical Latin from the Epicurean tradition, especially as passed down by Cicero, was understood as a negation of *cura*, "care." Thus, in a positive sense, it meant "carefree" as *tranquillitas animi;* in a negative sense it meant the "careless" in the sense of culpable negligence.[125] In the seventeenth century, following tentative Renaissance forerunners, "Sicherheit" moved in phases from a rare linguistic transference to a sociopolitical guiding principle—initially, in the context of territorial statism.[126] The above passage from 1665 points to the newness of this semantic development: "Sicherheit" is used not as "security/safety," in its sense as a political guiding principle, but still in its etymologically negative sense, as "carelessness." Indeed, disaster and security—or more precisely fire and security—appear here in a precisely reciprocal relationship, which is the central theme of the current study; however, this point in history reflects the very beginning or even the direct precursor of the incipient epochal development and transformation of this framework.

A Hamburg sermon from 1637 is a representative example which demonstrates how fire and security were bound up in a blending of both the religious function of admonition as well as a new political function related to the production of security in roughly simultaneous but very different ways. At this point there had been several major fires in Hamburg and the magistrate of the city had just adopted the first significant fire ordinance. The head preacher (*Hauptpastor*) of St. Michaelis, Jodocus Edzard, demanded the authority to

125 John T. Hamilton, *Security—Politics, Humanity and the Philology of Care*, Princeton 2013; Cornel Zwierlein, "*Se-curare, sine cura, se-curitas, assecuratio*: Innovationen der Sicherheitsproduktion in der Renaissance," in Gert Melville, Gregor Vogt-Spira, and Mirko Breitenstein (eds.), *Sorge*, Cologne 2015, pp. 109–35; a similar use with Erasmus, *De libero arbitrio ΔΙΑΤΡΙΒΗ Sive Collatio Per Desiderium Erasmum Roterodamum*, ed. J. v. Walter, Leipzig 1910, IV 16, p. 190. For quotes matching this observation from the political language around 1600 cf. Hannes Ziegler, "Im Schlaf der Sicherheit. Die Endlichkeit des Religionsfriedens und die Zukunft des Reiches (1606–1618)," in: Christoph Kampmann, Angela Marciniak, Wencke Meteling (eds.), *'Security turns its eye exclusively to the future'. Zum Verhältnis von Sicherheit und Zukunft in der Geschichte*, Baden-Baden 2018, pp. 189–206.

126 Cornel Zwierlein, "Sicherheitsgeschichte. Ein neues Feld der Geschichtswissenschaften," *Geschichte & Gesellschaft* 38 (2012), pp. 365–86.

interpret the meaning of the fires: he delivered a kind of systematic fire homiletics, which essentially asserted the framework of the fire sermons based on the theology of divine punishment popular during the sixteenth and seventeenth centuries, but placed them in close proximity to the practical production of security measures for the future.

Edzard bases his thesis on Matthew 22:7, "The king burned their city." He develops the theme in the following steps: 1) How was the fire started? 2) Why did He burn the city? 3) Which signs are given indicating that God would want to set fire to Hamburg? 4) What sort of "fire ordinance" is offered by the scriptures, "so that fire distress is either averted by our city, or if it has already started, that it can be quenched"?[127] For his first point, Edzard exaggerates a distinction and couples it with Biblical references: God ignites cities partly "immediately" via fire from heaven (Sodom and Gomorrah), partly "mediately," using various means: 1) through soldiers in a legitimate war, 2) through *Mordbrenner* ("murder burners"; i.e. mass arsonists), 3) through negligence, and 4) through the devil and his tools (which however are naturally empowered by "God's vengeance"). The purpose of fire disasters was therefore to warn and to console, because "it is consoling that it is done by the Lord, who takes away, and then gives again."

Regarding this second point, Edzard is brief and decisive: "Concerning the cause, why the Lord God would set cities on fires, it is simply sin alone." The relevant sins typically included a) not keeping the Sabbath day holy, b) taking the name of the Lord in vain, c) mocking His commandments, d) idolatry, e) falling away from that which is pure and accepting false teachings, f) the injustice of rulers, g) injustice in trade (the attempt to unjustly enrich oneself), h) arrogance, i) adultery and being unchaste, and j) obstinacy against just punishment. As his third point, Edzard asserts that one could argue that these are all punishments from the Old Testament imposed on Jews and heathens and question what this had to do with a Christian city like Hamburg; in response, Edzard subsumes the activities of Hamburg within the categories he established. Denizens of the city, for example, often gambled and drank on

127 Jodocus Edzard, *Feuer-Predigt über den Evangelischen Spruch: Text. Matth. 22. v. 7. Der Herr zündet ihre Stadt an. Darinn gehandelt wird, was von denen, in dieser guten Stadt, unlängst leyder oftmahls entstandenen Feuersbrünsten zu halten, und was hinferner zu Abwendung derselben in acht zu nehmen von nöthen sey, gehalten am XX. Sonntage nach Trinit. in der Kirchen St. Michaelis, und auf etlicher Bußeiferigen Begehren in Druck gegeben*, Hamburg 1637 [sic], added is: Bernard Vaget, *Wahrhaftiger Bericht von dem gräulichen und erschrecklichen Brande des Thurms St. Nicolai in Hamburg* [*im Jahr 1589*] (all together reprinted Hamburg 1750), p. 4f.

Sunday instead of going to church and were thus epicurean pleasure seekers rather than attentive churchgoers; other sins ranging from adultery to usury also occurred in Hamburg. In addition to these sins, he cites the flood of 1625 (which was followed by an outbreak of the plague) and the flood of October 11, 1634 as further portents of the impending incineration of the city. The reported observation of "fire in the sky" as well as destructive fires in other cities also served as warning signs. Finally, he mentions the conflagrations that had already occurred "in many particular houses, which were partly filled with the most precious wares." He condemns the arrogance of the merchants:

> As if they wanted to say, why should fire come in our houses, we will simply build more carefully; let the priests' blood boil, in the meantime we will also boil, and cook and fry, and eat and drink good things. Our government and rule will remain intact. Thus, the burned merchant homes are like a boiling pot of the Lord, suggesting what the Lord our God might have in mind for the merchant commerce and splendor of this city: Far be it from us to throw caution to the wind after such signs![128]

Edzard invokes the recently published fire ordinance of the city magistrate, but then refers to a second, metaphorical, fire ordinance: "In the scriptures the Holy Spirit has also prescribed a good fire ordinance." He goes so far as to compare individual articles: in the "burgher fire ordinance article 6, it is decreed that designated fire surveyors should diligently make [their rounds] and watch [for fires]." In this way, the Lord God himself goes through the cities as a surveyor of fire; the watchtowers that sound the alarms are like the preachers who warn and admonish on a spiritual level. "Who would mind if the preachers would start blasting the penitentiary trumpets again in these current dangerous times."[129] The translation and mirroring of spiritual Christian behavior within the newly adopted "fire ordinance"[130] of 1637 is thus one way (albeit in a slightly protesting tone, advocating penitence, church service, and the elimination of the first cause of threatening fire—i.e., sin) of uniting practical and spiritual fire prevention. The fact that Edzard makes reference to the newly adopted fire ordinance here at all shows how he welcomes practical considerations on the one hand, while on the other he wants to couple them with a call to penitence.

128 Ibid., p. 14f.

129 Ibid., p. 17.

130 For more on this see Ch. 3, 3.1.2, p. 199f.

Edzard's sermon, with its etiological systematization and its attempt to make penitence and divine orientation at least as important as activities related to fire *policey,* is typical for the seventeenth century. Even though ever now and then a document might be found that supports the rhetorically oriented demand of some preacher exhorting city residents to pray rather than fight fires, we should not think that the practical world of fire *policey* stood in conflict and opposition to the world of religious penitence and the faith of the city's community per se. The opposition always concluded with the admonition that one should pray *and* put out the fire.[131] The analysis of the changes within the apparatus of theological interpretation is what is important here.

One might identify a general secularization movement through examining the origins of the modern notion of conscience found within sermons on fires and lightning strikes.[132] In the sixteenth century, the interpretation of the theology of divine retribution remained absolute and maintained that thunderstorms—beginning with Luther's conversion experience—were sent by God in order to strike the sinner in his moral conscience. Gradually, other interpretations also became viable. The conflict between scientific and theological interpretations of thunderstorms and lightning smoldered in the physicotheological trend dominating the first half of the eighteenth century until lightning conductors were finally implemented, and thus a clear "purely" scientific explanation for the origins of lightning stripped the theological interpretation of much of its legitimacy.[133] This strictly teleological explication has little relevance as an ideal-type construction, but nonetheless it endures as a framework. One could certainly cite countless solid examples to the contrary, indicating that even in the sixteenth century significant scientific interpretations were already present in sermons and, conversely, that even into the nineteenth century crude tirades invoking the necessity and reality of divine punishment still existed.

In order to contrast the situation of the seventeenth century with a second historical moment period, I will single out one body of work here—the sermons written in 1750 following the lightning strike and ensuing fire which completely destroyed the St. Michaelis Church in Hamburg. Although the fire did not destroy an entire city, the exceptional quantity and quality of these sermons is particularly appealing, in that, like a prism, they reflect a very particular turning point in theologico-philosophical disaster perception, comparable only to the post-1755-Lisbon discussion.

131 Cf. Allemeyer, *Fewersnoth und Flammenschwert.*

132 Heinz D Kittsteiner, *Die Entstehung des modernen Gewissens,* Frankfurt am Main 1991.

133 Cf. Andreas Bähr, *Furcht und Furchtlosigkeit. Göttliche Gewalt und Selbstkonstitution im 17. Jahrhundert,* Göttingen 2013, pp. 192–228, who also touches on *Brontotheologie.*

The fire of the so-called "Michel" church in Hamburg resulted in the largest public discussions about the fear and threat of fire until those after the great Hamburg fire of 1842. On Thursday, March 19, nine days after the lightning strike, the city designated three days after the fires as a special repentance and prayer ceremony for the entire city. Services of worship were to be held in all churches, and the city council strictly regulated the format of the services by stipulating which Biblical texts should be included and which songs from the hymnal should be sung.[134] Admittedly, this concerted effort towards a major sermon event not only had spiritual purposes, but also served as a means of collecting donations for the reconstruction of the church. Indeed, the very impressive sum of 120,122 Courant Marks was gathered from the collection plates—in total, 439,737 Courant Marks[135] were donated during memorial collections held every year in March between 1750 and 1765 (Figure 21)

The texts to be included in the sermons to be preached at the first gathering of the Hamburg community in 1750 when these collections were done, were as follows:

1. For the morning sermon (5:00–6:00 am)
   Nahum I: 2, 3
   2. God is jealous, and the LORD revengeth; the LORD revengeth, and is furious; the LORD will take vengeance on his adversaries, and he reserveth wrath for his enemies.
   3. The LORD is slow to anger, and great in power, and will not at all acquit the wicked: the LORD hath his way in the whirlwind and in the storm, and the clouds are the dust of his feet.
2. For the main sermon (8:00–9:00 am)
   Amos IV: 11, 12
   11. I have overthrown *some* of you, as God overthrew Sodom and Gomorrah, and ye were as a firebrand plucked out of the burning: yet have ye not returned unto me, saith the LORD.

134 *Verordnung* [i.e. regulation, edict] *wegen des, bey Gelegenheit der, am 10ten Märtz, des 1750sten Jahres, durch einen Wetter-Strahl, geschehenen Entzündung des Thurmes zu St. Michaelis* [...] *auf den 19ten dieses Monats, besonders angesetzten Buß-Fast- und Bet-Tages*, Hamburg 1750.

135 At the end of the miscellaneous volume *Samlung* [*sic*] *der Obrigkeitlichen Verordnungen, Predigten, Gedichte und Kupfer welche bey Gelegenheit der durch einen Wetterstrahl eingeäscherten herrlichen Haupt-Kirche St. Michaelis in Hamburg zum Vorschein gekommen sind* [...] (StAHamburg, Bibliothek A 640–77) there are handwritten notes about the successive donation campaigns between 1750 and 1752.

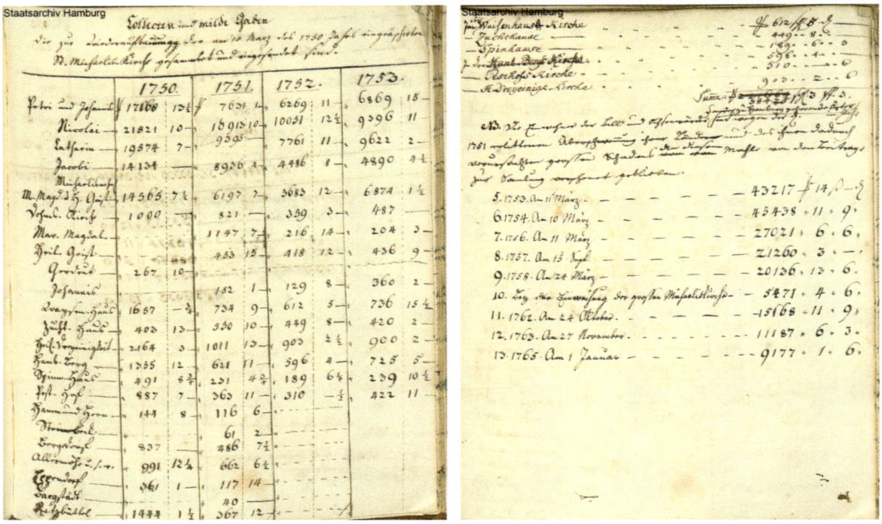

FIGURE 21 Manuscript notes by the church wardens on the sums collected after the destruction of St Michaelis 1750 in Hamburg by lightning, 1753 to 1765. On the following pages at the end of this volume, we find precious information about the Hamburg collections from 1703 to 1797 for foreign damages, cf. below, Ch. 5, 1.1; HStA Hamburg, Bibliothek A 640–77

12. Therefore thus will I do unto thee, O Israel: *and* because I will do this unto thee, prepare to meet thy God, O Israel.

3. For the afternoon sermon (2:00–3:00 pm)

Lamentations III: 39–42

Wherefore doth a living man complain, a man for the punishment of his sins?

Let us search and try our ways, and turn again to the LORD.

Let us lift up our heart with [our] hands unto God in the heavens.

We have transgressed and have rebelled: thou hast not pardoned.[136]

The morning sermon comes from the beginning of the prophecy of Nahum to the inhabitants of Nineveh, that their city would be destroyed; the main sermon comprises the destruction of Sodom and Gomorrah; and the afternoon sermon concludes with a penitence text from Lamentations.

Fifteen sermons by thirteen preachers—the preachers and deacons of the main churches—can be identified. The preachers Friedrich Wagner (1693–1760, preacher at St. Michaelis from 1736), Erdmann Neumeister (1671–1756,

136 *Verordnung* (1750), pp. 8–11 (King James Bible translation).

preacher at St. Jacobi from 1715), Johann Ludwig I. Schlosser (1702–1754, preacher at St. Katherine beginning in 1741 and the predecessor of Goeze), Ernst Friedrich Mylius (1710–1774, preacher at St. Petri from 1744), Henrich Hoeck (1700–1779, preacher at Holy Trinity from 1729), Christian Friedrich Schaub (preacher at the Pesthof [plague court]), Hinrich Wordenhoff (1709–1765/66, preacher at St. Pauli) all preached on the main text and also at the most popular times in the churches of St. Jacobi, St. Katharine, St. Petri, Holy Trinity, St. Job (Church of St. Hiob hospital), St. Pauli, and the Pesthof. Only Matthias Lüttmann (preacher at Holy Spirit), the Archdeacon of St. Michaelis (Daniel Rücker), and the deacons preached in the afternoon. All sermon texts were printed as memorabilia, accompanied by poems and historical commemorative texts on the history of St. Michaelis. There were also many other related sermons—for example, the sermon for the construction of St. Michaelis of 1646, the sermon for the groundbreaking ceremony of 1649, and the sermon honoring the tower completion in 1669, accompanied by a description of the July 16, 1589 lightning strike at the St. Nicolai church. In 1750, Edzard's fire sermon of 1637 was even reissued.[137]

In this way the 1750 destruction of St. Michaelis became embedded within Hamburg church and fire history and etched itself into the cultural memory of the city. Through the annual special penitence days held in the Hanseatic city until at least 1765, the lightning strike of 1750 founded a disaster memory that was fed by a corresponding type of sermon. These sermons anticipated a content and framing that appeared again five years later in the sermons of four Hamburg preachers, Goeze, Mylius, Hoeck, and Zimmermann on the occasion of the Lisbon earthquake.[138] The fact that the Lisbon earthquake sermons were presented more than four months after the actual event on March 11, 1756, the same date that Telemann's *Donnerode* (Thunder Ode) was first heard and that

137 Jodocus Edzard, *Tempel-Predigt HAGGAI, Erkläret und wiederholet, die Christliche Gemeine in der Newen- oder Vor-Stadt, und andere Christliebende Hertzen anzumahnen dem HERRN Salvatori daselbst sein Haus zu bawen* [original date 5 April 1646], reprint Hamburg 1750; Jodocus Edzard, *Grundlegungs-Predigt, da der erste Stein zur Neuen Kirchen in der Neuen Stadt Hamburg geleget worden, welche dem grossen Nahmen deß Heil. Michaels, deß Groß-Fürsten und lebendigen Sohns Gottes, zu Ehren soll gebauet und consecriret werden* [original copy 26 April 1649], reprint Hamburg 1750; Johann Surland, *Gott ist ein starcker Thurm, und der beste Schirm. Das ist: Eine Christliche Danck-Predigt, aus den Sprüchen Salomonis 18 Cap. Vers. 10. den 9 Martii, Anno Christi 1669. Nach glücklicher Verfertigung des Neuen Thurms an der Kirchen St. Michaelis* […], reprint Hamburg 1750; Edzard, *Feuer-Predigt* [1637/1750]; Vaget, *Wahrhaftiger Bericht* [1589/1750].

138 Ulrich Löffler, *Lissabons Fall—Europas Schrecken. Die Deutung des Erdbebens von Lissabon im deutschsprachigen Protestantismus des 18. Jahrhunderts*, Berlin 1999, pp. 438–554.

the community of Portuguese Jews in Hamburg set their day of repentance,[139] is connected to the tradition that began in 1750 of holding a special day of repentance in Hamburg in honor of the lightning-strike. This repentance day was held on the Thursday or Sunday before, on, or after the day of the lightning strike of March 10 (March 7, 1751; March 12, 1752; March 11, 1753; March 10, 1754; March 11, 1756).[140] Just as the Lisbon earthquake is later conflated with other catastrophic events such as the Seven Years' War,[141] 1750 was a critical year in the formation of Hamburg's disaster discourse and simply subsumed the later Lisbon memorials.

Whether a specific collection was held for Lisbon's victims is not clear.[142] Interestingly, other findings in related research indicate that the strongest rise of interest in the earthquake as a scientific object is documented exactly one decade *before* the 1755 Lisbon earthquake, and that news coverage of regional earthquakes was already well established by that date.[143] Lisbon was only a catalyst event that fit into an already existing horizon of expectations. Similarly, there is some evidence that Hamburg was at the height of contemporary international scientific developments, while in other German cities there may have been some lag in the reception of ideas and technological innovations such as those reported in Benjamin Franklin's then epochal publications and the subsequent French experiments that occurred two years after the Hamburg lightning strike (in 1752). This is due to the fact that central figures in Enlightenment science had already demonstrated deep interest in the problem of lightning and electricity: Hermann Samuel Reimarus, for example, had conducted

139 Martin Warnke, "Das Erdbeben von Lissabon 1755—eine Bewährung der Aufklärung," *Zeitschrift des Vereins für Hamburgische Geschichte* 95 (2009), pp. 1–22, 15.

140 Cf. donation entries in *St.Michaelis-Sammelband* StA Hamburg, Bibliothek, A 640–77, p. 5, 8; see also Warnke, "Das Erdbeben," p. 8 n 15 where he makes reference to the Michaelis fire in the council decision of 20 February 1756, but otherwise does not mention the date coincidence between the Michaelis fire and the Lisbon earthquake.

141 Ulrich Löffler, " 'Gott wollte verschonen, nun tödteten wir' Das Erdbeben von Lissabon und der Siebenjährige Krieg im Spiegel protestantischer Predigt," in Lauer and Unger, *Erdbeben*, pp. 437–58.

142 Whether a separate donation campaign was held for the Lisbon victims is not entirely clear: In the donation register for 1756 there is only a sum of 22,536 marks courant listed "Für die hiesigen Marschländer nach der hohen Wasserfluth" (StA Hamburg, Bibliothek, A 640–77, p. 22) and a not precisely categorized sum of 27,021 marks courant for the day of repentance (*Bußtag*) dedicated to the earthquake, 11 March 1756 (ibid., p. 5).

143 Grégory Quenet, *Les tremblements de terre aux XVIIe et XVIIIe siècles. La naissance d'un risque*, Paris 2005, pp. 274, 279f. On earlier parts of the learned investigations and debate about earthquakes cf. e.g. Louis D. Gerdelan, "The Royal Society, Port Royal and the Great Trans-Atlantic Earthquake of 1692," *Studi Storici* 60, 4 (2019), pp. 845–874.

experimental research and collected international news about the topic.[144] Thus, the impression arises that the Hamburg preachers of the Michaelis fire of 1750 had already largely exhausted their argumentative arsenal when the Lisbon earthquake happened. Of the thirteen authors who preached and wrote in 1750, only three were still active in 1756, in addition to the newly appointed Goeze.

This broader textual basis also invites certain comparisons with Ulrich Löffler's analysis of Lutheran preaching concerning the Lisbon earthquake, which emphasizes that the authors of the Hamburg disaster sermons of 1756 represented four different approaches standing side by side: eschatological, creationist, orthodox, and ascetic. Common to all of them, however, is an anti-Deistic, often physicotheological orientation.

144 This comes from a later publication of his son Johann Heinrich Reimarus: after the observation of lightning incidences between 1705 and 1760 the question of whether lightning always strikes in a certain direction, or if it is "different due to various causes" was introduced, and this is a question that the son inherited from his father (Johann Albrecht Heinrich Reimarus, *Die Ursache des Einschlagens vom Blitze, nebst dessen natürlichen Abwendung von unsern Gebäuden; aus zuverläßiger Erfahrung von Wetterschlägen vor Augen gelegt* [Langensalza: 1769], 39n, see also the lightning strike descriptions on pp. 33–38n). Concerning the natural science research on electricity, the role of the older Reimarus seems to have been ignored heretofore in research. In the *Allgemeinen Betrachtungen über die Triebe der Thiere*, ed. Jürgen von Kempski, Göttingen 1982 the animal/medical electricity is not mentioned; however from the quotations, it is clear that Reimarus also was familiar with the *Hamburgische Magazin*, the *Philosophical Transactions*, and the *Histoire de l'Académie* with its annual reports of the *Académie des sciences*. The literature on electricity cited by the younger Reimarus, including the publications in the *Philosophical Transactions*—there was no interruption in the series of the volumes received from the first decade of the eighteenth century—, were, even during his father's lifetime, consistently integrated into the personal family library. There are indications such as this regarding the exact chronology of the progression of knowledge—"[…] since these thoughts have been known in Europe in 1751, it follows that in 1752, Mr. Dalibard and Mr. Delor would begin the first real experiments in France […]," (Reimarus, *Ursache*, 72)—thus suggestive of a bio-bibliographical memory and an activity that had been going on for quite some time. In general, it is well-known that Reimarus had a deep understanding of (also called "atheistic") research related to the Newtonian and post-Newtonian discussion of physics in the circles of the Royal Society and the Académie des Sciences. On the prehistory and the genesis of his physicotheological *Vornehmsten Wahrheiten* (1754), see Dietrich Klein, *Hermann Samuel Reimarus (1694–1768). Das theologische Werk*, Tübingen 2009, pp. 200–66; Ulrich Groetsch, *Hermann Samuel Reimarus (1694–1768): Classicist, Hebraist, Enlightenment Radical in Disguise*, Leiden 2015 and Id., "The Devil in the Details: The Case of Hermann Samuel Reimarus (1694–1768)," in Martin Mulsow and Asaph Ben-Tov (eds.), *Knowledge and Profanation. Transgressing the Boundaries of Religion in Premodern Scholarship*, Leiden 2019, pp. 273–296, 282 n 49: Reimarus' interest in magnetic demonstrations by the Dutch physician and mathematician Nicolas Hartsoeker during his Holland travels.

Friedrich Wagner, the head preacher at the affected Michaelis church, warned at the beginning of his sermon against any suggestions that questioned God's destructive purpose from a theodicean perspective:

> I can easily imagine that many righteous souls in our community, which through this great loss ... have been painfully touched and stand deeply bowed, because of the inscrutable wrath of God prevailing here on earth ... or have begun to hold a grudge against God. ... I too am worried, and in part I know that with this great accident others have [opened] their mouths grumbling against God. ... O that none should open his mouth against God.[145]

Wagner emphasizes that Amos's comment on Sodom and Gomorrah indicates that even Israel could be directly hit by the fire of the Lord, since fire was specifically created to exact God's punishments.[146] Wagner then recalls afflictions from recent years: the very hard winter of 1740, the subsequent lack of grain and the livestock disease epidemic; the subsequent war that arose and the "insecurity of the seas" which suppressed trade; finally, a strong "stormy wind" on December 12–13, 1747. Even though these conditions had all affected other cities and countries as well as Hamburg, the destruction of St. Michaelis was a clear sign of "the burning zeal [of God] for our city."[147] It is important for Wagner that it was "God himself" who ignited the fire: lightning was seen simply as an extension of the divine. God specifically chose the newest and the only post-Reformation church. For the first time in the history of the city, and according to Wagner at least until the twelfth century, a major church had been destroyed in a fire.[148] Wagner nevertheless heavily invokes the still significant

145 Friedrich Wagner, *Der gerechte Feuer-Eifer Gottes über Hamburg. Als die neue St. Michaelis-Kirche den 10 Merz 1750 durch einen Blitz in die Asche geleget war, am ausserordentlichen Buß- Fast- und Bet-Tage den 19 Merz zu St. Marien Magdalenen aus Amos IV. 11. 12 vorgestellet: Nebst einigen Historischen Nachrichten*, Hamburg 1750, p. 4.

146 Ibid., p. 9f.

147 Ibid., p. 12. This was also repeated in the sermon Daniel Rücker, *Das durch ein zu Zion vom Herrn angesteckte Feuer zur Buße erweckte Israel wollte zur Erweckung einer hertzlichen Buße, als vorher am 10ten Martii 1750. die herrliche St. Michaelis-Kirche durch einen feurigen Wetter-Strahl entzündet und in die Asche geleget war, an dem ausserordentlichen Buß- und Bet-Tage aus Amos IV. 11, 12. seiner lieben und wehrten Gemeine in der Heiligen Geist-Kirche vorstellen*, Hamburg 1750, p. 19; and Johann Andreas Gottfried Schetelig, *Die von Hamburg empfundenen Rührungen des Zorns und der Güte Gottes, bey Gelegenheit der am 10. Merz 1750. durch einen Wetterstrahl geschehenen Entzündung und Einäscherung der Haupt-Kirche St. Michaelis, in gebundener Rede erwogen*, Hamburg 1750, p. 13.

148 Wagner, *Der gerechte Feuer-Eifer Gottes*, p. 20: Wagner recalls the major fire of 1684, but emphasizes that no church was affected before or after. The St. Nicolai church tower

goodness of God even in this act of punishment. After all, due to the simultaneous rain and favorable wind, the fire destroyed only the church; furthermore, the burnt tower fell in an especially benign way, which prevented the fire from spreading to other houses and ensured that nobody died. The fire is a small wound cut with scalpel-like precision from the flesh of the city, which in its *precise limitation* illustrates the nature of divine will—functioning here primarily symbolically to remind the Deists that God's direct and present intervention is still possible in the here and now in this world.

He then compares the question concerning the significance of the church's destruction to the people of Sodom and Gomorrah through the exegesis of Amos, and asks: "Oh Hamburg! Hamburg! What is now the cause? Why has the fiery zeal of the Lord been kindled upon you, that he has so painfully called you home with fire from heaven?"[149] The answer is, of course, the sins of Hamburg, which he further specifies. In particular he cites:

> the brash and irrepressible free spirits, the shameless mockers and blasphemers of the Christian religion, indeed the open despisers of all religion and of all God's services: those who reject God's Word and defile or blaspheme the name of Jesus, and who in general want and seek nothing more than to tear all belief in and fear of God from the hearts of all people and to lead an unbridled and unconscienced life; and who also have not refrained from teaching their un-Christian, godless, and abominable opinions against God and His word, not only in societies [i.e., reading groups] to others, but even by spreading their teachings through public writings and printed broadsheets.[150]

Wagner does not specify which writings and which reading groups or societies he means in particular, but one can gather that he probably interprets the lightning strike as a type of anti-Enlightenment divine punishment. The concept of the "free spirit" (*Freygeist*), also invoked by Joachim Plahn (see below), is probably linked to the idea of "free thinkers," a term that had become a negatively charged epithet—used, particularly in Hamburg, as a specific denunciation

lightning strike of 16 July 1589 only destroyed the tower, not the church; the lightning strike that hit the St. Catherine church tower on 3 March 1622, penetrating all the way to the baptismal font, did not start any kind of destructive fire; while the windstorm of 15 February 1649 knocked the spire off the tower of St. Catherine, the church remained standing.

149 Ibid., p. 28.

150 Ibid., p. 29.

of Deists and recipients of Deistic thought and loosely linked to the title of Anthony Collins's famous *A Discourse of Free-Thinking* (1713).[151] The older Reimarus's *Apologie,* the "purest" German Deistic original text, which circulated for a long time in hand-written form only and then through Lessing's partial edition, which led to the famous *Fragmentenstreit*,[152] cannot be the object of reference here because of Wagner's choice of words: "in published writings printed even here." More generally, he probably had in mind the translation of English writings by Johann Albert Fabricius or the Deistic translation of the Wertheim Bible, first printed in 1735 by Johann Lorenz Schmidt, who lived in Altona and Hamburg with Reimarus in the 1740s.[153]

In any case, Schlosser, the head preacher at St. Katharine's, also used his Michaelis sermon of 1750 against all manner of refutations of the Wertheim Bible and operated on a physicotheological basis, using the various Brontotheologies.[154] He asked:

151 The German loan-translation first appears in a review of Collins's work in 1714. Christopher Voigt, *Der englische Deismus in Deutschland. Eine Studie zur Rezeption englisch-deistischer Literatur in deutschen Zeitschriften und Kompendien des 18. Jahrhunderts*, Tübingen 2003, p. 44. The discussion between Toland, Collins, Woolston, and Tindal reaches its highpoint between 1710 and 1750. On Hamburg as the "center of reception and popularization of physicotheology and Deism" see Johann Anselm Steiger, "Ist es denn ein Wunder? Die aufgeklärte Wunderkritik. Oder: Von Spinoza zu Reimarus," in *500 Jahre Theologie in Hamburg*, ed. Johann Anselm Steiger, Berlin 2005, pp. 113–30. There are differing evaluations of the reception of Deism in the literature: Wayne Hudson, *Enlightenment and Modernity. The English Deists and Reform*, London 2009, pp. 146–48 emphasizes that its influence was more prominent in Germany than in England; Voigt emphasizes, in contrast, an earlier, purely bibliographic approximation of the reception, which evaluates it as quantitatively immense, but less prominent in its qualitative effect, and becoming even less significant after 1750; for a particular case study, see Jan van den Berg, "English Deism and Germany: The Thomas Morgan controversy," *The Journal of Ecclesiastical History* 59 (2008), pp. 48–61.

152 William Boehart, *Politik und Religion: Studien zum Fragmentenstreit* (*Reimarus, Goeze, Lessing*), Hamburg 1987. On the role of the elder Reimarus's children, such as Johann Heinrich Reimarus, author of the important treatise on lightning cf. note 158; on what concerns the administration of the parental inheritance and the *Apologie* manuscript, cf. Almut Spalding, "Der Fragmenten-Streit und seine Nachlese im Hamburger Reimarus-Kreis," *Aufklärung* 24 (2012), pp. 11–28.

153 Cf. Ursula Goldenbaum, "The public discourse of Hermann Samuel Reimarus and Johann Lorenz Schmidt in the *Hamburgische Berichte von gelehrten Sachen* in 1736," in Martin Mulsow (ed.), *Between Philology and Radical Enlightenment. Hermann Samuel Reimarus (1694–1768)*, Leiden 2011, pp. 75–101.

154 Johann Ludwig Schlosser, *Die Stimme des Herrn im Feuer: am Buß-Tage wegen der durch einen Wetter-Strahl verursachten Einäscherung der S. Michaelis-Kirche in Hamburg zur Erweckung vorgestellt*, Hamburg 1750, p. 9f.: works of the orthodox Johann Georg Hagemann (1684–1765), superintendent and preacher primarius at Blankenburg,

> Or would we therefore, since tempests can also have their natural origins, deny the presence of the nature's Lord and His government in them? Who created the order of the nature? And who is it that maintains it [nature] in its essence? It is the Lord, who does all of this: Who would deny Him the ability to guide that which His hand has created and which He still maintains, according to His will, in line with His commandments?[155]

Daniel Rücker, archdeacon of St. Michaelis (1692–1773, preacher at the church from 1729) addressed the problem of the relationship of God and nature in the same way: "Of course the works of God within the empire of nature have such qualities that no man can comprehend their very basis with his understanding … who among us knows and grasps how lightning works, when it strikes and when it doesn't strike; how thunder works." He then adds a footnote, in which the concept of electricity is articulated:

> When gunpowder was discovered, one believed that it was now possible to fully explain thunder on the basis of its function. Now with the electrical experiments, one has discovered many new things, also about lightning, as well as thunder, which creates a whole new understanding of thunderstorms. And perhaps neither of these two things can truly clarify the matter for us—perhaps it all behaves still very differently. *Quantum est, quod nescimus*! John III: 8.[156]

preacher for the Duke Wilhelm von Braunschweig-Lüneburg-Wolfenbüttel during his travels; Johann Dietrich Winkler (1758–84, became the head preacher at St. Nikolai); of Johann Thomas Haupt (died 1785, Prussian church and school inspector at Templin in der Mark); the Biblical commentary of Romanus Teller, as well as the *Cosmotheologischen Betrachtungen* of Johann Gottlieb Walpurger (1690–1765), were referred to (see also Löffler, *Lissabons Fall*, pp. 215–30). Schlosser also quotes (p. 13) Peter Ahlwardt, *Bronto-Theologie, oder Vernünftige und theologische Betrachtungen über den Blitz und Donner* […], Greifswald 1747; Anders Olofsson Rhyzelius, *Brontologia theologico-historica, thet är enfaldig lära och sanfertig berettelse: om åske-dunder, blixt och skott; bestående af fyra delar* […], Stockholm 1721 (a work that offers lightning strike statistics for Sweden's church towers, in particular); Johann Jakob Schmidt, *Biblischer Physicus, Oder Einleitung zur Biblischen Natur-Wissenschaft und deren besondern Theilen* […], Leipzig 1731; 2nd ed., 1748, Ch. 7, p. 113f.; Gottlieb Müller, *Einige Predigten über verschiedene Materien aus der Natur- und Geisterlehre an das Licht* [… *gebracht*], Wittenberg 1749. On the influence of physico-theology in Hamburg, with particular focus on the Derham translation by Fabricus (1730, further editions 1732, 1736, 1741, 1750, 1764); cf. Steiger, "Von Spinoza zu Reimarus," p. 120.

155 Schlosser, *Die Stimme des Herrn*, p. 12.

156 Rücker, *Zion*, p. 4.

Scientific research is defined as a human activity encompassed by a far larger frame of ignorance. Although the younger Reimarus, as a physician and a scientist, followed his father's highly rationalistic path of Enlightenment and did not include any theological reflection in his publication on lightning in 1768, we find very similar wording here too: "we still have much to learn about the whole nature of the subject of electricity or thunderstorms."[157] The reference to scientific research on electricity within the sermon as well as the admission of partial ignorance within what was one of the best informed treatises on electricity and lightning in Germany at that time demonstrate that Hamburg was at the forefront of scientific trends—the electrical experiments with the Leiden bottle as well as Franklin's correspondence with the Royal Society about electricity from 1747 were both well-known in Hamburg. From its very first issue in 1747, the *Hamburgische Magazin* regularly published excerpts from the *Philosophical Transactions* and from French and Italian scientific publications.[158] Reimarus's lightning writings were thus based on the steady

---

157 Reimarus, *Vom Einschlag*, p. 53. A Socratic topos, to be sure, which in this context, however, had more specifically become a topos of early Enlightenment science in the wake of Boyle, and thus was further developed in the Deism discourse; cf. Jeffrey R. Wigelsworth, *Deism in Enlightenment England: The Politics, and Newtonian Public Science*, Manchester 2009, p. 25f.

158 Concerning the documentation of the ever-present reception of Western European works in Hamburg's Englightenment journals, see *Hamburgisches Magazin* 1, no. 1 (1747): for example "Nachricht von einigen magnetischen Versuchen" (1744; from *Philosophical Transactions* 474), pp. 46–351; "Des Abt Nollet Versuch über die Electricität der Körper" (1746; from the Paris printing)," pp. 412–418; "Nachricht von einem italiänischen Werke: Von der Elektricität" (*Dell'Electricismo* [Venice: 1746]), pp. 154–71; *Hamburgisches Magazin* 2, no. 2 (1748): "Nachricht von einer Sammlung physikalischer Briefe des Marchese Maffei" (on the theory of the origin of lightning) (*Della formazione de' Fulmini* [Verona: 1747]), pp. 346–51; "Auszug aus einem Briefe, die Wirkung der Electricität auf Pflanzen betreffend" (i.e., translation of a letter from John Browning to Henry Baker, 11 Dec 1746), pp. 629–31; *Hamburgisches Magazin* 3, no. 1 (1748): "Schreiben von Kästner an Hrn Joh. Ellicot von der Art die Stärke der elektrischen Ausflüsse abzuwägen" (from *Philosophical Transactions* 479, 2), pp. 25–29; "Auszug von des Herrn Knights magnetischen Versuchen," pp. 94–103; Réaumur reception, pp. 104–07. The French/English debate on electricity in the years between 1746–50 between Nollet, Watson, Baker, Ellicot, in particular, is addressed in John L. Heilbron, *Electricity in the 17th and 18th Centuries. A Study of Early Modern Physics*, Mineola 1999, pp. 296–304, 309–43 and Paola Bertucci, "Sparking Controversy: Jean Antoine Nollet and Medical Electricity south of the Alps," *Nuncius* 20 (2005), pp. 153–87. The reception of that debate in Göttingen and Hamburg is not mentioned; on the *Hamburgische Magazin*, in fact edited by the Göttingen professor of mathematics and physics A. G. Kästner (1719–1800), cf. Martina Lorenz, "Physik im Hamburgischen Magazin (1747–1767)—Publizistische Utopie und Wirklichkeit," *Zeitschrift des Vereins für Hamburgische Geschichte* 80 (1994), pp. 13–46, on the editor Rainer Baasner, *Abraham Gotthelf Kästner, Aufklärer (1719–1800)*, Tübingen 1991.

and uninterrupted analysis of all relevant articles from the journal of the *Royal Society,* beginning with volume 47 (1750) and extending to Watson's magnet experiments.

Like Wagner, Schlosser, and Rücker, other preachers repeatedly and explicitly emphasize that lightning strikes demonstrated that the "old God, who ruled and dominated this world still lives; that he still has thunderbolts"[159]; that for all systematic scientific distinctions of fire in the earth's interior, earthquakes through fiery inflammation, volcanos, and fires caused by lightning, the question of who created these fires must always be answered: "God."[160] In line with this, Joachim Nicolaus Friedrich Plahn (1724–1787), who worked at this time in Hamburg (1751–1753 private minister for the Duchess of Holstein, preacher at Grabow beginning in 1753), wrote a physicotheological poem for the occasion. In it he presents first the scientific doctrine of the generation of lightning and thunder by rubbing together dense sulfur particles and the capacity this creates to drive fire along air that is rich in potassium nitrate. However, in the end, he emphasizes that Hamburg's punishment was related, above all, to the "free spirit in the art of ridicule/who only mocks hell and heaven", and that God continues to guide natural events, even in individual cases, in the here and now:

> He Himself ordered the fates of the world
> Suffering, fear, and accident
> He guides through His limitless gaze
> The greatest claps of thunder ...
>
> No fate, that originates from itself
> First created that flame
> God alone, who is over us all
> God alone ignited them.[161]

159 Joachim Johann Daniel Zimmermann, *Zwo Predigten nach der am 10ten März 1750 durch einen Wetterstrahl verursachten Einäscherung der Hauptkirche zu St. Michaelis in Hamburg*, Hamburg 1750, p. 52.

160 Johann Peter Gericke, *Der Herr im Feuer wurde der Gemeine zu St. Michaelis, am Palmsonntage 1750. in einer Predigt, nachdem das prächtige Gebäude der Michaelis Kirche durch einen am 10. März erfolgten Wetterstrahl angezündet und bis auf den Grund verbrannt worden, aus Ps. 50, 1. 2. 3. in der Kirche zu St. Marien Magdalenen vorgestellet, und auf Verlangen dem Druck übergeben*, Hamburg 1750, p. 20f.

161 Joachim Nicolaus Friedrich Plahn, *Das verbrannte Heiligthum Hamburgs. Eine Erzehlung der am 10ten März 1750. durch einen Wetter-Strahl geschehenen Entzündung und gäntzlichen Einäscherung des Thurms und der Hauptkirche zu St. Michaelis* [...], Hamburg 1750, f. C2v- C3r.

More so than in the sermons related to the Lisbon earthquake—perhaps because it concerned Hamburg itself, and thus the cause for the punishments could be found in the city—a strong unifying element can be seen here: the Hamburg clergy preached vehemently against Deism. To a greater or lesser degree, lightning was interpreted in the sermons as the manifestation of a "particular providence" (to use more theologically precise terminology) in contrast to the "universal providence" generally expressed through the creation of the world within its natural laws. It was not interpreted as a "miracle," but rather as a discrete action of God, who does not always obey the simple movement of the world according to laws already in place.[162]

It is not possible to identify with certainty a precise group, a text, or an author related to the repeated "free spirit" allusions in the 1750s sermons. It was a term that could be equally applied at this time to anything from materialistic philosophy, such as that of La Mettrie, to various types of Deists. The majority of preachers who published their sermons used the particular lightning bolt to illuminate a narrow distinction between a more traditional physicotheology, in which a place continues to be cleared for God's immediate causation of an event in the here and now, and a more Deism-oriented physicotheology, similar to that of Reimarus and his circle. In any case, the Hamburg Masonic Lodge, founded in 1737, the Spinozaists (in a stricter sense), and more extreme thinkers ranging from radical Enlightenment to atheism, attracted countless opponents who need hardly be named, and were excluded entirely from this distinction that was preached from the pulpit.

Even if there are enduring themes and constants in the theology of punishment, such as the recurring topoi of repentance, penitence, and conversion that were applicable to any catastrophic event from the sixteenth into the nineteenth century, we nevertheless see here a precise functionalization and development of this genre, which can be categorized according to a very specific epistemic and spatiotemporal context. The lightning strike that destroyed the church alone with laser-like specificity was the best possible argument for a conservative-orthodox interpretation of physicotheology, because it could be easily depicted as a precise symbolic speech-act of God, in contrast to the

162 The systematic differentiation, for example, was relatively clearly articulated by Chubb (1738)—see Jeffrey R. Wigelsworth, " 'God always acts suitable to his character, as a wise and good being': Thomas Chubb and Thomas Morgan on Miracles and Providence," in Wayne Hudson, Diego Lucci, and Jeffrey R. Wigelsworth (eds.), *Atheism and Deism Revalued. Heterodox Religious Identities in Britain, 1650–1800*, Aldershot 2014, pp. 157–72, 161. The German, and in some cases Latin, terminology would, of course, be different. Cf. for example, Reimarus's doctrine of predestination in Klein, *Reimarus*, pp. 234–47 and passim.

case of Lisbon, which with its undifferentiated obliteration of a huge mass of people, made for a much more difficult argument. Perhaps for this reason, the majority of Hamburg clergy were silent in the face of the Lisbon event.

The Michaelis strike occurred before the publication of Franklin's *Experiments and Observations*. On August 6, 1767 lightning again struck a bell tower, that of the St. Nicolai church; on February 17, 1768, Johann Albert Heinrich Reimarus pleaded[163] in front of the Patriotic Society for the installation of a lightning conductor at St. Jacobi. Reimarus was granted his request—against the wishes of the head preacher, Christian Samuel Ulber—but the introduction of this safety precaution met with conflict in Germany, as it did throughout the world.[164] The Hamburg constellation of the (older as well as younger) Reimarus circles, open to Deism and rationalist physicotheology groups on the one hand, and Lutheran orthodoxy and more traditional physicotheology on the other, thereby replicated in a specific Hamburg milieu the intellectual context in which even the Deist Franklin could witness the spread of anti-Deistic protests in Philadelphia and around the world.[165] Neither the Europe-wide discussion of theodicy after Lisbon nor the introduction of lightning conductors, however, silenced the preachers of their admonishments and atonements, or to a radical rethinking that would have mandated religion-free explanations of causation or, in the wake of Voltaire, a redefinition of evil in the world as merely a proof against the existence of God.

This new trend can be seen in the fire sermons which were held in memory of the total destruction of Gera in a fire on September 18, 1780.[166] It becomes

163 Reimarus, *Die Ursache.*

164 For its diffusion after Franklin and the Marly Experiment see Peter Heering and Oliver Hochadel (eds.), *Playing with Fire. Histories of the Lightning Rod*, Philadelphia 2009; two local studies: Oliver Hochadel, "'Hier haben Wetterableiter unter den Augsburger Gelehrten eine kleine Revolution gemacht': Die Debatte um die Einführung der Blitzableiter in Augsburg, 1783–91," *Zeitschrift des Historischen Vereins für Schwaben* 92 (1999/2000), pp. 139–64; Fritz Dross, "Gottes elektrischer Wille? Zum Düsseldorfer 'Blitzableiter-Aufruhr' 1782/83," in Jörg Engelbrecht (ed.), *Landes- und Reichsgeschichte. Festschrift für Hansgeorg Molitor zum 65. Geburtstag*, Bielefeld 2004, pp. 281–302.

165 Joseph Waligore, "The Christian Deist Writings of Benjamin Franklin," *The Pennsylvania Magazine of History and Biography* 140, 1 (2016), pp. 7–29: There is ample evidence of these positions for the years 1730–35, while for the late Franklin there was indeed a continuity of the basic notion of a Christianity based on a general natural religion; however, a passage from a letter from 1784 is mentioned, wherein it is made clear that the "miracles of God" in the context of the revolutionary wars, were clearly still relevant (ibid., p. 27f.).

166 *Summarischer Entwurf wie der zweyte große Fast-Buß- und Bethtag, welcher Donnerstags den 19 Oct. 1780. In den Hochgräflichen Reußischen Herrschaften zu Gera, mit Singen, Lectionen und Predigten allhier gefeyert und gehalten werden soll*, Cuba bei Gera 1780 [Cuba was a small town outside of the Gera walls, not burned, serving therefore in this moment

clear that certain patterns of reasoning were now barely even possible: all sermons now consistently spoke of the *Unglücksfall* ("accident")—the German translation of the Latin *terminus technicus* "casus fortuitus." In the seventeenth century, for Edzard around 1637, there is still no mention of causes of fires other than God. There is talk of the "court of fire" that God holds with people, but not of any classification as simply an "accident," other than in the sense in the sense of "*malum*." Even in the Hamburg sermons of 1750, the word "accident" hardly appears at all. In 1780, however, the city fire of Gera is consistently described as a "sad accident."[167] One can interpret this as an indication of the influence of secular thinking, however, the sermons hardly present a general theodicy debate. Even Johann Friedrich Teller (1736–1816), the orthodox-minded brother of the better-known neologist,[168] Abraham Wilhelm Teller, went against the tendency described as the "Sadduceean seeing of ghosts" (alluding to Acts 23:8) to explain disasters such as this fire in entirely natural ways.

In any case, the argumentative pattern stands: through such events, the earthly, direct effects of God were tangible and inspired repentance and contrition. In particular, when a fire caused the destruction not only of a few houses,

as place of retreat and printing, it became later part of the city]; Johann Friedrich Teller, *Gera in der Asche. Eine Predigt am 18. Sonntage nach Trinitatis 1780*, Zeitz 1780; Christoph Ernst Hertel, *Die erste Predigt nach dem am 18. Sept. 1780. Ausgebrochenen Geraischen Brand Von dem gottgefälligen Verhalten bey den Unglücksfällen unserer Nebenmenschen*, Greiz 1781; Johann Samuel Gottlob Gräf, *Predigt am XVIII. Sonntage nach Trinitatis als dem nächsten Sonntage nach dem 18ten September, an welchem die ganze Stadt Gera durch Feuer verwüstet worden war* […], Cuba bey Gera 1780; Johann Gottfried Senf, *Eine Predigt von der ernstlichen Entschlüßung eines unter dem Zorngerichte des verzehrenden Feuers liegenden Volks: am 18. Sonntage nach Trinit. 1780. über 2 Sam. 24, 14. nach dem unglücklichen Brande der den 18. September vorher die ganze Stadt Gera in einen Schutthaufen verwandelt hatte*, Cuba bei Gera 1780; Friedrich Wilhelm Streit, *Thränen und Entschließungen der Christen bey dem Aschenhaufen einer durch Feuer gänzlich verheerten Stadt. Eine Predigt am 18. Trinitatis-Sonntage 1780 in der Kirche zu Ronneburg bey Gelegenheit des großen Brandes in der benachbarten Stadt Gera gehalten*, Cuba bei Gera 1780; Justus Christoph Krafft, *Eine Predigt über Luc. 13, 1–5 mit Beziehung auf die Feuersbrunst, welche den 18. Sept. 1780 die Stadt Gera im Voigtlande, in die Asche geleget hat. Nach derselben folgen einige Nachrichten und Briefe, welche diese traurige Begebenheit betreffen*, s.l. 1780; Johann Gottlieb Lummer, *Ermunterung an die Geraischen Abgebrannten: Jn ihren Trübsalen Gott und ihren Nebenmenschen zu geben, was sich gebühret. Eine Predigt am 23. Sonnt. nach Trin. 1780 in der Gottesackerkirche zu Gera gehalten*, Gera 1781.

167 Teller, *Gera in der Asche*, p. 13.

168 Not 'Neologism' in the sense of coining a new word, but 'neology' as branch of eighteenth-century German Protestant theology which was foremost characterized by a strong historical-critical approach to the Bible text.

but an entire city, it was interpreted as directly caused by God: major disasters of the early modern period were clearly His monopoly.[169] Indeed, there was no need to waste time looking for the immediate causes in this context (in the case of Gera, farmers who carelessly burned pig stalls that had been contaminated by a rampant case of swine disease) because the "mediate causes" were of no consequence—only God, the primary cause, mattered.[170] The problem introduced by the Lisbon earthquake, that the event involved both the punishment of the guilty and the suffering of the innocent, was addressed, but presented argumentatively, emphasizing that human observers were not in a position to judge whether, for example, the Gera fire justly punished its sinful residents—only God knew how to differentiate the guilty from the innocent.[171] Johann Samuel Gottlob Gräf explained the phenomenon with the help of a "corporate law" theory of sin: "a city must always be considered as a whole, or as a person." God operates

> [as He does] with individual people, not destroying it immediately all at once; rather, He carries [the city] for centuries, sparing it. When He has warned it enough and spared it enough, the sins are then taken as a whole, finally ripe for judgement, and when judgement falls, [the city] is not only punished for the sins of the people living at that time, but rather for the entire sum of the sins of the city.[172]

The current residents bear what might be described as a medium-term "original sin" or an accumulated collective burden of sin because of the sins attributed to the earlier inhabitants of the city. The main accent of all sermons fell on a focused emotional appeal for sympathy for the victims and the extension of Christian love towards those among them affected by the disaster. Lisbon was never directly named in the sermons because there were ample major disasters to refer to in the vicinity. A quarter of a century later the earthquake might perhaps have caused a rupture in the discourses of major literary works; however, as before, the theology of everyday sermons continued to be based primarily on repentance and divine retribution. In addition to the competing explanatory forces of natural sciences, the other new tendency was how natural sciences became embedded into the tendencies of late eighteenth-century

169 Ibid., 20f.

170 Hertel, *Die erste Predigt*, p. 11.

171 Ibid., p. 16.

172 Gräf, *Predigt am XVIII. Sonntage*, p. 16.

sentimentalism (*Empfindsamkeit*) as a secular functional equivalent to religious admonition.

A number of sermons suddenly began to switch to verse form in 1780 and the proportion of "secular" reports, poems, and memorials in the case of Gera seems to have increased in comparison to earlier examples. An anonymous poem sent by "an unnamed humanitarian (*Menschenfreund*)" from Saxony attempts to dramatize the Gera fire in almost balladlike fashion:

> Woe is me! Abandoned and naked I sit walled in by ruins.
> The ruins of my fallen fortune […]
> My spirit! Consider the misery—look at the terrifying scene,
> The sorrows of an ignited city:
> Here a landlord screams in distress, because not far from his head
> The smoky flames are reaching toward the heavens.
> He calls a neighbor in vain, who, deafened from his own terror,
> Does not hear the call of his fearful friend. […]
> There quivers a powerless old man, pleading with a trembling voice
> For the help of one stronger;
> His call, weak as it sounds, does penetrate sympathetic hearts,
> One comes to his rescue.
> But, alas! The fire has already filled the narrow passage,
> Through which one climbs to his chamber
> No man is bold enough to sacrifice his life for a service,
> That will not help the old one.
> One stands around, wanting to help—not helping—charity fights here
> With love of self a harrowing battle. […]
> As far as the eye can see, there is nothing but hungry flames—
> They curve terrifyingly down the alleys.
> The wind heats up the anger of the fire, already so angry,
> And plays with the horrific smoke. […]
> The damage is as great as an ocean—who can bandage these wounds?
> With whose sorrow can you be consoled?
> Only he—whose compassion knows no limits,
> And who remains a father even in his punishment,
> Only he devours in his sea of immeasurable omnipotence
> Your sea of pain, like a drop.
> He—he shall breathe new life into the dust of your fallen walls!
> Rise up anew from the ashes like a phoenix!
> Then will return back with words of praise to the bosom of the comforting mother.

> The errant horde of children.
> Now hope stirs in my eye—strengthened it sees through the mist
> Of the dark future, you, Gera, renewed [...].[173]

While the beginning sounds almost like an Old Testament lamentation, as the poem continues, it emancipates itself completely from this sacred preface and offers a focused subjective perspective, emotionalizing perception on the level of personal experience in an unusual verse form and presenting, above all, the reenactment of terror and suffering. The increasing popularity of such texts in the context of fire disasters indicates a tendency toward the neutralization of catastrophe and the endowment of meaning through empathetic suffering. This was in line with the ever-intensifying trend toward affect-driven poetics as a response to social individualization that had occurred since the age of Sentimentality (*Empfindsamkeit*). An increasingly sophisticated system of literature obviously begins to fulfill certain functions of religion.[174]

Thus, a transformation in the field of affective-religious security production arises: the city fire shocks the citizens' sense of security on a religious level. In the words of Daniel Rücker: "And thereby they should first find their security disturbed, and be brought to humility before God, which is the beginning of true conversion."[175] Similarly, Johann Ludwig I. Schlosser states, regarding the relationship between fire and security: "Because Israel goes forth in its security,

173 Anon., *Klagen eines Ungenannten über Gera am 18.ten des Septembers 1780*, ed. Friedrich Wilhelm Streit, Altenburg 1781.

174 Poem in Gräf, *Predigt am XVIII. Sonntage*, p. 19ff.; Lummer, *Ermunterung*, pp. 54ff.; Anon., *Gedanken über den Brand zu Gera*, s.l. 1780. Reports and descriptions: Anon., *Das Geraische Tabeera oder Beschreibung der erschrecklichen Feuersbrunst welche im Herbst des Jahres 1780 die hochgräflich Reußische Kauf- und Handelstadt Gera ohne alle Rettung völlig in die Asche legte* [...], s.l., 1780; Anon., *Beschreibung der bedenklichen Verwüstung der berühmten Stadt Gera im Voigtlande* [...], Basel 1780 (with several interjections of religious lyrics, pp. 41ff.); Anon., *Einige Nachrichten von dem undenkbar grossen Unglück welches die Handelsstadt Gera, am 18. September 1780 betroffen hat*, s.l. 1780; Anon., *An die hin und her zerstreueten vormaligen Einwohner der durch Feuer zerstörten Stadt Gera und an ein mitleidiges Publikum*, Cuba bey Gera 1781; Anon., *Gute Gedanken eines Bürgers bey dem höchst traurigen Schicksale seiner geliebten Vaterstadt Gera*, Gera 1781; Christian August Behr, *Auf das Gera am 18. September 1780 betroffene traurige Schicksal*, Cuba bei Gera 1781; Carl Friedrich Uhrlandt, *Klagelied eines geraischen Stadtkindes bey dem Verlust seiner Mutter*, Cuba bei Gera 1781; Carl Friedrich Uhrlandt, *Kurze Beschreibung von dem grossen Brande, der am 18. September 1780. Die hochgräfliche Reußplauische Residenz und Handelsstadt Gera ganz in die Asche geleget hat*, Cuba bei Gera 1780; B., *Die Größe Gottes in der Zerstörung von Gera zu zeigen und nähere die unglückliche Stadt betreffende Nachrichten* [...], s.l. 1781.

175 Rücker, *Zion*, p. 10.

therefore, says the Lord, continue to do as Israel does!"[176] Older and newer concepts of security coalesce in order to provide citizens with a consistent sense of security in their towns and cities—the territorialized concept of political security production begins to merge now with *se-curitas*, in its positive sense of being "care-free." God becomes a conscience shaker, powerfully evoking the awareness of that which is not self-evident and suggesting the culpable impropriety of such a carefree feeling of safety and security. Through the destruction of the physicality of the city, the feeling of safety and security, which was taken for granted, was also destroyed; religion again reared its head powerfully in the newly opened space of fear. Indeed, to use the metaphor of Edzard's fire *policey*, the "watchtowers" eagerly champion this space of fear in order to offer a feeling of safety through a spiritual substitute for the lost sense of material security. Fires, like other disasters, start to act as a link between regimes of material and spiritual security. A certain emotionalizing of the impact of perception asserts itself. This phenomenon will be further examined below in the context of images of urban fires.

2.2 *Visualization and Affect*

As has been seen above, sermons, and later, disaster poetics, can be understood as responding to "the reality" of the fire threat, eventually also mirroring, to a certain extent, the higher and lower frequencies in fire trends. On that level, they sometimes even reflect the development of actual safety provisions—ranging from metaphorical references, to *policey* ordinances, to reflections on the newest electricity research of the Enlightenment era. On a different level, they respond to particular development in theology. It would be far too easy to think of that discursive level as simply and passively dependent on and stimulated and shaped by the reality of fire destruction. As is true for all discourses, and for literature and art in particular, specific developments obey their own rules. It is important to note that this relationship of loose coupling also functions in reverse—cohesive congregational calls to repentance and the emotionalized mobilization of Enlightened churchgoers, while certainly not the only decisive factors, were also important factors in the development of disaster management.

On a similar level, this section will explore the correlation between developments in the visualization of major fires and the fire trends that have been identified in the context of the epochal threshold around 1700 in relation to the history of security. The intention here is not to demonstrate the representational

176 Schlosser, *Die Stimme des Herrn*, p. 19.

relationship between reality and the image in any simple sense—rather, of primary interest here is an investigation into the function of visualization in the framework of the process that led to the expansion of the aforementioned fire gaps. The beginning of this spread, which can be localized in Europe at an earlier point than that identified by Frost and Jones in America, appears more or less in tandem with what is referred to here as the "normal secure society": a set of conditions ("constellation" in the meaning evocated at the beginning of chapter 4), beginning approximately 1680–1700, in which natural disasters were conceptualized as exceptions to a counterfactual-normative horizon of expectations—i.e., the idea that towns and cities, even entire territories, were thought to be basically safe (or it was expected that they would be safe). An understanding of how these counterfactual expectations developed can be assisted by an investigation into how major fires were visualized and which developments appeared from the late Middle Ages until the end of early modernity. A special type of disaster-oriented visualization is devoted to the actual result of the disaster—i.e., the scope of destruction, the "ground zero." In terms of ground zero representations, in particular, there is a very direct relationship between visual representations and practical administrative security provisions implemented before and after disasters. I will discuss this relationship at the end of this section as a transition toward the treatment of *policey* developments of actual pre- and postfire provisions.

2.2.1 The Modernity of City Fire Images and Their Dutch Provenance

Art history research specializing in urban fire visualizations—beyond studies concentrating on individual paintings or painters—is limited to a few important works.[177] Since 2011, conferences have begun to address this subject,[178] but we are still lacking a thorough study that examines the general

177 Nancy A. Corwin, *The Fire Landscape: Its sources and its development from Bosch through Jan Brueghel I with special emphasis on the Mid-Sixteenth Century Bosch 'revival,'* PhD diss. University of Washington 1976; Victor I. Stoichita, " 'Lochi di foco'—La città ardente nella pittura del Cinquecento," in Ulrich Pfisterer and Max Seidel (eds.), *Visuelle Topoi. Erfindung und tradiertes Wissen in den Künsten der italienischen Renaissance*, Munich 2003, pp. 439–51; Victor I. Stoichita, " 'Lochi di foco.' La ville en flammes dans la peinture des XVIe et XVIIe siècles," in Dimiter Daphinoff and Edgar Marsch (eds.), *Das Feuer. Beiträge zu einem interdisziplinären Gespräch,* Freiburg in Switzerland 1998, pp. 91–116; Johan Hanselaer, "Aeneas' vlucht uit het brandende Troje. Virtuositeit en Allegorie in de zuidnederlandse schilderkunst van de 16de en van de eerste helft van de 17de eeuw," *Gentse bijdragen tot de kunstgeschiedenis en oudheidkunde* 28 (1989), pp. 105–18; Vitor Serrão, "O mito do herói redentor: a representação de Eneias na pintura do Portugal restaurado," *Quintana* 1 (2002), pp. 71–81.

178 Koppenleitner, Rössler, and Tiemann (eds.), *Urbs incensa*.

topic of fire motifs and the iconography of city fires for the period between late antiquity and the late Middle Ages—though a contribution on the Sodom motif does exist.[179]

It is important to note that most of the representations of interest for this study lie beyond the material basis of canonical major works of fresco- and oil painting and higher pictorial art. The aesthetic transformations occurring at the apices of the respective developments in art history (which, of course, are always more visible today) only operate as early indicators of new image codes that often took a long time to spread into lower levels of handicraft works of art, which are particularly important for the observation of fire trends. However, at some point, these image codes start to become generally accessible, manifesting themselves in common, practical pictorial forms, such as printed broadsheets, church paintings, illustrations in chronicles, and other minor artworks, including maps and drawings found within administrative files.[180] Occasionally the reverse cycle of innovation was also at work—for instance, methods of visualizing movement, speed and motion in pictorial art were only later integrated into certain forms of oil painting (see below). In order to make a representative statement about this development, a systematic inquiry was undertaken in state and municipal museums, communal archives, and other cultural institutions in the cities afflicted by fires as registered in the fire database discussed above (Ch. 3, 1). In total, this inquiry involved 1880 cultural institutions. 529 illustrations of 294 fires in 252 cities were collected in an image databank that was made publicly accessible.[181] Consequently, the following statements do, to some extent, proceed from an analysis of the iconographic developments in higher art and reflect the early development of certain patterns and representational forms. This development will then be contextualized against the

179 Ulrike Koenen, "Die brennende Stadt. Sodom in spätantiken Darstellungen," *Actes du Xie Congrès international d'archéologie chrétienne* [1986], *Collection de l'Ecole Française de Rome* 123, 3, Rome 1989, pp. 1355–70.

180 In art history research the distinction between fine art and craftmanship is, of course, a traditional research topic; for the Nürnberg artists see, for example, Martin Tacke et al's prosopography *"Der Mahler Ordnung und Gebräuch in Nürnberg": Die Nürnberger Maler(zunft)bücher ergänzt durch weitere Quellen, Genealogien und Viten des 16., 17. und 18. Jahrhunderts*, Munich 2001.

181 The database is included as one of the thematic sub-databases into the meta-archive "prometheus", for licensed institutions accessible at http://prometheus.uni-koeln.de/. In order to access the 529 images directly, rather than by cross-referencing them using city names or fire dates, it is neccessary to click on "92 Databases" in the entry portal, and then on the names themselves ("Die Ikonographie der brennenden Stadt"). The work of collecting and corresponding about the images with the institutions was handled in part by Anna Meissner and Max Ammareller, whom I gratefully acknowledge again here.

background of the great number of low and medium-quality images included in the database. This level of "average"-quality representations is more important for the impact on the development of security production, and, vice versa, the effect of the fire threat and the development of a new standard of safety on the development of visual representations.

In light of this research, it becomes increasingly clear that little visual evidence has been found of contemporary/historical (versus Biblical or mythological) city fires from the Middle Ages, outside of chronicle illustrations: there are always hell fires in the context of the last judgement, or fires in Sodom and Gomorrah, for example, represented in churches, but only very rarely do you see the precisely dateable burning of a neighboring city or the church's city itself beyond representations of wars and sieges. Exceptions can be found—primarily in surviving fresco depictions from the early Italian Renaissance—but this was an incidental, not a prominent motif.[182] The image of the Frankenberg fire of 1476 is a well-known chronicle illustration,[183] but there are many other similar depictions. One example of the enduring tradition of chronicle illustration is the watercolor drawing of 1707 in a Mühlhausen chronicle (Figure 22).

The image was placed in the middle of a running text about inflation, the price of grain, and the whitewashing of the Blasius church. The city council's response to the fire itself is also reported: a new regulation was introduced requiring citizens to have their chimneys cleaned every year by a city-sanctioned chimney sweep (previously this aspect of home maintenance was left to the discretion of the homeowners).[184] Similarly, an illustration of the Mühlhausen fire of 1689 was embedded within the chronicle and characterized as an eye-witness account ("The great fire looked terrifying like this when one stood on the city mountain"). The fact that color images of the fire were included in a chronicle narrative demonstrates the particular significance that visual impressions had in the recollection, and with good reason: the chronicle's description of the measures of the fire *policey* ordinance in reaction to the fire is thereby tied to the horrific image of the fire and is thus accorded memorial emphasis. On the other hand, since the image is buried in this codex, it did not reach large number of viewers, even in Mühlhausen itself. It almost seems as if people did not want

182 Marco Folin and Monica Preti (eds.), *Wounded Cities: The Representation of Urban Disasters in European Art (14th-20th Centuries)*, Leiden 2015 have followed the same order and chronology of examples.

183 Gerhard Fouquet, "Für eine Kulturgeschichte der Naturkatastrophen. Erdbeben in Basel 1356 und Großfeuer in Frankenberg 1476," in Andreas Ranft and Stephan Selzer (eds.), *Städte aus Trümmern. Katastrophenbewältigung zwischen Antike und Moderne*, Göttingen 2004, pp. 101–31.

184 Stadtarchiv Mühlhausen Th. 61/38, p. 318.

FIGURE 22 City fire of Mühlhausen/Thuringia 1707 in a manuscript chronicle, StadtA Mühlhausen Th. 61/38

to see their own city damaged by fire—an image that would openly reveal the possibility that God had punished them. This great city-destroying event was, on the one hand, revered in the memorialization of city culture, while on the other hand, it was also hidden—banished to the "dark side" of the city's history. This reluctance to depict the fire was manifested in forms of self-censorship, evidenced by the quantity of images in the findings. These invisible barriers of *decorum* forbidding the representation of such an *anti-laus urbis* and the destruction of the city were broken down in spurts and starts through the use of various media in the course of early modernity, as will be shown.

The representation of urban fires in oil paintings—i.e., in the highest form of available early modern visualization media—is a subgenre of landscape painting and, as such, a child of modernity. However, scant attention in the relevant literature has been paid to images of city fires or to images of natural disasters in general, which, in part, are not even recognized as belonging to the genre of landscape painting—even though there are very specific defining elements that can be recognized in such works.[185] If the prevailing

185 See Nils Büttner, *Die Erfindung der Landschaft. Kosmographie und Landschaftskunst im Zeitalter Bruegels*, Göttingen 2000; Catherine Levesque, *Journey through Landscape*

presupposition is true, then like landscape painting itself, paintings of city fires could be considered a primarily Dutch invention that also attracted imitators on the Iberian peninsula in the seventeenth century. The difference from Italian art is striking here. While Raphael and Caracci do not actually depict fire itself in their "fire" paintings, only people (Aeneas/Anchises, fleeing the burning Troy), in this new genre of (natural) disaster painting of Dutch provenance, one finds the exact opposite: in such paintings, fire and flames are the central, exclusive subjects. If Dutch painters did place figures in the foreground, it was invariably in the painting context and technique of "Italianate" work (e.g. Daniel van Dyck, David Teniers the Younger).[186] Conversely, there was at least one painting that appeared in Italy in memory of the fire in the Venetian Doge Palace of 1577, in which people were not the main subject—unsurprisingly, the painter, Lodewijk Toeput, was Dutch (and it was not a painting of a whole city).[187]

This overwhelmingly Dutch body of works concerning city fires can be quickly summarized in terms of quantity. Nancy Corwin has pointed out that in 10 of the 28–30 known Bosch paintings, large fire scenes can be seen;[188] beyond these, the burning of Troy paintings by Lancelot Blondeel[189] and Adam Elsheimer are well-known. Additionally, there are Gillis Moestart's three Troy fire paintings,[190] Jan Breughel the Elder's two Troy fire paintings, Pieter Schoubroeck's six fires of Troy and a painting based on the Nero fire of Rome.[191]

---

in Seventeenth-Century Holland, Pennsylvania 1994; Lawrence O. Goedde, "Naturalism as Convention. Subject, Style, and Artistic Self-Consciousness in Dutch Landscape," in Wayne Franits (ed.), *Looking at Seventeenth-Century Dutch Art*, Cambridge 1997, pp. 129–43. On the topic of city-landscape painting generally see Ralph Andraschek-Holzer, "Topographische Ansichten Österreichs (in den Grenzen der Republik) 1500–1800," in Josef Pauser, Martin Scheutz, and Thomas Winkelbauer (eds.), *Quellenkunde der Habsburgermonarchie (16.-18. Jahrhundert). Ein exemplarisches Handbuch*, Vienna 2004, pp. 1048–59.

186 Hanselaer, "Aeneas' vlucht," p. 115f.

187 Cf. Donatella Calabi, *Venezia in Fumo. I grandi incendi della città-fenice*, Bergamo 2006, p. 46.

188 Corwin, "The Fire Landscape," p. 51.

189 Hanselaer, "Aeneas' vlucht," p. 106.

190 Hanselaer, "Aeneas' vlucht," pp. 108–11.

191 Margaretha Krämer, "Der Brand von Troja und von Rom—Sieben Versionen von Pieter Schoubroeck," in Edgar J. Hürkey (ed.), *Kunst, Kommerz, Glaubenskampf. Frankenthal um 1600*, Worms 1995, pp. 97–102; Martin Papenbrock, "'Das Brennende Troja' in den Bildern von Pieter Schoubroeck. Zur Revision des Vaterlandsbegriffs in Kunst und Philosophie um 1600," in Koppenleitner, Rößler, and Tiemann (eds.), *Urbs incensa*, pp. 45–60.

FIGURE 23 Kerstiaen van Keuninck: *Lot and his daughters in front of the burning Sodom* (ca. 1620), Kortrijk, Stedelijk Museum, Inv. Nr. 924

Daniel van Heil and Kerstiaen de Keuninck painted at least seven fires of Troy and five fires of Sodom (for an example see Figure 23),[192] Enghelbert Ergo painted four Troy fires and two fires of Sodom,[193] and Monsu Desiderio painted one Troy fire painting. According to Vítor Serrão, there were also ten Troy paintings by the Portuguese painter Diogo Pereira (who was probably significantly influenced by the Dutch tradition) that position the city skyline, the Trojan horse, and the night/fire contrast quite strongly in the foreground.[194] When one considers the number of probable lost paintings (for example, the twenty "burning landscapes" that Federico Gonzaga supposedly purchased in 1535, the current location of which is unknown),[195] the fire motif can be described as virtually mass-produced in the formal sector of oil painting between 1500 and 1650.

---

192 Hans Devisscher, *Kerstiaen de Keuninck*, Freren 1987 (= *Vlaamse schilders uit de tijd van de grote meesters*, deel 3), catalogs A3, A23, A 24, A26, B 45, B46, B47, B48, B49, B50, B51, B52. Five others paintings can be added to this not included here by Devisscher or not attributed by him to Keuninck: Kortrijk, Stedelijke Musea, Inv.Nr. 924—reproduction in *Die Flämische Landschaft, 1520–1700*, ed. Kulturstiftung Ruhr Essen/Kunsthistorischen Museum Wien (Ausstellungskatalog: Essen, Villa Hügel, 23. Aug.—30. Nov. 2003), Lingen 2003, p. 319, Fig. 116; Juan Miguel Serrera, Otro, "'Incendio de Troya' de Christian de Keuninck," *Archivo Español de Arte* 59, 236 (1986), pp. 420–23.

193 Hans Devisscher, "Enkele schilderijen van Enghelbert Ergo," in Arnout Balis (ed.), *Rubens and His World*, Antwerp 1985, pp. 235–44.

194 Serrão, "O mito do herói redentor," p. 77: Historical sources point to the existence of since lost fire paintings also by painters such as El Greco, Vicente Carducho, and the Hispanicized Flamish immigrants Juan de la Corte, Francisco de Collantes, and Roque Ponce.

195 Corwin, "The Fire Landscape," p. 69.

There is, perhaps, a very simple reason for the Dutch dominance in this area. One could certainly make the claim that since the modern cities of Italy were more often built in stone, they were therefore less vulnerable to fire, as evidenced by the still largely preserved architecture.[196] Perhaps Italian painters simply had less personal experience with catastrophic fires than the Dutch.[197]

2.2.2 A Discrepancy: Mythological/Biblical City Fire Paintings vs. the Low Number/Quality of Paintings of "Real" City Fires

In any case, this reasoning alone is not sufficient to clarify the character of early modern fire representations. There is, in fact, no surviving record of a single contemporary "real" city fire painted in oil by any of the Dutch and Iberian painters specializing in the *mythological* fire paintings mentioned above, even though there was no scarcity of such events in northwestern and central Europe during the lifetimes and in the immediate environments of these artists. Thus, as "realistic" or naturalistic the Troy and Sodom fires might seem at first glance, they nevertheless always present a fictitious subject. Until the eighteenth century, even the larger fires in Europe attracted often little interest from famous (or even from decent second- and third-rate) painters. While one of the first drawings of Pieter Brueghel concerns the fire caused by the Ottoman warship shelling of Reggio in 1552, there are no such references to current events within his oil paintings.[198]

Is there a deeper reason underlying the rarity of paintings of real (i.e., not Biblical or mythological) city fires and their late emergence, beginning only in the middle of the seventeenth century? The collection and recording of urban fire representations in the image database further confirmed the finding that the number of representations in the early modern period was amazingly

---

196 D. Ballestracci, "La lotta contro il fuoco (XIII-XVI secolo)," in *Città e servizi sociali nell'Italia dei secoli XII-XV*, Pistoia 1990, pp. 417–38 at first gives a contrary impression; however, all the major series of fire originate in the Middle Ages, for early modernity there is no comparable attempt at a survey, but reports of large fatal fires in Italian cities are almost nonexistent in the early modern media.

197 But this hypothesis cannot be grounded on firm evidence since Italian urban history has not yet produced a synthesis or representative overview on the reality and frequency of late medieval city fires. For one of the largest Italian cities, see Calabi, *Venezia in Fumo*, although even for Venice one has to state: "The history of fires in medieval Venice is still to be written" (Kiril Petkov, *The anxieties of a Citizen Class: The Miracles of the True Cross of San Giovanni Evangelista, Venice 1370–1480*, Leiden 2014, p. 49 n 4). For other cities and territories there is not even as much as a study identifying the gaps in research.

198 Cf. Büttner, *Erfindung der Landschaft*, pp. 28, 318 (Fig. 4).

small relative to the modern era, which was particularly surprising considering the precisely inverse relationship regarding the number of fires: almost 8,200 major city fires were situated primarily in the premodern period. In many cases, when a city was affected by a very large fire, there was no surviving illustration at all, even hen the fire was thoroughly documented in local-historical memory. For example, there is no oil painting of the large Erlangen fire of 1706, to which the city owes its present appearance;[199] there are only engravings or broadsheets illustrating the large city fires of Passau in 1662 and Stralsund in 1678/80; similarly, there are no paintings memorializing the fires in Aalen in 1634, Cottbus in 1671, Offenburg in 1689, or the Nordhausen fires; for the Copenhagen fire of 1728 there is only a copper engraving of the cityscape made in 1717, to which someone subsequently added clouds of smoke.[200] This list could go on and on. Concerning probably the largest northern European fire of all, in the 'secret' Finnish capital of Turku in 1827, which destroyed 2,543 houses and burned down a cathedral, only a single post-fire color lithograph has been found.[201]

The paintings we do have can be categorized into different types. One type is an almost paradoxical disaster-idyll, in which the standardized aesthetics of landscape painting and idylls seem strangely overemphasized in light of the disaster that occurred and/or its consequences. Such is the case with some rare, more elaborate ruin/post-fire paintings, such as those related to the Göppingen fire of 1782, by the Ulm city painter Nikolaus Kleemann (Figure 24).[202]

The total destruction of the city down to the foundations is presented, with wood from broken beams still glowing red. In sharp contrast is the intact front yard and garden belonging to the physician Dr. Gottlieb Friedrich Oetinger. The wall enclosing the yard creates a space of security and tranquility here as if it belonged to another world. Spaces of security and insecurity are sharply

---

199 Cf. Andreas Jakob, "'Der Ort stieg aus seiner Asche viel schöner empor'—Der Brand der Altstadt Erlangen am 14. August 1706 und ihr Wiederaufbau als moderne Planstadt bis 1712," *Erlanger Bausteine zur fränkischen Heimatforschung* 51 (2006), pp. 9–47.

200 Kåre Lauring, *Byen brænder. Københavns brand 1728*, Copenhagen 2003, p. 11; on the fire of 1807, which was, of course, quite spectacular due to England's siege as an act of war, see Thomas Munch-Petersen, *Defying Napoleon. How Britain bombarded Copenhagen and seized the Danish Fleet in 1807*, Stroud 2007, Fig. 22ff., pp. 138ff.

201 Gustaf Wilhelm Finnberg (1784–1833), *After the Great Fire of Turku, 1827*, colored litograph, height 35.70 cm, width 55.00 cm. Finnish National Gallery Ateneum—with appreciation to Merja Strandén, Turun maakuntamuseo for her assistance.

202 On the Göppingen fire, cf. Dieter Kauß, "Der Göppinger Stadtbrand vom 25. August 1782 und seine Folgen," *Hohenstaufen* 13 (1986), pp. 135–55; *"Das schöne nahrhafte Göppingen in Asche"—Die Brandkatastrophe von 1782 und der Wiederaufbau der Stadt*, [Stadtarchiv Göppingen], DVD, Göppingen s.d.

FIGURE 24 Nikolaus Kleemann: *Göppingen after the city fire 1782*, Stadtmuseum Göppingen

separated by a dividing line. Both areas are deserted, except for a few viewers and others depicted in conversation. The extinction of the city after the disaster is captured in this large-format image: the memory of the potential complete obliteration of the city is emphatically preserved (the painting was carefully restored in 1871). It was created at a time when disaster imagery had already been developed and when paintings of present ruins—not only of ancient ruins—also belonged to the repertoire of European painters.

130 years earlier there are scarcely any representations at all of the fire of Aachen in 1656, the largest early modern German fire before the Hamburg fire of 1842—nevertheless there is one contemporaneous surviving Dutch school painting (Figure 25).[203]

The representation primarily remains mired in the genre of landscape and cityscapes, which was still relatively new and in which, due to the wide-angle perspective, most people, and even dynamics or movement, play only minor roles. The image shows the city at dawn, with the Lousberg mountain in the background—even this is an idealization.[204] Were there no flames, this would be

203 On the city fire see Kraus, "Der Aachener Stadtbrand", Fig. 12, p. 60.

204 The fire took place between eight and nine o'clock in the morning; the sun would have actually been much higher in the sky at this time.

FIGURE 25 City Fire of Aachen 1656, Dutch school, Museum Burg Frankenberg, Aachen, Photograph Anne Gold

a pleasant image showing the city in tranquil peace and beauty with the majestic cathedral in the center and the solid walls in the foreground, so that reverence, quiet strength, and security are expressed.[205] Through the tongues of fires depicted on top of the houses and spilling down the sides of the cathedral, this painting reverses itself: the strength of the city walls featured in the foreground, which are otherwise emblematic of the security against outside threats, underline here *ex negativo* only the severity of the disaster inside the city, which is presented as deeply vulnerable and defenseless. The fire seems to draw a line through the city and split it; the town center and the cathedral seem to gather around it, as if the fire almost emerges from there. The dark cloud of fire, fed by impossibly thin curls of smoke, hangs ominously above the city, even though it in no way completely conceals the beauty of the mountain range and the reddish sunlight.

Thus the image captures the reversal of the spaces of security and insecurity at the moment of the disaster: citizens flee the city, rescuing a few possessions

205 The city walls and the fortification of cities were, as is well known, a constitutive element of urban identity and security. Tearing down the walls and the defortification of the cities, responding to new war techniques as well as to new demands caused by the growth of urban population have been major events in the history of many cities. This was addressed already in a famous chapter by Walker, *Home Towns* and now analyzed for the crucial period of the Napoleonic Wars by Yair Mintsker, *The defortification of the German city, 1689–1866* (Cambridge 2012).

and goods. In the foreground to the left, brightly illuminated people express little obvious sorrow and a woman in the foreground dressed in red and blue with a white apron and holding a child by the hand even smiles. Other than in a park-like area in front of the city, few people can be seen in the city itself—the protagonists here are the burning main buildings of the city. There is a certain beauty in the whole ensemble and in the color composition, which ranges from dawn-red to fire-orange to the bright darkness of the foreground. It is a dynamic scene: while the fire is front and center, the demise of one of the oldest and most important ceremonial churches in the entire empire occurs more or less in the background. Thus, the demise of the historical is juxtaposed with the departure of individuals. Along the right edge, a legend written in Dutch testifies to the external view on the demise of the ancient imperial city, eschewing any emphatic compassionate subjective perspective. As a whole, the painting shows a borrowing from the natural and city-landscape genre and the bucolic idyll, rather than from the aforementioned mythological fire-disaster paintings.

At almost the same time as the Göppingen painting, a depiction of the great city fire of Gera of 1780 by an anonymous artist (Figure 26) appeared.

This painting is an example of the *locus-amoenus* tradition and is as similarly undramatic as the Aachen painting: because of the azure sky and the

FIGURE 26 City fire of Gera 1780, Stadtmuseum and Stadtarchiv Gera

fairly innocent-looking clouds of reddish smoke hovering over the city, an impression of tranquillity rather than terror is invoked. Gentle hills are visible in the foreground and a few people are shown with their backs turned away from the viewer, grouped around a makeshift shelter made from barrels and boards, clutching a few rescued possessions. Once again, the viewer is first drawn to the surviving remains of the city: on the right are two infirmaries surrounded by white city walls, which show no traces of the fire; here again one sees the reversal of spaces of security and insecurity as the city of ruins enclosed by the wall is the unsafe space. The disaster is pushed so far into the background of the painting that the ruins actually start to disappear in a bluish haze, so that one must closely examine the painting in order to distinguish it from a normal city silhouette complete with rising chimney smoke. In contrast to the disaster descriptions included in contemporaneous poems and sermons, the viewer here is shielded from the 9,000 newly homeless residents, from their complaints and suffering. The ruin-landscape painting, intended for extended remembrance, appears to be focused more on contemplation and even on hope, allegorized by embedding the remnants of the city within nature.

The fact that city views like these are considered landscapes follows a pictorial tradition but also suggests the sublimation of the disaster within the idyll. The countermodel to such depictions of disaster denial demonstrates a development in aesthetic processing, first attempted, as has been shown, by the Dutch around 1600, but only in respect to mythological images of fire: there are two anonymous paintings of the major Goslar city fires of 1728 and 1780 that esteem night-time fire representations as evidence of virtuosity (Figure 27 and 28).

The dark image dominated entirely by fire-red is something of a secularized representation of hell—no devil figures emerge, but the urban landscape painted entirely in black, the red reflection on the faces and bodies of the people in the foreground, and the pale yellowish-edged moon on the borders of the image, convey an unvarnished impression of hopelessness and helplessness. In contrast to images like the Aachen painting, the people in the foreground are also clearly in a state of fear and panic—many are depicted in the foreground on their knees offering prayers toward heaven. The painting itself avoids a strongly religious interpretative framing, its caption records only the time and range of the destruction, and the victims in prayer are depicted rather objectively, lending a distanced perspective to their great consternation. The disaster and its effect on the people remains as if caught in an illustration of an experiment. A place of safety is not seen in this image at all—the fleeing people do not find themselves beyond this disaster;

FIGURES 27 AND 28 City fires 1728 and 1780 of the Free Imperial City of Goslar, Oil 34x23cm and 113x87cm, Goslarer Museum, photography: Bernhard Heinze

FIGURE 29 Votive picture in memory of the 1709 fire of 11 buildings in Bad Tölz, Leonhardikapelle Bad Tölz (property of the church foundation Maria Himmelfahrt Bad Tölz, photography: Claus Janßen)

they have, perhaps, outrun physical death, but cloaked in the dark of night, now as before, they are part of the disaster situation.[206]

Compared to these rather rare oil paintings, composed with care and craftsmanship, votive paintings become fairly common in the Catholic arena beginning in the sixteenth century—for example, the Bad Tölz image (Figure 29) of the house fires of 1709, featuring a precise depiction of the homeowner with

206 The caption states: "Eigentlicher Prospect der Kayserl: Freyen Reichs Stadt Goslar mit numerirter Anzeige der Situation und unterschiedener Bemärkung, sonderlich wie dieselbe Anno 1728 von 26 bis 27 April des Nachts zwischen 12 und 1 Uhr mit einer entsetzlichen Feuers=Brunst heimgesuchet, wodurch dero unter-Theil von Westen gegen Osten auff die 186 vor-Häuser worunter 40 Brau-Häuser und also beynahe in die 500 Gebäude sammt der Stephanschen Kirchen und ihren 9 Glocken, Pastorat, Diaconat, Wittben und Küster-Hause item das breite Thor mit seinen 5 Zwingern und allen Zubehöhr, auch auff beyden Seiten ein gut Theil des Bedecks der Stadt=Maur etc alles innerhalb 8 Stunden gäntzlich verhaeret und verzehret worden."

Mary, Christ, and five saints hovering above him. The sponsor was not one of those affected by fire, but rather Joseph Boier of Brabant, whose house had been narrowly spared. This image also bears a notice of restoration in 1888, which had been financed by the current residents of the once-burned house. The fire was thus maintained in the collective memory.

There are many such instances which suggest a heightened interest in conscious memory work at the end of the nineteenth century.[207] Votive paintings are mostly found in churches or in local museums and primarily serve local memory, the historical remembrance of the respective city, or local Mary or Florian worship. There are also city fire memorial images found on the target plates of marksmen guilds—a rather unexpected motif. For example, a Feldkirch target plate from 1697 depicts the city fire that had just taken place and the annual winners of the shooting festivals were engraved on the edges (Figure 30). A similar plate of Mittweida 1793 shows how, 100 years after the large city fire of 1683, the shooting festival was devoted to the memory of that catastrophe (Figure 31).

The center of the image, depicting the city fire, shows the arrow shot marks. Sometime every year, and sometimes just in a special year of commemoration, there was a target shooting in (dis)honor of the city fire, which must be interpreted as an apotreptic gesture, a gesture of survival, recovery and re-empowerment ex-post in eventually remarkable conviviality. The often poor quality of these local fire images also weighs on and partially reduces their ability to invoke emotions of terror, fear, and disaster angst related to the fire: while traces of such fear are evident in the Biblical and mythological fire images, they also appear—as will be shown below—in other visual media of the time, and particularly in printed broadsheets and engravings.

After the relative absence of city fire illustrations in the Middle Ages, various types of images arose in early modernity— even in oil painting, which, on the one hand could communicate the virtual negation of the disaster within the idyll and the depiction of remaining safe places (e.g., nature, new vast spaces, and hope for the future), but on the other hand, could also portray the hopelessness and the absolute nature of the disastrous moment from a superhuman perspective. Votive and memory images, especially (but not only) in Catholic areas, also flourished during this time. Overall, there seems to be

207 This is also the case, for example, for Bieberach/Rieß, Bobingen, Braunau, Giengen/Br. [Neubronner Epitaph], Hechingen, Klosterneuburg, Knittelfeld, Oberndorf 1757/Laufen 1864, Murau.

FIGURE 30 Feldkirch target plate (*Schützenscheibe*), Fire of 1697 (Property of the Hauptschützengilde Feldkirch, reproduction by kind permission of Oberschützenmeister Mathei)

a glaring disparity between the number of elaborate *Biblical* and *mythological* city fire images in Dutch painting between 1600 and 1650 in comparison to the very limited number of paintings devoted to *real* fires. Excluding, perhaps, images devoted to the London fire of 1666 (the most famous of which were painted by Jan Griffier),[208] if such oil paintings are found, they are almost invariably painted by a relatively unknown artist, or otherwise escape the attention of art history.

208 Cf. the contibutions of Koppenleitner and Heyl in Koppenleitner, Rößler and Thimann (eds.), *Urbs incensa*. The Museum of London has devoted a whole section and website to the event in 2006: http://www.fireoflondon.org.uk/.

FIGURE 31 Burning Mittweida from south-east on a target plate (*Schützenscheibe*), oil on wood, diameter: 169 cm, Stadtmuseum Mittweida

2.2.3 Early Modern Image Theory and Disaster Images

There is thus a strange distance between the representational *realism* of mythological or Biblical locations versus the underrepresentation of *real* medieval city fires in paintings and the *idealization* of disasters, as if representing them was considered somehow inappropriate for this upscale medium. At this point, a rereading of contemporaneous image theory, beginning with Cristoforo Sorte, could be of further help; as Victor Stoichita points out, Sorte was the first theorist of fire representation and one of the few who dealt with this topic at all in early modern painting theory.

According to Stoichita, beginning with Erasmus, art theory classified fire and burnings within Pliny's notion of "the unrepresentable." Sorte was the first to incorporate the unrepresentable into the context of theoretical reflexion. According to Sorte's conception of a city fire image, "nature alone makes the 'story' rather than the [human] figures." Thus in comparison to classical painting, there is a "complete upheaval of the concepts: the background (nature) becomes the main character, while the figures (the mass of Veronese citizens in the fire image described by Sorte) are transformed into the background."[209] When he then brings up as a counterexample the figure-centric Italian tradition of the Troy fire representation in the form of Raphael's *Incendio del Borgo*, he invokes the first unexplained curiosity: why should an Italian, rather than a Dutch theorist be the first to reflect upon city-fire landscape painting? Let us therefore consider again the *Osservazioni nella pittura* and its author.

While, as is well-known, Sorte was a pupil of Giulio Romano, he was also a cartographer, geographer, chorographer, and architect.[210] In fact, there are no surviving oil paintings by him;[211] however, there are seven larger maps and seventy-four cartographical-chorographical landscape drawings attributed to him. He had written the *Osservazioni* because a nobleman from Desenzano del Garda, Bartolomeo Vitali, asked him for further clarification of his painting methods—a request which was related to the reflections on a "corografia" that Sorte had given him. It is not known if Sorte worked for Vitali, as he did for many other "private" noble clients, as well as monasteries, bishops, and the Signoria herself in the surveying and drafting of the many small territories and places in the *terraferma*—but he probably did.[212] This was obviously a (not surviving)[213] "descrizzion del Veronese," as it was later called, and in it Vitali

209 Stoichita, "Lochi di foco, 448.

210 Cf. Jürgen Schulz, "Cristoforo Sorte and the Ducal Palace in Venice," *Mitteilungen des Kunsthistorischen Instituts in Florenz* 3 (1962), pp. 193–208; Jürgen Schulz, "New Maps and Landscape Drawings by Cristoforo Sorte," *Mitteilungen des Kunsthistorischen Instituts in Florenz* 20 (1976), p. 107—Italian translation of these essays: Jürgen Schulz, *La cartografia tra scienza e arte: carte e cartografi nel Rinascimento italiano*, Modena 2006, pp. 91–140. See also the newer collection: Silvino Salgaro (ed.), *Cristoforo Sorte e il suo tempo*, Bologna 2012 which examines the Sorte´s background from many perspectives.

211 Cristoforo Sorte, *Osservazioni nella pittura, in: Paola Barocchi* (ed.), *Trattati d'arte del cinquecento fra manierismo e controriforma*, vol. 1, Bari 1960, pp. 271–301, 276 n 1.

212 On Sorte's clients and on his way of working see the study on the archival tradition at this point by Silvino Salgaro: "'Conoscere i luoghi senza leggere le lettere de' loro nomi.' Cristoforo Sorte cartografo," in Salgaro, *Cristoforo Sorte*, pp. 305–52, 309–14. This contribution also contains information about Sorte's surveying techniques, his notation techniques, and his impact.

213 Denis Cosgrove, "Prospect, Perspective and the Evolution of the Landscape Idea," *Transactions of the Institute of British Geographers* 10 (1985), pp. 45–62, 52: "The work that Vitali refers to is sadly unknown."

already admired the craftsmanship with which Sorte painted the "incendii." Taking the comments of Sorte and Vitali together, it is evident that in his *Osservazioni,* Sorte always had his own work in mind as the object of commentary—i.e. the watercolor chorograph that he gave to Vitali, identified by Cosgrove on the basis of its description as "part-map, part landscape drawing."[214] Supposedly, this chorograph contained an illustration of the Verona fire of 1541.[215]

This "private" chorograph is said to be yet another side project in addition to Sorte's main occupation at the end of the 1570s (Vitali's letter dates to 1578): the manufacture of six chorographs (maps of the Venetian terra ferma) for which he was commissioned in 1578 by the *Sala della secreta* for the Doge's Palace in Venice on the occasion of its reconstruction after the fire of 1577.[216] Some similar watercolor maps by Sorte still survive. Stoichita's comments on the pictorial topos of the burning city on the basis of Sorte's work thus appear somewhat misleading: Sorte does not stand in the tradition of those theorists who, like Varchi quoting Pliny, conceive of fire as "unrepresentable;" rather, he is aligned with those who revived the artistic and simultaneously technical surveying description and collection of countries and cities, particularly Venice, after the rediscovery of Ptolemy's *Geography/Chorographie*.[217] The chorographic was thereby the qualitative descriptive element that completed geography, and thus also counted as the *storia* of the country. In this way it is not surprising that Sorte believed that a decisive event in city history, like the Verona fire, belonged to the *corografia* and should be represented as such. From this perspective, it becomes clear why Sorte coupled his *Osservazioni* in the second edition of 1594 with a medieval chronicle of Verona. Written and painted *storia*—descriptions of space and time—intermingled in chorography as qualitative descriptions of the place.

214 Already in Cosgrove, "Prospect;" also see Denis Cosgrove, *The Palladian landscape. Geographical Change and its Cultural Representations in Sixteenth Century Italy*, Leicester 1993.

215 Cf. Bartolomeo Vitalis's letter to Cristoforo Sorte, written in Desenzano [at the Garda Lake] on 28 February 1578, in Sorte, *Osservazioni*, p. 275f.

216 Gian Maria Varanini, "Governo del territorio e raffigurazioni cartografiche. La Terraferma veneta tra Quattrocento e tardo Cinquecento," in Salgaro, *Sorte*, pp. 87–107, 106f. He places Sorte's chorographic activity within the then 150-old Venitian tradition of creating a visual record of the successively conquered areas of the *terraferma* for administrative purposes, with reference to the rich tradition of determining borders, including the diverse *relazioni* and the related military defensive and fiscal purposes of tax extraction.

217 Instead of multiple references, see: Numa Broc, *La géographie de la Renaissance (1420–1620)*, Paris 1980; Marica Milanesi, *Tolomeo sostituito. Studi di storia delle conoscenze geografiche nel XVI secolo*, Milan 1984.

His image of the Verona fire of 1541, which has not survived, was one of many documents and memorial images, preserved in local historical sources; typically, better painters did not choose this subject for elaborate paintings. In any case, Sorte does briefly argue at the end of this textual passage that past and mythological city fires such as those of Troy or Corinth should also be painted *in just the same way*.[218] This short note is remarkable, because it succinctly bridges the disjunction expressed at the beginning of this section between *realistic* representations of fictional fires and the underrepresentation of real fires in paintings: Troy fire representations thereby reflect real fires, which means conversely that, according to Sorte, representations of mythological city fires should basically be read chorographically.

Sorte's categorization of such images in the chorographic tradition suggests city fire painting as a distinct subgenre of landscape painting, and thereby connects it to the older discussion (related to Svetlana Alpers' compelling argument) emphasizing the affinities between Dutch painting aesthetics with chorography and with the new scientific methods of the sixteenth- and seventeenth-century world.[219] While the structures of Troy, Sodom, or Rome presented in the fire paintings are fictitious, the important consideration was to make the representation of the fire "seem real."[220] However, a secret message was rarely coded into a second level of meaning.[221] At the most, a certain generalization was intended: the representation of the fictitious town stands for the city fire in itself, for the phenomenon known to all, for the fire of *every city*. There is no strong religious or moral emblematic meaning accorded to the paintings and a punitive God is not the represented, even in paintings of Sodom. Instead, Sodom and Troy are almost interchangeable—they represent the structure of the breathtaking tragedy of the outright destruction of a city through flames and the ensuing forced flight of the denizens of the city. Dutch paintings, such as those

---

218 Sorte, *Osservazioni*, p. 292.

219 Svetlana Alpers, *The Art of Describing: Dutch Art in the Seventeenth Century*, Chicago 1983, esp. 119–68.

220 Eddy de Jongh, "Realism and Seeming Realism in Seventeenth-Century Dutch Painting [1971]," in Franits, *Looking at Seventeenth-Century Dutch Art*, pp. 21–56, 28: "They are non-existent landscapes that nevertheless could have existed" (not in relationship to moving fire images). Cf. also Eddy de Jongh, *Questions of Meaning: Theme and Motif in Dutch Seventeenth-Century Painting*, trans. and ed. Michael Hoyle, Leiden 2000, pp. 167–91.

221 Even de Jongh, "Realism and Seeming Realism," p. 27 concedes, in relation to the landscape painting, that in this case it is possible that only "the form remains while the allegorical or moralizing contents evaporate"; cf. Goedde, *Naturalism*, pp. 129–43, 132.

by Schoubroek—himself a religious refugee in Frankenthal—perhaps thereby simultaneously capture the fire and exile experience associated with the wars of religion, although this is not specified in the paintings. Each and every fire was implied in these paintings—whether caused by war or not. The emblematic surplus concerning the choice of fictional Biblical/mythological subjects reflected this general applicability and thereby drew the focus and feelings of the observer entirely toward the fire itself.[222] This notion is also supported by the fact that in the case of Schoubroek, the paintings were painted on copper, probably so that they would shine in a small, dark chamber: the virtuosity of the material illustration and the effect of the fire's glow were also important.[223] If Biblical and mythological images of fire in oil painting before 1650 stand in for depiction of real fires in painting, then this substitution can be interpreted as an element of the *decorum* of the first half of the early modern period already touched upon above: on the one hand, the representation of non-sacred and non-mythological subjects was, of course, effectively not allowed; on the other hand, representing actual cities as destroyed may have also contradicted the dignity of the cities as bearers of their own culture of identity and memory.

2.2.4 Temporalization, Eventfulness and Affect Control

There is one key difference that distinguishes landscape paintings from city fire paintings, which are otherwise classic objects of art history research concerning Dutch painting: in Dutch paintings, the landscape tends to be calm, motionless; still lifes are also influenced by this timeless, contemplative

222 In contrast to Alpers's theory (that the Dutch visual culture owes its character of proximity to the descriptive/empirical sciences and their instruments) Goedde emphasizes that the artistic landscape image is characterized by a "contrived informality and apparent spontaneity of realistic landscapes," and presupposes a "sophistication of purpose and taste on the part of both artists and buyers of art." Landscape painting functions precisely based on the paradox that it is an illusion of reality and the appearance of an unmediated, natural experience that could be accomplished only through the practice of the highest artistic abilities, not just the application of cartographic and measurement craftmanship transferred from administrative and technical purposes to art (Goedde, *Naturalism*, p. 142f.). Both approaches are not incompatible: artistic play combined with the enjoyment of the now realized instrumental culture of visuality that has become so familar is in fact derived from this, even as it presupposes it.

223 For the material specifications, see Krämer, *Brand von Troja*, p. 101f. (e.g., note 53). In Karel van Mander's treatise, fires are treated in Ch. 7 ("Van de Reflecty, Reverberaty, teghenglans oft weerschijn"), with precise information how the artistic representation of sparks, firelight can be realized.

mood.[224] In Vermeer's paintings, it is often momentary movements that are recorded—a stream of water that has just been poured from a jug, a glass from which someone has just drunk, a letter that is just being read.[225] But they always concern small movements in the framework of general peace. Images of ruins might be endowed with a certain reference, pregnant with meaning, to a disastrous event, from which they had been generated; however, as image-objects they express a "temporal continuity."[226] While city fire painting can be classified, also chorographically, as a subgenre of the landscape tradition, it is the momentary quality, the emotion of the catastrophic moment that stands at the center in these paintings. Chorography does not appear to exclude a temporalization of "landscape" and the specific affect control associated with it:[227] the landscape becomes not only the main autarchic subject, but also a threatening actor generating affect control and a point of reference for its impact.

This trend, noticeable in the Dutch Troy and Sodom paintings, appears less often in contemporaneous oil paintings of actual city fires than in corresponding engravings. News broadsheets reporting on sieges and other war events often depicted fires "in action," as did the less common depictions of city fires not related to war—the Passau fire of 1662 and the Annaberg fire of 1664 are easily accessible examples in this regard.[228] In addition there was a wide variety of independent broadsheets.

224 "Nothing ever happens" (Alpers, *Art of Describing*, p. 147 with regard to Dutch landscape painting). On the content, for example, of van de Velde's landscape engravings, cf. Levesque, *Journey*, p. 105: "throughout the seasonal and peasant prints, man—simple and natural, and specifically Dutch—is shown in harmony with his environment, and with time, cosmic and agricultural."

225 Cf. Markus A. Castor, "Augenblick und Zeitfluß—Bemerkungen zu einigen Konstanten in der Malerei Jan Vermeers," *Pantheon* 54 (1996), pp. 84–101.

226 "They commemorate past troubles, celebrate present-day abundance, and encourage future vigilance." (Levesque, *Journey*, 59).

227 In order to avoid misunderstandings, it should be pointed out that the concept of temporalization has already been addressed in the context of landscape painting by Josiane Rieu in her article "La temporalisation de l'espace" in *Le Paysage à la Renaissance*, ed. Yves Giraud (Fribourg: 1988), 297–310—albeit in a much different sense. She discusses the reintroduction of a neomedieval pluritemporality from temporally distant action- and biographical elements in the landscape sphere of the mannerism school of Fontainebleau from around 1540, after the common representations from the Middle Ages (e.g., of saints in a tableau) had been repressed during the Italian Renaissance.

228 Both representations are easily accessible through the databank "Prometheus" and the broadsheet digitalization of the state library of Bavaria: http://www.bsb-muenchen.de/Einblattdrucke.178.0.html. Cf. Horst Carl, "'Brüssel 1695': Kriegszerstörungen und ihre Visualisierung im späten 17. und 18. Jahrhundert," in Birgit Emich and Gabriela Signori (eds.), *Kriegs/Bilder in Mittelalter und Früher Neuzeit*, Berlin 2009, pp. 295–315.

FIGURE 32 Engraving of Baden-Baden burning during the Palatinate war of 1689, as in Wahrmund, *Die Jammer-gedrückte* [...] *Neckar-Pfalz* (cf. n. 229), Stadtmuseum und -archiv Baden-Baden, Inv. Nr. 317

For example, images of the fires of Baden-Baden in 1689 (Fig. 32)[229], Esslingen in 1701 (Fig. 33)[230], and Goslar 1728 (Fig. 38) featured remarkably vivid

229 Theophil Wahrmund [pseud.], *Die Jammer-gedrückte, Hülffleistend erquickte, Und Kronen-beglückte Rhein- und Neckar-Pfalz* [...] *Wie solche, samt andern benachbarten Ländern und angränzenden Flüssen,* [...] *von denen alles-verheerenden Franzosen, so übel zugerichtet, und jämmerlich verwiestet worden*, s.l. 1691, between p. 238 and 239. I thank Heike Kronenwett of the Stadtarchiv/Stadtmuseum Baden-Baden for providing me with a copy.

230 Entsetzliche Feuers=Brunst/ Welche den 25. Octobr. des zuruckgelegten 1701. Jahrs/ die Reichs-Stadt Eßlingen troffen/ und leider! solche mehrertheils eingeäschert (from Stadtarchiv Esslingen—with gratitude to Ursula Kümmel).

FIGURE 33 Engraving on single-leaf broadsheet of the Esslingen fire, October 25, 1701, Stadtarchiv Esslingen Karten und Pläne Q 15, printed in Nürnberg, Felseckerische Erben

FIGURE 34 Engraving on single-leaf broadsheet of the Reutlingen fire September 23, 1726, Historisches Museum Reutlingen, Nr. 12, printed Augsburg: Elias Bäck

representations of fire, as did images of the Reutlingen fire of 1726 (Fig. 34 to 37, 39)[231]—clearly one of the most frequently illustrated fires.

231 Cf. Werner Ströbele, "Der Stadtbrand von 1726—ein Ereignis schafft Bilder," in *Stadt—Bild—Geschichte. Reutlingen in Ansichten aus fünf Jahrhunderten. Katalog und Ausstellung*, Reutlingen 1990, pp. 49–56. On the fire see P. Schwarz, "Der große Reutlinger Stadtbrand im September 1726," in *Reutlinger Geschichtsblätter* 14 (1976), pp. 7–43.

FIGURE 35 Hand colored engraving of the Reutlingen fire of 1726, Historisches Museum Reutlingen, Nr. 994

FIGURE 36 French Single-leaf woodcut of the Reutlingen Fire of 1726, Städtisches Museum Ludwigsburg Nr. 1131

FIGURE 37 Engraving of the Reutlingen fire of 1726 by Gabriel Bodenehr, *Prospect der Hl. Röm. Reichs Statt Reutlingen, wie solche A.° 1726, d. 23. 24. u. 25. Sept. durch eine entsezlich wütende Feuers-Brunst bis auf etliche wenige Häusser u. Gebäude in die Asche geleget worden,* Städtisches Museum Ludwigsburg Nr. 1132

FIGURE 38 Engraving of the Goslar city fire of 1728 as published in the *Verbesserter und Neuer Kriegs- Mord- und Tod-Jammer- und Noth-Calender* (Nürnberg 1729), Stadtarchiv Goslar

FIGURE 39 Small anonymous engraving of the Reutlingen fire of 1726, Städtisches Museum Ludwigsburg Nr. 1145 (Schefold 6462)

One engraving in particular (Figure 39) should be emphasized here, in which the forms of people, full of suffering and reaching toward heaven with outstretched hands, are mirrored in the tongues of flame reaching upward into the heavy curtain of smoke. The group of horrified people reinforces the line of the town wall: the contours of the elemental violence and the contours of the expressions of suffering, fear, and horror are isomorphic. This is a different, new imagery, which combines representations of disaster with the suffering of mankind. The meadow and barn in the foreground are not primarily depicted as spaces of safety in contrast to the inferno of the city; rather, the disaster situation neatly comprises the image as a whole.

At about the same time, the French engraving of the biggest early modern city fire in France (Rennes, 1720) appeared, capturing the situation, the moment, and the perspective of the dramatic inferno in very different ways and

FIGURE 40 Engraving by Thomassin after a drawing by Jean-François Hughes of the Rennes city fire December 22 to 29, 1720, Musée de Bretagne, Inv. Nr. 86.2684.E.2

means than the listless aerial-view oil paintings of the seventeenth century (Figure 40). The view of the crashing avalanche of flames is from the town square itself and the observer of the engraving feels just as suffocated by the clouds of smoke and just as oppressed by the wall of flames as the residents depicted fleeing the fire. We can see the details of people throwing down their possessions at the last moment and clearly discern beams and roofs in the midst of the fiery explosion. This is unquestionably a much more strongly affect-oriented representation than seen in most surviving oil paintings—it comes closer in effect to a Troy or Sodom painting.

There also seems to be a developmental step between the aesthetics of the single-leaf news broadsheets and representations such as this one, which is most famously associated with Jan van der Heyden. His *Brandspuitenboek* (Fire Hose book) always presents the scene of the fire itself and above all the attempts at fighting the fire from the perspective of someone directly involved (Figures 41, 42)—Susan Donahue Kuretsky has aptly described this as taking the perspective of the fireman, a particularly appropriate description

FIGURE 41 Jan van der Heyden, *Fire men using fire equipment*, pen and wash drawing, plate 1 of the 1690 *Beschrijving*

FIGURE 42 Jan van der Heyden, *Firemen fighting a fire after a lightning struck four rows of adjoining houses between Elands Canal and Elands Street in Amsterdam, July 27, 1679*, plate 8 of the 1690 *Beschrijving*

in this context as Jan van der Heyden was indeed chief of the Amsterdam fire department.[232]

As realistic and gripping as his fire representations are, they are also influenced by his particular perspective and also by the specific context of their publication: Van der Heyden intended to use these images as advertisements

232 The original title is Jan van der Heyden, *Beschrijving der nieuwlijks uitgevonden en geoctrojeerde Slang-Brand-Spuiten* [..], Amsterdam 1690. Cf. Jan van der Heyden, *Fire Engines with Water Hoses and the Method of Fighting Fires now used in Amsterdam*, Canton 1996; Hans Wildeboer, "De ontwikkeling van de brandspuit in de zeventiende en achttiende eeuw," in *Jaarboek voor de Geschiedenis van Bedrijf en Techniek* 10 (1993), pp. 9–41, 23–35; Peter C. Sutton (ed.), *Jan van der Heyden (1637–1712)*, New Haven 2006, pp. 73–81; Susan Donahue Kuretsky, "Jan van der Heyden and the Origins of Modern Firefighting. Art and Technology in Seventeenth-Century Amsterdam," in Bankoff, Lübken and Sand (eds.), *Flammable Cities*, pp. 23–43, 35.

for his invention of the fire hose (1671–1676), which he first distributed throughout Europe on a large scale. Especially in western German municipal archives near the Netherlands, such as Aachen and Cologne, the engravings, which Van der Heyden distributed as advertising brochures with instructions for using his fire hoses, can still be found today (see below Ch. 3, 3.1-2). As eighteen fire hoses are shown in use in the depiction of the burning of one Amsterdam canal house (plate 2 of his *Fire hose book*), this can hardly be considered a "realistic" example—outside of Amsterdam, hardly any European city possessed that many large fire hoses around 1700 and they were of course never simultaneously employed to fight a single house fire. In this case, the goal was obviously the presentation of the formidable operational capability and manoeuvrability of the hoses and what might be called an "super-realistic" representation.

Whether the engravings and similar drawings prepared for the *Brandspuitenboek* were sent also, in a similar fashion as the fire hose advertisement engravings and broadsheets themselves, to city magistrates and princes, and whether the innovative and dramatic style of the images, geared toward producing affects of fear, worry, and anxiety about possible future fires in one's home town, should therefore be put into the context of this purpose of advertising of the product, is not clear from current archival documentation.[233] In any case, in examining the remaining Van der Heyden œuvre, it can be determined that there is not a single fire image among his color oil paintings—neither a Biblical or mythological image nor one related to a real fire, though he is known

233 Recent research just on this question of how advertisement for fire hoses was done in small Lower Rhine and Westfalian cities has not revealed any further information on this particular question concerning van der Heyden's sketches and engravings, Christine Brauner, "Recommendation und Reklame. Niederrheinische Brandspritzenmacher und Praktiken der Werbung in der Frühen Neuzeit," *Zeitschrift für Historische Forschung* 46, 1 (2019), pp. 1–45. For the (mostly eighteenth century) history of advertising in the Anglo-American context cf. Christine Y. Ferdinand, "Selling it to the Provinces: News and commerce Round Eighteenth-Century Salisbury", in John Brewer and Roy Porter (eds.), Consumption and the World of Goods, London 1993, pp. 393–411; Timothy H. Breen, *The Marketplace of Revolution: How Consumer Politics Shaped American Independence*, New York 2004; Emma Hart, "A British Atlantic World of Advertising? Colonial American 'For Sale' Notices in Comparative Context, *American Periodicals* 24, 2 (2014), pp. 120–127; Jon Stobart, *Sugar & Spice. Grocers and Groceries in Provincial England, 1650–1830,* Oxford 2016, chapter 7 ('advertising'). However, this is embedded into the framework of a history of (individual, private) consumption like the history of the *Bureaux d'adresse*, while the diffusion of technological innovation in a semi-public network of governmental and inter-magistrate communication is still different on what concerns the distinction between the private and public realm.

as one of the most important city landscape painters of his time.[234] In fact, there are hardly any surviving Troy or Sodom paintings in the style of Breughel, Mostaert, Schoubroek, or Keuninck from the second half of the seventeenth century on. This genre seems to have gone extinct in the oil painting sector. With Van der Heyden and Ruisdael, the city itself as a landscape object is discovered,[235] but even more so, there seems to be a functional differentiation at work: while landscape oil painting is devoted to more contemplative, quiet observation, engraving and drawing captures moments and dynamic situations of movement.[236]

The aesthetics of the dynamic image are first realized only in images that capture discrete moments. The almost comic-like aesthetics known today, of the movement of firebombs marking flight lines, originates in the news medium of the news broadsheet or "flyer."[237] Even the constant linking of representation and commentary in the form of the legend that not only memorialized the names of people and places, but also situational moments in the image, fixed a specific moment of time in the representation. This informative, temporalized aesthetic of event reporting was broadcast so far into the realm of painting that legends were embedded into almost all known oil paintings of contemporary fires.[238]

The next step was perfecting the representation of the disaster as a temporal moment itself, in particular regarding explosions. This is comparable to the high-speed footage of Harold Edgerton, in particular his now-iconic and almost universally recognized image from 1939, in which a water droplet hits the water—a phenomenon that is invisible to the naked eye.[239] In the same

234 Helga Heikamp-Wagner, *Jan van der Heyden*, Amsterdam 1972; Sutton, *Jan van der Heyden*.

235 On this concept see Jan Peeters, "De stad als landschap de stad als stilleven," in *Kunstschrift/Openbaar Kunstbezit* 36 (1992), pp. 20–24.

236 Cf. Arthur K. Wheelock, "Accidents and Disasters," in Susan Donahue Kuretsky (ed.), *Time and Transformation in Seventeenth-Century Dutch Art*, New York 2005, pp. 73–82.

237 Cf. Wolfgang Harms (ed.), *Deutsche illustrierte Flugblätter des 16. und 17. Jahrhunderts*, 7 vols., Stuttgart 1980–2005, vol. 2, Nr. 163; on the 1610 siege of Jülich, vol. 2, Nr. 402; on the 1631 conquest of Magdeburg, vol. 2, Nr. 546; on the 1641 siege of Hohentwiel, vol. 2, Nr. 613; on the siege of Neuhäusl, vol. 2, Nr. 664; on the 1664 siege of Kanischas; vol. 4, Nr. 369, on the 1675 conquest of Trier.

238 That relates somewhat to the above-mentioned fire images of Aachen 1656, Göppingen 1782, and Goslar 1728.

239 See Gunnar Schmidt, "Splashes & Flashes. Über High-Speed-Visualisierungen und Wissensformen," in Inge Hinterwaldner and Markus Buschhaus (eds.), *The Picture's Image. Wissenschaftliche Visualisierung als Komposit*, Munich 2006, pp. 180–95.

way, an engraving, such as that of 1750, showing the explosion of the burning St. Michael's Church Tower in Hamburg, fixes visually a moment that could never have actually been seen to *that* degree of detail, and could not have been remembered as such. It is this temporalized aesthetic that constructs the visual event— that stupendous, threatening moment—in order to render the event as a whole (the fire of the Hamburg Church), capable of being experienced in a visually sublime form (Figure 43).

Through the technical possibilities of exposing the actual moment, photography capitalizes on the possibility of creating simultaneity between the event and image production;[240] however, the fascination with the event-ness itself and the search for its adequate representability and the specific possibilities of affect control associated with that are much older, as this eighteenth century image proves.[241] In graphics and in single-leaf engravings and woodcuts this temporalization of disaster representation was increasingly perfected.

Even after the fire, these media served not only the purposes of remembrance, like oil paintings, but were also part of post-fire communications, in which the temporalized images that focused on the natural catastrophe as an actual affect-controlling subject contributed to a specific function of appeal intended to trigger various acts of solidarity—not least of which were donations.[242]

2.2.5 Threat Perception, Security Requirements, and Emotionalization

Embedded within representations of city fires is thus an appellative and affective script that can be interpreted according to their relationship with and,

240 The invention of the snapshot takes place around 1880, as the collodion method was replaced by a high-resolution, momentary exposure process (improvement of the bromine silver-gelatin method) and thus made on-site photographs captured with a portable camera possible. Cf. Bibliothèque nationale de France (ed.), *La Révolution de la photographie instantanée 1880–1990*, Paris 1996; André Gunthert, "Esthétique de l'occasion. La naissance de la photographie instantanée comme genre," *Études Photographiques* 9 (2001), pp. 64–87.

241 Van der Heyden's images were compared with photography as early as the nineteenth century—see Alpers, *Art of Describing*, p. 29.

242 Like the St. Michael's church tower explosion image and the fire sermons in the context of the massive donation efforts (see above p. 116 and n. 145), similar collection contexts could often be inferred from the texts on the broadsheets. Most broadsheets today have, however, been separated from their original contexts and included in collections of images or engravings, while collection books and donation notations, if they are preserved at all, are preserved again at different places, to be found in parish documents and state records of collections, charity briefs, donations, etc. Historically, however, both are important parts of a *one single* function and activity, even though the images that were later detached also serve as a remembrance.

FIGURE 43 Explosion of St Michael's church tower after lightning strike in 1750. Anonymous engraving in the miscellaneous volume on the 1750 event, HStA Hamburg, Bibliothek, A 640/77

perhaps, causal contribution to change and the bridging of an epochal threshold of security history in the *longue durée*.[243]

Art theorists of the Renaissance and the seventeenth century were already engaged in complicated reflections concerning affect control through art and in relation to the conceptualization of the role and perspective of the observer via the painter, primarily in recourse to the theory of affect, which first originated in rhetoric. This link to rhetoric, however, meant that these theories were functionally primarily tied to figural representations. Observers are conceptualized as audience members listening to a speech, who are emotionally engaged by the expressive capacities of the represented persons. For example, the sixth chapter of Karel van Manders' *Schilder-boeck* on the representation of affect ("Wtbeeldinghe der Affecten, passien, begeerlijckheden, en lijdens der Menschen") specifically concerns representations of people, whose suffering or joy has the capacity to invoke corresponding empathy in observers; in contrast, the chapter on landscape (Ch. 8) primarily emphasizes practical instructions on how to paint things effectively, with no mention of associated emotions.[244]

In any case, a general tendency for visualizations of city fires was that people increasingly disappeared and that the fire itself became the main subject. In the overall schematics of Troy and Sodom in paintings around 1600, the minor figures of fleeing people, of Aeneas and Lot in the mostly dark foregrounds, appear so superfluous that they seem hardly capable of operating as a focal point for generating an emotional response; even if they were, they are only schematically perceivable as a group. This dynamic was, of course, different in the Italian tradition, and it was with good reason that affect in Raphael's *Incendio del Borgo* was linked to the then new reception of the Aristotelian theory of tragedy.[245] However, in Dutch and Portuguese painting, as well as in the

---

243 For further theoretical discussions see William M. Reddy, "Against Constructionism. The Historical Ethnography of Emotions," in *Current Anthropology* 38, no. 3 (1997), pp. 327–51; William M. Reddy, "The Logic of Action. Indeterminacy, Emotion, and Historical Narrative," in *History and Theory* 40 (2001), pp. 10–33; Katja Mellmann, *Emotionalisierung. Von der Nebenstundenpoesie zum Buch als Freund. Eine emotionspsychologische Analyse der Literatur der Aufklärungsepoche*, Paderborn 2006; J. Plamper, "The History of Emotions: An Interview with William Reddy, Barbara Rosenwein, and Peter Stearns", *History and Theory* 49 (2010), pp. 237–65.

244 On the reconstruction of Dutch art theory see the relevant work: Andreas Hahn, "... *dat zy de aanschouwers schynen te willen aansprecken"—Untersuchungen zur Rolle des Betrachters in der Niederländischen Malerei des 17. Jahrhunderts*, Munich 1996.

245 J. Rasmus-Brandt, Pity and Fear. "A note of Raphael's 'Incendio del Borgo,'" *Acta ad Archaeologiam et Artium Historiam Pertinentia. Institutum Romanum Norvegiae Roma* N.S. 1 (1981), pp. 259–74.

"temporalized" fire representations in engravings and woodcut broadsheets, the observer directly faces the fire. This perspective and the possibilities for identification are based on the small and powerless-seeming individuals in contrast to the forces of nature. In fire representations in higher art (i.e., beyond votive pictures and pure local memorializations), the emergence and function of the aesthetic category of the sublime in the eighteenth century in conjunction with the classification of "natural disasters" as particularly terrible manifestations of the sublime is significant:[246] the English aesthetic theories of the sublime that emerged in the late seventeenth century and culminated with Edmund Burke (1757), and later those of Schiller and Kant, mark the development of a theory exploited by the history of art, which was particularly significant for the absolutization of "terrifying nature," and which developed in tandem with the history of natural sciences in the eighteenth and nineteenth centuries.[247]

When, in the eighteenth century, volcano representations, especially Vesuvius images, like those of Jakob Philipp Hackert and Pierre-Jacques Volaire, became a popular genre, it was because they particularly clearly manifested this notion of terrifying nature.[248] Most were fashioned using elemental symbolism—sea and moonlight against volcano fire—which formed the

---

246 For the parallel development in German literature, see Carsten Zelle, *"Angenehmes Grauen." Literaturhistorische Beiträge zur Ästhetik des Schrecklichen im Achtzehnten Jahrhundert*, Hamburg 1987.

247 Cf. Werner Busch, "Die Naturwissenschaften als Basis des Erhabenen in der Kunst des 18. und frühen 19. Jahrhunderts" in *Jahrbuch des Historischen Kollegs* (2004), pp. 83–110. For the early reception of Pseudo-Longinus *peri hypseous* see Hana Gründler, " 'La più orrenda pittura'—Zu einer Ästhetik des Erhabenen in Vasaris *Vite*, in Koppenleitner, Rößler, and Timmann (eds.), *Urbs incensa*, pp. 61–79.

248 Sean Cocco, *Watching Vesuvius: A History of Science and Culture in Early Modern Italy*, Chicago 2012 counter-intuitively treats not this image tradition, but rather the visual-scientific development of Vesuvius in the seventheenth century and, along with Hamilton (pp. 209–25), barely touches on the post-Hackert/Volaire era. Current discussions of the relationship between arts, literature and vulcanism have concentrated more on the perception of the bad weather situation after the Tambora eruption ('the year without a summer', 1816) Gillen d'Arcy Wood, *Tambora. The eruption that changed the World,* Princeton 2015, pp. 1–33; Adriana Craciun, "Writing the Disaster: Franklin and *Frankenstein*", *Nineteenth-Century Literature* 65, 4 (2011), pp. 433–480; Siobhan Carroll, "Crusades Against Frost: Frankenstein, Polar Ice, and Climate Change in 1818", *European Romantic Review* 24, 2 (2013), pp. 211–230—here, the relationship between the diegetic level of the narrative, the production process and the perception of a real weather anomaly caused by the eruption forms a less coherent line of causal or conditional argument, cf. Zwierlein, "Klimageschichte und Kulturgeschichte". For the development of the aesthetics of the sublime and the iconography of volcanic eruptions, this is a logic internal to the artistic episteme, therefore less problematic.

FIGURE 44 Johann Christoph Jünger: Burning of Dresden's Pirna suburb by the Prussian soldiers when the Austrians were approaching to besiege the city, November 10, 1758 (Staatliche Kunstsammlungen Dresden 1978/k 248)

background of a natural-scientific geognostic discussion of time; the earth's age and the origins of its form also invoked the still-open debate regarding Neptunism vs. Vulcanism. In most cases, small depictions of a few "disaster tourists" appeared in the foreground, so that the observers of the painting—through the doubling effect of observers observing—could reflect their own external position regarding the disaster in accordance with theories of the sublime.[249]

Johann Christoph Jünger's 1758 depiction of the fire set by Prussian occupiers in the Dresden suburbs (instigated by the Austrians' advance during the Seven Years' War) is executed in precisely this way and demonstrates that in the Enlightenment schematization of visual perception, the war-induced city

249 For the period around 1750, see the small synthesis by Jörg Trempler, *"Katastrophen." Ihre Entstehung aus dem Bild*, Berlin 2013. For Volaire cf. Émilie Beck-Saiello, *Pierre Jacques Volaire, 1729–1799: dit le Chevalier Volaire*, Paris 2010, pp. 113–146 for the interpretation of the several Vesuvius pictures he painted during his stay at Naples after 1768; for Hackert (most important Vesuvius picture of 1774) cf. Monica Leonardi, "Der Vesuv und die phlegräischen Felder. Feuer, Wasser, Schwefel und Gestein", in *'Kennst Du das Land'. Italienbilder der Goethezeit*, eds. Frank Büttner, Herbert W. Rott, Cologne 2005, pp. 255–271.

FIGURE 45 Pierre-Jacques Volaire, "Outbreak of the Vesuvius" 1774, Compton Verney Manor, Warwickshire, England

fire is also categorized as a *natural* disaster, just like a volcano eruption, and clarifies why absolutizing the fire event stands in the center of such paintings as the actual affect-directing mover (Figure 44, 45). The Dutch school had, in practice, already anticipated this sublime aesthetics of disaster in some respects before the theory arrived. In print graphics, capturing the danger and horror of the fire, the stupendous moment itself, becomes increasingly central. If there were a role that the observer could assume, it would be David in a battle against the inhuman Goliath of the fire inferno—not the compassion for a suffering individual human depicted.

What actually occurs is the always increasing attempt to confront the spectators emotionally, and always more powerfully, with the danger of fire *in itself,* without taking a detour through empathy—empathy for the suffering, the danger, the fear of other people—as was the primary aim of the theory and practice of most poetic tragedy. Even in the engraving of the Reutlingen city fire of 1726, in which the composition almost produced a symmetry between the high-reaching tongues of flame and the raised arms of the terrified observers, the people remain depicted as small and unindividualized, with their backs facing the viewer. This demonstrates the attempt to directly address the simple base emotions of fear in the face of an uncontrollable fire, in contrast to the derivative, specifically culturally coded emotional program.[250] Thus, in this way, a pictorial language for "disaster" develops in tandem with the linguistic and philosophical concept of disaster in the eighteenth century.[251]

Unfortunately provenance research has not yet come far enough to construct a true social history of the owners of "realistic" landscape and fire paintings and to thereby answer the question of "why" they owned these landscape paintings.[252] Concerning the broadsheets and engravings of city fires in near or distant cities, however, it has been repeatedly documented that they were used in correspondences requesting support between authorities or to directly solicit donations—thus as media used to appeal for solidarity, rather than merely as remembrance and memorial images like some oil paintings.

The development of the visualization of city fires and the mobilizing power of fire images is thus at least correlative, since temporally speaking, and perhaps even relating causally to a small extent, it contributed to the general development beginning in 1650–1700, which since the eighteenth century, and irreversibly since the nineteenth century, led to the flattening of the peaks in fire frequencies and the spread of the *fire gap*. It is thus a parallel phenomenon to the somewhat astounding ethically and emotionally manipulative topoi of advertising for fire insurance in the numerous smaller cameralistic tracts and treatises of the eighteenth century, which led to the implementation of the fire funds. The founders of fire funds and fire insurance companies attempted to weld together potential members (cities, villages, and their individual citizens) in institutional associations of risk and solidarity, which were incommensurate and neither spatially nor socially coextensive with traditional relationships because they far exceeded conventional local networks (cf. below Ch. 5, 1.1). In order to do this, they appealed to the moral responsibility of the individual, the importance of charity and brotherly love, and even the common good of all mankind. Also, it is noteworthy that one transformative effect of insurance institutions—the inauguration and cultivation of more abstract, impersonal relationships than in the old forms of social solidarity—was that

---

250 J.H. Barkow, L. Cosmides, and J. Tooby (eds.), *The Adapted Mind*, Oxford 1992.

251 The history of concepts and the history of visualization are two components that should be taken together to create an integrative "History of perception". On the first topic see Mercier-Faivre and Thomas, *L'invention de la catastrophe*; Olaf Briese and Timo Günther, "Katastrophe. Terminologische Vergangenheit, Gegenwart und Zukunft," *Archiv für Begriffsgeschichte* 51 (2009), pp. 155–95.

252 Cf. Goedde, *Naturalism*, p. 142: "Until more work is undertaken on ownership and provenance, it will be difficult to reach conclusions about who owned realistic landscapes and why."

they necessitated a stronger form of emotional discourse. The technical side of insurance projects—which invokes personal responsibility and collective self-interest according to a probability calculation involving large numbers, and speaks to the reader's "rational" side through various tables and calculation examples—bills itself as the engine of a new form of supralocal passion related to humanity and mankind, which not only has a life of its own in the literature and art of the time, but also operates on a practical level (cf. Ch. 5, 1.1-2).

Thus the demonstrated tendency in fire visualizations is understandable in the following way: on the one hand there was a development toward a much more direct confrontation with the threat, and on the other hand, due to lack of an offer of intermediary human identification, a move toward a more abstract perception of fire. This is analogous to the somewhat dialectic combination between an emerging individual concept of personal responsibility for future losses and anticipating care for one self (*Eigenvorsorge*) with the far more abstract deindividualization of social bonds that the insurance institutions effected. Above all, the exceptional state of the disaster as an avoidable phenomenon first revealed itself in concrete contours through developments in artistic representations of catastrophes—as a contrast to the expected normal state of safety and security, disasters became comprehensible and capable of being experienced visually. Such visual (in addition to linguistic and semantic) conceivability was necessary so that the "normal secure society" always had its abhorrent opposite placed before the eyes: a new quality of the sphere of security could only arise by refining and emotionalizing the image and the concept of the threat to security.

2.3 Ground Zeros: *Visualization and Time Horizons*

The previously discussed developmental lines regarding urban fire visualizations addressed the autonomy of the disaster as an image object and its related impact on affect. Even here there were close links to the area of practical fire prevention and also to notions of resilience and disaster management—just as within the aesthetic development itself, a striving toward the chorographic-geographical, scientific form of landscape and city visualization was apparent. As will be discussed below, the analysis of the visualization of the burned cities themselves—the ground zeros—relates even more closely to the areas of fire *policey*, resilience and reconstruction management:[253] even before the autonomy of an aesthetics of ruins, based not only on Classical ruins, but also

253 Cf. the quick response to the 2001 catastrophe Joan Ockman (ed.), *Out of Ground Zero. Case studies in urban reinvention*, Munich 2002.

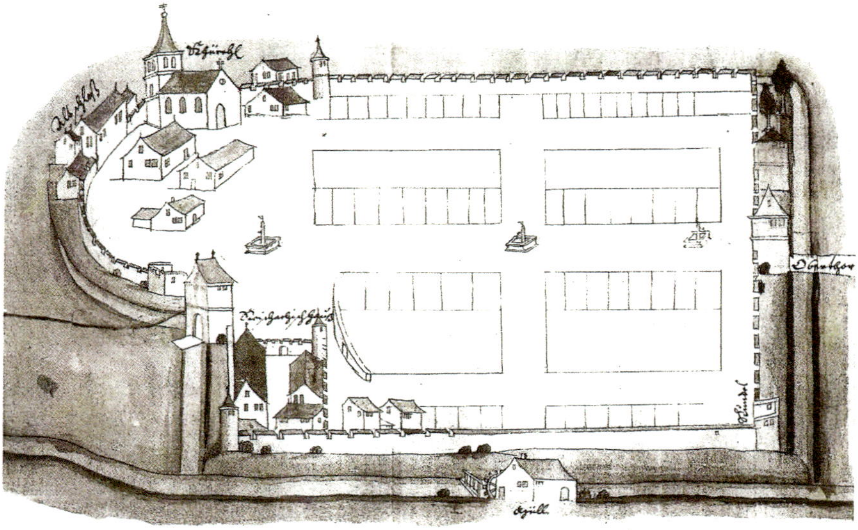

FIGURE 46 Map of the city of Elzach indicating the burned part, Landesregierungsarchiv Innsbruck, Ferdinandea 31, Freiheiten, Faszikel 35 Elzach

on actual ruins, the urban technical/architectonic view of the destroyed area of the city had been long established. A comprehensive longitudinal study is, however, still absent. Studies on urban cartography and on hazard mapping in the nineteenth century do exist,[254] as do studies of individual city fires and postfire cartography, zoning, and mapping along with other map types used in the Prussian bureaucracy of reconstruction.[255] For the purposes of the argument unfolding here concerning a long-term epochal change, however, I can only attempt to sketch some aspects and vertices of this development, focusing on the question of how certain temporal horizons and images of danger and safety have been inscribed onto images of the city. In addition to the development of the affect-related side of the disaster image and the temporalization of its representation, the question thus arises: what images of a city are created in a post-disaster situation and, potentially, in the anticipation of disasters?

---

254 Mark Tebeau, *Eating Smoke. Fire in Urban America, 1800–1950*, Baltimore 2003, pp. 89–125; Mark Monmonier, *Cartographies of Danger: Mapping Hazards in America*, Chicago 1997; Niels van Manen, "Les plans d'assurance incendie de Goad: Cartographie des risques d'incendie et normalisation des risques industriels (1885–1903)," *Le Mouvement Social* 249 (2014), pp. 163–85.

255 See below Ch. 5, 2.

One of the earliest representations (of the type that later became common) of a burned area that I am aware of comes from the city of Elzach in the Breisgau area of Anterior Austria—a territorial city under the rule of Archduke Ferdinand at the time of the fire (September 27, 1583) (Fig. 46). The fire destroyed almost the entire building stock of this small town of forty houses, five towers, and forty-three barns; as was often the case, these structures were still contained within the city walls. The fact that such a map was created here was probably due to the great distance between the city and its government (Elzach was located on the right bank of the Rhine and northeast of Freiburg/Br., while the Anterior Austrian government was in Ensisheim [on the left bank of the Rhine and north of Mulhouse/France] and the seat of the ruling Archduke was in Innsbruck).[256] The map was sent to the Archduke along with a request for donations to support reconstruction, and was intended to serve as documentation of the reconstruction plan.[257] The name of the artist is not known but its type corresponds to the aerial-view cityscape, specifically to that of the southern German late Renaissance painting of the handicraft milieu, as could also be found in Augsburg or Nuremberg. Elzach is located in the immediate vicinity of the Lordship of Triberg, whose ruler was Lazarus von Schwendi. Schwendi also pledged reconstruction aid for Elzach.[258] He was known to have read Machiavelli's early writings and was probably the most important military and tactical late-humanist advisor of Emperor Maximilian II. He led the siege of Magdeburg in 1550, and had, at this point, just organized the Anterior Austrian and Alsatian militia according to a late-Renaissance military model.[259] Without knowing more details about this specific fire among the many hundreds of city fires

256 On the politico-institutional context of the Austrian estates in this area, see Kurt Gustav Speck-Nagel, *Die vorderösterreichischen Landstände im 15. und 16. Jahrhundert*, 2 vols., Karlsruhe 1991.

257 Landesregierungsarchiv Innsbruck, Ferdinandea 31, Freiheiten, Faszikel 35 Elzach, attached to the letter of the council to Ferdinand on November 17, 1583, Josef Weber, *Zur Geschichte der Stadt Elzach*, Elzach 1978, pp. 52–59. The map is decidedly intended for illustrative purposes "welchermassen dies Stättlein Elzach wiederumb in guter Ordnung und Creuzgassen damit auf chünftiger Füersnot, die der Allmächtige allerorts verhüten wolle, bessere und gelegenere Rettung angeschafft und zu erbauen sein möchte." (ibid., p. 53).

258 Weber, *Elzach*, p. 55.

259 On Schwendi's reception of Machiavelli, see Lina Baillet, "Schwendi, lecteur de Machiavel," *Revue d'Alsace* 112 (1986), pp. 119–97; Thomas Nicklas, *Um Macht und Einheit des Reiches. Konzeption und Wirklichkeit der Politik bei Lazarus von Schwendi (1522–1583)*, Husum 1995; on Schwendi's contribution to the *Landrettung* (defense organization of the country) of Upper Austria, see Cornel Zwierlein: *Discorso und Lex Dei. Die Entstehung neuer Denkrahmen im 16. Jahrhundert und die Wahrnehmung der französischen Religionskriege in Italien und Deutschland*, Göttingen 2006, pp. 738–50.

and reconstruction organizations, it is not surprising, in this context, that the mapping of the empty reconstruction area, including designating foundations for rebuilding, reveals the influence of a rudimentary chorographically and geographically trained form of a rather "southern" Renaissance provenance—the artist was presumably trained in mathematical measurement practices. The map, in light of its finishing technique, is certainly comparable to works by surveyors—for example, those by Tilemann Stella during this time in southern and central Germany.[260] These reflections are necessary to contextualize the map's exceptionality: it is surprisingly early example of this form of recording. More elaborate aerial-view depictions of burned areas and the subsequent mapping of burned sections and the parceled plans for reconstruction, which thus point toward the future, are otherwise common only at the beginning of the seventeenth century—at least for central Europe and according to all the related literature of southern and western Europe regarding maps and legends.[261]

A map some decades later, which shows a clear developmental step and essentially already reflects the type common throughout the seventeenth and eighteenth centuries, is from the Lower Saxon town of Herford in 1638. Although Herford had formally declared its independence from the empire in the wake of the Thirty Years' War, it was quickly subject to loss of sovereignty and occupation again. In the middle of the war, a major fire broke out on July 25 *stylo veteri* (August 4/5 Gregorian) which, uncharacteristically, was not war related. The city issued a letter of appeal to the Duke of Braunschweig-Lüneburg, who had served since 1637 as patron, asking him to intercede with the emperor to request aid. Along with this letter, a map was included, clearly marking the areas of the burned buildings in red, apparently in order to visually illustrate for the Duke the damage that was later notarially estimated at

260 The classical case study for a German territorial "Landeserfassung" (survey) is Kersten Krüger, *Finanzstaat Hessen 1500–1567. Staatsbildung im Übergang vom Domänenstaat zum Steuerstaat*, Marburg 1980, which analyzes the developments even prior to the "Ökonomische Staat" (economic state [however meaning 'budget' or 'survey']) of Wilhelm IV, the first German survey that could compete with western and southern European comparative cases. For the cartographic work that was produced in the same contexts and with similar purposes, see Axelle Chassagnette, "Mettre en carte: fonctions pratiques, intellectuelles et sociales de la cartographie allemande au XVIe siècle," in Isabelle Pantin and Gérald Péoux (eds.), *Mise en forme des savoirs à la Renaissance. À la croisée des idées, des techniques et des publics*, Paris 2013.

261 One might presume that something similar has been already in use in Renaissance Italy —as the idea and sketches of (undestroyed) planned cities (*Planstadt*) were common there since at least the early sixteenth century—however, I have found no hint of this either in sources or in the literature related to the city, art, or the history of disaster until today.

FIGURE 47 *Plan der Stadt Hervordt Mitt Ihren Strassen vnd Gebäwen, Auch Was daran verbrandt, so geschen den 25 July A.o 1638*, Map of Herford after the fire, showing the part destroyed in red, Staatsarchiv Hannover, Notariatsinstrument des Hermann Schmackepepfer von 1643

150,000 Rtl. —just as the Elzach map served to illustrate the damage for the distant Archduke Ferdinand. Unlike in Elzach, the Herford map is not characterized as a blueprint for reconstruction. However, the Herford map was apparently able to draw on an earlier, recently completed town survey, which was created by the engineer Antonius Bose during the reconstruction of the defensive city wall in 1626; thus, the mapping of the burned areas corresponded to the then most modern techniques of the art of military surveying (Figure 47). It was not until 1827 that another survey-based city map was produced; this gap in surveying was not atypical for such small towns.[262]

This type of fire damage map is absolutely identical to maps sketching areas of fire damage known from news broadsheets concerning the most famous early modern city fire—London 1666. In the case of London, the aerial-view maps with markings of the areas damaged by fire are essentially derived from

262 Johannes Kretzschmar, "Zur Geschichte Herfords im 30jährigen Kriege. Mit einem Plane der Stadt von 1638," *Westfälische Zeitschrift* 58 (1900), pp. 5–29.

engravings made directly after the fire.[263] The engraving made by Wenceslaus Hollar after a very quick survey in December 1666 (printed in 1667) is most well-known[264] and served as the starting point for the competing reconstruction plans by Wren, Eveyln and Knight.[265] These engravings were based on accurate surveys and visual maps, which already existed for a big metropolis like London. Measurements of the burned areas were taken during the reconstruction process by Robert Hooke and others, but since this process took over two decades,[266] these measurements did not form the basis for the ground zero images immediately published after the fire in 1666.

Initially, the function of these early visualizations of the damaged areas in aerial-view city maps was primarily to illustrate the massive damage, in particular, for the purpose of advertising for support and preparing to solicit donations—they served only secondarily as the rough basis for mapping sketches for future planning in the London case. In small towns such maps emerged as a response to the distanced communication between city magistrates and ducal patrons: cities needed to present an image of themselves and their damaged state to the lords who were "looking at them." It was no accident that the first of these maps appeared, in the case of London, on a Dutch broadsheet, thus within the framework of early modern news and solicitation communication. As late as 1726 the ground zero image for the Reutlingen fire of 1726 was expressly formulated as a "begging letter" by Augsburg engravers—thus detached from the city itself. In contrast to contemporaneous Prussian situation maps, it also had no relation to the reconstruction; it was thus intended purely for the purpose of providing a visualization of the dramatic range of destruction in order to solicit donations.[267] The presence of earlier such examples always seems to be related to the random, or in the case of a metropolis

263 Probably the earliest "ground zero" of the London fire is illustrated on a Dutch broadsheet by Marcus Willemsz Doornick, printed in 1666, Michael Cooper, *"A More Beautiful City": Robert Hooke and the Rebuilding of London after the Great Fire*, Stroud 2003, p. 102.

264 Repeatedly reproduced, for example in Derek Keene, "Fire in London: Destruction and Reconstruction, A.D. 982–1676," in Körner, *Wiederaufbau*, pp. 187–211, 190 and Tinniswood, *By Permission of Heaven*, Fig. 17.

265 These are reproduced in Stephen Porter, *The Great Fire of London*; Cooper, *A More Beautiful City;* and Tinniswood, *By Permission*, figs. 18–21.

266 For surveying work after the fire, see, especially, Cooper, *A More Beautiful City*, pp. 95–220 and Michael Cooper, "Robert Hooke's work as surveyor for the City of London in the aftermath of the Great Fire," *Notes and Records of the Royal Society of London* 51 (1997), pp. 161–74; 52 (1998), pp. 25–38, 205–220.

267 The electoral/Bavarian councilor Johann Thomas von Rauner, whose own feudal property made him a neighbor to the city of Reutlingen, commissioned the engraving with the Augsburg engraver Gabriel Bodenehr for the purposes of inspiring donations; see Schwarz, "Der große Reutlinger Stadtbrand," p. 21.

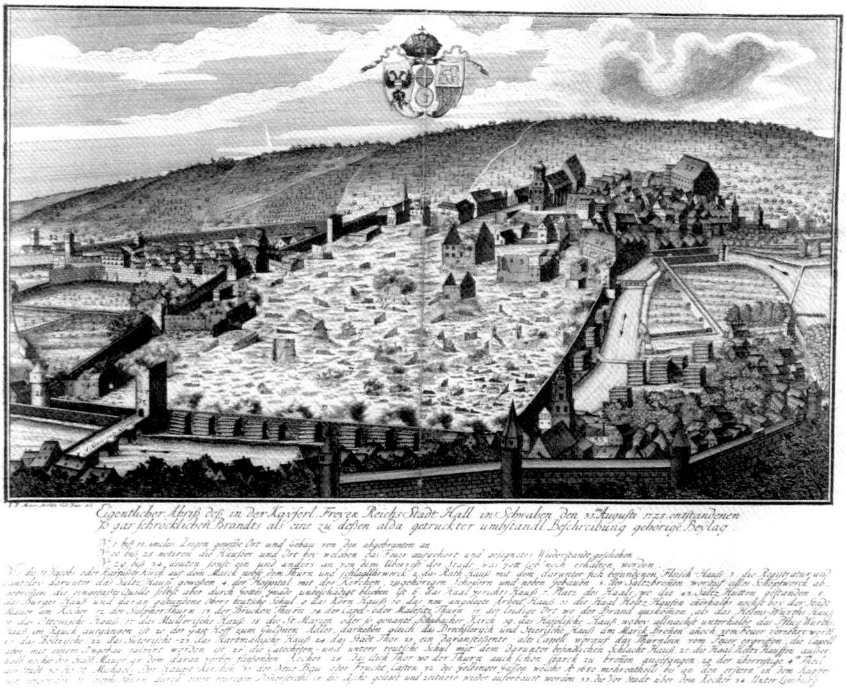

FIGURE 48 Engraving of Schwäbisch Hall's *Ground Zero* after the city fire of 1728 by the architect I.P. Meyer (Stadtarchiv Schwäbisch Hall, Grafiksammlung S10/ 2340, thanks to Sara Wiest for sending a reproduction)

like London easily explainable, availability of a current survey associated with mapping the city for some purpose.

In addition to such aerial-view renditions marking damaged areas as ground zeros in survey-like city maps and representations, there were also prospectus-type representations of burned cities, which appeared with any great regularity only after the mid-seventeenth century (early eighteenth century for central Europe). These featured ruins in copper engravings and in oil paintings—beginning, in this sense, with Hollar's London prospectus of the city silhouette.[268] The best examples for Germany for our purposes are Schwäbisch Hall (1728, Figure 48), Gera (in 1686 as well as 1780, Figure 49, 50), and Göppingen (1782, see above Figure 24).

268 This is currently in the British Museum; the woodcut has been reproduced in Timothy Baker, *London. Rebuilding the City after the Great Fire*, London 2000, p. 2f.

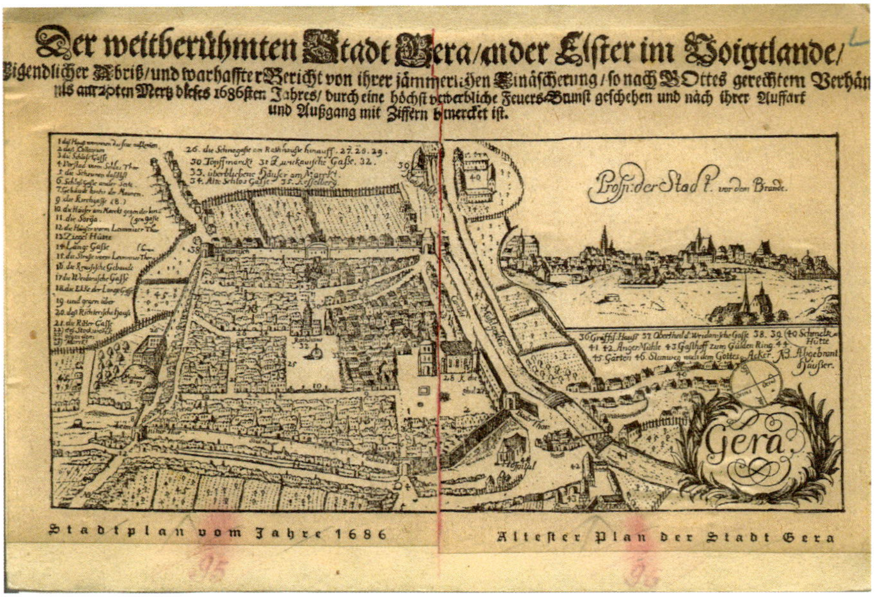

FIGURE 49 Engraving on a single-leaf broadsheet *Der weitberühmten Stadt Gera/ an der Elster im Voigtlande/ Eigendlicher Abriß* [...] Leipzig: Justinus Brandt 1686, Stadtarchiv Gera, A 1505b (facsimile reprint twentieth century, A0105)

In Schwäbisch Hall, we see the city's coat of arms, with the imperial crown and the double-headed eagle, both signs of the Holy Roman Empire and of the city's immediate status, above the disastrous scene of destruction. The picture has, evidently, no immediate technical function for the rebuilding of the city as in other cases discussed below, but it serves as a memory of that sorrowful moment of complete annihilation of one of the most important south-western imperial cities—and this also in the form of a deep-strata memory: while now the center is completely destroyed, the engraver remembers the part to the left which had resisted the flames this time (number 32, the *Gelbinger Gasse*), when it had already been destroyed in 1680. Thus, the city consists nearly more or less of nothing, nothing that concerns material existence on earth is firm and stable, even the parts that still seem to be strong and stable, built of stone, are just built on older ground zeros: a city consists not of deeper and deeper strata of stone, earth, memories and life, but of a succession of layers of ground zeros.

However, it seems typical that such sophisticated representations of the zeroness of the city, its reduction to the immateriality of its living community by force of the material destruction of its stock of houses, is to be found perhaps most in such small and mid-size cities that retained their immediate status and remained

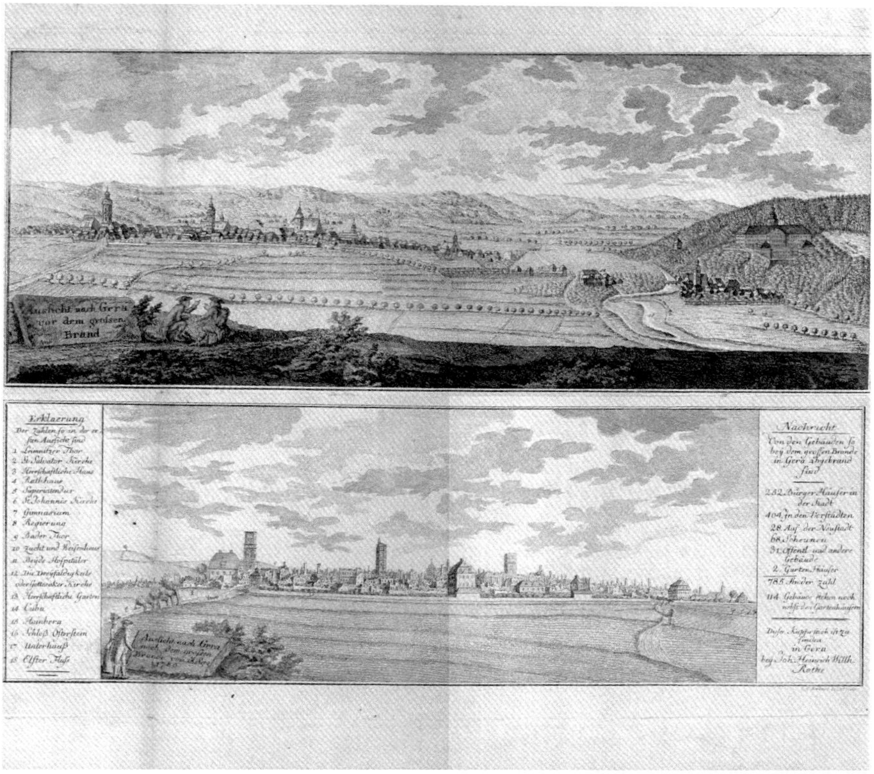

FIGURE 50 Engraving *Aussicht nach Gera vor dem grossen Brand—Nachricht von den Gebäuden so bey dem grossen Brande in Gera abgebrand sind*, Gera 1780, by G.G. Endner (sculp.), printed by Johann Heinrich Roth, Stadtarchiv Gera, A5630

quite proud and self-conscious as is the case of the southwestern imperial and proto-bourgeois cities, who since the Middle Ages had a tradition of cultivating their individual identity as a collective body of citizens. In contrast, there are no such (or far less or only sometimes exceptionally) equivalent representations for the many hundreds of small towns of northern and northeastern Germany that burned after 1650: Gera distinguishes itself here from other country towns as a city of residence for the Count of Reuss—it was not one of the small-scale territorial cities of the large Prussian or Saxon territories that were administrated more or less in an equalizing manner. This representational form, which has no functional relationship to reconstruction planning, but rather only served memorial purposes by invoking consternation and dismay and horrific memories of the past disaster, appears to arise only in those municipal communities which practiced an "early-civic" cultivation of city consciousness and collective memory.

The elaborate ground zero memorial belongs to those cities, which Walter Mack has characterized as "hometowns", going beyond the constitutional typology of imperial, free, and rural towns and cities. These were often located in the southwestern region;[269] the almost generic, state-governed and -engineered small towns in the north and northeast were fully subject to the lords of that region.

In the north-east, however, other developments were special and to some extent were advancing and innovative with regard to the home-towns of the south-west. In Prussia, earlier than in the largest south-western territory of Württemberg, a kind of professionalization and bureaucratization of the reconstruction administration emerged in the cameralistic management that was not found elsewhere in western Europe. This led to a standardization of the map-like representation of cities in case of fire, which transformed a possibility that was only a happy accident before 1650 into standard and normal practice: after every fire, the extent of damage was surveyed and entered onto a situation map, which was then combined with a reconstruction map for future planning, either in a single map using several colors or in two separate maps. This will be discussed in the next section (Ch. 3, 3) as an example of yet another aspect of the reconstruction process in Prussia, which later took place in other German territories, especially in the second half of the eighteenth century. Due to the high cost, this process was more effectively implemented through the overarching state in territorial cities than it was in the many small, poor, southwestern German towns of the seventeenth and eighteenth century, which were not supported by any rich country lord. Concerning the standard application of fastidious engineering surveys of such burned cities, these otherwise hardly significant communities paradoxically possessed a higher status in terms of implementable innovations and administrative techniques than many of the traditional imperial cities of the South. Curiously, there is also no comparable standardized bureaucratization of these operations leading to a standardization of these ground zero images known for France[270]—even though France was recognized as the leader, or at least a very active player, in terms of both engineering and surveying technology[271] and also regarding tendencies toward the increasing intervention of the state in post-disaster aid during the

269 Mack Walter, *German home towns: community, state, and general estate, 1648–1871*, 2nd ed., intro. James J. Sheehan, Ithaca 1998; Cf. the special issue *Central European History* 47, (2014), ed. Robert Mark Spaulding.

270 Similarly, in France there was also no state administrative system of 'prophylactic' anticipating fire funds. Cf. below Ch. 5, 1.1.

271 Josef W. Konvitz, *Cartography in France, 1660–1848: science, engineering, and statecraft*, Chicago 1987; Hélène Vérin, *La gloire des ingénieurs. L'intelligence technique du XVIe au XVIIIe siècle*, Paris 1993.

eighteenth century.[272] It is thus not the cities that abstractly created their own images of damage and sent them to the city rulers, but rather the sovereign who ruled those cities (approximately 78 of which belong to the Kurmark and Neumark areas, to mention the Prussian central provinces alone) as objects to be managed who created these maps—the ground zero was seen from this high and central administrative perspective, which surveyed the urban network of the entire territory. The aerial perspective,[273] which has been common since the Italian Renaissance and particularly since the rise of Dutch city cartography (extending even to aerial views with the painting in of cloud shadows long before airplanes could indeed take such photographs in reality)[274] in the seventeenth century, was popularized by and for self-assured city republics. Insofar as the thriving city could be captured as a whole, the standardized form of the ground zero image as the mapping of the aggregate area of destruction is found precisely where the view from above corresponds to a bureaucratic-planning view of the ruler of *his* cities and his welfare-like management of them.[275]

Whether images of ruins and fires served a memorial function or served an administrative and city planning function as a map of the situation and

---

272 Quenet, *Tremblements*, pp. 265–305.

273 For the corresponding history of the development of the concept of the metropolis, see Cornel Zwierlein, "Durée, stabilité et grandeur urbaine: de la cité humaniste à la métropole moderne," in Élisabeth Crouzet-Pavan, Denis Crouzet, and Philippe Desan (eds.), *Cités humanistes, cités politiques (1400–1600,)*, Paris 2014, pp. 237–51.

274 Peeters, "De stad".

275 Jaeckel presents a number of postfire maps in the appendix of his unfortunately unpublished thesis. These maps, are based, in part on earlier surveys of the mediate and immediate cities of Kurmark and Neumark starting with the 1720s for Crossen in 1721, Luckenwalde in 1724, and then resuming in midcentury. Jaeckel's dissertation advisor Reinisch explored the case of Neuruppin in his own monograph, which also reproduces the relevant maps from 1787 (Ulrich Reinisch, *Der Wiederaufbau der Stadt Neuruppin nach dem großen Brand von 1787*, Worms 2001). Thomas Spohn's " 'Sollen recht ordentlich bauen sonder Resoniren. F W'—Hausbau und Stadtplanung unter preußischem Einfluß, dargelegt am Wiederaufbau der märkischen Städte und Flecken im 18. Jahrhundert," *Der Märker* 39 (1990), pp. 191–206, 250–61 presents similar maps following the same pattern for Prussian administrated Westfalian cities, including Schwelm 1722, Unna 1723, Plettenberg 1724, Breckerfeld 1727, Hagen 1724, a four-colored, relatively carefully drawn map by mathematics student Nicolaus Stell for Hamm 1734 and a rough idealized map of reconstruction for Lüdenscheid 1724. Additionally, one could mention the destruction and reconstruction maps for Lebus 1723, conceived according to the same bureaucratic pattern (rendered as an engraving in Johann Christian Wellmann, *Das von der göttlichen Regierung an denen Mordbrennern, welche in der Nacht zwischen den 19. und 20. May 1723 die Lebusische Vorstadt zu Franckfurt an der Oder* [...] *angesteckt bewiesene Denk-Mahl* [...], Frankfurt an der Oder 1725, f. b3V); on the Lebus fire of 1801 the files in BLHP Rep. 2 S 5526 contain information about the preparation of the situation map by Conducteur

reconstruction, the premoderns were only aware of the depiction of *past* damages, and future images of cities to be built anew—the destruction was in the past and the future was worthy of being illustrated as a complete city. In the modern era, in contrast, a further, basically new type of image developed where different forms of potential future destruction are inscribed into the city and its representations. *Risk* or *hazard mapping* is a visualized anticipation of the possible ground zero of a city and was first developed in its embryonic form by English fire insurers in the last third of the eighteenth century.[276] Examples of such map types will be introduced in the last chapter, when the global expansion of English insurers will be addressed in order to conceptualize this new epochal threshold concerning the relationship between fire and safety in modernity. If one would not explain the colors differently, it would be possible to judge the 1638 Herford map with red markings based on a previously existing map of the city (see above, Figure 47), and for instance the far later 1858 map of the Istanbul district of Pera with red markings of relevant quarters on a printed map engraved by Stolpe through Agent Wood of the Sun Insurance Company, as belonging to exactly the same type of map (see below, Ch. 6, 2, Figure 69). In one case, however, the damage shown had already occurred; in the other case, future prognostics were sketched on and into the map: houses judged to be definitively less risky and better-built are shown, which, in the anticipation of a possible disaster, thus appear to be more resistant to damage. Risk mapping quickly assumed more sophisticated forms, and cities were separated into zones of different levels of risk—that is, in terms of their potential vulnerability to destruction.

Using immensely more powerful visualization techniques, today's forms of electronic hazard mapping capture a variety of risks and future scenarios that can be projected onto city spaces as well as other spaces—concerning not only fire or flood damage risks, but also zones denoting a higher risk of racist violence and other forms of risk. The potential of the disaster and the ground zero state is always visually inscribed within the conception of the present and future state of a city.[277] There is little research specifying which cartographic

Lietzmann; from the context of the Seven Years' War regarding Küstrin 1758 the damage reports serve as preparatory work for the survey in BLHP Rep 3 Nr. 10603/1; for Peitz 1758 there is still a map in a file archiving a similar procedure in BLHP Rep 3 Nr. 10376. For other German regions there are very similar maps; for example, Tübingen 1789, Tuttlingen 1803 (fire reconstruction and geometrical reconstruction maps), the Electoral Trier-Nassau-Oranienburgian village of Camberg 1798, and the Electoral Bavarian-Sulzbachian Markt of Erbendorf 1796. In the late eighteenth century this becomes the common pattern in all regions and also the smallest dominions. The temporal development and the special position of the Prussian administration in this area is, in any case, clear, in spite of incomplete source preservation and further gaps in the responses to the inquiry of 1800 communities.

276 Van Manen, *Les plans*, p. 171f.

277 Cf. note 261 above.

models were used to create the first fire insurance risk maps—in central Europe there are no surviving maps from the time of the cameralist fire funds: cameralist administrations did not produce such anticipations of zones that were more likely to be damaged in future than others. However, it can be shown that an intermediate, transitional element of this development is obviously a differentiation and refinement of reconstruction maps, which were crafted *at the moment* of the destruction and aimed at consequent reconstruction—there was no projection of future scenarios onto the map *independently* of the destruction. The degree of detail of the damage measured and plotted on the map increased, however, and the way to represent the current state of ground zero after the disaster became more sophisticated. Nicolaus Stell, a student of mathematics, drew a map of Hamm after 43% of the 456 houses of the city had burned down on September 22, 1734, distinguishing with four colors the burnt houses (red), the remaining (blue), the zone of garden and unbuilt zones (yellow) and the public church places, the cemetery and the streets (white).[278] He did not distinguish the types of building, however, nor degrees of destruction, nor did he integrate the measurements of the estates (Figure 51).

W. Dornstein drew a plan of the town of Peitz, destroyed on December 24, 1758 by order of the Kurmark War and Domaine Chamber (*Kriegs- und Domänen-Kammer*), the mediate administrative institution existing for each Prussian province depending on the central *Generaldirectorium*: he distinguished between houses in parts not at all touched by the fire (in Latin numbers) and those hit by the fire (in Arabic numbers). He marked those houses better preserved than the others within the zone hit by the fire with red ink. He measured all the ground-plots of each estate, listing both types of houses in a large legend. A picture of the city emerged which offered everyone one quick look at a comprehensive representation of the damage (Figure 52).

Some of the more sophisticated situation maps distinguish building materials/types quite precisely, and in particular the roofing material (straw or stone), and thereby point out a fundamental fact of fire risk in the buildings still standing. Reconstruction often ordered or encouraged stone buildings in the place of older half-timbered and thatched-roofed construction, and thus a lower or higher future fire risk for the remaining part of the city was visualized—as, for example, in 1756 for the Mecklenburg Grevesmühlen in 1756 (Figure 53):[279]

278 Heinz Stoob, "Grundrißbild und Entwicklung der Altstadt Hamm bis 1830," in Herbert Zink (ed.), *750 Jahre Stadt Hamm*, Hamm 1971, p. 21.

279 The fire has nothing to do with the Seven Years' war as the date might suggest: on March 25, 1756 the former shoemaker Christian Miltz, who maintained himself as a distiller of brandy, had caused the fire by careless handling of his distillery, 59 houses, 8 barns and many stables were burnt. By order of the Duke of Mecklenburg, the detailed plan of the

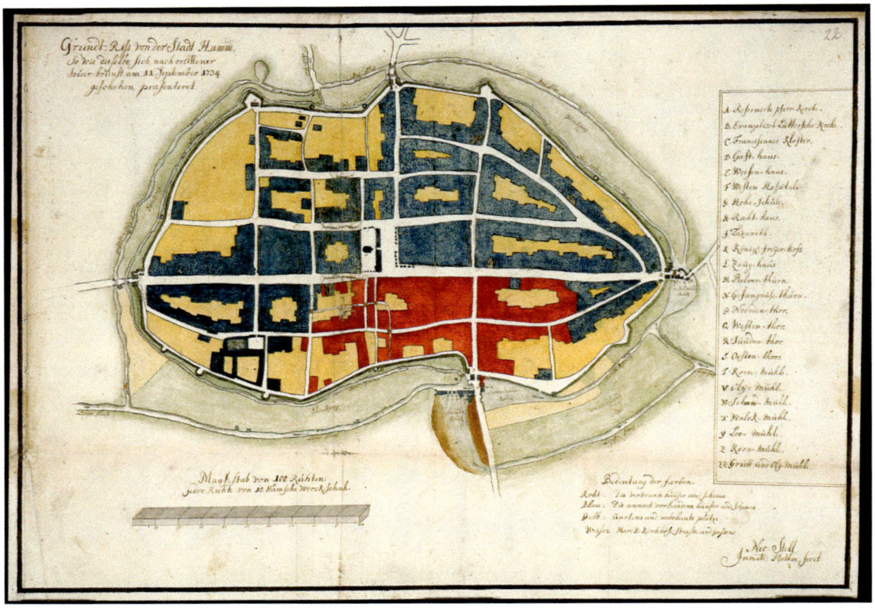

FIGURE 51 Groundplot of the city of Hamm (*Grundt-Riss von der Stadt Hamm, so wie dieselbe sich nach erlittener Feuer-brunst am 11. September 1734 geschehen praesentiret*) drawn by Nicolaus Stell, cf. Spohn "Sollen recht ordentlich bauen", p. 201

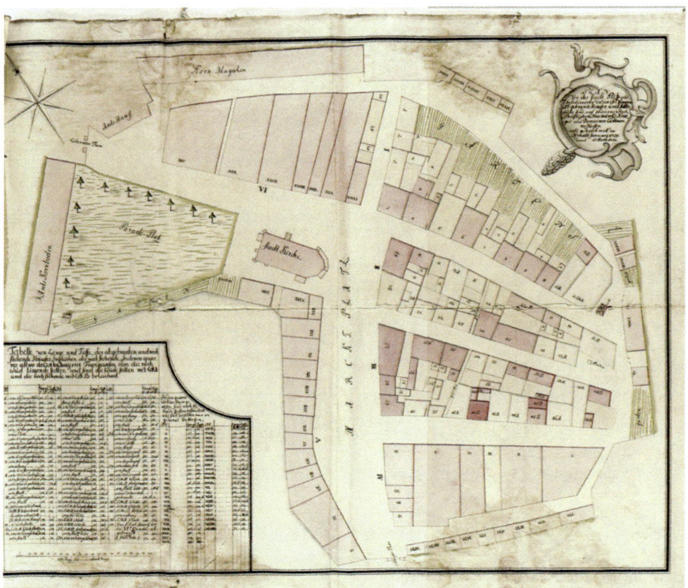

FIGURE 52 Plan of zones of destruction and possibility of rebuilding by W. Dornstein of the Fortified City (*Festungsstadt*) of Peitz after the fire of June 23, 1758, BLHA Potsdam, Rep. 3, Nr. 10376

FIGURE 53 Map of Grevesmühlen (*Plan der Stadt Grevismühlen*), 80x60 cm, drawn by military surveyor and artillery lieutenant J.D.F. Blume, early June 1756, Landeshauptarchiv Schwerin 2.21-1 Geheimes Staatsministerium und Regierung, Nr. 2009

The mapping expert from the ducal army's artillery distinguished six types of houses next to the zones of burned houses ("die Brandstellen"): 1) living houses, 2) barns, and 3) stables with each either "stone roof [*Steindach*]" (tiled or brick roof), or houses with thatched roof ("*Strohdach*"): although this was a post-catastrophe map, it was still mapping—somehow and in a very rough way—the present and future degree of risk for given zones of the city.

Narrative and tabular explanations of past destructions are, of course, common throughout the eighteenth century, but visualizations are by no means common and not yet systematized—even less so for small independent cities or those in territories which had not yet developed a systematic form and bureaucratic scheme of recovery and rebuilding as was the case in Prussia since the early eighteenth century (cf. below Ch. 5, 1 and Ch. 5, 2). However, within the cameralist state management of central Europe, starting points do appear for the emergence of such a new administrative image of risk for cities and of a new level of complexity and interconnection of the time horizons concerning the visualization of "disaster." The image of the *possibility* of future disasters is beginning to be constantly co-present.

In addition to the previously addressed development of the autonomy of the disaster as the autarchic image-subject itself (Ch. 3, 2.2), and to the refining of the *affect* control on the side of *appeal* and *expression*, this important development at the level of disaster consciousness and the simultaneous continual visual processing of "disaster potential" thus emerges as a further decisive element on the *cognitive* side. Taken together, a striking change in the visualization of city fires and fire disasters manifests itself from the late Middle Ages into the early nineteenth century, once again with a particular break or phase of concentration in the developmental trajectory at the end of the seventeenth and the beginning of the eighteenth century.

3 Developmental Trends of "Real-Assecuration": Fire *Policey*, Construction

The previous section offered an important level of analysis that involved tracing the emergence and changes associated with how the threat of fire was perceived. We will now shift our attention to the ways in which this threat was approached and managed practically. Fire *policey* constitutes the classical operational field of these practical efforts. As mentioned at the start, Leibniz ingeniously labeled this area of security production as "real-assecuration," (i.e., security of and in physical reality) in contrast to "verbal-assecuration" (i.e., security granted by insurance institutions) (cf. Ch. 4, 1.2). The procedures related to the production of security before and after a putative disaster are both quite distinct and also systematically tied to each other. The major sections of

city was drawn by the artillery lieutenant Blume who was paid 27 Reichstaler and 43 shillings, Landeshauptarchiv Schwerin 2.21-1 Geheimes Staatsministerium und Regierung, Nr. 2009.

Chapters 4 and 5 will address the transformation of fire prevention and firefighting specifically in this sense, in the context of the development of a new insurance-based system of security, which appeared around 1680–1700.

3.1 *The Security Regimes between the Late Middle Ages and the Enlightenment and between the City and the Territories*

From the smallest of villages to the largest of cities, all premodern communities were keenly interested in fire prevention as one of the most important matters influencing basic survival. In principle, the fire *policey* measures implemented were always the same: there was a specific set of firefighting equipment—a triad of leather buckets, hooks, and ladders. While the first rudimentary precursors of city fire ordinances had emerged by the late fourteenth century, in the fifteenth century, they began to acquire a highly detailed form—for example, the Cologne ordinances of 1452, which served as the foundation of the entire early modern development of fire *policey* in that imperial city. Primarily, this ordinance divided the city into areas of responsibility, often according to parish districts. It also detailed storage locations for firefighting equipment and the duties of the citizen for the maintenance thereof.

Policey research has paid relatively little attention to fire *policey* and in more recent times it has investigated it almost entirely in respect to the territories, in spite of the fact that the most significant innovations in this area until well into the seventeenth century clearly concerned cities.[280] The main tasks regulated by fire *policey* were the constant evaluation of the functional capacity of firefighting equipment (i.e., eliminating buckets with holes and broken hooks and ladders), replacing it when necessary and, beginning in the eighteenth century, maintaining large fire hoses, the so-called "serpentine fire hoses" (*Schlangenfeuerspritzen*). Fire ordinances were often related to building ordinances, watchman ordinances, fire fund ordinances, and insurance ordinances. While in earlier years they were frequently integrated into one single ordinance, in

280 M. Ventzke, "Fürsten als Feuerbekämpfer: Handlungsmotive einer sich wandelnden Hofgesellschaft am Ende des 18. Jahrhunderts," in Id. (ed.) *Hofkultur und aufklärerische Reformen in Thüringen. Die Bedeutung des Hofes im späten 18. Jahrhundert*, Cologne 2002, pp. 223–35; André Holenstein, "'Gute Policey' und lokale Gesellschaft. Erfahrung als Kategorie im Verwaltungshandeln des 18. Jahrhunderts," in *"Erfahrung" als Kategorie der Frühneuzeitgeschichte*, Munich 2001, pp. 433–50; Achim Landwehr, *Policey im Alltag. Die Implementation frühneuzeitlicher Policeyordnungen in Leonberg*, Frankfurt am Main 2001, pp. 262–273; André P. Holenstein, *"Gute Policey" und lokale Gesellschaft im Staat des Ancien Régime. Das Fallbeispiel der Markgrafschaft Baden(-Durlach)*, Tübingen 2003, pp. 727–49.

later periods they were differentiated according to specific subject matter. The ordinances reflected different levels of expectation and horizons of concern—for example, concern about fires caused by tobacco smoking in the eighteenth century.[281] In order to provide a reasonably representative impression, I will compare two cities, Cologne and Hamburg, and a territory—the Prussian central province of Kurmark.

3.1.1 Cologne: An Important Medieval Imperial City[282]

In Cologne, the first fire ordinance, which seems to have had relatively little impact, dates from the year 1360[283] and was written not on parchment, but already on a piece of paper.[284] However, the foundational fire ordinance was not written until June 21, 1451, and it was this ordinance that ultimately ruled the entire early modern period—its effects were still relevant in the city's Prussian era of the nineteenth century.[285] This fire ordinance, originally written in Low German, formed the initial part of the so-called "fire scrolls" (*Brandrolle*), as translated into official High German—the collections of norms, council decisions, and instructions issued to the four city "fire lords" (*Brandherren*). Further decisions made by the magistrate relating to fire prevention were added later in accordance with article 17 of the old fire ordinance.[286] The structure of this ordinance was based on the sixteen, and later twenty, Cologne parishes.[287] With the church heads as leaders, each parish was required to provide and maintain firefighting devices (282 buckets, 66 ladders, 35 hooks). This number was increased in the eighteenth century (as of 1729: 744 buckets, 72 ladders, 43 hooks). From 1711 on, city militia captains were also obliged to provide 12

281 Nassau-Weil fire ordinance 1751.

282 The archival research for this section was carried out in 2007/08; verifying some details was no longer possible due to the collapse of the Cologne city archives.

283 Karl Härter (ed.), *Repertorium der Policeyordnungen der Frühen Neuzeit*, vol. 6, *Reichsstädte 2: Köln*, 2 vols., ed. Klaus Militzer, Frankfurt am Main 2005, no. 81; Walter Stein (ed.), *Akten zur Geschichte der Verfassung und Verwaltung der Stadt Köln im 14. Und 15. Jahrhundert*, 2 vols. Bonn 1893–95, vol. 2, pp. 26–28.

284 Paper was introduced just around that time in Cologne, the first paper document in the archive dates from 1343. Thus, fire *policey* administration obviously belongs to the new segments of municipal administration that needed frequent communication and that were handled now by help of the new writing material, cf. Zwierlein, "Gegenwartshorizonte im Mittelalter," p. 34.

285 Stein, *Akten*, vol. 2, pp. 365–69.

286 For example, in HStA Köln V+V N17, f. 11v-12v; N21, f. 12v-13r; N23, f. 11r.

287 In the fire scrolls of Hieronymus Schophorn (1776–79) the numbers for 20 parishes are significantly altered in contrast to the 16 parishes of the old order, HStA Köln V+V N 23; in the fire scrolls of Johann Frantz Hyacinth Canot of 1776, the old numbers are listed in f. 3r, as in N 21, f. 7v.

buckets each (at 54 captains, 648 additional buckets).[288] From 1741 on,[289] the 25 guilds were also required to provide 30 buckets, one ladder, and one hook (thus 725 buckets, 25 ladders, 25 hooks). Visitation logs and notes from 1729, 1742, 1753, 1763, 1769, 1787, and 1794 show that these target values were, for the most part, not attained; either because the leather buckets were rotten and unusable or because there were simply not enough present.[290]

The maintenance of firefighting equipment was flanked by many, in part continually repetitive, edicts. For example, there was an edict against possessing uncontrolled amounts of gunpowder (April 26, 1583),[291] and one regulating the possession of gunpowder (not more than 12 pounds per private residence, March 29, 1647, renewed on August 20, 1787. Regulations changed under French occupation (dated 6 Vendemiaire an III), returning to the 12-pound regulation in the Prussian ordinance of January 11, 1815).[292] There was also an edict prohibiting the dangerous Cologne custom of a fire on the night and day of the "St. Peter ad vincula" (St. Peter in Chains) celebration on August 1 of every year to commemorate the liberation of St. Peter from prison (Acts 12: 1–11).[293] In periods of great drought and heat, the council also regularly ordered every house to provide a water barrel in front of the house for fire-extinguishing purposes, imposing a fine of one gold florin for noncompliance.[294]

Beyond these regulations, the fire ordinance was also flanked by a published "alarm" (*Auflauf*) ordinance, i.e. an ordinance regulating the behavior of fire watchmen, as well as the organization and logistics of fire extinguishing methods—while not dated, it is probably from about 1580. Accordingly, the fire bell at the city hall was supposed to be rung "henceforth no longer for a single fire but only when multiple fires break out at the same time."[295] As in every city,

288 HStA Köln V+V N 12, p. 18.

289 HStA Köln V+V N 860, f. 6v.

290 In 1729 150 buckets were lacking; in 1742 there were a total of 576; in 1753 355; in 1769, 716 buckets, 101 ladders, and 82 hooks were lacking; in 1787 there were 348 buckets, 41 ladders, and 25 hooks lacking; HStA Köln V+V N 336, N 860, N 862.

291 HStA Köln Edikte 6 Nr. 44.

292 HStA Köln Edikte 6, Nr. 45–48.

293 July 21, 1581, July 25, 1590, July 29, 1680, July 20, 1681, July 26, 1684, July 20, 1689, July 28, 1698, HStA Köln, Edikte 6, Nr. 138, 139, 172, 178, 179; Edikte 27 f. 38r-v, 196r-v. Cf. Bächtold-Stäubli (ed.), *Handwörterbuch des Aberglaubens*, 10 vols., Berlin 1929–42, vol. 6: 1526.

294 Other than the St. Peter ad vincula edicts and also November 11, 1580, June 27, 1659, July 1, 1684, July 28, 1688 (HStA Köln, Edikte 6 Nr. 171, 173, 174, 175.).

295 HStA Köln Edikte 6, Nr. 140, Art. 7; Härter and Militzer, *Repertorium*, no. 2067—dated 1550 without any clear justification; through comparision with other ordinances in this group of edicts, I would estimate the date of printing to be closer to the 1580s. Further printings: January 5, 1604, March 25, 1633, March 12, 1635, January 30, 1642, March 14, 1659, October 6, 1688 (Härter and Militzer, *Repertorium*, no. 141, 142, 143, 157).

an especially important developmental threshold in fire *policey* was the acquisition and installation of the new "serpentine fire hoses" (*Schlangenfeuerspritzen*). In an edict from June 10, 1665, there was already talk of "water sprayers" (*Wassersprützen*), and an admonition that "in every parish one of those commonly useful sprayers should be placed, for which donations should be collected."[296] Of course, leather fire hoses cannot yet be intended here. When in a decree of February 1, 1679, the advice of the parishes was given to establish "for common necessity, a water sprayer" through donations, it appears only as a repetition of the older ordinance, probably regarding pipe sprayers without leather hoses. This was, however, obviously not heeded, since on September 9, 1722, an edict was released that it "would be highly necessary that the fire and water sprayers should be purchased with accompanying leather channels" and, accordingly, the council once again decreed "that one should be bought for each parish and militia unit"[297]

The decisive input of practical knowledge came in 1682: on April 27 it is noted in the council protocol that "the model of a large hose, useful in times of fire … from Amsterdam up to now described and engraved in copper [has been] brought forward." This discovery was then conveyed to the treasurer and the council-affiliated Fire Chief Owell for evaluation.[298] In fact, these engravings, used by the brothers Jan and Samuel van der Heyden to extol the virtues of their invention and to advertise the pressure-pump water sprayer with movable leather hose ("snake"), can still be found in the Cologne city archive (Fig. 54-57). Though the air chamber that allowed the continous flow of water while pumping was an earlier invention,[299] the combination of the whole together with the movable hose was a major technological step forward at that time mastered only by the van der Heydens and only by them distributed in such a professional way. The invention had first been tested in 1672 in Amsterdam and in 1682, when the model finally reached Hamburg, it was already being widely distributed throughout Amsterdam.[300] This advertising prospectus can also be found in Aachen.[301] In Aachen as well as Cologne, the council had the instructions translated into High German; thus it can be seen how carefully the engraving was examined:[302] The van der Heydens sold

296 HStA Köln Edikte 6, Nr. 177.

297 HStA Köln Edikte 6, Nr. 159.

298 HStA Köln, Rpr. 129, f. 95r.

299 Cf. below Ch. 4, 1.2.2, p. 299.

300 Cf. for literature above, Ch. 3, 2.2.4 after note 232.

301 In the otherwise identical copy in the Stadtarchiv Aachen RA II Allg. Akten 1068; cf. the reprint in Kraus, "Der Aachener Stadtbrand", p. 78, which does not include the price tag.

302 HStA Köln, V+V N 864: the above report (*Berigt*) still includes the printed "Instructie, rakende't Onderhoud en Conservatie der Slang-brand-spuiten" and the printed price tag

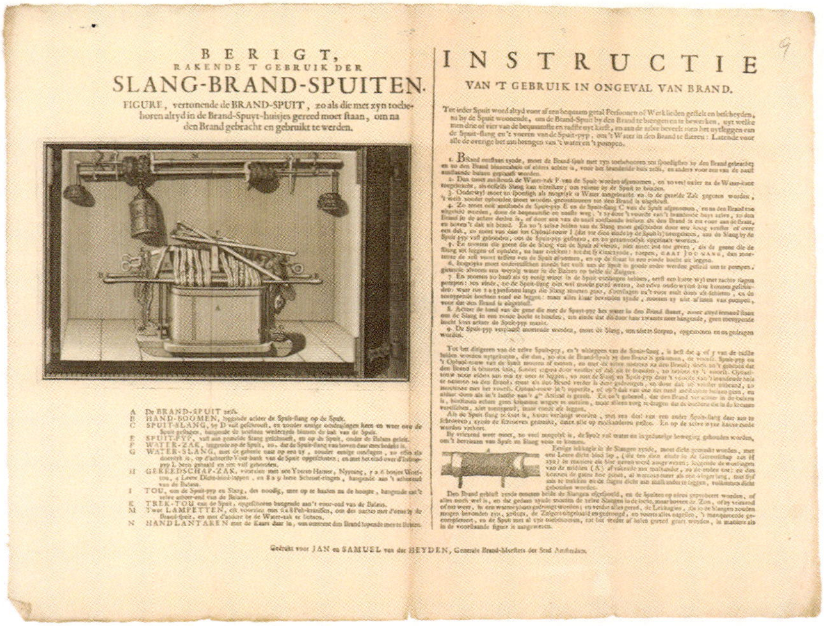

BERIGT,
RAKENDE 'T GEBRUIK DER
SLANG-BRAND-SPUITEN.

FIGURE, vertonende de BRAND-SPUIT, zo als die met zyn toebehoren altyd in de Brand-Spuyt-huisjes gereed moet staan, om na den Brand gebracht en gebruikt te werden.

INSTRUCTIE
VAN 'T GEBRUIK IN ONGEVAL VAN BRAND.

Gedrukt voor JAN en SAMUEL van der HEYDEN, Generale Brand-Meesters der Stad Amsterdam.

FIGURE 54

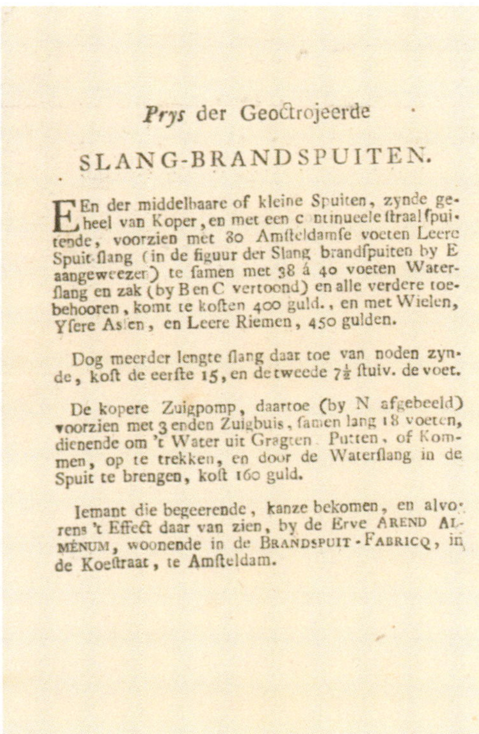

Prys der Geoctrojeerde

SLANG-BRANDSPUITEN.

EEn der middelbaare of kleine Spuiten, zynde geheel van Koper, en met een continueele straal spuitende, voorzien met 80 Amsteldamse voeten Leere Spuit-slang (in de figuur der Slang brandspuiten by E aangeweezen) te samen met 38 á 40 voeten Waterslang en zak (by B en C vertoond) en alle verdere toebehooren, komt te kosten 400 guld., en met Wielen, Ysere Assen, en Leere Riemen, 450 gulden.

Dog meerder lengte slang daar toe van noden zynde, kost de eerste 15, en de tweede 7½ stuiv. de voet.

De kopere Zuigpomp, daartoe (by N afgebeeld) voorzien met 3 enden Zuigbuis, samen lang 18 voeten, dienende om 't Water uit Gragten, Putten, of Kommen, op te trekken, en door de Waterslang in de Spuit te brengen, kost 160 guld.

Iemant die begeerende, kanze bekomen, en alvorens 't Effect daar van zien, by de Erve AREND ALMÉNUM, woonende in de BRANDSPUIT-FABRICQ, in de Koestraat, te Amsteldam.

FIGURE 55

FIGURES 54 TO 57
Jan and Samuel van der Heyden: Advertisement sheet for the serpentine fire hoses, sent to German city magistrates (1682) with details of the construction, and how to use (Fig. 54), a printed price note (Fig. 55), care and maintenance instructions (Fig. 56, p. 194), and with a German translation of the legend (Fig. 57, p. 195), HStA Köln V+V N 864

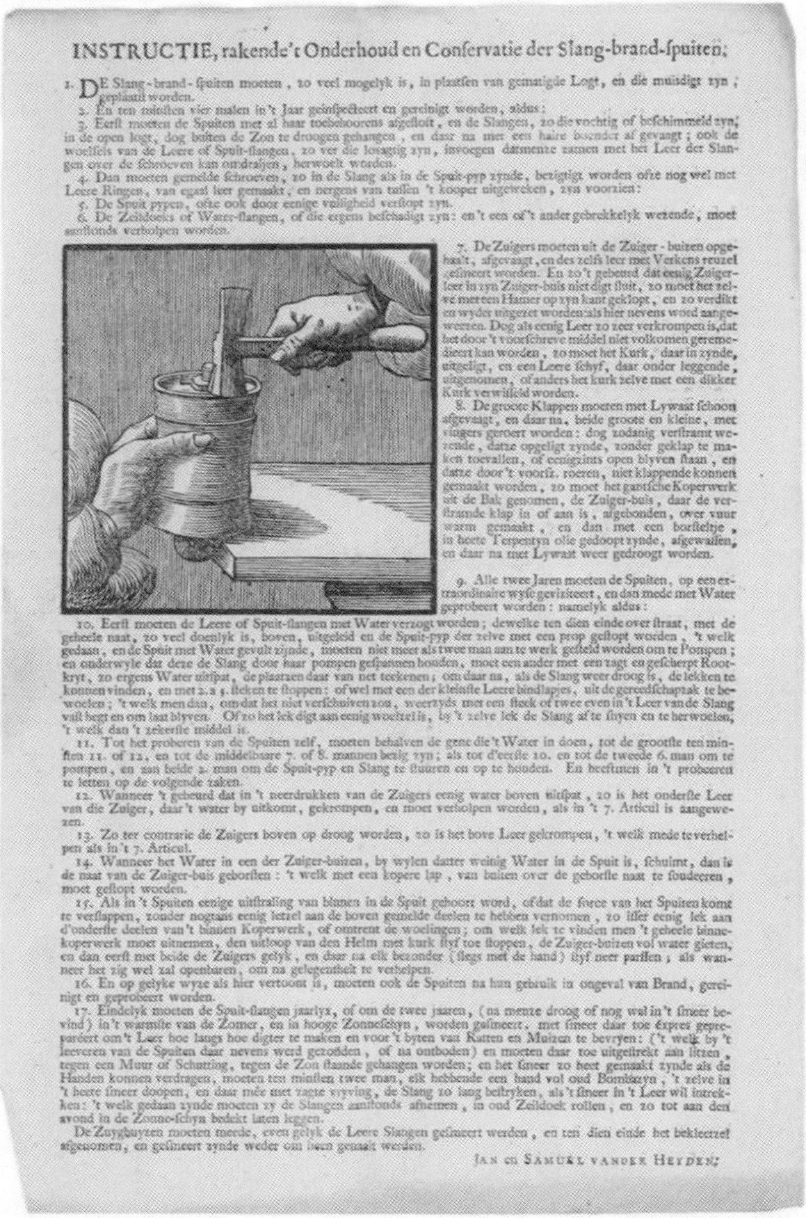

INSTRUCTIE, rakende 't Onderhoud en Conſervatie der Slang-brand-ſpuiten.

1. DE Slang-brand-ſpuiten moeten, zo veel mogelyk is, in plaatſen van gematigde Logt, en die muisdigt zyn, geplaatſt worden.

2. En ten minſten vier malen in 't Jaar geinſpecteert en gereinigt worden, aldus:

3. Eerſt moeten de Spuiten met al haar toebehoorens afgeſloſt, en de Slangen, zo die vochtig of beſchimmeld zyn, in de open logt, dog buiten de Zon te droogen gehangen, en daar na met een haire boender af gevaagt; ook de woelſels van de Leere of Spuit-ſlangen, zo ver die losagtig zyn, invoegen datmenze zamen met het Leer der Slangen over de ſchroeven kan omdraijen, herwoelt worden.

4. Dan moeten gemelde ſchroeven, zo in de Slang als in de Spuit-pyp zynde, bezigtigt worden ofze nog wel met Leere Ringen, van egaal leer gemaakt, en nergens van tuſſen 't kooper uitgeweken, zyn voorzien:

5. De Spuit pypen, ofte ook door eenige vuiligheid verſtopt zyn.

6. De Zeildoeks of Water-ſlangen, of die ergens beſchadigt zyn: en 't een of 't ander gebrekkelyk wezende, moet aanſtonds verholpen worden.

7. De Zuigers moeten uit de Zuiger-buizen opgehaalt, afgevaagt, en des zelfs leer met Verkens reuzel geſmeert worden. En zo 't gebeurd dat eenig Zuiger-leer in zyn Zuiger-buis niet digt ſluit, zo moet het zelve met een Hamer op zyn kant geklopt, en zo verdikt en wyder uitgezet worden als hier nevens word aangeweezen. Dog als eenig Leer zo zeer verkrompen is, dat het door 't voorſchreve middel niet volkomen geremedieert kan worden, zo moet het Kurk, daar in zynde, uitgeligt, en een Leere ſchyf, daar onder leggende, uitgenomen, of anders het kurk zelve met een dikker Kurk verwiſſeld worden.

8. De groote Klappen moeten met Lywaat ſchoon afgevaagt, en daar na, beide groote en kleine, met vingers geroert worden: dog zodanig verſtramt wezende, datze opgeligt zynde, zonder geklap te maken toevallen, of eenigzints open blyven ſtaan, en datze door 't voorſz. roeren, niet klappende konnen gemaakt worden, zo moet het gantſche Koperwerk uit de Bak genomen, de Zuiger-buis, daar de verſtramde klap in of aan is, afgebonden, over vuur warm gemaakt, en dan met een borſteltje, in heete Terpentyn olie gedoopt zynde, afgewaſſen, en daar na met Lywaat weer gedroogt worden.

9. Alle twee Jaren moeten de Spuiten, op een extraordinaire wyſe geviziteert, en dan mede met Water geprobeert worden: namelyk aldus:

10. Eerſt moeten de Leere of Spuit-ſlangen met Water verzogt worden; dewelke ten dien einde over ſtraat, met de geheele naat, zo veel doenlyk is, boven, uitgeleid en de Spuit-pyp der zelve met een prop geſtopt worden, 't welk gedaan, en de Spuit met Water gevult zijnde, moeten niet meer als twee man aan te werk geſteld worden om te Pompen; en onderwyle dat deze de Slang door haar pompen geſpannen houden, moet een ander met een zagt en geſcherpt Root-kryt, zo ergens Water uitſpat, de plaatzen daar van net teekenen; om daar na, als de Slang weer droog is, de lekken te konnen vinden, en met 2. a 3. ſteken te ſtoppen: of wel met een der kleinſte Leere bindlapjes, uit de gereedſchapzak te bewoelen; 't welk men dan, om dat het niet verſchuiven zou, weerzyds met een ſteek of twee even in 't Leer van de Slang vaſt hegt en om laat blyven. Of zo het lek digt aan eenig woelzel is, by 't zelve lek de Slang af te ſnyen en te herwoelen, 't welk dan 't zekerſte middel is.

11. Tot het proberen van de Spuiten zelf, moeten behalven de gene die 't Water in doen, tot de grootſte ten minſten 11. of 12, en tot de middelbaare 7. of 8. mannen bezig zyn; als tot d'eerſte 10. en tot de tweede 6. man om te pompen, en aan beide 2. man om de Spuit-pyp en Slang te ſtuuren en op te houden. En heeftmen in 't probeeren te letten op de volgende zaken.

12. Wanneer 't gebeurd dat in 't neerdrukken van de Zuigers eenig water boven uitſpat, zo is het onderſte Leer van die Zuiger, daar 't water by uitkomt, gekrompen, en moet verholpen worden, als in 't 7. Articul is aangewezen.

13. Zo ter contrarie de Zuigers boven op droog worden, zo is het bove Leer gekrompen, 't welk mede te verhelpen als in 't 7. Articul.

14. Wanneer het Water in een der Zuiger-buizen, by wylen datter weinig Water in de Spuit is, ſchuimt, dan is de naat van de Zuiger-buis geborſten: 't welk met een kopere lap, van buiten over de geborſte naat te ſoudeeren, moet geſtopt worden.

15. Als in 't Spuiten eenige uitſtraling van binnen in de Spuit gehoort word, of dat de force van het Spuiten komt te verſlappen, zonder nogtans eenig letzel aan de boven gemelde deelen te hebben vernomen, zo iſſer eenig lek aan d'onderſte deelen van 't binnen Koperwerk, of omtrent de woelingen; om welk lek te vinden men 't geheele binne-koperwerk moet uitnemen, den uitloop van den Helm met kurk ſtyf toe ſtoppen, de Zuiger-buizen vol water gieten, en dan eerſt met beide de Zuigers gelyk, en daar na elk bezonder (ſlegs met de hand) ſtyf neer parſſen; als wanneer het zig wel zal openbaren, om na gelegentheit te verhelpen.

16. En op gelyke wyze als hier vertoont is, moeten ook de Spuiten na hun gebruik in ongeval van Brand, gereinigt en geprobeert worden.

17. Eindelyk moeten de Spuit-ſlangen jaarlyx, of om de twee jaaren, (na menze droog of nog wel in 't ſmeer bevind) in 't warmſte van de Zomer, en in hooge Zonneſchyn, worden geſmeert, met ſmeer daar toe éxpres geprepareert om 't Leer hoe langs hoe digter te maken en voor 't byten van Ratten en Muizen te bevryen: ('t welk by 't leeveren van de Spuiten daar nevens werd gezonden, of na ontboden) en moeten daar toe uitgeſtrekt aan lijnen, tegen een Muur of Schutting, tegen de Zon ſtaande gehangen worden; en het ſmeer zo heet gemaakt zynde als de Handen konnen verdragen, moeten ten minſten twee man, elk hebbende een hand vol oud Bombazyn, 't zelve in 't heete ſmeer doopen, en daar mée met zagte vryving, de Slang zo lang beſtryken, als 't ſmeer in 't Leer wil intrekken: 't welk gedaan zynde moeten zy de Slangen aanſtonds afnemen, in oud Zeildoek rollen, en zo tot aan den avond in de Zonne-ſchyn bedekt laten leggen.

De Zuyghuyzen moeten meede, even gelyk de Leere Slangen geſmeert werden, en ten dien einde het bekleetzel afgenomen, en geſmeert zynde weder om heen genaait werden.

JAN en SAMUEL VANDER HEYDEN.

FIGURE 56 Jan and Samuel van der Heyden: Advertisement sheet for the serpentine fire hoses, sent to German city magistrates (1682), care and maintenance instructions, HStA Köln V+V N 864

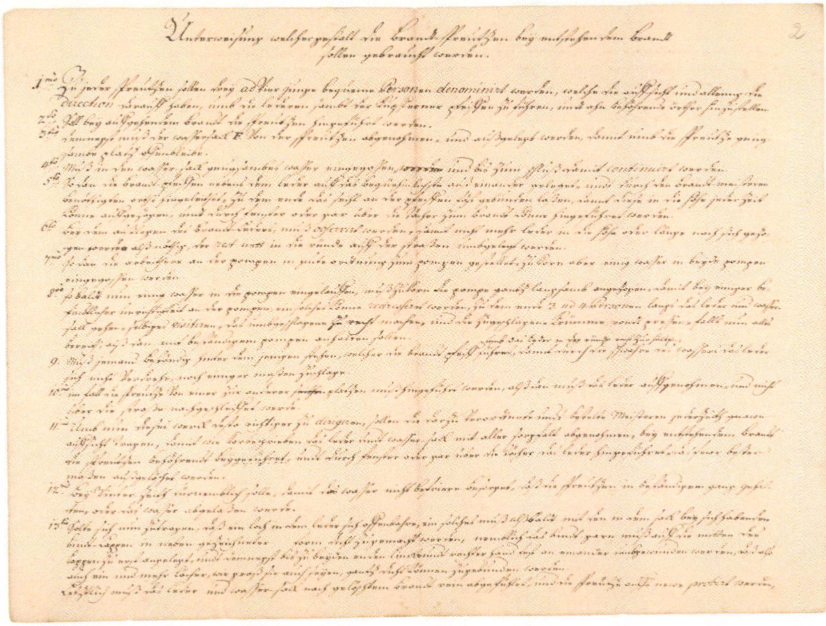

FIGURE 57 Jan and Samuel van der Heyden: Advertisement sheet for the serpentine fire hoses, sent to German city magistrates (1682), German translation of the legend, HStA Köln V+V N 864

the "middelbaare of kleine Spuiten [...] met een continueele straal spuitende [with a continous water stream]" for 450 gulden and the pomp made of copper for 160 gulden, all to be bought at the "fire hose factory (*Brandspuit-Fabriq*)".

Exactly how many "serpentine hoses" were bought and precisely when they were employed in Cologne cannot be clearly determined. In the supplementary articles of the fire scrolls in 1709 there is talk of the completion of a new hose, in 1722 there is talk of the task of inspecting the hoses, and in 1727 talk of an additional discovery—in addition to leather hoses, leather pulls which could draw water from distant springs had also been invented in Holland. In

"Prys der Geoctrojeerde Slang-Brandspuiten." Christine Brauner ("Recommendation und Reklame") has undertaken now a more extensive survey of this advertising communication in the early eighteenth century in the small cities of the region like Herford, Duisburg, Bad Salzuflen, Detmold, Xanten. Cf. above n. 233: Advertising was an integral part of what became a network of risk communities, replacing a world of loosely connected solidarity communities (see below Ch. 5, 1).

1728 a new hose was reportedly completed and another was bought from Holland.[303] In 1729 there were probably 16 large sprayer-type extinguishers in operation, some "without leather."[304] In 1801 there were 22,[305] and in 1833 only 20 such extinguishers remained.[306]

The council stimulated engagement in the area of firefighting through performance-based incentives: the first person to arrive with a horse-drawn fire extinguisher after a fire broke out received two Talers (Rtl.), the second 1.5 Rtl., the third 1 Rtl.,[307] and the "fire lords" were compensated for every firefighting effort which they actively led. Certain irregularities arose within this system: for each fire, the fire lords basically received a specified number of *Ratzeichen* ("council tallies")—a Cologne inner-administrative currency or valuation that could be exchanged for actual money.[308] Between 1758 and 1761, the four "perpetual" and six rotating fire lords participated in a total of 80 fires and earned 1600 *Ratszeichen* that were fully paid out. However, contradictory statements led to a report on the matter, which revealed that the fire lords often took credit for firefighting efforts employed in their absence— typically fires extinguished by the homeowners themselves without any assistance from firefighting forces or equipment. Sometimes the lords even took credit for fires that did not happen at all. The anonymous reporter estimated that "hardly one quarter of listed [fires] likely consisted in documented and dangerous fires."[309]

Though relatively rare, such cases documenting the unreliability of some fire figures can certainly shake confidence in archivally demonstrable fire statistics—but in the case of Cologne, there is no significant reliable material available for a longer period. Descriptions in the chronicles are insubstantial[310] and no fire statistics at all are recorded in the administrative files related to fire *policey*. This is striking, especially in comparison to Hamburg, where administrators continually worked on writing the "fire biography" of the city. Certainly, one reason for this is the fact that major fire disasters were rare in Cologne, probably because stone architecture was already quite well-developed there,

303 HStA Köln V+V N 21, f. 16r-22v.

304 HStA Köln V+V N 862, f. 13r-14v.

305 HStA Köln V+V, N 862, f. 41r: List of 10 Fructidor IX/August 28, 1801.

306 Revised fire ordinance for the city of Cologne, August 1/August 26, 1833, HStA Köln Bibliothek Ec 127, § 37.

307 HStA Köln V+V N 23, f. 11r.

308 Cf. Clemens Graf von Looz-Corswarem, *Das Finanzwesen der Stadt Köln im 18. Jahrhundert*, Cologne 1978, pp. 34–38.

309 HStA Köln V+V N 862, f. 22r-26r, auch V+V N 12, f. 41–46.

310 Several hints are found in Leonard Ennen, *Geschichte der Stadt Köln: meist aus den Quellen des Kölner Stadt-Archivs*, 5 vols., Cologne 1863–80, vol. 2, p. 507f. and vol. 3, p. 1049f. for the fourteenth and fifteenth century; in contrast, there is no mention of fires whatsoever for the sixteenth and seventeenth centuries (vol. 6).

although thatched roofs were still common.[311] However, the trend toward reform in the aftermath of fires was also apparent in Cologne. For example, the fire of the Jesuit School in 1727 was followed by a definite impulse toward renewal, particularly regarding the acquisition and care of fire extinguishers.[312] Quantitative material about the Cologne fire ecology is scarce, doubtlessly due to the absence of a fire fund in the city until 1778, when one was finally established by the archbishopric. A corresponding secular city fire fund was rejected sometime after 1815 by the Prussian government, and it was not until 1818 that the Bergisch Fire Insurance Company, headquartered in Düsseldorf, Germany expanded into the upper presidential district (*Oberpräsidialbezirk*) of Cologne. Indeed, it was the founding of private fire insurance companies (the Vaterländische of Elberfeld 1823, Aachener 1825, Kölner 1839—known as "Colonia" after 1841) that constituted the important developments in this regard during the nineteenth century.[313] In early modernity, there was still a distinct lack of coherent technical expertise regarding the quantification of risk and fire losses.

As was generally the case elsewhere, city building ordinances were also of central significance for fire *policey* in Cologne—a fire ordinance proposal from 1760/61 (which was never enforced), for example, includes several relevant prescriptions in the first part of the document.[314] However, all early modern regulations are relatively vague when compared with the intense precision of the *Réglement pour les incendies de la ville de Cologne*, enacted in 1801 in the then French occupied city.[315] This document stipulated that chimney flues were to be built out of "bricks of the best kind," and that "the interior space of ordinary chimneys should measure 431 millimeters deep and 574 millimeters wide. Every flue should also protrude one meter over the roof. … For those chimneys that must be swept with bristles, there should be at least 287 millimeters in the corners and in the interior space. …" These and other specific architectural elements, accurate to the millimeter, could only be implemented after obtaining preliminary permission from the city—an "inspector of public building places" assessed these preliminary permits and granted final permission.[316] These millimeter prescriptions clearly reflect the autosuggestion of revolutionary and

311 See the edict of July 21, 1610 against straw-roof construction in HStA Köln Edikte 6, Nr. 176.

312 HStA Köln V+V N 21, f. 6r-22v.

313 Hans Pohl, "Wirtschaftsgeschichte Kölns im 18. und beginnenden 19. Jahrhundert," in *Zwei Jahrtausende Kölner Wirtschaft*, ed. Hermann Kellenbenz, vol. 2: *Vom 18. Jahrhundert bis zur Gegenwart*, Cologne 1975, pp. 9–162, 111f. and Klara van Eyll, "Wirtschaftsgeschichte Kölns vom Beginn der Preussischen Zeit bis zur Reichsgründung," in Kellenbenz (ed.), *Zwei Jahrtausende*, vol. 2, pp. 163–266.

314 HStA Köln V+V N 12, f. 1–13.

315 HStA Köln Edikte 6 Nr. 170 (1. Fructidor IX = August 19, 1801).

316 HStA Köln Edikte 6 Nr. 170 (1. Fructidor IX = August 19, 1801).

then Napoleonic rationalism imposed within the occupied territories— these regulations appeared only eight years after the establishment of the new measure of the "meter" through the National Convention, and two years after the determination of the value of the meter by Delambre and Méchain, embodied in the platinum prototype meter.[317]

The fact that these regulations did not actually lead to any kind of radical changes in architecture is not unique to Cologne. All in all, in comparison to other German cities, fire *policey* in Cologne can be characterized as absolutely exemplary for the Middle Ages—the precision of the fire ordinance of 1451 coupled with its large number of regulations is impressive. There was no comprehensive revision of it in early modernity, which on the one hand attests to its inherent strength, while on the other, to the fact that the impulses of administrative Enlightenment were almost wholly absent in Cologne. Beyond the rather sluggish implementation of the large "serpentine" fire hoses, the introduction of improved inspection techniques seems to be the single point of progress in the area of firefighting technology from about 1750 on. Thus, the impression that Peter Kissling formulated of the Upper German region is confirmed here too: the cities were initially the leaders of innovation, but later they definitely lagged behind the territories.[318]

In Cologne this lag can even be seen in the frequency of related *policey* regulations. During the main period of *policey* regulation activity (between 1568

---

317 Research on the French influence in the neighboring countries is large, cf. only Michael Rowe, *From Reich to State: The Rhineland in the Revolutionary Age*, Cambridge 2003; there is also a good tradition of studies on the history of internal French state statistics (Marie-Noëlle Bourguet, *Déchiffrer la France: la statistique départementale à l'époque napoléonienne*, Paris 1988) and certainly on the introduction of the metric system (cf. only Maurice Crosland, "The Congress on Definitive Metric Standards, 1798–1799: The First International Scientific Conference?", *Isis* 60, 2 (1969), pp. 226–231). For Germany cf. Wolfgang von Hippel, "Von der Mühsal der Modernisierung—die Einführung des metrischen Systems in der linksrheinischen Pfalz", Aus südwestdeutscher Geschichte: FS Hans-Martin Maurer, ed. Wolfgang Schmierer et al., Stuttgart 1994, pp. 580–594, here 582f.: it usually took quite a while to implement it, so this is an early example. Contributions to the political, social and economic history of the Rhineland tend often to take the metric system as an already established condition; for the relevant literature cf. Robert Mark Spaulding, "Changing Patterns of Rhine Commerce in the Era of French Hegemony, 1793–1813", *Vierteljahrschrift für Sozial- und Wirtschaftsgeschichte* 100, 4 (2013), pp. 413–431.

318 Peter Kissling, "Reichsstädtische und territoriale Policeyen. Ein Vergleich der Reichsstadt und der Grafschaft Kempten im 16. Jahrhundert," in Peter Blickle, Peter Kissling, and Heinrich Richard Schmidt (eds.), *Gute Policey als Politik im 16. Jahrhundert. Die Entstehung des öffentlichen Raumes in Oberdeutschland*, Frankfurt am Main 2003, pp. 229–83, here 231: "[…] Die erste Differenz zwischen Stadt und Stift besteht demnach in einem beinahe 200jährigen Vorsprung der Stadt in der Gesetzgebungspraxis."

and 1711) there were, on average, about 0.6 ordinances or council decrees per year regarding fires; the average for the entire time period of more or less continuous *policey* lawmaking (1437–1796) was 0.45; and in the late phase of the Old Reich (1711–96), the average was only 0.39—the intensity of regulation thus decreases substantially after 1711.[319] Because of its favorable location, Cologne enjoyed access to and a full reception of western innovations such as the fire hose from the Netherlands. However, after the transition to what we are used to call early modernity, a rather leisurely pace reigned within the walls of Cologne. A relatively ample feeling of safety dominated—a feeling of safety that was never shaken by a great disaster. This is why Cologne serves as a good example of this type of large medieval imperial city—as do Augsburg, Nuremberg (last larger fire in 1130), and, at least for the time between the Thirty Years War (last fire in 1642) and the Napoleonic Wars (burning in 1809), Regensburg. Such cities initially entered modernity in a position of strength and, in comparison to smaller cities, they were less vulnerable to fire. Among the large imperial cities, only Frankfurt am Main experienced multiple large fires at the beginning of the eighteenth century. The others were mostly spared: they were relatively fire-safe, but then stagnated and failed to innovate in the area of fire safety.

3.1.2 The German Center of Security Innovations: Hamburg

This was all quite different in the case of Hamburg, as I have illustrated in the earlier discussion of Hamburg's "fire ecology" (Ch. 3, 1.6). In Hamburg, the frequency of fires and the amount of fire losses were confirmed by the exact observation and documentation of contemporaries beginning in the seventeenth century: the city thus offers a singularly precise case study for Germany. There are surviving early Hamburg fire ordinance texts dating to 1462,[320] but they had no lasting effect to speak of. A series of printed fire ordinances first begins with the watchman and fire ordinances of 1626,[321] which formed the basis of the 1637 fire ordinance.[322] Thus, Hamburg enters the scene of fire *policey* regulation almost three hundred years later than Cologne. According to the ordinance of 1637 (which inspired the preacher Edzard's parallelization of secular and spiritual fire

319 Statistics on the content analysis and thematic classification in Härter and Militzer, *Repertorium*.

320 HStA Hamburg Cl. VII Lit. La 1, vol. 1, Nr. 3c (note in Fe Nr. 2 vol. 2).

321 A handwritten "Fuer [sic] Ordenung," at the end of the sixteenth century in HStA Hamburg Cl. VII Lit. Fe Nr. 2, vol. 3, Nr. 10.

322 Heinrich Werner, *Der Stadt Hamburgk Fewer-Ordnung/Wornach Dero Bürgere/ Einwohnere/Vnterthanen vnd jedermänniglich daselbst in begebenen Fewersnöthen sich richten vnd verhalten sollen*, Hamburg 1637.

ordinances[323]), various pieces of equipment should be kept with the "artillery" as well as in the parishes and with the captains of the city militia. Large water tanks should be placed in building yards, in limestone yards, and elsewhere. There should be annual inspections and citizens should have their chimneys cleaned every year. In case of fire, the three watchmen on the towers of the Cathedral, St. Nicolai, and St. Jacobi, should sound a storm warning—during the day, they should raise a red flag in the direction of the fire's location, and at night, a lantern. The firefighting work should be undertaken by all citizens under the leadership of court administrators, parish leaders, and artillery leaders.[324]

In this regard, the watchman ordinance, which divided the entire citizenry of Hamburg into five so-called *Colonell* districts (parishes) and into main teams, was relevant.[325] After the fire of 1676 and the great fire of June 23, 1684, a revised version of the ordinance was issued, featuring only minor content changes.[326] A new version was also issued in 1697; however, major changes were not implemented until the revision of Sept 10, 1750. The preparations for this reform date back to 1748. The shock of the St. Michaelis fire (discussed above, in Ch. 3, 2), which led to the proliferation of disaster theology, also stimulated the ordinance reform.[327] This new version of the watchman ordinance included many detailed stipulations that went beyond the spatial division of the city, such as precise instructions for dealing with fire in different circles of society; it also included extremely detailed instructions for firefighting in general and for keeping equipment operable. As a reaction to the St. Michaelis fire, it was suggested that the comparable Leipzig ordinances concerning watchmen on Leipzig's St. Nicholas church tower should be employed,[328] and there were eight new paragraphs devoted to the handling of tower fires, as well

323 Cf. Ch. 3, 2, p. 113.

324 Per parish: 100 buckets, 1 large and 1 small fire extinguisher, 6 buckets, 4 strong fire hooks, 2 large runners; also, the societies and fraternities should maintain the buckets and runners; the city captains should ensure that there are 50–100 buckets and 2 hand extinguishers available for each company.

325 *Der Stadt Hamburg Wacht- und Feuer-Ordnung* [...], Hamburg 1643 (HStA Hamburg Cl. VII Lit. Fe Nr. 2, vol. 3).

326 *Der Stadt Hamburg Neu revidirte Feuer=Ordnung* 1676; *Der Stadt Hamburg Anno 1685. Neu Revidirte Feuer=Ordnung* [...], Hamburg 1685 (HStA Hamburg Cl. VII Lit. Fe Nr. 2, vol. 3).

327 Cf. the first versions of the revision of 1748 and then, "Lit. B Zusätze zu der neuen Feuer-Ordnung in Betracht der Thürme und Kirchen, welche derselben behüfigen Orten zu inseriren wären" (Lect. d. 26. Aug. 1750), HStA Hamburg Cl. VII. Lit. Fe Nr. 2, vol. 7, f. 42ff., here 141–46.

328 Instructions from the Leipzig city council for the tower guard of St. Nicolai, Johann Luschküen, September 16, 1720 in the files of the fire ordinance revisions included in HStA Hamburg Cl. VII Lit. Fe Nr. 2, vol. 7, f. 165–69.

as instructions for those responsible for church towers during thunderstorms. As in Ahlwardt's *Bronto-Theologie*,[329] these tower watchmen were primarily responsible for the observation of thunderstorms and defending against their effects.[330]

However, the decisive change had already been implemented in Hamburg—namely the purchase and installation of large fire hoses. On December 12, 1677, Jan van der Heyden's famous fire hose was used for the first time in Hamburg.[331] In 1720, the restructuring was complete—25 city fire hoses were placed at fixed specific locations and, under the authority of the two head fire chiefs, the watchmen ordinance prescribed exactly how the fire watchmen should make their rounds in an exact order, leaving "fire signs" at specified points along their routes to document that they had done their work.[332] Hamburg thus had greater affinity to the Dutch model—Amsterdam (and its environs) used an even greater number of large fire hoses. By the nineteenth century, Cologne still did not have as many large fire hoses as Hamburg had at the beginning of the eighteenth century. In 1748 Hamburg already had six ship-based fire hoses, located in the Alsterfleet, the kleine Alster, and two channels.[333]

As such, Hamburg was exemplary in the Holy Roman Empire. Hamburg also modeled Amsterdam's organization of fire prevention and firefighting, as established by Jan van der Heyden, earlier than Cologne; this was accomplished not only through purchasing fire hoses and by copying the fire-hose extinguishing ordinance in place since 1677,[334] but also through deploying officials to

329 Kittsteiner, *Entstehung*, pp. 74–79.

330 *Neu-revidirte Feuer-Ordnung der Stadt Hamburg*, Hamburg 1750 (HStA Hamburg Cl. VII Lit Fe Nr. 2, vol. 3, nr. 9): according to article 30 every church tower should provide a fire extinguisher, supplies of sheets, leather buckets, axes, as well as a vessel full of water. Cf. esp. article 38, clearly dictated following the St. Michaelis fire, which prescribed careful inspection of roofs for fire during thunderstorms.

331 Johann Ehlert Bieber, *Verzeichniss aller Feuersbrünste in Hamburg und der Umgegend, von 1281 bis 1830*, Hamburg 1830, p. 10.

332 Twenty-four fire hoses are mentioned in HStA Hamburg Cl. VII, Lit. Fe, Nr. 3, vol. 3, fol. 2–6. "'Reglement der Hamburgischen Brandwachten' und 'Tabellen' jn welchen verzeichnet sind die gassen und Gänge, wo die von den 25 Stadt-Feuer-Sprützen genommene Brand-Wächter zu patrouilliren haben; wobey zugleich angezeiget die Corps de garde in welchem dieselbe ihre kupferne Brandwacht-Zeichen abgeben müssen" [1740] (HStA Hamburg Cl. VII Lit. Fe Nr. 2, vol. 7, nr. 2 u. 3). The number of twenty-five hoses remained constant until 1834 (HStA Hamburg Cl. VII Lit. Fe, Nr. 2, vol. 7, fol. 6–21; Nr. 3, vol. 20, nr. 5).

333 HStA Hamburg Cl. VII Lit. Fe Nr. 2, vol. 7, f. 68v.

334 "Entwurff Wie es in Amsterdam mit den bey den Sprützen benöthigten Arbeitsleuten gehalten wirdt. Ist Johan van der Heyden, Haubtman, und hat von allen Provincien die octroye, daß keiner die Schlangen-Brandt-Sprützen ihm nachzumachen oder auß den Lande zuführen sich unterstehn muß" (HStA Hamburg Cl. VII Lit. Fe Nr. 2, vol. 1b, n. 4).

Amsterdam.[335] In 1719 at the latest, an inspection trip to Amsterdam by Hamburg Fire Chief Thomas Simon Ammon is documented.[336] While Hamburg went to Amsterdam for firefighting knowledge during the eighteenth century, surrounding territories looked to Hamburg. Friedrich Karl von Hardenberg, for example, requested Fire Chief Ammon's expertise in October 1733 for the royal government in Hanover, requesting that he dispense his knowledge in Hanover and Celle for a total of three weeks. Hamburg approved this request only reluctantly, because the absence of the local specialist was always a liability.[337] Harburg also asked for Ammon's help in replicating Hamburg's firefighting techniques.[338]

In 1750, immediately after the lightning strike had destroyed St. Michael's church, the witty project maker Andreas Liedel was proposing a fire alarm clock to be positioned in each house and church, functionally similar to what is today used as fire and smoke detectors (Figure 58 and 59):

The machine was working like a clock with some gear wheels that would store enough tension for an alarm bell; through brass wires it was connected with different parts of the house supposed to conduct the eventual heat of a fire from all parts of the building to the machine where a release mechanism, hindered only by a portion of wax that would melt away by the heat conducted through the

335 An undated *Ordnung der Schlangen Sprütze: Ambsterdam* may have already originated in the time after the major fire of 1684. It countains thirteen articles, of which the first five concern: *Entwurff Wie es in Amsterdam mit den bey den Sprützen benöthigten Arbeitsleuten gehalten wirdt.*

"1. Ist Johan van der Heyden, Haubtman, und hand [= hat] von allen Provincien die octroye, daß keiner die Schlangen-Brandt-Sprützen ihm nachzumachen oder auß den Lande zuführen sich unterstehn muß.

2. Hat daß Commando über 300 Mann, welche, wann ein Brandt entstehet, sich sofort an den Ohrte einfinden müßen.

3. Von diesen 300 Mann werden 50 zu rettung der jenigen Güther in deren Häußer der Brandt entstehet gebraucht, und müssen dieselbe einen eyd, daß sie die güther gehörigen Ohrtes restituiren wollen abstatten, auch das Wapen der Stadt auff den Armen und einen Lang roten Schleyer zum abziehen auffm Huet tragen.

4. Daß Hauß, in welchen der brandt außkömbt, wirdt von denen Bürgern in negster Capitain Hoff besetzet, damit nicht der gemeine Man so heuffig hinzulaufft und die arbeits Leute in ihrer arbeit verhindern, denen anverwandten aber so die Güther mit retten helffen wollen, wirdt ein sonderlich zeichen, welches die Capitains darzu haben gegeben.

5. Werden 50 beeydigte Zimmer und 50 Maurerleute die nichts alß was die högste Nottdurfft erfordert, herunterreisen mussen gehalten [...]." (HStA Hamburg Cl. VII Lit. Fe Nr. 2, vol. 1b, Nr. 4).

336 HStA Hamburg Cl. VII Lit. Fe Nr. 2, vol. 2, nr. 7/8.

337 HStA Hamburg Cl. VII Lit. Fe Nr. 2, vol. 10.

338 HStA Hamburg Cl. VII Lit. Fe Nr. 15, vol. 8, fasc. 2 (1736–1738).

Staatsarchiv Hamburg

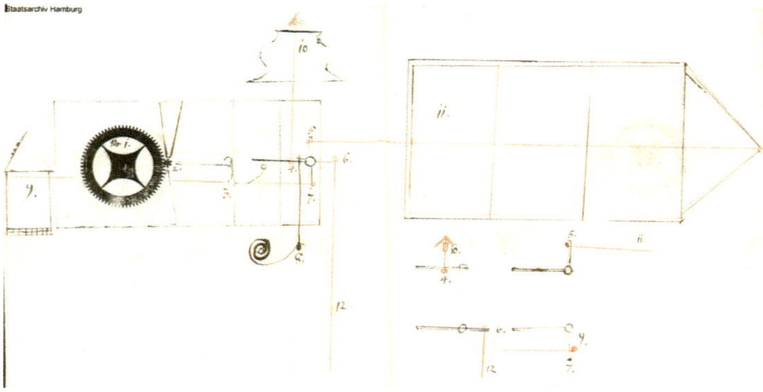

Staatsarchiv Hamburg

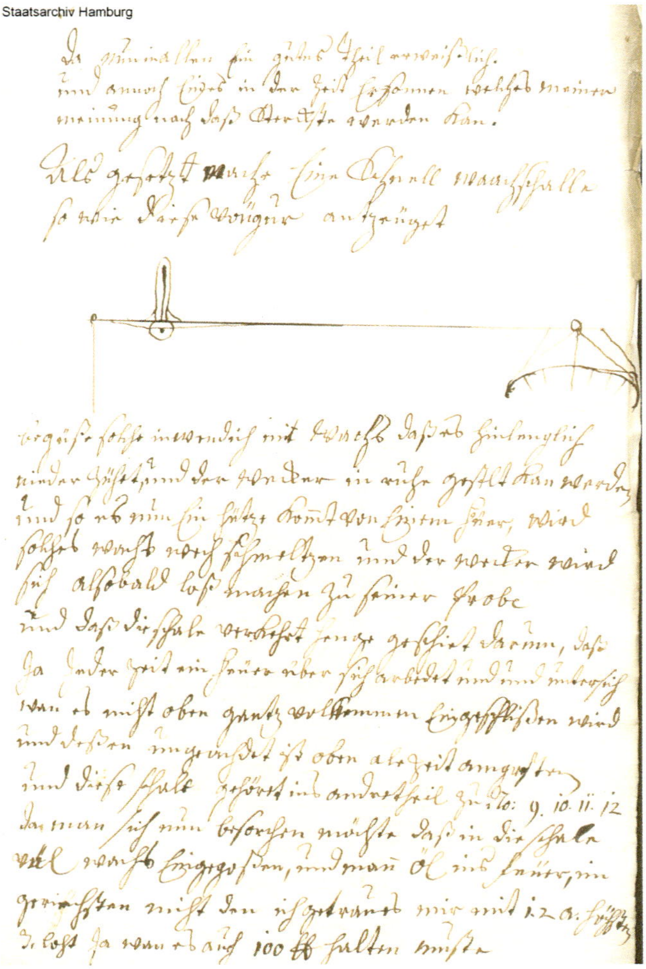

FIGURES 58 AND 59 Andreas Liedel explains to the magistrate of Hamburg by letter of May 1, 1750 his project for a *Feuer-Wecker* (fire alarm clock or detector)

brass wires, would then enable an alarm signal, indicating immediately also the exact place in the building where the fire had broken out. Apparently, this was not yet ready for serial production and the proposition remained in the archives.[339]

Because there were constant increases in fire danger in Hamburg during the late eighteenth/early nineteenth century, there were calls for reform following every fire. While old fire *policey* techniques were still in use—such as mandates advocating caution from the city pulpits, as on Dec. 10, 1802,[340]the most important intervention involved the interaction between city fire specialists (Fire surveyors [*Brandschauer*], head fire chiefs, fire fund deputies) and the Patriotic Society (*Patriotische Gesellschaft*), which with input from the city council, sought "scientific" avenues of reform and improvements in security production systems.

A main focus was the sector of early industrialization. For example, after a fire broke out in a tobacco factory (of the owner *Flickwier*) in 1793, general evaluations and a special fire *policey* ordinance proposal outlined the conditions warranting particular caution in tobacco factories.[341] The Patriotic Society was also automatically engaged in such problems. Senator Johann Arnold Günther constructed the relevant proposals and asked the society to appoint experts to examine them; the discussion also extended to sugar refineries and other factories vulnerable to fire. His proposal was then published in the proceedings of the company and also in a separate format.[342] Alderman Tonnies, the fire surveyor Ulrich Moller, Professor Brodhagen, the sugar factory owner J.F. Dunker, and the pharmacist Thorey participated in the review process and contributed to the development of a "theory of the prevention of fire risk." Peter Heinrich Christoph Brodhagen deemed all factories and manufacturing sites that he identified as "chemical industries" to be particularly vulnerable to fire—thus "laboratories, pharmacies, sugar works, lime kilns, dyeworks, breweries and spirit distilleries, metalworks of all kinds, but particularly those dealing with the smelting of metals, calico printing, tobacco factories, wool manufacturing,

---

339 HStA Hamburg Cl. VII Lit. Fe Nr. 2, vol. 13.

340 Mandate due to increased instances of fire, Hamburg, December 10, 1802, HStA Hamburg Cl. VII, Lit. Fe Nr. 2, vol. 28a.

341 "Einen Hochedlen und Hochweisen Radh dieser Kayserlich freien Reichs Stadt unumgängliche Vorstellung und gehorsamste Bitte, abseiten Stielcke und Oberdorffer, Supplicanten. Zur Untersuchungs Sache wegen der bey Carl Siemon Flickwier am 8ten Januar dieses Jahres abends nach 10. Uhr ausgebrochenen Feuersbrunst" (HStA Hamburg Cl. VII, Lit. Fe Nr. 2, vol. 19, nr. 1); "Nothwendige Vorsichtigkeits Regel zur Verhütung von Feuers-Gefahr in Tobacks-Fabriken 1793. Febr." (HStA Hamburg Cl. VII, Lit. Fe Nr. 2, vol. 19, nr. 2, nr. 4).

342 Johann Arnold Günther, *Ueber die anwendbarsten Vorschläge zur Verhütung von Feuers-Gefahr in Fabriken, und über die Gefahren von Selbst-Entzündungen*, Hamburg 1797 (HStA Hamburg Cl. VII, Lit. Fe Nr. 2, vol. 19, nr. 5).

ropemakers, leather tanneries, silk boilers."[343] The proposals were sometimes quite unspectacular: tobacco factory workers, for example, were instructed not to smoke, to use specially protected lamps, and to unpack and transport tobacco only in the safety of natural light.

More interesting is the category of "spontaneous combustion," which included an extended overview of the inflammatory response of simple and composite materials, according to the then-current knowledge of chemistry[344]—for example, "roasted, ground, and heated coffee" was thought to spontaneously combust in about half an hour. These reports were based on personal experience, but also on the observations of spontaneous combustion collected in journal and magazine reports that suggest that older theories of combustibility were still relevant in Hamburg: "According to chemistry, the basis of spontaneous combustion lies in substances of heat and light, which are dispersed in all bodies, and are released in one way or another, and which through the influx of respirable air the various mixtures of bodies often break out in violent flames. With mineral substances this combustion is due to decomposition; with vegetable and animal substances, it occurs when they are exposed to fermentation."[345] Lavoisier's refutation of the Phlogiston theory was not yet widely known in Hamburg; in the context of practical questions of fire *policey* and careful storage, this fact, of course, made little difference. More significant here is the fact that academic chemistry was deemed relevant to practical considerations.

A proposal advocating "care for the rescue of movable possessions during sudden fire outbreaks" emphasized in a similar context of administrative/semipublic discussions that during fires unauthorized persons often stole items from burning domiciles under the pretext of helping to save belongings. The proposed answer to this was to hire a troop of particularly competent persons supported by the fire funds, who would patrol the city in pairs, wearing a specific uniform. These people alone would be authorized to carry personal possessions out in case of fire. The proposal was discussed by the Patriotic Society and in the city government between 1791 and 1795; however, it was eventually implemented by a private enterprise: Georg Elert Bieber's *Association Hamburgischer Einwohner zur Versicherung gegen Feuers-Gefahr* (Association of Hamburg Residents for Insurance Against the Threat of Fire) employed 50 rescuers, 5 supervisors, and 45 simple carriers who earned 4–6 Reichstaler per year as well as a certain compensation for each fire and for services rendered.

343 Günther, *Ueber die anwendbarsten Vorschläge*, p. 41f.

344 Phosphorus, fulminating gold, and pickling lime ignite in contact with air or water. A multitude of mixtures (iron filings with concentrated mineral acids, vitriol, nitric and hydrocloric acids with turpentine, etc.) are mentioned.

345 Günther, *Ueber die anwendbarsten Vorschläge*, p. 60f.

Each rescuer had to wear a special bracelet at all times, which identified him accordingly in the event of a fire.[346] In 1797 a black jacket and a cap with a flame painted on it were also introduced as the trademarks of a rescuer.[347]

In 1820 a comparable project was introduced: a squad of 36 "people rescuers" (*Menschenretter*) were employed to rescue people during fires.[348] This project was clearly lifted from a Danzig provision of 1818, the statutes of which were known in Hamburg.[349] The rescuers were paid with funds from the General Fire Fund (*General Feuer-Kasse*) and were instructed to wear a black jacket and pants with white lacing and a badge on the left arm inscribed with the words "Feuer-Casse" (Fire Fund). Twelve of them were posted every evening at six strategic posts, equipped with a ladder, crowbar, lantern, winch, and rope. In case of fire, they were to concern themselves solely with "rescuing people, and nothing else," and this "without shying away from danger to themselves," until the point of obvious futility was reached. They were paid and otherwise rewarded for both their everyday service and for special efforts, and in case of injury or death, the rescuer or his survivors received a pension. "Only a man totally committed to the importance of the task of human rescue entrusted to him can be and remain a rescuer. Zeal for service, in addition to physical strength and ability, detached from lowly profit-seeking, are thus the essential conditions of retention."[350] Through some effort, this new service was made known to the city public. A civic culture supporting safety and security was inaugurated and widely advertised.[351] However, since these rescuers often arrived late to fires, and since people were harmed in spite of them, this relatively expensive but short-lived endeavor was absolved in 1834, following unsuccessful rescue attempts during three fires.[352]

Once again, the pervasive sensitivity in Hamburg concerning the increase in fire danger (statistically confirmed earlier in this study) clearly resonates in this context: "The frequent and saddening accidents occurring during the

346 Cf. "Jnstruction für die Retter der Brandversicherungs-Association" (HStA Hamburg Cl. VII, Lit. Fe Nr. 2, vol. 27, nr. 12).

347 HStA Hamburg Cl. VII, Lit. Fe Nr. 2, vol. 27, nr. 13.

348 Files in HStA Hamburg Cl. VII Lit. Fe Nr. 3, vol. 14.

349 HStA Hamburg Cl. VII Lit. Fe Nr. 3, vol. 14. Following the print attached to nr. 17: *Statut des im Jahre 1818 gestifteten Vereins zur Rettung bei Feuers-Gefahr*, Danzig 1819.

350 HStA Hamburg Cl. VII Lit. Fe Nr. 3 vol. 14, nr. 11a: "Instruction für die zur Menschen-Rettung bestimmten Retter," for exact wording, see HStA Hamburg Cl. VII Lit. Fe Nr. 3 vol. 14, nr. 11b.

351 "Bericht der General-Feuer-Cassen-Deputation über die zur Beförderung der Menschen-Rettung bey entstandenem Feuer zu treffenden Einrichtungen," *Privilegirte wöchentliche gemeinnützige Nachrichten von und für Hamburg* 126, Monday, May 29, 1820, 1–5.

352 HStA Hamburg Cl. VII Lit. Fe Nr. 3, vol. 20.

fires occurring here in recent years, have, as is well known, engaged the general attention of the public, and it is not easy to have made many and diverse, albeit also often very premature, suggestions for remedies."[353] This quote was written in a supplicating tone by the fire deputies Chateauneuf, Fersenfeldt, Sibertz, Nagel, Repsold and Wenke in 1835 when they suggested that these few paid rescuers ought to be replaced with a community of approximately 120 supervisors and 500–600 assistants coordinated by a few higher city authorities, who would be distributed across the entire city, creating a much denser network of potential rescuers. The notion of a volunteer fire department and of the cohesion of an active citizenry are clearly perceptible in the background here. However, this plan was rejected by the senate; instead, two reserve workers were posted at each of the 25 fire hose locations and equipped as people-rescuers, with the advantage being that everyone knew where to find them. The "many projects of artificial machineries, which would be brought from outside to the burning houses," were also rejected, because "one cannot not even send people there quickly, machines even less quickly."[354]

Thus a long-lasting series of proposals of various types of "rescue machines" (*Rettungsmaschinen*) ended; models, which unfortunately no longer exist, were presented to the Patriotic Society. There were various proposals from Copenhagen and other cities concerning the possibility of improving or replacing traditional fire ladders in order to improve access to tall buildings. Johann Georg Repsold, who served as fire chief in 1800, deemed a hinge-like construction that would hook into a high-lying windowsill with a pulley construction especially worthy of support (Fig. 60).[355]

According to Repsold, sixty such "machines" costing 25–30 Courant Marks each would have been necessary. This model was just as unsuccessful in gaining support as many other ladder models, however, it does document the existence of a constant discourse concerning projects and innovations intended to improve the methods and means of security production.

Repsold, as the resident expert, is simultaneously invoked as a major figure within the Hamburg security "laboratory". Repsold was a scientist serving as "head fire chief." While firefighting operations had generally been supervised by related professions—especially smiths and foundries within the fire *policey* administration, since they could also provide for the construction and maintenance of firefighting equipment—over the course of

353 HStA Hamburg Cl. VII Lit. Fe Nr. 3, vol. 20, nr. 11.

354 HStA Hamburg Cl. VII LIt. Fe Nr. 3, vol. 20, nr. 20.

355 Recommendation of Johann Georg Repsold of May 22, 1806, in HStA Hamburg Cl. VII Lit. Fe Nr. 2, vol. 28a, Nr. 45.

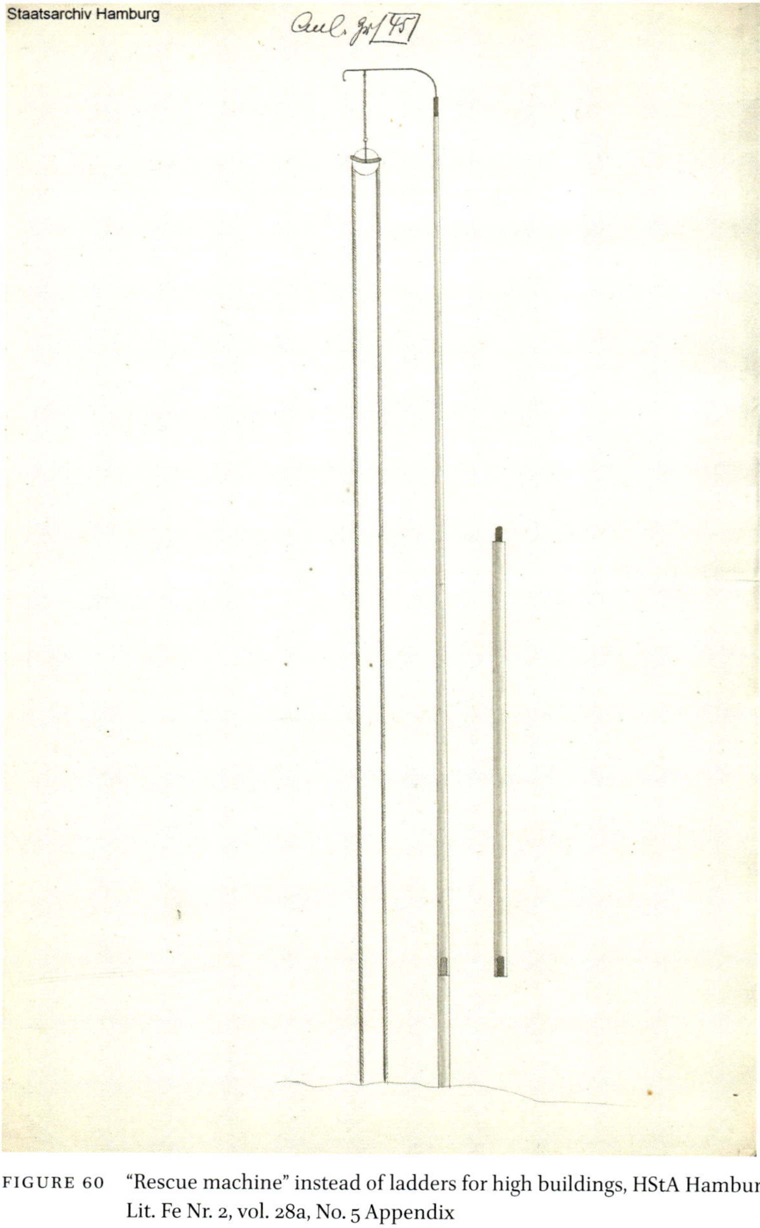

FIGURE 60 "Rescue machine" instead of ladders for high buildings, HStA Hamburg CL. VII Lit. Fe Nr. 2, vol. 28a, No. 5 Appendix

the eighteenth-century, a process of scientification and specialization of the profession evolved in Hamburg. The members of the Bieber family are one example of this; Johann Georg Repsold is another. He was not a simple, subordinate public servant; rather, he belonged to a circle of practical-scientific citizens. Repsold is perhaps better known today as the founder of the astronomical observatory in Hamburg, as a scientist and precision mechanic, who communicated with Fraunhofer and Gauß;[356] but his main occupation was deputy fire chief, and, from 1808, head fire chief of the Hanseatic city, in which role he succumbed to a hero's death due to a falling beam during a firefighting operation in 1830. A statue erected in his honor still stands in front of the Museum of Hamburg History where the observatory was previously located.[357] There are probably very few nineteenth-century firefighters who were honored with their own statue. He was given a complete state funeral with an entourage of one hundred carriages led by Mayors Bartels and Sillem, the syndic Sieveking, and the senators. Several pieces of verse devoted to the event, punctuating the early bourgeois culture of the time, appeared in the newspapers: in the *Hamburger Nachrichten* of January 30, 1830, Theodor Körners's "Auf dem Schlachtfelde von Aspern—1812" (On the battle field of Aspern—1812") was quoted: Repsold had died for the fatherland.[358] Two weeks earlier, a poem devoted to the hero, actually entitled "Repsold," had appeared in the same newspaper.

> Only he, the bold rescuer in danger
> The protection in the time of need, never to return again!
> No cry for help will wake him! Now and forever
> His shattered head fell to the earth!

---

356 Jürgen W. Koch, *Der Hamburger Spritzenmeister und Mechaniker Johann Georg Repsold (1770–1830), ein Beispiel für die Feinmechanik im norddeutschen Raum zu Beginn des 19. Jahrhunderts* (Hamburg: 2001).

357 "Eine Skizze des Denkmals in Skizze des Denkmals für Joh. Georg Repsold 1830" (StAH Best. 622-1 Repsold 7); One might take this as a starting point of patriotic nationalizing the heroic deeds of firemen, cf. today the commemoration statues and monuments built for the 347 firemen killed while fighting the fire and rescuing the remaining people in the World Trade Towers on September 11, 2001: *Zipporah Lax Yamamoto*: After 9/11: Transformations of Memory into History. Diss. UCLA 2011; Cornel Zwierlein, "Vom Londoner Brand 1666 bis 9/11. Feuergefahr und Feuerexperten seit dem 17. Jahrhundert," *Geschichte in Wissenschaft und Unterricht* 67, 11/12 (2016), pp. 711–730.

358 "D'rum soll es die Nachwelt laut erfahren, \\ Wie auch Deutsche Bürger dankbar waren, \\ Wie wir der Gefallenen That erkannt, \\ Daß ihr Tod uns Lebende ermuthet, \\ Daß sie für Unwürd'ge nicht geblutet, \\ Das beweise, Deutsches Vaterland!—\\ Deine Sänger laß in Liedern stürmen, \\ Und zum Steine füge kühn den Stein \\ Und die Pyramide laß sich thürmen \\ Der gefallnen Brüder werth zu seyn." (according to a transcription in StA Hamburg best. 622-1 Nachlass Repsold, S. 9). For further literature on early nineteenth century firefighting Zwierlein, "Vom Londoner Brand 1666"; for the Berlin fire department cf. Olaf Briese, *Für des Staates Sicherheit. Das Löschwesen im 19. Jahrhundert und die Gründung der ersten Berufsfeuerwehr Deutschlands in Berlin 1851*, Berlin 2018.

> His eye will never look up again, calculating,
> To force, from the freely chosen paths of terror,
> The wild run of increasing doom
> Into its more bounded limits.[359]

In stark contrast to the old imperial city of Cologne (the largest German city until 1600), Hamburg demonstrated a much more advanced development in the theory and practice of security production in terms of fire safety from the seventeenth century onwards. This is not only exemplified by the existence of the General Fire Fund (see below Ch. 4, 1.2), but also in terms of fire *policey*, which was no longer limited to counting buckets and ladders, but fully integrated within the discourse of the popular Enlightenment.[360] Hamburg was home to the first Enlightenment society in Germany definitively oriented toward practical, economic improvement—the Hamburg Society for the Promotion of Arts and Useful Industries (*Hamburgische Gesellschaft zur Beförderung der Künste und nützlichen Gewerbe*), also known as the Patriotic Society, was founded in 1765. The society served as a context for expert dialogues between the senate, aldermen, and fire chiefs—projects were started there, and everything was carefully evaluated by experts like Repsold. In the formative phase of the late Enlightenment, the scientification and specialization of this area was an elementary part of the process of transformation within society: the production of security became a state function, shaped for the benefit of all citizens by a few exemplary researcher-citizens. The ways in which the specific Enlightenment world of Hamburg was intertwined with national Enlightenment discourse will be further explored in section 2 of this chapter. Among other major cities assuming a leadership role during the seventeenth century for old Rhenish and southern German imperial cities—the "winners" in terms of population growth and importance—none played as important a role as Hamburg, even in supraregional terms, as a laboratory for the mechanisms of security production. In the major territorial capitals of the empire—Berlin, Dresden, Munich—the primary stimulus for potential changes typically originated at the state level, rather than emerging from independent circles of the local elite. Concerning the territorial system of fire *policey* and security production, Prussia offers the best third paradigmatic case study.

---

359 StA Hamburg best. 622-1 Nachlass Repsold, p. 10.

360 On this see Franklin Kopitzsch, *Grundzüge einer Sozialgeschichte der Aufklärung in Hamburg und Altona*, 2 vol. (Hamburg: 1982).

3.1.3 The State's "Images" of the City Collective: An Approach to Disaster Memory and to Learning from Disaster (e.g., Prussia)

Since most territorial cities were ruled by a superordinate authority at the territorial level, as was the case in Prussia, Saxony, Hanover, and other regions, focusing on the territorial administration yields the most relevant insights into the development of security production mechanisms in these small territorial cities. In the "Kurmark," the Prussian central province, the supreme authority was the Prussian General Directorate, established in 1723, which operated as the central decision-making body regarding all provincial management issues—including fire *policey* and fire insurance societies. While these administrative bodies were directly managed by the Chambers of War and Domains (CWD; Kriegs- und Domänenkammern), the General Directorate made all final decisions. There were distinctions made between Kurmark and Neumark cities, between the rural areas (*plattes Land*), and Berlin—which because of its special importance was not subordinate to rule by the Kurmark CWD. The fire regulations for the royal residence cities of Berlin, Cölln, and Friedrichswerder took precedence over other cities and were later deemed exemplary.[361] In particularly urgent matters, the General Directorate could attend to a case directly—thus, for example, the reconstruction of Neuruppin after the fire of 1787 and that of Zehdenick after the fire of 1801 were facilitated directly by the highest authorities rather than funneled through the CWD.[362] Above all, the General Directorate dealt with questions concerning decrees of edicts and ordinances in the cities and in rural areas—thus with fire ordinances in particular[363]—but it was also responsible for final approval of larger fire equipment acquisitions and for the supervision of provincial fire societies.

The CWD in Kurmark and Neumark had the oversight to regulate all fire *policey* measures. In rural areas, i.e. in the villages, this was basically managed

361 Christian Otto Mylius (ed.), *Corpus constitutionum Marchicarum oder Königlich preußische und churfürstlich Brandenburgische in der Chur- und Mark Brandenburg, auch incorporirten Landen publicirte und ergangene Ordnungen, edicta, mandata, rescripta, Teil 1–6, nebst Anhang und Continuatio 1–4*, Berlin 1737–55, V, I, II, Nr. 3, col. 143–56: fire ordinance July 15, 1672 for Berlin and Cölln; no. 4, col. 155–68: fire ordinance Novemer 8, 1672 for Friedrichswerder; no. 7, col. 169–72 fire ordinance January 26, 1701 for the rural areas (*plattes Land*); no. 13, col. 191–214: revised fire ordinance 3 May 1707 for Berlin and its suburbs.

362 GStAPK Berlin HA II, Abt. 14, tit. CLXIV Stadt Ruppin und tit. CXC Stadt Zehdenick.

363 Cf. approbation files for the fire ordinances of the cities of Frankfurt an der Oder, Rathenow, Brandenburg, Potsdam, Beelitz, Trebbin, Werder, Liebenwalde, Teltow, Zossen, Saarmund, the Prignitz cities, Büsenthal, Bernau, Köpenick, Havelberg, Liebenwalde, Mittenwalde, Müncheberg, Oderberg, Oranienburg, Strausberg, Zinna and other Kurmark cities (GStAPK Berlin HA II, Abt. 14, tit. CCXVII, Nr. 7, 8, 11–17, 20–22, 29, 31, 33).

by the district administration (*Domänenamt*). In the cities, fire *policey* was managed, in principle, by the city magistrates themselves, regardless of whether it concerned mediate or immediate cities (*Mediatstädte* or *Immediatstädte*): mediate cities were subordinate to one *Domänenamt* while immediate cities were subordinate only to the CWD. But the Chamber of War and Domains always prescribed the framework of rules and ordinances and exercised overall control—always subject, of course, to the General Directorate. In reality, therefore, the cities adopted laws and ordinances that were imposed or validated by the central authorities, and which always required approval.

Previously, individual cities still had the authority to enact small, unsophisticated fire ordinances—an example being the fire ordinance of 1716 from the small town of Lenzen, which was forcefully imposed on all citizens.[364] However, in 1718 the Berlin government imposed a collective fire ordinance on all Kurmark cities, which then became exemplary for many other Prussian provinces, perhaps, in part, because it was recorded in the ordinance compilations by Christian Otto Mylius.[365] This fire ordinance was then widely distributed through administrative channels: around 1719, Steuerrat (commissioner) Lütkens of the Uckermark and Niederbarnim district forwarded 450 copies of the new fire ordinance to the magistrate of Strasburg—almost every single homeowner must have been thereby provisioned.[366]

Fire inspectors were sent to inspect individual cities and were instructed to pay attention to: barns still standing in the city, remaining wood and straw roofs, as well as other architectural entities; the existence and quantity of highly flammable material (gunpowder, pitch, tar) in the houses; whether the fire extinguishers were in good condition and were stored in a publically accessible manner; whether every homeowner also possessed a copy of the city fire ordinance and whether an excerpt thereof was also posted in the home.[367] The inspection reports provide an overview of the entire scope of the territorial firefighting system: in 1723 it was determined that 34 large wheeled fire hoses, 88 simple hand sprayers, 2290 leather buckets, 122 fire hooks, and 194 fire ladders were required to fully equip 46 administrative areas;[368] however, in 1742/43 it was found that there were hardly any large fire hoses in the territories at all. In 1756 a general overview was created, the *Extract from those received*

364 "Revidirte Feuer-Ordnung Der Stadt Lentzen, Stendal [s.d., 1716]" (BLHA Potsdam Rep 8 Lenzen 40A, unfol.).

365 "Allgemeine Feuer-Ordnung in denen Städten der Chur-Marck," Berlin, November 1, 1718, Mylius, *Corpus constitutionum Marchicarum* V, I, II, Nr. 28, Sp. 241–46.

366 BLHA Potsdam Rep. 8 Strasburg 829 (July 20, 1719).

367 BLHA Potsdam Rep. 2 D 721, f. 62r-63v.

368 BLHA Potsdam D 680, f. 21v-23r.

reports and negotiated fire-instrument acts, what fire-instruments exist among all Kurmark districts and related outpost farms, and what must still necessarily be acquired (*Extract aus denen eingekommenen Berichten und verhandelten Feuer-Jnstrumenten-Acten, was an Feuer-Jnstrumenten auf sämtlichen Churmärckh. Aemtern und denen dazu gehörigen Vorwerkern vorhanden, und was an dergleichen noch nothwendig anzuschaffen nöthig ist*). This document covered 64 administrative districts and listed the following equipment as either existing or still needed (the latter figure in parentheses): 32 (81) large serpentine hoses, 86 (268) "Wasser-Thienen" (i.e., water tanks for the hoses), 663 (612) small sprayers, 659 (122) hooks, 823 (94) ladders, and 1370 (832) leather buckets.[369]

In general, the diffusion of large fire hoses in the territories and rural areas was very slow and halting. Including ship-based hoses, the city of Hamburg had as many or more large hoses than the entire Kurmark area at this point. As late as 1783 it was noted that "the fire hoses in the territorial cities are extremely inadequate, which basically means that they are equipped only with turning pipes but not with hoses."[370] Nevertheless there was obviously some progress made during the reign of Friedrich II. Quantitatively, an overview by the Kurmark authorities investigating the existing fire hoses in the territorial cities and associated villages, which began on August 4, 1789 and ended in 1795, demonstrates a distinct difference from 1756: among the 59 administrative districts that submitted information, covering 585 outposts and villages, 252 existing fire hoses were reported and an additional 145 were listed as needed.[371] But even so, there was often only one large fire hose per area, or sometimes none at all, and many villages shared a single one, which, in case of fire, was difficult to transport: "The lack of fire hoses is often a cause of the size and increase of a fire, so that even when a fire hose is brought from a neighboring region to help only hours after a fire has started, it no longer has its intended effect. It would therefore be very good if every village, outpost, and colony, could have its own fire hose." This passage was written by the Beeskow watchmaker Johann Christian Lube in 1806, who submitted a model and description of a fire hose to the Kurmark CWD. In his opinion, it was an inexpensive alternative since it could be built in stages used many common parts, thus cutting costs. The project was however skeptically and immediately rejected.[372] Another proposal in 1743 from the third mayor of Salzwedel, Jacob Friedrich Annisius, to replace the mostly wooden

369 BLHA Potsdam D 680, f. 21v-23r.

370 BLHA Potsdam Rep. 2 D 721, f. 93r (Chamber of War and Domains [Kriegs- und Domänenkammern; hereafter CWD], Berlin, November 26, 1783—to Old March [*Altmark*]-Prignitz Deputation).

371 BLHA Potsdam Rep. 2 D 686, f. 77r-84r.

372 Lube an die KDK, Beeskow 4.10.1806, BLHA Potsdam, Rep. 2 D 706.

chimneys with stone seemed like a more long-term solution, but there was also no discussion of any widespread implementation of this plan.[373] The forester Hartmann was obviously more successful with his pro memoria, which he submitted directly to the General Directorate: he proposed a new ladder, based on the simple idea that through creating a means of connecting multiple ladders, one could construct appropriately long ladders out of inexpensive wood from common trees, instead of relying only on expensive "split tall fir trees." This successfully persuaded the evaluating authorities and the General Directorate forwarded the concept and a sample drawing to the CWD with instructions to use these new connectable ladders for any new or replacement acquisitions (Fig. 61).[374]

These examples notwithstanding, there are significantly fewer projects and reform proposals in the CWD and General Directorate files than there are in Hamburg. Furthermore, the projects that are documented are targeted less at novel technical, chemical, or logistical-organizational innovations than at ways to reduce costs and improve the deployment of large quantities of firefighting equipment. The primary energies were clearly directed toward attaining a uniform basic level in all places, on widespread coverage and penetration into all areas. In some sense, comparing Prussia to Hamburg or other large western European cities as innovation centers is like comparing China of the 1990s to the West: there were cheap imitations in great quantity instead of original inventions. However, this was itself a highly important, even decisive, contribution to the development and distribution of the new order of security viewed as a whole; had they remained limited to Amsterdam, they would have failed to create much of an impact. After all, a developmental biotope or laboratory of modern security innovations would always remain historically marginal were it not for widespread diffusion.

Another unusual feature unique to Prussia, at least in its central provinces, in the context of eighteenth-century central Europe, is the way in which city reconstruction was organized after a disaster—especially after a fire disaster. In Prussia, reconstruction always served as a means to improve the technology of security. However, in contrast to most territories at this time, reconstruction was always directed by central authorities. Indeed, as the following examples indicate, this trend started much earlier than the date of 1855 reported in the literature.[375] On August 2, 1708, after the fire of the Prussian town of Crossen, King Friedrich I issued a reconstruction order[376] that became an exemplary

373 BLHA Potsdam Rep. 2 D 721, f. 1r-27v.

374 BLHA Potsdam Rep. 2 D 721, f. 121a,b, GD to CWD 29.11.1786.

375 Spohn, "Der Stadtbrand als Chance", p. 52.

376 Mylius, *Corpus constitutionum* Marchicarum, V, I, II, Nr. 16: "Verordnung, wegen Wiederaufbauung der abgebrandten Stadt Crossen, und wie es damit gehalten werden soll" (August 2, 1708).

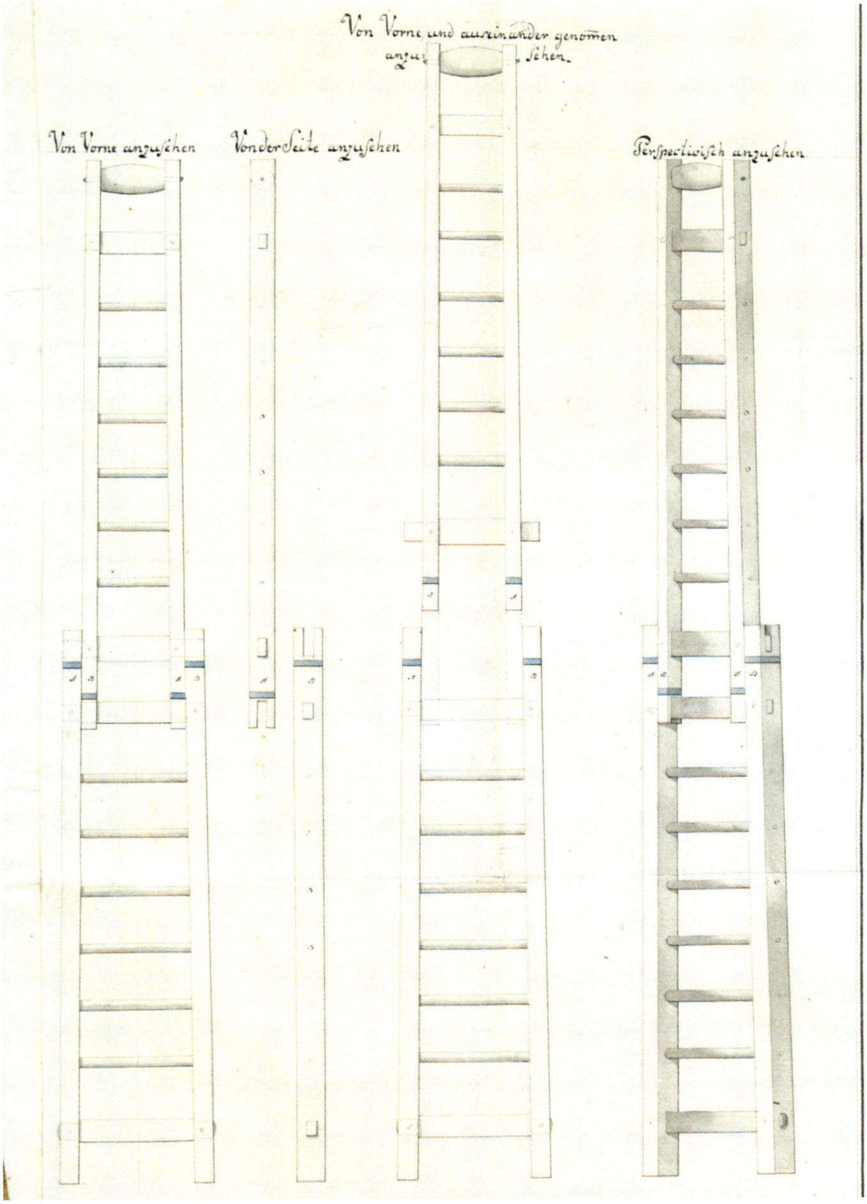

FIGURE 61 Connectable ladder system proposed by the forester Hartmann 1786, BLHA Potsdam Rep. 2 D 721, f. 121b

model for city reconstruction operations in the eighteenth century.[377] While the reconstruction of Cottbus after the fire of 1671 was relatively chaotic and dragged into the 1730s,[378] by the eighteenth century, at least after the Luckenwalde reconstruction of 1723, the reconstruction process invariably followed the pattern of a 15-point process, which was also presumably applied earlier during the 1708 reconstruction of Crossen.[379] This process was clearly employed in the reconstruction of Templin after the fire of August 24, 1735,[380] in Neuruppin after the fire of August 1787,[381] in Zehdenick after the fire of the May 27, 1801,[382] and certainly also in a majority of Wesphalian cities studied by Thomas Spohn, to the extent that they were already Prussian in the eighteenth century.[383]

These centrally organized reconstruction activities patterned themselves after the ideal city planning notions of the Renaissance, still popular at the time,[384] which involved the idea of designing cities according to aerial perspective maps on a drawing board and then imposing them *en bloc* onto the landscape.[385] Beginning in the sixteenth century there are several examples of

377 On this topic, see Reinisch, *Der Wiederaufbau der Stadt Neuruppin*; Ralph Jaeckel, '... *daß so wohl der Regularität nach, und zum Zierath der Strassen, als vor Feuers-Gefahr sicher gebauet werde ... '—Die Erneuerung der Städte in Brandenburg-Preußen im Zeitalter des Absolutismus zwischen 1640 und 1806/11. Untersuchungen zu den städtebaulichen Retablissements und Etablissements in Städten der südlichen und östlichen Kurmark sowie der Neumark Brandenburg*, PhD thesis, Humboldt University Berlin 1999; Christof Baier, *Bürgerhäuser in Templin und Zehdenick. Die Theorie vom bürgerlichen Wohnhaus und die Praxis des provinzialstädtischen Bürgerhausbaus in Brandenburg-Preußen im 18. Jahrhundert*, PhD thesis, Humboldt University Berlin 2006.

378 Jaeckel, "Erneuerung," pp. 436–466.

379 On the reconstruction of Crossen, see Jaeckel, "Erneuerung," pp. 466–507.

380 Baier, *Bürgerhäuser*, pp. 61–88.

381 Reinisch, *Wiederaufbau*.

382 Baier, *Bürgerhäuser*, pp. 153–83, the reconstruction regulations appeared in print in 1802 and are reproduced in Jaeckel, "Erneuerung," appendix 2.4.

383 Spohn, "Stadtbrand als Chance"; Thomas Spohn, "Sollen recht ordentlich bauen," *Der Märker* 39 (1990), pp. 191–206, 250–61.

384 On this Hanno-Walter Kruft, *Städte in Utopia. Die Idealstadt vom 15. bis zum 18. Jahrhundert zwischen Staatsutopie und Wirklichkeit*, Munich 1989; Eva-Maria Seng, *Stadt—Idee und Planung. Neue Ansätze im Städtebau des 16. und 17. Jahrhunderts*, Munich 2003; Kersten Krüger, "Die Idealstadt der frühen Neuzeit, insbesondere in Nordeuropa," in Frank Braun (ed.), *Städtesystem und Urbanisierung im Ostseeraum in der frühen Neuzeit. Wirtschaft, Baukultur und historische Informationssysteme*, Münster 2004, pp. 11–47; City planning and city concepts even beyond the more narrow focus of architectural history are treated in Élisabeth Crouzet-Pavan, Denis Crouzet, and Philippe Desan (eds.), *Cités humanistes, cités politiques: 1400–1600*, Paris 2014.

385 There is discussion about whether cities in the Middle Ages were already conceived according to a "plan"—for example, the cities founded by the House of Zähringen, for which a

this implementation in Europe, but many more are to be found in overseas colonies. In Europe, these visions of urban design were primarily realized only in the new construction of fortress cities. Typically, during the reconstruction of even totally destroyed major cities, the inhabitants demanded reconstruction according to the precise boundaries of their properties; this led to cities typically being rebuilt without any major changes in road and parcel layout. For this reason, the grid-layout planned city was fully realized mostly only in the colonial arena.[386] Although Brandenburg attempts to found their own oversea colonies had limited significance during this time (the *Africa-Compagnie*),[387] a different, inwardly directed idea of colonization became increasingly important in Brandenburg-Prussia from the second half of the seventeenth century on. The settlement of Huguenot and Salzburg religious refugees resulted in "colonies;" and as David Blackbourn has vividly recounted,[388] certain land reclamation projects, such as the draining of the wetlands surrounding the Oder and Warta rivers, instigated settlement policies that operated explicitly as policies of colonization.

The administrative terminology in Prussia, which constantly refers to "Retablissement" or "Melioration," demonstrates that the influence of French absolutism—imported perhaps by the French Huguenot immigrants—was also at play here. By the late sixteenth/early seventeenth century, the *mélioration* of the state and the empire was counted among the primary tasks and duties of the French king. In a text called *L'Horoscope du Roy*, which was composed around 1622 and describes the future and the duties of Louis XIII, François Goujon of Lyon included under the concept "mélioration" the king's duty "to

---

rudimentary compass technique was probably employed in the planning. See Klaus Humpert and Martin Schenk, *Entdeckung der mittelalterlichen Stadtplanung. Das Ende vom Mythos der "gewachsenen Stadt"*, Stuttgart 2001; the history of architecture has always traditionally maintained, in contrast, that for early cities there is no evidence whatsoever of any plan in written form, other than certain exceptions such as the St. Gallen convent plan; cf. Günther Binding, Susanne Linscheid-Burdich, and Julia Wippermann, *Planen und Bauen im frühen und hohen Mittelalter nach den Schriftquellen bis 1250*, Darmstadt 2002. It is undisputed that in terms of the city planned along a grid from an aerial viewpoint, the abundance of city planning utopias of the Renaissance are something qualitatively very different.

386 Cf. Robert Home, *Of Planting and Planning: The Making of British Colonial Cities*, London 1997, who analyzes the idea and practice of city planning throughout the emerging British Empire beginning with the Plantation of Ulster (Londonderry, Nordirland 1622). Cf. Jay Kinsbruner, *The Colonial Spanish-American City: Urban Life in the Age of Atlantic Capitalism*, Austin 2005.

387 Well-known is the Brandenburgisch-Afrikanische Compagnie attempt of 1682.

388 David Blackbourn, *The Conquest of Nature: Water, Landscape and the Making of Modern Germany*, London 2006.

restore what weather and bad fortunes have destroyed [decreased]."[389] Repair and recovery after accident-based destruction is precisely what the Prussian administration systematically implemented. Regarding French colonies, Vauban wrote a famous opinion in 1699 that employed the concept of "rétablissement", the *Moien de retablir nos colonnies de l'Amérique & de les accroistre en peu de temps.*[390] "Rétablissement" and "melioration" are thus the French equivalents of that what was called "improvement" within English interior and colonial development and reform politics.[391]

This is not the place to take up the multifaceted discussion of absolutism beginning with this aspect of reform politics.[392] While it is unclear how one should understand the impulses of centralizing government, of its achievement, or lack thereof, due to resistance by estates, nobles, juridical, or clerical counterbalancing powers, a decisive point is found in the new epistemic framework of early modern ("absolutist") governance, characterized by the methodization of empiricism, reflexivity, and long-term planning located within the interplay between rulers and advisors.[393] It is precisely this "long-term planning"—within the overall empiricist aim of shaping the world and its future—that functioned in Prussia as the "Retablissement" and the "Melioration" of territories and cities, which appeared there in the purest possible form within the obviously limited context of early modern realities. Of course, there is always deviation between the vision for and the actual realization of a project; nonetheless the attempt and aspiration, the basic positioning and reflexive "aiming toward" in decision-making processes and long-term management are of great significance. How then was building safety integrated into concrete cases of reconstruction?

Ralph Jaeckel compared the Crossen regulations with the reconstruction practices in the Kurmark and from that he distilled a 15-point plan comprising the common Prussian administrative system of the eighteenth century. This 15-point plan demonstrates the impressively standardized manner in which a bureaucratic routine for the planning and realization of a unified idealized city was established:

389 Qtd. Yann Lignereux, *Lyon et le roi. De la ‚bonne ville' à l'absolutisme municipal (1594–1654)*, Seyssel 2003, pp. 540–42.

390 On this see Michèle Virol, *Vauban. De la gloire du roi au service de l'état*, Seyssel 2003, pp. 175–79.

391 Cf. J.H. Elliott, *Empires of the Atlantic World. Britain and Spain in America 1492–1830*, New Haven 2006, pp. 88–97; Richard Drayton, *Nature's Government: Science, Imperial Britain, and the "Improvement" of the World*, New Haven 2000.

392 See the overviews of Lothar Schilling (ed.), *Der Absolutismus—ein unersetzliches Forschungskonzept?*, Munich 2008 and Dagmar Freist, *Absolutismus*, Darmstadt 2008.

393 Zwierlein, *Discorso und Lex Dei*, pp. 25ff., 298f., 310–13.

1. Report of the *commissarius loci/Steuerrat* responsible for the area to the CWD including a list of victims, damage estimation, information about causes of the fire, etc.
2. Determination of the legitimate claims to payments from the fire insurance funds according to the number of insured houses burnt down, and determination of the financial circumstances of the affected citizens.
3. Creation of temporary shelters for the victims (wooden huts in front of the city gates).
4. Negotiations on the granting of state/direct royal support funds and reductions (freedom to build funds [*Baufreiheitsgelder*], *beneficia*)
5. Preparation of a reconstruction plan by the district's building inspector according to the general specifications for fire-resistant buildings of the whole territory.
    a) Conception and justifications for major changes, rough planning of the construction processes.[394]
    b) Situation map of the area with damaged areas marked in red in contrast to the undamaged building stock marked in black, planned regulatory measures (according to the Kurmark construction regulations of July 26, 1741) sketched in with red lines.
6. Approval of the reconstruction plan by the building department, the CWD, and the General Directorate.
7. Begin concrete planning, in particular the establishment of plans for the construction of residential buildings according to appropriate types for different social classes (houses for craftsmen, traders, beer and brandy brewers, farmers, middle and upper-class burghers, cottage inhabitants, etc.)
8. Negotiations with contractors and craftsmen.
9. Negotiations for the procurement of construction materials.
10. Direction of the construction processes by the responsible building inspector, periodically through the *commissarius loci/Steuerrat*; periodic reporting to the General Directorate and the CWD.
11. After rebuilding the homes, reconstruction of public and religious buildings.
12. Revision and acceptance of the reconstruction by the chamber's construction department council.
13. Final improvements, correction of defects.
14. If necessary, report on punishment of those who caused the fire.

---

394 A corresponding tabular reconstruction plan for Luckenwalde 1723 is reproduced in Jaeckel, "Erneuerung," Fig. 68.

15. Final settlement of the *Retablissements*, completion of legal issues (compensation negotiations because of changes to building foundation lines, etc.).[395]

These points were not always performed in this exact order, but there was basically a a scheme, and a procedural form that was astounding for the early modern period. We will also encounter the same impressive organizational will (if not always force) in regard to fire insurance, as will be discussed in a later chapter. To illustrate just point nr. 7, we might take a look at the model sketches drawn for the rebuilding of the small Kurmark town of Seelow after the fire of 1788.

The rebuilding administration was planning with 10 different types of buildings (Figure 62), starting with "a preacher widow's house", with its interior and stable, a "farmer-citizen house" with its stable including space for horses and cattle—the term "farmer-citizen [*Ackerbürger*]" denoted the upper class citizen who had rights of lease or property of land outside the city which they were effectively cultivating like a large farmer but were recognized as citizens of the city, living there –, houses for "middle class citizens [*Mittelbürger Wohnung*]", for "Lower Class citizens [*Kleinbürger Wohnung*]", and for shepherds [*Hirtenwohnung*].

On the one hand, transforming the procedural steps into the written form of reconstruction ordinances facilitated the direction of local activities from the far away headquarters in Berlin (or from the respective "subheadquarters" of the seats of the district administration and the local commissioners [*Steuerräte*]); on the other hand, this written form also functioned at the headquarters to create an archivable objectification of the process, which could be used again in subsequent cases. Pressure imposed by reconstruction ordinances and the inclusion of the Crossen ordinance in the *Corpus Constitutionum Marchicarum* also reinforced this process. In communities that independently organized reconstruction efforts, this process was quite different: such communities rarely successfully created "an image of themselves" as an object of observation. Thus, there is no existing evidence of such administrative operations for most smaller and midsized imperial cities in Germany, nor for many territorial cities that were not yet tightly integrated within the administrative network by the eighteenth century. Once in a while, one encounters reconstruction plans for these types of cities, but often they are entirely absent.

395 These scheduling patterns are somewhat abbreviated and based on Jaeckel, "Erneuerung," pp. 400–02.

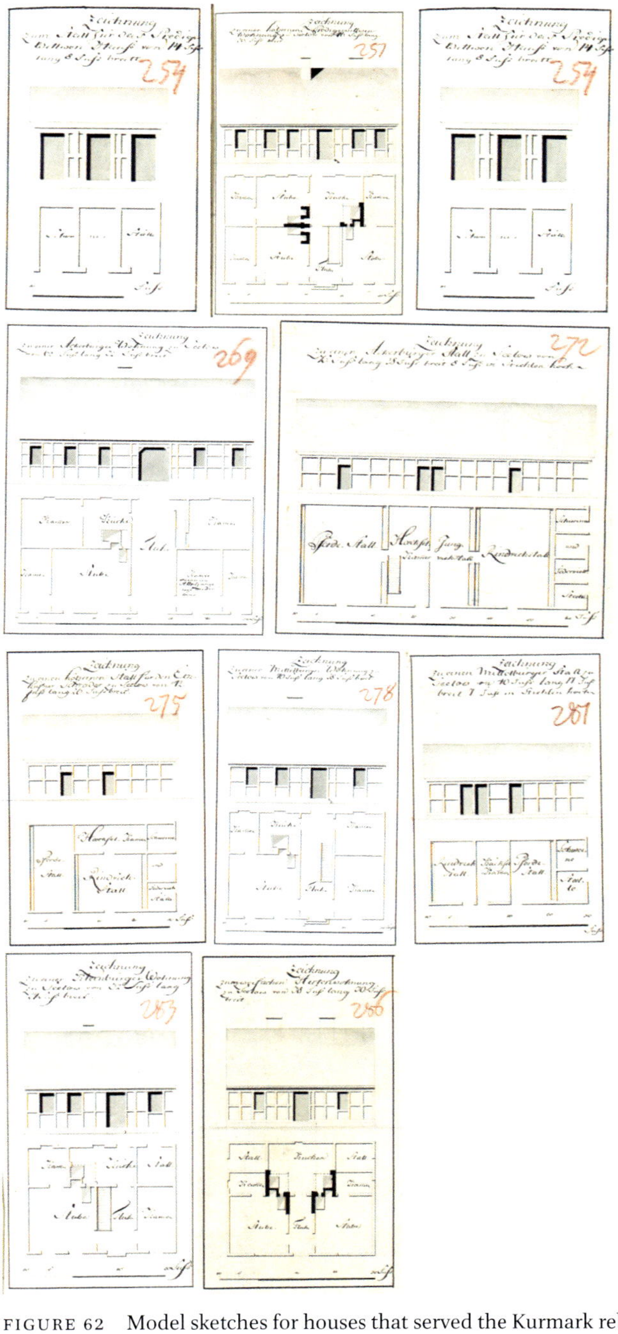

FIGURE 62 Model sketches for houses that served the Kurmark rebuilding administration, and the architects as patterns for reconstructing the town of Seelow in 1788: each citizen had a right to a given type of building according to their "class" and as they were taxed; there was no individual rebuilding and freedom of choice about how to rebuild (BLHP Potsdam Rep. 2 S7247)

In Prussia, a type of self-sustaining discourse arose within the country, involving the Prussian territorial city administration, the building department, the CWD, and the General Directorate. A "virtual image" of the country circulated, which was manipulated and constantly changed, but which still aimed at a certain ideal, and which was fostered and framed by the various "improvement" campaigns. Likewise, though to a far lesser degree, imperial city administrations also possessed a self-generated view of their cities as malleable objects capable of being shaped. Within these circles of state communication, every city fire was perceived as a chance, as a new opportunity.

Let us examine one such process in detail:

> On May 8, 1801 a fire broke out in the small town of Lebus, located a few kilometers north of Frankfurt an der Oder, which destroyed approximately 40 per cent of the town of 1160 inhabitants, including the church and schoolhouse: 41 burgher homes and 28 cottages along with many barns were reduced to ashes. Lebus, at this point, was a largely insignificant town. Nevertheless, it traditionally belonged to the 34 "mediate" cities of the Prussian central province of Kurmark.[396] The inhabitants of Lebus were divided into three social groups: farmers, burghers, and poorer people living in cottages (*Büdner*). The farmers possessed arable land, burghers possessed only meadows for raising livestock, and the *Büdner* possessed no land at all and worked either as skilled craftsmen (*Profeßionisten*) or as day laborers for the merchants.[397] As was usual in premodern times, the fire spread quickly due to the straw and thatched roofs and could not be extinguished.[398]

Following the fire, the reconstruction machinery asserted itself: after the city magistrate and Oberamtmann (district head) Gülle had informed the responsible commissioner, Steuerrat Zöller,[399] the damage was assessed and reported

396 Klaus Vetter, "Lebus," in *Städtebuch Brandenburg und Berlin*, ed. Evamaria Engel, Stuttgart 2000, p. 280; Vetter, *Zwischen Dorf und Stadt*, pp. 19–24; Jaeckel, "Erneuerung," pp. 154–59.

397 Cf. corresponding notes in the input of the lower class (*Kleinbürger*, petite bourgeoisie) and of the peasants of Lebus to the the king, Lebus 16 June 1801, BLHA Potsdam Rep. 2, S 5526, unpag.; Vetter, *Zwischen Dorf und Stadt*, pp. 73–93.

398 Steuerrat Zöller to the Kurmark CWD, Frankfurt an der Oder, 16.1801, BLHA Potsdam Rep. 2, S 5526, unpag.

399 Regarding the local commisioners (*Steuerräte*), see Otto Hintze, *Der Commissarius und seine Bedeutung in der allgemeinen Verwaltungsgeschichte* (*1910*), reprinted in Otto Hintze, *Staat und Verfassung. Gesammelte Abhandlungen zur allgemeinen Verfassungsgeschichte*, ed. Gerhard Oestreich, Göttingen 1970, pp. 242–74.

to the appropriate CWD authorities on May 16.[400] According to the report submitted to the highest authority, the General Directorate, Zöller's proposals were approved in the name of the king on July 15.[401] With little regard for their unpleasant living conditions, the inhabitants of Lebus were, up to this point, left in temporary accommodations in stables and barns, leaving livestock with no roof over their heads. Over two months after the fire, on July 17, there was finally a thorough site inspection, from which Conducteur Lietzmann generated a "situation map" based on the state of the city before the fire with the fire-destroyed sections sketched in, as well as a "reconstruction map" for future building development. The core of the transformation would result in a straightening of the roads, the construction of buildings in a straight line to achieve "more regularity," as well as the reclamation of additional space effected by banning barns within the city limits. Building Inspector Berger created model sketches for burgher houses, *Büdner* cottages, and stables, which served as the standard for all future buildings.[402]

Since reconstruction using solid materials was deemed unfeasible due to cost, half-timbered buildings with solid fire gables and tiled roofs were recommended. The costs for reconstruction were only partially borne by the residents themselves: of the estimated total construction costs of 66,596 Rtl., 47,496 were provided by the fire fund, 743 Rtl. were paid through construction aid, 9,176 Rtl. were paid by the homeowners themselves, and 9,176 more were added to that through state funds. This suggests that the fire fund was actually relatively effective, as the affected individuals only had to contribute 14 per cent of the total sum in actual costs at the time of reconstruction.[403] When questioned if the church could be built on its old foundations, the General Directorate approved this as an exception, but complained about the previous reconstruction of 13 houses and 25 stables.[404] The preferences of the owners to rebuild their land 1:1 were taken too seriously, thereby impeding the opportunity to implement more effective measures:

400 BLHA Potsdam Rep. 2, S 5526, unpag.

401 General Directorate to the Kurmark CWD, Berlin, 15 July 1801, BLHA Potsdam Rep. 2, S 5526, unpag.

402 BLHA Potsdam Rep. 2, S 5526, unpag.; Gerhard Fehl, "Perspektivischer Stadtraum. Modellmäßiger Hausbau. Zum planmäßigen Stadt- und Wohnungsbau zwischen 15. und 18. Jahrhundert," in Tilman Harlander (ed.), *Stadtwohnen. Geschichte, Städtebau, Perspektiven*, Munich 2007, pp. 19–45.

403 Zöller to the Kurmark CWD, Frankfurt an der Oder, July 17, 1801, BLHA Potsdam Rep. 2, S 5526, unpag, also Frankfurt an der Oder, August 12, 1801.

404 Zöller to the Kurmark CWD, Frankfurt an der Oder, September 17, 1801, BLHA Potsdam Rep. 2, S 5526, unpag.

> In the meantime, even you must see in your unbiased judgement, that a *Retablissement* plan involving the preservation of the old foundations of burned buildings hardly deserves to be called a *Retablissement* plan, and that if this is the basis for reconstruction, a proper *Retablissement* will never be actualized.[405]

Several builders had built stables as instructed, but then added unsanctioned details, such as large main doors and "hay floors covering the stables"—as in a barn, residents could thus store a "necessary supply of feed for [their] livestock" in the stables, because in the winter it was impossible to transport hay from the barns on the far bank of the Oder River.

> For one thing, as so-called citizens, we have nothing else in common with the citizens of other cities than the name, because we neither operate in an urban economy nor do we live the same lifestyle. Our revenues are based only in the meadows given to us that we can use to our greatest advantage not otherwise than by keeping a proportionate number of livestock on them. How could we possibly responsibly care for them and ensure their protection, if their feed must kept on the far side of the Oder in the designated barns, and if we are only allowed to fetch a supply that is sufficient for only a few days from there?[406]

The Lebus residents therefore circumvented the official *Retablissement* plan by building stables with state subsidy funds that could also be used as barns. The abstractly envisioned ideal image of the city was not realized according to plan: the long-term ideas of the planners collided with local inertia. The residents of Lebus argued for their lack of "authentic urbanity"—an argument, which in other contexts, had it concerned the preservation of their city status, would certainly never have arose.

War losses were handled in the same way as the fire losses in the tiny town of Lebus—examples being the small fortress town of Peitz and also the more important fortress city of Küstrin, both of which were bombarded and set on fire in 1758. In these cases, in addition to the usual processes in place since the beginning of the century, a precise reconstruction map was drawn (See above fig. 52: Reconstruction map of Peitz after the Fire of June 23, 1758) and detailed

---

405 General Directorate to the Kurmark CWD, Berlin, September 25, 1801, BLHA Potsdam Rep. 2, S 5526, unpag.

406 Meeting minutes, Lebus, November 26, 1801, conveyed to the Kurmark CWD by the magistrate, BLHA Potsdam Rep. 2, S 5526, unpag.

reconstruction regulations were created and printed.[407] These regulations focused particularly on the exact designation of the prices and labor costs that could be charged for materials and construction work. This was legitimized by the fact that the largest part of the money made available was of a public nature: it either came from the fire funds or—as in the case of Küstrin—from a specific royal grant of 200,000 Rtl.

The general planning approach at work here clearly falls into the perceptual schemes and procedures described by James C. Scott—admittedly with somewhat teleological and ideological critical undertones—as "state simplifications." They have already been partially evaluated for the eighteenth-century from an environmental-historical perspective in the field of forestry science and in relation to Prussian land reclamation (drainage) projects.[408] The long-term orientation expressed in the current context is clearly reflected in the administrative language: "Melioration," the improvement of the entire country, is the goal, the *terminus technicus* embodied by "Retablissement." As the above quote illustrates, the goal is not the restoration of what was lost, but rather new construction according to an idealized plan through the erasure of old structures. "Retablieren" refers only to the restoration of value, however the goal of growth is de facto implied here (which is why the insistence on the exact re-construction of the city, is, somewhat counterintuitively, deemed *not* to be a "Retablissement"). The administration thus implicitly negates the distinction between *Établissement* and *Rétablissement*, which is otherwise linguistically implied.[409] These reconstruction protocols were part of a wealth of large government programs leading to "Retablissement plans for Kurmark cities" (1767) and to the "general plan for the Retablissement of the healthy state of Kurmark cities" (1770) following the great destruction caused by the Seven Years' War. These large *rétablissement* projects were projecting somehow the Prussian cities, eventually even the Frederician state as a whole, into an ideal future yet to be realized.

In later periods, one can see the emergence of truly temporal (rather than spatial) utopian visions in the zero hour, imposing their influence on the process of reconstruction. After the Hamburg fire of 1842, for example, the manufacturing entrepreneur Harel wrote to the senate from Paris, calling for the

407 Building regulations for the reconstruction of the city of Küstrin in 1758, BLHA Potsdam Rep. 3 Nr. 10445.

408 See James C. Scott, *Seeing Like a State: How Certain Schemes to Improve the Human Condition have Failed*, Princeton 1998, pp. 1–52; Blackbourn, *The Conquest of Nature*, pp. 33–96.

409 Too strictly divided with Jaeckel, "Erneuerung," pp. 66–89.

concord and communality of all civilizations particularly in times of misfortune, and advertised the plan to rebuild Hamburg with houses equipped with metal roofs on which gardens would be planted: the soil on the roofs would prevent any spread of fire and the city would be transformed into a paradise of clean air. The project of the garden-roof city actually turned out to be verbatim implementation of passages from Mercier's *L'an 2240*: "Oh if only my century could see a new Babylon and marvel at the gardens of Semiramis!"[410] Utopian thinking is often sparked by disaster and the chance it affords to create new beginnings. Once again, however, these utopian ideas conflicted with the actual reconstruction work that had already begun.

3.2 *Panaceas: from "Local Knowledge" to Science and Back Again to Popular Enlightenment*

Questions concerning fire *policey* were highly important for Enlightenment science, due in no small part to their central importance for municipalities. This is indicated by the strong representation of the problem in corresponding media as well as its emergence as a topic for academic prize essays, and, similarly too, by its presence in the cameralist compendia of the academic field of "*policey* science" (*Policeywissenschaft*). "Accidents" (*Unglücksfälle*)—as natural disasters and extreme events were called, from the Latin "casus fortuitus"—represented the most striking break in the sought-after state of enduring happiness in a well-managed society.[411] Under the influence of the eudemonistic idea of the state, administrators were constantly searching for ways to eliminate accidents through inventions and improvements. Thus, there emerges the sense of a sort of laboratory of the economic Enlightenment existing within the network of public business and administrative discourse across the German states and territories. In the field of scientific discourse there were, for example, several academic prize questions dedicated to the fire problem, which were sponsored by scientific academies and societies—for example, the 1756 prize question of the Royal Society of Sciences in Göttingen: "How to fix a specific humidity in lumber through moistening it and coating it, so that it does not easily catch fire; and how to make this application not too expensive and effective for at least

410 HStA Hamburg Best. 323-1 (Rat- und Bürgerdeputation von 1842), Nr. 142, Harel to the senate, Paris, May 24, 1842, unpag.

411 As a broad overview and without a precise evaluation of conceptuality, it has been suggested by Torsten Meyer, "Sicherheit durch Erfindungen. Zur sozialen Bedeutung der Technologie im ausgehenden 18. Jahrhundert," in Gerhard Banse and Hans-Peter Müller (eds.), *Johann Beckmann und die Folgen. Erfindungen—Versuch der historischen, theoretischen und empirischen Annäherung an einen vielschichtigen Begriff*, Münster 2001, pp. 191–200.

several years; or where required, how to renew it without incurring too great a cost."[412] Another prize question was advertised by the same Göttingen society in 1772: "Demonstrate interventions that are cheap and correspond to local (Göttingen or even all of Hanover) state constitutions, which in their execution function 1) to improve the fire extinguishing procedures in small towns and in rural areas, and 2) to quickly gain control over potential conflagrations."[413] In Prussia, there were functional and labor divisions between the Academy of Sciences and the General Directorate in matters of fire safety (a distinction of tasks that was established shortly after the founding of the Academy in spite of Leibniz's opposition). While there is no evidence of a single prize question concerning fire *policey* offered by the Academy (1745–85), in the 1770s alone, the General Directorate advertised several related prize questions.[414] Between 1760 and 1795, in addition to the already mentioned Göttingen and Berlin prize essays, nine additional prize questions divided between three academies in Germany have been identified.[415] In Denmark, which was recognized as a contributor to the cameralist German Enlightenment discourse because in part, it shared a common scientific language, a prize question appears in 1795 in the Royal Society of Sciences in Copenhagen: "What is required to form a fully functional administration of fire prevention and resilience in larger cities?"[416]

412 *Göttingische Anzeigen von gelehrten Sachen* 141 (1756).

413 *Göttingische Anzeigen von gelehrten Sachen* (1774), part 144; The next four submitted prize writings were judged negatively by the prototechnologist Johann Beckmann: "Alle vier Abhandlungen scheinen mir sehr mittelmäßig zu sein, nichts wesentliches Neues zu enthalten, und auch in Anwendung bekannter Sachen nichts lobenswürdiges zu haben." (Beckmann to the society November 3, 1773)—the prize question was thus re-published once more in 1773. Glaser and Heinemann were chosen from the relatively numerous submissions (Göttingen, Archiv der Akademie der Wissenschaften, Scient. 196, vol. 10, Fasz. 29; cf. Rebecca Knapp, *Eine Wissensgeschichte der Feuersicherheit. Kommunikation über Brandbekämpfung zwischen 1600 und 1800*, PhD Diss., Ruhr University Bochum 2013, p. 70). The winning contributions were then published in the *Hannoverischen Magazin* 44 (1775). On Enlightenment academic prize contests in general, see for France Jeremy L. Caradonna, *The Enlightenment in Practice. Academic Prize Contests and Intellectual Culture in France, 1670–1794*, Ithaca 2012.

414 Thus in 1772–73 there were two prize writings relating to fire extinguishing machines, which would be suitable for serial distribution in rural areas and relating to fireproof roof construction, vgl. Knapp, "Wissensgeschichte," p. 67f.

415 They concern the wood coating, the construction of fireproof roofs, the improvement of fire extinguishers, the improvement of firefighting practices in general, rescuing property items during a fire, the planning and construction of more fire resistant villages; cf. Knapp, "Wissensgeschichte," appendix 1, p. 255f.

416 Prize question from June 11, 1796 of the Royal Society of Sciences in Copenhagen, in *Schleswig-Holsteinische Provinzialberichte* 9 (1795), pp. 224–27.

These prize competitions represent only the tip of the iceberg of an economic Enlightenment discourse which both preceded and extended beyond the prize essay discussion. Significantly, this discourse also operated outside of publicly printed materials, for example, in the correspondence of administrators and in projects proposed to the fire *policey* authorities.[417] Continually improving models and corresponding sketches of fire hoses were submitted to magistrates and governments everywhere. There were also somewhat adventitious advertisements for legendary "fire extinguishing machines" which unfortunately were accompanied by no exact descriptions or drawings: either those have not survived or they were not communicated at all, as the inventors usually tried to keep the secret for a long time as this was the only asset they could sell.[418] Such projects and proposals were generally concerned with two major themes: first, improvements in the extinguishing capacity of water, and second, the possibilities of a fire-protective coating for wood. These projects consistently attracted the attention of city and territorial administrations and some of the project proposers may have perhaps written more cautious applications, had they known that the administrations had already slowly accumulated a series of similar such proposals within their archives.

In late October 1698, the former imperial, later Hanoverian, artillery captain, Johann Franck of Nordhausen/Harz, appeared in Hamburg. Supposedly, "he had muddled around in the art of fire and had learned many sciences, even *secreta* and *arcana* … over the course of many wars," including in the war against the Ottomans for the emperor of the Holy Roman empire, and was thus in a position to offer the production of a cheap "extinguishing water" for the municipal water tanks positioned at strategic points around the city, which could then be used in the extinguishers (fire hoses with and without movable pipes)

417 Cf., for example, the separate fonds "'[Ökonomische] Projekte und Erfindungen' des 1660 gegründeten preußischen *Consilium Oeconomicum*" (GStAPK Berlin Hauptabt. I, Rep. 9, Nr. C6a1).

418 Hinrich Christian Schultz, for example, wrote an introduction while advertising his invented machine, addressing it to the members of the council, but he did not reveal enough details to enable the addressees to grasp the core of the project: "Wird zweifels ohne annoch in Jhren andencken schweben, wie vor einiger zeit [ich] die genug zu rühmende gnad gehabt, die Probe der so geschwind feüerlöschenden Machine vor dero erläuchtigste Augen, untertähnigst darzu legen. [Die Maschine wäre von nur drei Personen zu bedienen, er habe sich damals, wie es] einjeder Mechanicus zu tun pflegt, der nicht allemahl so gleich seiner bewegenden ursachen halber alle Vortheil so gleich entdeckt." He wanted to "reserve" such details, and requested from the council the promise of a priviledge, so that nobody other than he might be able to build the machine ("niemand als ich solche Machine verfertigen möge") (Hamburg, June 26, 1722, HStA Hamburg Cl. VII Lit. Fe Nr. 2, vol. 12a).

in case of fire.[419] Similar, half-alchemical recipes, billed as secret "arcana" of the "virtue of water," were adopted by fire *policey* writers.[420] They amounted, more or less, to recipes of how to make water more effective as an extinguishing agent through the addition of specific chemicals and elements, and occasionally, the same water mixture was also touted as a coating to make wood virtually fireproof. Johann Gerlach Wilhelmi, a captain in a Prussian grenadier regiment until 1716,[421] presented an "antipathy against fire" in 1721 in Hamburg with a relatively exact recipe for a water mixture that could serve as an extinguisher as well as a coating for wood.[422]

An "A. Bräef or Brach" advertised in Cologne in 1736 for a similar miraculous substance, which he also praised as originating in England—after the London fire of 1666, one could presume that the English possessed particular expertise concerning fire. He praised the "Veritable *Secrets*, from the widely famous English fire water, and how it is prepared in England." Supposedly, "noble people and persons of position kept this [fire water] in their homes as a precaution (against outbreak of fire)" and it was used "in the sudden outbreak of fire" before an alarm had even been sounded in the city and before one could count on assistance from the fire hoses or even from any helping hand, and it is established by many experiments, that this prepared fire water, in addition to divine providence, has saved many beautiful homes and

---

419 HStA Hamburg Cl. VII Lit. Fe Nr. 2, vol. 11.

420 Ibid.

421 See Wilhelm Werner Kurt von Priesdorff, *Offizier-Stammliste des Grenadier-Regiments König Friedrich Wilhelm IV. (1. Pommerschen)*, no. 2, Berlin 1906, no. 511, p. 184: born around 1680, Fähnrich (ensign) since 1711, retired in 1716.

422 "Antipathie wider das Feuer

R. Kreyde	10 pfundt
Alaune	20 pfundt
Potasch	5 pfundt
Bleiasch	5 pfundt
Vitriol	8 pfundt

Diese sachen werden sehr pulverisieret und kommen zu einer Tonne Wasser, man nimpt einen großen Kessel, wo eine Tonne Wasser kan eingekochet werden worinnen nun eine halbe Tonne Wasser siedendt heiß werden lässet, dan must man die Kreyde, jhns selbige algemötlich eine Handvoll nach der andern hierein im Wührmchen ['worm-like' meaning stirring, forming a funnel in the liquid for receiving small portions of the chalk one after another] hienein thun uber diese sachen muß alle zeit einer rühren damit alles wohl zergeht, und sich nichts am grundt setze. [...]." Later the addition of 12 pounds of clay was also recommended: "Wo dieses Wasser rauff kömpt brendt das Holtz in keiner halben stunde, wan es auch in völliger glut lieget"— Hamburg, 17 Jan. 1721, dorso: "Recept das Wasser zu kochen" (HStA Hamburg Cl. VII Lit. Fe Nr. 2, vol. 11).

buildings. Though they had sometimes already completely burst into flames, through this fire water the flames were fortunately put out before the fire hoses had even arrived.[423]

Until the early eighteenth century, it is clear that many of these "projects" were proposed by "project makers" (*Projektemacher*), or even by "quacks [*Quacksalber*]"—individuals who probably did not have regular work, such as soldiers and officers who had been released from service. The proposals circulated only in this administrative context, and while they appear in the archives of the fire *policey* authorities, they rarely appeared in print.

Proposals submitted after 1750 were similar in content, but now they were proposed by physical-economic Enlightenment scholars. Among them was perhaps the most active and best-known contemporary self-declared expert, doctor of medicine and physician of the Thuringian city of Suhl, Johann Friedrich Glaser (1707–83), who was also a member of the Leopoldina and an honorary member of the Electoral-Saxon Economic Society in Leipzig. Glaser, the son of an executioner, studied medicine in Erfurt and Altdorf and completed his doctorate in Hardwijk.[424] He was personally affected by the fourth major fire in the city of Suhl (following those of 1509, 1590 and 1634) on May 1, 1753, and this experience seems to have left a deep impression: from that point onwards, he dedicated his life's work from then on to the improvement of fire *policey*. In 1756 he published the first edition of his burgeoning fire *policey* compendium; the fourth edition appeared in 1772 and became almost a standard handbook for every little administrative office in the German territories and cities.[425] In the same year, the Göttingen Society of Sciences published the prize question mentioned above soliciting proposals for a fire-resistant wood coating, and Glaser's submission was awarded the prize.[426] He had tried various techniques, such as soaking the wood in salt water, soaking it in vitriol, and painting the wood

423 HStA Köln, V+V N 863, f. 1r-v: the water would then have the positive effect, of preventing pests, worms, spiders, etc. in the wood.

424 "Beytrag zur Lebensgeschichte des vor kurzen verstorbenen Arztes, D. Johann Fridrich Glasers," *Journal von und für Franken* 6 (1793), pp. 411–23; on him now Ruth Schilling, *Johann Friedrich Glaser (1707–1789). Scharfrichtersohn und Stadtphysikus in Suhl*, Vienna 2015.

425 Johann Friedrich Glaser, *Nützliche und durch die Erfahrung bewährte Vorschläge, bey heftigen und geschwinden Feuersbrünsten Häuser und Mobilien sicher zu retten* [...], 4th much expanded and improved edition, Hildburghausen 1772.

426 First published in *Hannoverische Beyträge zum Nutzen und Vergnügen* 3 (1761), 84. In part, 19 October 1761, pp. 1321–80, then as a separate edition: Johann Friedrich Glaser, *Preisschrift, wie das Bauholz in den Gebäuden zu Abhaltung grosser Feuersbrünste, leicht, wohlfeil, bewährt und dauerhaft zuzurichten, daß es nicht leicht Feuer fange oder fortbrenne*, with notes, clarifications, and improvements, Hildburghausen 1762.

with a "delicate" slurry of glue and clay. He documented experiments (e.g., the experiment of the "21st of August in the year 1760") and finally sent the society a piece of wood covered with his recipe in order for it to be tested.

After this successful endeavor, his advertising efforts for the recipe intensified: in 1773 he finally found two sponsors, the Economic Society in Leipzig and, interestingly, the Society for the Promotion of Arts and Useful Industries in Hamburg. It is not surprising that this issue resonated particularly well with the practically oriented Enlightenment milieu of Hamburg. Glaser had wrested the financing from his sponsors for several experiments—in particular, the comparative burning of three full houses with and without his miracle solution. As typical in such matters, a list of reliable witnesses was published with the findings and the exact elapsed time of the burning process was included; a copper engraving further clarified the operation.[427] Although his wood application spent nearly twenty years basking in the light of academic honor, a critic finally emerged: an author "W.," who remained anonymous, published a tiny miscell in a Hanover magazine in 1773, reporting that he tried out the recipe comprising "a slurry of glue, with one part slurry of clay, and one part flour paste … but there was not a single trial in which it proved to be reliable." Furthermore, he claimed, the flour paste functioned more as a fire propellant than a fire retardant.[428]

Glaser felt challenged by this brief objection and published defenses and also further works on firefighting.[429] In 1772, he also won the Göttingen learned

427 Johann Friedrich Glaser, [*A*]*usführliche Beschreibung der glücklich abgelaufenen großen Feuerprobe, welche mit seinem erfundenen brandabhaltenden Holzanstriche* […] *öffentlich gemacht worden* […], Leipzig 1773.

428 W., "Ueber die Abhaltung des Feuers vom Bauholze," *Hannoverisches Magazin* 11 (1773), p. 1631f.

429 Johann Friedrich Glaser, *Beantwortung und Widerlegung verschiedlicher wider seinen erfundenen und in der damit an etlichen kleinen Probehäusern öffentlich angestellten und glücklich abgelaufenen großen Feuerprobe bewährt befundenen brandabhaltenden Holzanstrich gemachten ungegründeten Einwendungen und erregten Zweifel*, Leipzig 1774. A sober review is found in *Allgemeine deutsche Bibliothek* 25, no. 1 (1775), p. 310; Johann Friedrich Glaser, *Ausführliche und auf richtige Erfahrung gegründete Abhandlung und Vorschläge, wie thunlicher Weise die meisten Feuerbrünste an und in den Gebäuden wohl verhütet* […] *werden können*, Dessau 1783; an annoyed review in *Allgemeine deutsche Bibliothek* 56, no. 1 (1783), pp. 301–05, which finds the proliferation of Glaser to be "bey nahe ekelhaft" (almost revolting); Glaser, *Beweisgründe, daß und warum die in seiner gedruckten Schrift: Abhandlung und Vorschläge, wie die meisten Feuersbrünste zu verhüten, viel wirksamer und gewisser sind, als die bisher gewöhnlichen*, Schleusingen 1784, newly reinvigorated by the annoyed tone of the review regarding Glaser's belief that he was the single, unchallenged fire expert of Germany, *Allgemeine deutsche Bibliothek* 61, no. 2 (1785), pp. 606–08.

society's second call for prize essays.[430] Here too, he had to defend himself against the second-place candidate, Johann Wilhelm Heinemann, who not entirely without reason, accused Glaser of excessive verbosity, complaining that he "bombarded his readers with a mass of minutiae, that one does not want to know because they are already long known."[431] Glaser stressed that there was a method behind his prolixity: his texts are "laboriously and clearly written with diligence ... so that they can and will be correctly understood and employed by everyone, even by common people, for whom knowing such things is particularly useful and necessary."[432] Glaser, himself of lower-class descendance, definitively considered himself to be a promotor of popular Enlightenment and took this as his mission. Reviewers decided his application had a certain effectivity: of course it would wash away in a heavy rainfall, and even if it did not actually ignite in flames, coated wood would still at least smolder; nevertheless, treating wood would help slow the spread of fire.[433] However, quite a few reviewers became increasingly weary of Glaser's publication mania: "Into his old age Mr. D. Glaser never stopped [talking] about fire deterrents and extinguishing methods, sending one treatise after another into the world."[434]

Beyond Glaser's recipe, other competing but only slightly different mixtures circulated, such as that of Ewald Friedrich Herzberg (1774)[435] and Lord Mahon

430 Cf. note 551: Johann Friedrich Glaser, "Von Verbesserung der Feuerlöschungsanstalten in den kleinen Städten und auf den Dörfern hiesiger Lande," *Hannoverisches Magazin* 13, no. 44, June 2 (1775), pp. 689–736; also published separately (Leipzig 1775); see reviews in *Allgemeine deutsche Bibliothek* 28, 2 (1776), pp. 515–18; Johann Friedrich Glaser, *Fernere Erörterung und Aufklärung seiner verbesserten Preisschrift, von Verbesserung der Feuerlöschanstalten* [...], Hildburghausen 1779.

431 Johann Wilhelm Heinemanns von Königl. Societät der Wissenschaften zu Göttingen, *zuerst gekrönete Abhandlung über die Feuerlöschungsanstalten in kleinen Städten und auf den Dörfern*, Lemgo 1777 qtd. Glaser, *Fernere Erörterung und Aufklärung*, p. 11. The prize essay itself was first published in 1775: Johann Wilhelm Heinemann, "Anzeige der Mittel, die in der Ausführung thunlich, nicht gar zu kostbar und der Braunschweig. Lüneburgischen Landesverfassung angemessen sind, wordurch a) die Feuerlöschungsanstalten in den kleinen Städten und auf dem Lande zu verbessern, und b) Dem schnellen Ueberhandnehmen einer Feuersbrunst zuvorzukommen sey," *Hannoverisches Magazin* 13 (1775), parts 14–16, pp. 211–44.

432 *Hannoverisches Magazin* 13 (1775).

433 J. H. Lambert, "Rezension der 'Feuerprobe,'" *Allgemeine deutsche Bibliothek* 21, 1 (1795/96), pp. 306f.

434 Review of Glaser's fire extinguishing attempts in *Allgemeine deutsche Bibliothek* 68, 1 (1786), pp. 302–04.

435 Ewald Friedrich Herzberg, *Vorschläge zur Verbesserung der bisher üblichen Dächer*, Breslau 1774; supplement Breslau 1779.

(1777).[436] It was obviously clear to publishers and writers in Enlightenment discourse that there were no revolutionary inventions to be discovered in this field within the context of an overall unchanged and even pre-Lavoisierian epistemic framework; at most, things that were already known could be clarified and widely communicated. This enactment of popular Enlightenment had an impact on urban and territorial fire safety. Hamburg, which had already contributed to the financing of Glaser's attempts in 1773, was confronted again twenty years later with a similar "invention," publicized by Swedish assessor Franz Joachim van Aken and sent to the senate of the city. Again logs from fire experiments on houses were featured; they had been painted with an alum-vitriol solution and therefore burned less quickly than the others or not at all. Senator Günther submitted the printed descriptions of the experiment, testified to by a large number of honorable observers,[437] to the Hamburg Society for the Promotion of Arts and Useful Industries, which in its 48th meeting (March 27, 1794) charged four individuals with its evaluation: Brodhagen, the pharmacist Thorey, Deputy Fire Chief Scharf, and border warden Reinke.[438] Thorey, in particular, dispensed very hesitant judgments:

> It has been known for centuries that solutions of different types of salts are excellent: alum, vitriol, and alkalis have been used not only to extinguish fires but also, through their application (particularly of alum and vitriol solutions), have been used to protect buildings against fire. Because these solutions are difficult and costly to produce and implement, the usefulness of them is not generally known.[439]

In particular, the chemically experienced evaluators warned that the brine would destroy the fire hoses made of canvas and leather and that the clay slurry would clog the valves and pipes. In light of these significant criticisms the senate decided against implementing the solution.

---

436 [Sholto Percy], *The Percy Anecdotes*, Revised Edition, 2 vols. (New York: 1845), vol. 1, p. 76f.

437 "Protocolles, concernant les expériences faites par le Sieur François Joachim von Aken, Assesseur, contre les progrès des incendies, Stockholm 1793" (HStA Hamburg Cl. VII Lit. Fe Nr. 2, vol. 12b n. 2); on this see the hand-written Swedish description in HStA Hamburg Cl. VII Lit. Fe Nr. 2, vol. 12b, n. 3; the first edition (Örebro 1793) in addition to the French and Swedish versions contains: "Försök til Eldsäckning och Brandskadors förekommande: Protocoller."

438 "Protocoll-Extract der Hamburgischen Gesellschaft zur Beforderung der Künste und nützlichen Gewerbe acht und vierzigste Versamlung, d. 27. März 1794" (HStA Hamburg Cl. VII Lit. Fe Nr. 2, vol. 12b, n. 4).

439 HStA Hamburg Cl. VII Lit. Fe Nr. 2, vol. 12b, n. 4.

The Hamburg specialists were revisiting a question that had been discussed in the city thirty years earlier. On August 6, 1767, the famous lightning bolt hit the Nicolai tower. It should result in Reimarus becoming a "prophet" of lightning rods.[440] On the day itself, the syndicus Faber sent immediately a request to the saltworks in Lüneburg and through them eventually questioned the Halle saltworks about whether it might be possible to store a certain quantity of Lüneburg source water, i.e. a strong salty brine, in the church towers, since this was supposedly more suitable for fire extinguishing than plain water. Syndicus Kraut at the salt headquarters of Lüneburg replied on September 25 that they did not use brine for this purpose because it "eats away at the leather on the fire hoses, and … doesn't appear to improve or change the efficacy in any way". Instead, they used a barrel of vinegar on the church towers;[441] but the request was forwarded to Halle's local salt count Johann Christoph von Dreyhaupt, who had eagerly experimented in this area and published in the same year as Johann Friedrich Glaser. In Halle, brine was indeed used for fire extinguishing, since one bucket of the salty water supposedly had the extinguishing power of 3–4 buckets of normal water. Halle brine was also supposedly superior to sea water because it contained 20 lots of salt, while sea water contained only 4 lots; Halle brine was therefore heavier and thus longer lasting during firefighting efforts. Dreyhaupt had designed a special fire extinguisher capable of spraying water especially long distances[442] and strongly argued that the "old superstitions were completely wrong": the notion that "fire caused by lightning could only be extinguished with milk is quite false, as we see from experience, … while the elemental fire of lightning [is] indeed very subtle in and of itself, when it ignites woodwork or other things, the resulting fire becomes a common fire, like a common hearth or kitchen fire."[443] Dreyhaupt also included a recipe, which again relied on salt and alum, for use if natural brine was not to hand. He disputed the idea, addressed so sharply by the pharmacist Thorey a few decades later in Hamburg, that saltwater could damage fire hoses, stating that it was safe

---

440 Cf. p. 127.

441 Letter exchange in HStA Hamburg Cl. VII Lit. Fe Nr. 2, vol. 14.

442 HStA Hamburg Cl. VII Lit. Fe Nr. 2, vol. 14, Dreyhaupt's letter to a professor of theology "Nosfelt?", Halle, 10 Nov. 1767—Dreyhaupt claims he also introduced these fire hoses in two articles in the *Göttingische Anzeigen von gelehrten Sachen* in 1753 eingestellt—he is in error here, they are found in the *Hannoverischen Gelehrten Anzeigen* 3, 92 (1753), pp. 1361–64. The particular fire extinguisher is also mentioned in another one of Dreyhaupt's publications: Johann Christoph von Dreyhaupt, *Pagus neletici et nvdzici, oder diplomatisch-historische Beschreibung des Saal-Creyses* […], part 2, Halle 1773, p. 388f.

443 Ibid.; HStA Hamburg Cl. VII Lit. Fe Nr. 2, vol. 14.

as long as the hoses were flushed with normal well water or river water after use with saltwater.

These sources from Hamburg and other places clearly demonstrate how the "popular Enlightenment" constructed itself and located its themes: the fire-retarding effects of alkaline solutions and saltwater had actually been known for a long time, and members of occupations that often encountered fire, for example artillery soldiers, were very well aware of these effects. Whether this historical-anthropological knowledge should be classified as "local knowledge" is still up for debate,[444] but it was not "tacit knowledge" in Polanyi's sense, because the project maker could easily publish it and widely communicate it.[445] As the older form of *policey* administration increasingly transformed itself into a scientifically informed cameralism, the development of a supply and demand market for projects and mechanisms intensified accordingly—thus proposals for recipes are found in the archives dating from the late seventeenth and early eighteenth century. From 1750 on, a reversal takes place so that "knowledge from below" no longer reaches the administration; rather, promoters of "popular Enlightenment" and specialist administrative officials like Glaser, Dreyhaupt, or administrative amateurs such as Aken, dispersed information about the supposedly new knowledge of the special extinguishing capacities of saltwater or the associated application of fire-retardant substances in countless writings. Their intentions were decidedly grounded in popular Enlightenment, as the above discussion about Glaser's verbose style demonstrates: it is a proliferation of "knowledge from above," the great enemy of which, as always, are the various superstitions—such as milk presumably being a special means of extinguishing lightning-induced fires. The methods and media of these publications were modeled after the recently established notion of the scientifically verified, documented, and repeatable experiment, which justified the burning of test houses supported by money from two learned societies.[446]

In the context of this transformation it was now scientist-administrators who worked on the problem of security production. Here the question is less

444 Clifford Geertz, *Local Knowledge: Further Essays in Interpretative Anthropology*, New York 1983.

445 For a reconstruction of the the adoption of Peirce's semiotics in the founding of Polanyis's concept of "tacit knowledge" see M. Zappavigna, *Tacit Knowledge and Spoken Discourse*, London 2013, pp. 1–43.

446 For an overview of the subtle variations and semasio logical-semantic multiplicity of the terminology of "experience/experiment" during the Englightenment (Shapin, Jasanoff), see Claude Rosental, "Toward a Sociology of Public Demonstrations," *Sociological Theory* 31, 4 (2013), pp. 343–65, 343–46; and Christian Licoppe, *La formation de la pratique scientifique. Le discours de l'expérience en France et en Angleterre (1630–1820)*, Paris 1996.

related to the effect that "tacit" or "local knowledge" had overall on the genesis of innovation; instead, it primarily concerns the relationship between various strata claiming to be carriers of "knowledge" and the *fashioning* of certain knowledge elements *as* innovations. The question of how much practical *tacit knowledge* and how much "real" scientific knowledge combine to make an innovation is not important; instead, we should note that information was now sold *as* (new) scientific knowledge. This was almost certainly not done in any consciously simulating, falsifying sense, but rather reflects a somewhat unconscious appropriation of the gesture of the Enlightenment.[447] If the supposed origin of the knowledge of the Enlightenment thinkers is not questioned, one easily runs the risk of assuming the perspective of Enlightenment thinkers themselves. The fact that these "experiments" were merely simulated rather than used to generate knowledge illustrates how *local knowledge* was translated into "scientific knowledge" and how popular Enlightenment was used to enlighten the "Volk" with their own knowledge.

4 Summary

Between the discussion of premodern maritime insurance as a purely spatially oriented mechanism of security production (Ch. 2) and the investigation

---

447 On the state of discourse about *tacit knowledge* in the technological process of innovation, see Paul Nightingale, "Tacit knowledge and Engineering Design," in Anthonie Meijers (ed.), *Philosophy of Technology and Engineering Sciences*, Amsterdam 2009, pp. 351–79; Ulrich Wengenroth, "Innovationspolitik und Innovationsforschung," in Gerd Graßhoff and Rainer C. Schwinges (eds.), *Innovationskultur. Von der Wissenschaft zum Produkt*, Zürich 2008, pp. 61–77. For a concrete example: Eberhard Wolff, *Einschneidende Maßnahmen: Pockenschutzimpfung und traditionale Gesellschaft im Württemberg des frühen 19. Jahrhunderts*, Stuttgart 1998 and further refined in the contribution: Wolff, Eberhard, "Der 'willkommene Würgeengel': Verstehende Innenperspektive und 'genaue' Quelleninterpretation—am Beispiel des erwünschten Kindertods in den Anfängen der Pockenschutzimpfung," in Martin Dinges (ed.), *Neue Wege in der Seuchengeschichte*, Stuttgart 1995, pp. 105–41, which more or less demonstrates that the popular Enlightenment discourse claiming that apparantly simple-minded people, affected by death of pox-infected children, declined the small pox vaccination for economic reasons, is pure speculation—rather, there were were likely other reasons at the basis of this decision. But see also Holger Böning, Hanno Schmitt, Reinhart Siegert (eds.), *Volksaufklärung. Eine praktische Reformbewegung des 18. und 19. Jahrhunderts*, Bremen 2007. Anne Conrad, "Aufgeklärte Elite und aufzuklärendes Volk," in Anne Conrad, Arno Herzig, and Franklin Kopitzsch (eds.), *Das Volk im Visier der Aufklärung. Studien zur Popularisierung der Aufklärung im späten 18. Jahrhundert*, Hamburg 1998, pp. 1–15, however rather reproduces the assessment of the Enlightenment thinkers themselves.

that will follow concerning the emergence of a new order of safety and security around 1700 (chs. 4, 5), this chapter 3 began with an attempt to trace the contours of the main issue that eventually led to the implementation of the new security mechanism of fire insurance: the structural problem of high fire vulnerability in the premodern building stock. In particular I focused on the greatest threat—major city fires. These fires represented, next to wars and epidemics, the greatest possible quasi-natural threat of the premodern era, exceeding the limits of conceivability and planning—assuming "planning" correctly describes the concepts of long-term future orientation accessible within a regional and small-town scope. The enormous conflagrations—the burning of 1700 houses in Dresden, of 4000 houses in Aachen or 13,200 houses in London—were not of primary interest to municipal fire funds, which had limited capacities. However, those long memorized mega-catastrophes served as symbolic representations of potential complete city destruction that became part of a general frame of thought and perception. It represented the threat of fire and stimulated developments regarding general safety measures. This also clarifies why many later theoretical contributions discussing the forms of the fire funds and of fire insurance companies referred to major fires as starting points, taken as the key data of the respective local and regional fire history (cf. below Ch. 5, 1.2, 2).

In this respect, these large fires were also invoked in the visualizations of fire danger: while in oil paintings these visualizations were initially still metonymically encoded by depictions of Biblical and mythological fires, later—especially in the new media of copper engraving and the broadly disseminated leaflets they illustrated—an increasingly affect-oriented pictorial language developed that strongly emphasized the confrontation between the observer and the disaster itself. Fire was presented in its entire dynamic explosivity, and representations focused on the materiality of the destruction of the city. The temporalization and emotionalization of fire danger reached a peak during the seventeenth and eighteenth centuries, belying the initial detour taken during the Renaissance concerning the emotional involvement of the viewer with a represented human fire observer or fire victim. This was caused by a refinement of but also a strong dependency on the conceptual framework of ancient poetics, rhetorics and aesthetics transferred to the realm of pictorial aesthetics. The realization of the temporalized and emotionalized totality of the fire as the main subject of the representation did not have its counterpart in a simple erosion of retribution theology in the sense of a unilinear secularization process. On the contrary, it is striking how unchanged the genre of the fire sermon remained—with its Biblical topoi and recourse to the Biblical Sodom and Gomorrah fires—until the end of early modernity. Rather, a differentiation and unfolding in different areas of the possibilities of reaction clearly occurs, so that, for example, literary emotionalized reactions appear *in addition to* the sermons, without completely replacing them.

All these forms would seem to give voice to a stronger power of stimulation within the cultural domain; however, this would still imply a simplistic model, placing image, graphics, sermons, literature, and other texts in a simple causal logic of stimulating action. Yet even if this is interpreted as correlation rather than cause, as an indicator for other fields (administrative and fire prevention practice), this is still significant enough: making the representation of fire absolute beyond all human reaction symbolizes the genesis of the modern concept of disaster. The manufacture of the totalizing exceptional case existing beyond the norm of societal safety and security—which is presupposed but which must also be created—is realized here. The fact that increasingly from 1700 on this notion of a "normal secure society" was constructed will be discussed later (see Ch. 5).

Thus, the perspective in the preceding chapter (Ch. 3, 3) focused on this development in the field of fire *policey*. Three administrative types were distinguished and compared: Cologne, as the largest medieval imperial city, was avant-garde in this area until the end of the late Middle Ages. Through comparison with other similar cities, it becomes clear that larger traditional, bureaucratically and administratively disciplined, successful cities had a higher degree of security at the end of the Middle Ages in relation to other hamlets, villages, and territorial cities—not insignificantly due to the large proportion of stone buildings in the city centers, but also because of fire *policey* ordinances and inspection practices. This is clearly reflected in the German fire statistics taken as a whole, in which there are hardly any major fires documented in large imperial cities at all—with few exceptions such as Aachen, whose building stock was heavily half-timbered.

Accordingly, because these large cities were mostly spared, the ensuing progress during the seventeenth and eighteenth centuries was also slow in these cities. Domination in the area of security innovation was quickly assumed by the fast-growing cities and territorial capitals, as well as those subsumed by the northern German territorial administrations—Prussia and its provinces in particular. This may seem strange at first glance, in light of the early modern research of the last few decades, which has celebrated the customs, structures, and administrative traditions of the Holy Roman Empire. While there are some tendencies within Prussian historiography which held the national state embodied by Prussia to be the teleological endpoint of all developments, the significance of the Holy Roman Empire's protofederalist and consensus-building qualities—its "different modernity"—has been emphasized in more recent literature. Nevertheless, this much deserved reevaluation of the empire has very little relevance for some historical topics: the administrative structures and institutions of the empire contributed next to nothing to the development of new territorial and urban measures of security production in the fields addressed here. The revisionist approach to the history of the Holy Roman Empire is quite legitimate for the sixteenth and seventeenth centuries. But emphasizing the vitality of the empire for large parts of the eighteenth century necessitates a focus on forms not realized (imperial reform projects), on discursive fields (imperial patriotism), or on regions characterized by small territories where the empire substantially limited its function to the consolidation of interterritorial cooperation and the support of border defense and other tasks that could not be fulfilled on a local level.

In terms of the questions at stake here, it is difficult to come to different conclusions regarding the impulses and stimuli of development. Hamburg and Prussia are forerunners in fire security innovation, although or because of the fact that Hamburg was home to some of the most destructive conflagrations in Germany during this time period—in the 1660s, 1683, and in 1842—and was therefore a fairly "unsafe" city. Apart from the generally rather fire-resistant eighteenth century, these fires were symptomatic of a riskier path paved by increasing trade volumes, by population expansion, and at the beginning of the nineteenth century, by early industrialization. This resulted in higher pressure for the development of security production.

The Prussian provinces, in contrast, constituted a laboratory of sorts for the institution of widespread territorialized coverage of cities and villages in that the innovation potential was mainly concerned with the area of infrastructure and the "objectification" of the small territorial cities as a whole. In Prussia, the externalized image of cities as "objects" to be organized and reformed was a prerequisite for the implementation of innovations, while the largely autonomous imperial cities obviously lacked the development of a functionally equivalent stance of reflective self-observation as a starting point for innovation. This small difference, which is actually a matter of different modes of perception rather than a material difference of structures, of the quality of stones and wood used for building the cities, seems to be the main difference clarifying why the cities of administrative states like Prussia possessed an entirely different motivation for the development of a widespread and far-reaching security culture.

Finally, the object of this last section was a reflection on the paths of innovation. One widely generalizable phenomenon in popular Enlightenment is particularly noteworthy: a number of technical innovations were passed down by generations of craftsmen or "project makers" and "quacks" in the seventeenth century. The knowledge-based methodization of fire prevention and firefighting then increased and was transformed around 1750 in a trend toward scientification. At this point, well-known *local knowledge* was at least partially transformed and sublimated into Enlightenment scientific discourse in a type of historical feedback loop. The "Enlighteners" simply translated such knowledge into the code of scientific "inventions," which were then experimentally validated and published in scientific media. Information coded in this way was then recognized as Enlightenment knowledge and disseminated among the supposedly clueless "Volk" with all the usual popular Enlightenment rhetoric, lamenting the uninformed and stubborn nature of ignorant people. In several other treatments of popular Enlightenment, parallel phenomena have been observed; in general, however, popular Enlightenment research still risks following the rhetoric and self-definition of the Enlighteners themselves, believing too strongly in their own innovativeness. Some inventions, such as the lightning rod, really were heretofore unimagined elements of knowledge that found their way from above down to the "Volk," but the paths were often much more complex.

This all leads to the realization that beyond the representation and perception of fire danger treated in this chapter, and beyond the development of an administrative-technical level of innovation concerning measures of prevention and firefighting, it is the construction of a new "normality," a normal world *conceived* to be *normally* free of danger, to be *normally* secure and safe, in which insecurity becomes the exception. This emerging conceptual configuration and reversal of that what was previously regarded as the "normal state of things," is probably at least as important as the back-and-forth in practical administrative communication concerning the details of the aforementioned measures: but both the steady communication and the emerging reversal of the fundamental assumption of what is or should be normal were interdependent. This interdependence will be the focus of the next two chapters.

CHAPTER 4

The Epochal Threshold of the Security Regimes 1680–1700

The previous chapter presented the threat of urban fire in the premodern era: on the one hand, it demonstrated the "reality" of this threat as reflected in the frequency trends concerning major fire events; on the other hand, it analyzed how theological concerns and pictorial representations, which were in flux and continuously developing, influenced how threats were perceived. The previous chapter also addressed another level of development in the methods of fire prevention—what Leibniz called "real-assecuration"—and the pecuniary restitution after an event through insurance, which he referred to as "verbal-assecuration." These developments all intensify around 1700, when an epochal threshold emerges. At this point in time, the spread of the "fire gap" (cf. Ch. 3, 1.1) had already begun in central Europe, due in part to developments in military technology that favored tactics involving battles rather than infrastructure destruction, which contributed to a decrease in the frequency of major city fires. Similarly, on the level of culture and governmental administration, an epochal threshold also appears around 1680–1700, which clearly coincides with the emergence of different forms of fire insurance in England and Germany.

As far as the question of perception goes, this appears to be the inaugural moment of the quickly growing assumption that city fires and fire disasters in general constituted a break from normality, an exception to the rule. An important topic of the current chapter will be to demonstrate how, by crossing this epochal threshold, this "new order" of security essentially established a new mode of perceiving the world and constructed safety and security as the new "normal" situation. In order to trace this development, the emerging insurance-like institutions provide an almost ideal research subject, since they are among the most important factors responsible for constructing this new "normal secure society." The following chapter will examine two laboratories of innovation culture[1]—London and Hamburg. I will account for the concrete

1 The concept of the laboratory is not used here in a technical sense, but rather as a metaphor for a narrow structural relationship—in terms of place and the individuals involved—generating both intended and unintended innovation. There were, of course, no scientists that intentionally constructed these cities as "laboratories" for the purpose of creating inventions—at

 | DOI:10.1163/9789004431225_005

conditions that lead to the emergence of fire insurance techniques (1) and then generate a hypothesis as to why the universalization of the insurance principle began at precisely this point in time and in precisely these places (2).

1 Laboratories of Innovation: Hamburg and Berlin—1680–1700

It is common knowledge that the Hamburg Fire Fund (*Feuerkasse*) was founded in 1676, shortly before the first private insurance agency was founded in London as a joint-stock company in 1681. However, these two cities, which were both important centers of development for the history of insurance, have seldom been analyzed in close comparison, and it will therefore be useful to enter into an investigation of these two historical "laboratories"—London and Hamburg around 1680—where specific simultaneous conditions arose, leading to the emergence of the new security product of institutionalized fire insurance, though in very different forms.[2]

In both cities, one central person is associated with innovation: Nicholas Barbon was key to the developments in London and Gottfried Wilhelm Leibniz to those in Hamburg. Barbon is definitively regarded as the "inventor" of the English model of insurance; Leibniz, in contrast, is only indirectly associated with insurance in Hamburg. In a certain sense, he fulfilled a historical function otherwise absent in Hamburg at this time given the lack of an already strongly developed culture of 'merchant-philosophers' as in England (Graunt, Misselden, Mun, Barbon, J. Child) and given the lack of a full university in the city: he provided a bridge between academic discourse and practical innovation. In contrast, Barbon united both of these elements within his own work.

In a certain sense, the "culture of innovation" important at both places developed in tandem with the many different projects and attempts at problem solving that emerged in different regions throughout Europe (as discussed

the most, it was the "invisible hand" of history that was responsible. This concept thus has nothing to do with Bruno Latour's or Karin Knorr-Cetina's concept of the laboratory and the question of how society is related to the laboratory as a locus of scientific production.

2 Martin Mulsow generalizes the concept of "constellation" in the tradition of Dieter Henrich for specific purposes related to the history of ideas in his *Konstellationsforschung*, Frankfurt am Main 2005. Cf. Martin Mulsow, "Qu'est-ce qu'une constellation philosophique? Propositions pour une analyse des réseaux intellectuels," *Annales HSS* 64, no. 1 (2009), pp. 81–109. In the following the emphasis is less on "ideas" as on a laboratory-like link between practical economic, political, and communicative conditions, according to which "ideas" in the sense of a learned discourse operates more retrospectively by way of categorization, interpretation, and legitimation, whereas the actual innovation is emerging earlier in the nexus of actions and reactions of administrative practices.

in the previous chapter). On the one hand, the transnational development of innovation is clearly at work—demonstrated by, for example, the spread of firefighting technology across Europe. Nevertheless, as made clear by the different mechanisms developed by Germany and England to manage losses through insurance-like institutions, we are also dealing with a local problem that emerged at the same time in many different places. Based on different societal frameworks, different regions responded to the problem by using different forms of innovation, and for quite some time, there seemed to be little or no reciprocal exchange of innovation technologies. The reflexivity of innovation was clearly stronger in England than in Germany—demonstrated by the conflation of theory and practice in England, versus the separation of these two realms in Germany.[3]

1.1 *London: Nicholas Barbon*

1.1.1 Protostatistics, Protoprobabilistic Reasoning, and the Conflict between State and Private Economies

As discussed above, the history of early fire insurance for England has been researched essentially through the lens of three large companies, founded after 1710, for which corporate archival information still exists: the Sun, the Royal Exchange, the late Phoenix and the Norwich Union.[4] However, the earliest phase (1670–1720) has not been thoroughly investigated since the 1960s.[5] The sources for this time period—other than some rare mentions in parliamentary debates and the London city records—consist primarily of short, ephemeral printed matter attesting to the founding, rivalry, and reciprocal competition among different insurance models. Since none of the first insurance companies survived, there are no individual corporate archival materials from this time.[6]

The earliest insurance project was submitted to the Privy Council by William Riley and Edward Mabb in 1635, and in May of 1638 they applied for a patent for

3 The concept of innovation culture is introduced somewhat differently for the twentieth/twenty-first century within the history of technology; cf. Ulrich Wengenroth, "Vom Innovationssystem zur Innovationskultur. Perspektivwechsel in der Innovationsforschung," in Johannes Abele et al. (eds.), *Innovationskulturen und Fortschrittserwartungen im geteilten Deutschland*, Cologne 2001, pp. 23–32.

4 Cf. R. J. Ryan, *A History of the Norwich Union Fire and Life Insurance Societies from 1797 to 1914*, PhD thesis, University of East Anglia 1983.

5 Peter G. M. Dickson, *The Sun Insurance Office 1710–1960. The History of Two and a Half Centuries of British Insurance*, London 1960, pp. 1–16; Harold E. Raynes, *A History of British Insurance*, 2nd ed., London 1964, pp. 70–83; Hugh A. L. Cockerell and Edwin Green (eds.), *The British Insurance Business. A Guide to its History and Records*, 2nd ed., Sheffield 1994, pp. 26–28.

6 Raynes writes simply, "the time was ripe" for this innovation (*History*, p. 74).

the idea of a transfer of the premium-based insurance principle to real estate. Information about that project is very scarce but it seems comparable to the continental Straßburg and Oldenburg cameralist precursors to the Hamburg project, which will be touched on shortly below. It was approved in October of that year; however, there is no trace of it in later years.[7] The first official Fire Office was founded in London in September 1681 as a joint-stock company; its founding rendered irrelevant a municipal project that had long been under discussion. A certain Benjamin de Laune is credited with presenting the first project devoted to insuring houses against fire to the magistrate in 1669–70.[8] In 1674, 1675, and 1679, the businessman Augustine Newbold repeatedly discussed the establishment of a municipal fire insurance with the successive mayors of the city. This idea was even employed for three months in an attempt to control private competition. While Newbold addressed his project suggestions to the city authorities, seeking an implementation of insurance "from above," private insurers organized a type of grassroots effort, based on advertising, attempting implementation "from below": they called a number of "general meetings" of all people interested in insuring their homes in order to discuss the ways and means of actualizing insurance among interested parties. The placard printed for the opening of the Fire Office explicitly punctuates the consideration behind this approach:

> The [Fire] Office must undergo the following test:
> This Scrutiny this Office was to pass: For though some Publick Designes, whose Fund have been Reputation, & Personal Credit, have Thriven best without Examination, & (of a suddain) Raised Themselves into great Reputation; yet sometimes (Mushroom-like) they have had their Duration and Rise equal: their Creditors having lost their Money, before they had the Leisure to consider why they Trusted them. So, on the contrary, those Designes, whose Fund are in Land, being Real and Certain, can have no greater Advantage, than to have their Foundations Examined by those that Trust them: And though such Deliberation may make their Beginnings and Growth slow; yet They will be like the Oak, Durable.[9]

Barbon and his colleagues aimed to build trust among future depositors through early involvement in the process; the importance of this basic element of "trust" for their project was clear to them. Four central issues were publicly debated and agreed upon by the majority: the amount of necessary capital,

7 Dickson, *The Sun*, p. 3; Raynes, *History*, p. 73f.

8 *Observations on the Proposals of the City, to insure Houses in case of fire*, London 1681, f. A1r.

9 *An Advertisement from the Insurance-Office for Houses*, September 16, 1681, s.l., p. 1.

the modality of insuring it, how to manage the capital, and how to manage the entire enterprise. They agreed on 40,000 pounds as the amount of basic capital necessary to insure 5,000 homes—with 10,000 pounds more per additional 5,000 buildings. To secure the capital, ground rents of 2,100 pounds were charged in the three quarters of the city that had been reconstructed yet again. Since many homes had still not been rebuilt and since the amount that could be charged was therefore still insufficient, they decided to initially insure only up to 3,000 buildings. The assets were held in escrow by a group of ten men selected for this purpose, and they elected an eight-person council, appointed to make decisions under the supervision of a committee. The form of the policy was fleshed out, and Samuel Vincent, Nicholas Barbon, Benjamin Bartlet, and Samuel Tookey were granted authority to issue the policies. It is thus safe to assume that it was, in fact, these four men who were actually involved with the everyday business of the Office located behind the Royal Exchange.[10]

After a number of arguments, otherwise known as the "general meetings of the shareholders," the Fire Office was equipped with a large amount of startup capital. From this position, it was able to deploy propaganda to fight the municipal competition that was introduced on December 1, 1681. The Fire Office advertised that due to its financial security, derived from trust-managed ground rent earnings, it could offer a higher level of security. Thus, for all subsequent 170 fire insurance agencies founded before 1850, the Law of Trusts, adopted in 1673, was customarily employed in this manner, so that the trust-managed assets could not be executed by the creditors of the trustee, which limited the liability of the shareholders to their contribution to the company.[11] Barbon's Fire Office diversified itself from the outset into two risk categories (though this term did not yet exist): brick buildings and wood buildings. The premiums for the latter were estimated at twice the rate of brick. In competition with the municipal insurance, Barbon & Co. published the rates and wrote insurance contracts for a maximum of 31 years (see Table 2).

There is no indication of how, exactly, the Fire Office first established these premiums for the fire insurance industry. The baseline value is based on 5 pence per pound, which was then extrapolated with the inclusion of certain discounts (5 years payment for 8 insured years, etc.). The fact that Barbon actively engaged

10 To entice the first customers, a rebate was offered until November 1 (pay for five years of insurance and get eight, etc.) Among the trustees and the members of the council were a large number of important individuals, such as Sir George Treby, Recorder of the City of London, Francis Pemberton, Lord Chief Justice, and Sir William Williams of Chester, former Speaker to the House of Commons.

11 Pearson, *Insuring the Industrial Revolution*, p. 235.

TABLE 2 Table of insurance rates for Barbon's Fire Office in 1681: the far right shows the building values to be insured, ranging from 10 to 1000 pounds. On the far left is the corresponding ground rent per year for a house that still needed to be paid off in pounds (l<ibrum>.), shillings (sh.), and pence (d<enarius>.). The four middle sections show the premiums for a corresponding brick house insured for 8, 12, 21, and 31 years (*Observations on the proposal of the city* [1681], p. 4)

Pound Rent	Insured for 8. Years	Insured for 12. Years	Insured for 21. Years	Insured for 31. Years	Houses Burnt, Money Paid
Lib.	l. sh. d.	l. sh. d.	l. sh. d. ob.	l. sh. d.	l. sh. d.
1	00 02 01	00 02 11	00 03 11 0,5	00 04 07	10
2	00 04 02	00 05 10	00 07 11 0	00 09 02	20
3	00 06 03	00 08 09	00 11 10 0,5	00 13 09	30
4	00 08 04	00 11 08	00 15 10 00	00 18 04	40
5	00 10 05	00 14 07	00 19 09 0,5	01 02 11	50
6	00 12 06	00 17 06	01 03 09 0,5	01 07 06	60
7	00 14 07	01 00 05	01 07 08 0,5	01 12 01	70
8	00 16 08	01 03 04	01 11 08 00	01 16 08	80
9	00 18 09	01 06 03	01 15 07 0,5	02 01 03	90
10	01 00 10	01 09 02	01 19 02 00	02 05 10	100
20	02 01 08	02 18 04	03 19 02 00	04 11 08	200
30	03 02 06	04 07 06	05 18 09 00	06 17 06	300
40	04 03 04	05 16 08	07 18 04 00	09 03 04	400
50	05 04 02	07 05 10	09 17 11 00	11 09 02	500
60	06 05 00	08 15 00	11 17 06 00	13 15 00	600
70	07 05 10	10 04 02	13 17 01 00	16 00 10	700
80	08 06 08	11 13 04	15 16 08 00	18 06 08	800
90	09 07 06	13 02 06	17 16 03 00	20 12 06	900
100	10 08 04	14 11 08	19 15 10 00	22 18 04	1000

in accounting and that he had access to other statistics regarding the number of houses destroyed by fire in London since the great fire during the period from 1666 to 1681 is confirmed in another pamphlet.[12] He calculated roughly

12 Nicholas Barbon, *An Enquiry, Whether it be the Interest of the City To Insure Houses from FIRE; And whether the Insured may expect Any Advantage thereby, more than from the Insurance-Office already Setled*, S.l. 1681, p. 3.

£400 of damages per house, which for the part of the city covered by insurance (about half) came to £19,400 (i.e., £1,300 per year on average), but he emphasized that larger fires (a 40-building fire had already occurred in the interim in the Temple quarter) would lead to strong fluctuations. Barbon explained in a pamphlet directed against his closest competition, the Friendly Society—founded in 1683–84 as a mutual society—that "the Insuring of Houses being a New Design, *it is impossible to make a certain guess*" regarding the accuracy of the assumptions. He also admitted that given the insecurity regarding this new innovation, the premiums were initially fixed on rough estimates, which may tend to be a bit high. Three years later, the first experiential data had arrived: in the space of three years, 4000 buildings were insured at an average premium of £4.5 per house, making £18,000, of which £7,000 had already been dispersed before only a quarter of the policies had expired, and that it was therefore very probable "that the Premium will not be sufficient to pay the Losses."[13] Barbon's pamphlet is the only seventeenth-century pamphlet that shows traces of consideration about fire insurance problematics in the form of probability calculations (in a strict sense). In contrast to life insurance theory, there appear to be no other such applications of explicit probability and "chance" calculations in a fire insurance context until well into the eighteenth century.

In this pamphlet, Barbon seeks to prove that the specific process of damage assessment practiced by mutual societies—i.e., that each individual is charged a share of damages as they occur—is ultimately superfluous, since, in time, the financial burden placed on the individual is identical to the cost of replacing his uninsured house after a fire on his own, because what is initially saved would have been "lost by increase of the hazard." To clarify, Barbon explains the basic principle of the probability of a house fire: if one estimated that there were 100,000 houses within a district in which the city published an annual mortality table,[14] and 1000 of these houses are insured by the Friendly Society, then the chances would be 100:1 that one of the insured houses would burn in a fire. At the same time the probability of each individual house not burning would be 100,000:1. Thus, there is the same probability that all of the 1000 insured houses would burn as that every uninsured house would burn: 1000x100:1=100,000:1. He compared this probability ratio explicitly to a game of chance that had just be introduced by two Italians, Francisco Corbetta and Francisco Finochelli, and which had been given a royal patent in London: the "Lottery of the Royal Oak," in which a die was played on a field of

13 Nicholas Barbon, *A Letter to A Gentleman in the Country, Giving an Account of the Two Insurance-Offices; the Fire-Office & Friendly Society*, London 1684, p. 2.

14 On this also see the discussion of Graunt below in notes 47 and 48.

different painted figures:[15] the probabilities that a quarter figure (frequency 8:1) would appear four times and that a full figure would appear one time (frequency 32:1) were basically the same as the house/fire example, since 4x8=32. Similar to the mathematical probability theory in relation to the mortality tables, this game of chance offers a context for analogous transferences. According to this basic illustration of probability distributions, Barbon then accounts for the fire insurance ratios. He assumes that there are 310 brick houses in a city, each worth 100 pounds, of which 300 are insured by the Friendly Society and 10 are not. If one assumes that 10 houses are destroyed by fire each year, and that in one month 4 houses insured at 300 pounds combined burned down, an outlay of 20 shillings (20 shillings = 1 pound) for each of the 300 insured houses would cover the damages. If during another month two houses, insured at 150 pounds, were to burn, each homeowner would have to pay 10 shillings to cover all damages, etc. Barbon concludes that in 7.5 years, one quarter of the city would be destroyed by fire and rebuilt, thus 10 houses per year on average. That would, however, mean that each homeowner would have also lost one-quarter of the value of his own home: "those that don't burn down are [lost] through the payment of premiums on those that do burn down." Thus, in terms of value, after 31 years, all 310 buildings would have been destroyed by fire. Although actual individual houses may burn many times and others may remain undamaged, based on the levy and in terms of value, there is actually complete equality between the insured houses that were destroyed by fire and those that remained intact, as well as—and this is the important argument at this point—between the 300 insured and the 10 uninsured houses. While the uninsured houses may have also been affected by fire, the only difference is that the insured houses gradually lost their value through the annual payment of damage shares, while the uninsured houses lost their value all at once when fire struck. However, since the uninsured individuals did not have to pay a proportion of the administrative costs of the mutual society, and because they retained the capital of premiums not paid, they had the advantage of more freely available assets and did not have to deal with the hassle of managing their insurance accounts. So, from this point of view, insuring a house in a mutual society would make no sense at all—a reaction which was, by the way, not rare to be found later on in insurance history when poor house owners in small cities or villages rejected the offer to pay yearly premium sums or to participate in the collective share of damages: better not to pay those small sums in cash

15 Cecil L. Ewen, *Lotteries and Sweepstakes: An Historical, Legal, and Ethical Survey of Their Introduction, Suppression and Re-establishment in the British Isles*, New York 1932, pp. 95–98.

and risking the house to burn down—a house, often of simple wooden material and walls which would be easy to rebuild.[16]

In contrast, Barbon maintains that the Fire Office solution is more favorable for the insured, because the insured and insurers are not identical, and also because the insurers would have achieved a certain amount of security and value reserve through the mortgaged properties. Obviously, Barbon's comparison of both modes of insurance is clearly motivated by his competition with the municipal institution and by his interest in acquiring customers—the Fire Office had similarly calculated that all damages must in effect be covered by premium income, because otherwise it could never become a profitable business. Thus, in the end, the premium payer of the joint-stock model also pays for the fire damages suffered within the insurance collective. In any case, through the regular prepayment of premiums it was possible to reinvest temporarily uncommitted capital, thus facilitating a more flexible individual employment of the capital.

It is significant that Barbon actually argues in terms of the emergent theory of probability. His promotional pamphlet was written at a time that Anders Hald has characterized as a period of "stagnation" in the development of probabilistic mathematics that Blaise Pascal, Pierre de Fermat, and Christiaan Huygens introduced between 1654–64, and which led to a new developmental leap between 1708–18.[17] In 1662, during the phase which Ian Hacking places at the center of his argument on the development of modern probability theory, the first printed edition of Cardano's 1564 *Liber de ludo aleae* appears.[18] Eight years earlier, in 1654, Huygens had composed a small treatise, *Van Rekeningh in Speelen van Geluck*; in 1657, he wrote *De Ratiociniis in Ludo Aleae* (Dutch 1660, English 1690). Huygens represents the one side of probabilistic thought—the mathematical-numerical—while John Graunt's treatise on the Bills of Mortality of London (1662) represents the other, empirical-statistical side. Barbon uses Graunt's work in his three larger treatises and quite thoroughly in the *Apology* where he also refers to Graunt's calculations. Graunt determined the population size and population development of London, but Barbon used this data differently in order to estimate the building stock of the city based on the increasing number of inhabitants. He already employs a transference here from the sphere of discourse centered around the object "life" to the object "fire" (e.g., of houses), a transfer similar to the first and principal transfer of

16 Barbon, *A Letter to A Gentleman*, p. 4.

17 Anders Hald, *A History of Probability and Statistics and Their Applications before 1750*, New York 1990, p. 4.

18 Ian Hacking, *The Emergence of Probability: A Philosophical Study of Early Ideas about Probability, Induction and Statistical Inference*, Cambridge 1976.

the premium insurance principle from maritime transport to other spheres of insurance. It is not clear whether Barbon was familiar with Huygens or other works of early probabilistic mathematics. It is possible, since his studies in Leiden coincided with the celebrated publication of Huygens's work; he received his doctorate in Utrecht in 1661, and his general familiarity with developments in population calculation is evidenced by his use of Graunt's treatise.

It is thus not unlikely that Barbon was familiar with both sides of probability discourse and that he therefore represented one of the very rare exceptions, situated at the very beginning of the development of fire insurance, in which these two aspects were combined—even if this combination was only used rhetorically in competition with the mutual societies, and not (at least not obviously) operationalized for the calculation of premiums and the practice of insurance itself. During the entire course of the eighteenth century and even into parts of the nineteenth century, it was far more common for *fire* insurers (as was previously the case in the maritime context) to base their premiums and business calculations on rough experiential values of trial and error.[19] Before the *Society for Equitable Assurances on Lives and Survivorships*, which drew on James Dodson's calculations of 1750, was founded in 1762, the early statistical thinking about mortality and population development had not even been connected to the practical business of *life* insurance. Although empirically based mortality tables had circulated in all manner of scientific journals for almost a hundred years, widespread implementation lagged far behind.[20] Barbon's mode of argumentation is hardly surprising: the reference to the game of chance mirrors the common conception in early probability discourse that insurance and gambling were closely related. This idea played a role from late medieval times to the era of continental legal codification during the nineteenth century and appeared as a moral/theological and legal point of discussion in fifteenth-century academic discourse. Of course, since the supposed connection between insurance and gambling did not seem to cause any substantial prohibitive effects on the practice of insurance itself, this point is more important to historians interested in theory; starting in 1820 probability

19 A protoprobalistic mindset is already present among Renaissance merchants. See Giovanni Ceccarelli, "Coping with Unknown Risks in Renaissance Florence: insurers, Friars and Abacus Teachers," in Cornel Zwierlein (ed.), *The Dark Side of Knowledge. Histories of Ignorance, 1400 to 1800*, Leiden/Boston 2016, pp. 117–38. However, there was no relationship between early *abbaco* teachers' mathematical theory and practice.

20 Cf. the classical perspective: Lorraine Daston, *Classical Probability in the Enlightenment*, Princeton 1988, passim, esp. pp. 137, 172–74. Pearson, *Insuring the Industrial Revolution*, p. 4 reconfirmed this "antistatistical bias" after having analysed most of the handwritten record of the major eighteenth century fire insurance business remaining for Britain.

discourse began to separate the mathematics of insurance from calculations oriented purely at games of chance.

The problem for Barbon, however, was that he had to assume a constant number (10 houses per year) in order to facilitate a numerical-probabilistic treatment of the problem of the unit—i.e., the annual frequency of fire—which, in reality, remained yet to be empirically determined. The whole question of how many houses per year, and more specifically, the question of the average value of damages per year, in relation to the scope, development, and quality of the building stock—thus a truly statistically based probabilistic calculation—is necessarily excluded from his argument. The *Apology for the Builder* of 1685 suggests that Barbon definitely intended to evaluate the building stock protostatistically in a Grauntian sense, and his 1681 pamphlet *Enquiry* also makes it clear that he collected data about houses destroyed by fires in London, or at least that he had a source for those data, which even included details concerning, for example, numbers of brick buildings versus wood buildings that burned down (Table 3).

TABLE 3 Table of houses burnt in different parts of London between 1666 and 1681 as empirical foundation for his argument for fire frequency and demands for building, Barbon, *An Enquiry*, p. 3.[a]

The Particular of the Houses Burnt within the City and Liberties, since the Great Fire, in Number Ninety Seven, are as followeth, viz.

Brick		Brick		Timber	
Warwick-Lane	1	Bartholomew-Lane	1	Jewen-Street	2
Cole-Harbour	2	Knight-Rider-Street	1	New-Cheapside	1
Budge-Row	2	King's-Bench-Office,		Barbican	2
Mincing-Lane	1	and White-Fryars	9	Cammimile-Street	2
St. Clements-Lane	2	Fetter-Lane	2	The Navy-Office	15
Cheapside	2	Temple	40	Bishopsgate-Street	6
Dyers-Hall	2	Pauls Church-Yard	1	Bishopsgate-Street	2
Bow-Lane	1				

[a] Barbon was well informed about the fire incidents in London, as he offered the service of fire-fighters for those who were insured with his Fire Office; other insurance companies followed that example (each with ca. 30 men). Then followed about 15 privately run 'waterwork' companies emerged which were active during the eighteenth century, the first such company having been established already in 1581 (Carry van Lieshout, "'The Most Valuable Means of Extinguishing the Destroying Fires': Fire-fighting and the London Water Companies since the Long Eighteenth Century", *The London Journal* 42, 1 (2017), pp. 53–69).

Collectively, these observations tentatively contribute to a reformulation of the idea of a complete breach between theory and practice: while the projects from the 1630s until Newbold's published project *London's Improvement* of 1679 already included many major elements of fire insurance, Barbon's project of 1681 distinguishes itself not only because it was the first project that was actually implemented in practice, but also because there was an actual protostatistical and protoprobabilistic liaison between theory and practice that he attempted to implement for the notion of fire risk, which most likely also affected the calculation of premiums—even if this detail is not precisely documented in *written* form. For the history of science and insurance, this means that an element that was not widely accepted until 1750 (and then only in terms of life insurance) had already been introduced in the context of fire insurance around 1680, and that the gap between insurance practice and probability theory actually began to widen again *after* Barbon. The gap could emerge again without consequence from this point on, because due to the proliferation of competing insurance practices, the calculation of premiums began to function in a quasi-evolutionary manner through modes of imitation, modification, and improvements discovered through the trial-and-error process within the market. Why should a fire insurance company concern itself with complex mathematical theories if the business was running and nobody else engaged with theory?

In another context, the situation around 1680 anticipates a further important topic: the pamphlets that the Fire Office and its opponents published in 1681–82, which were distributed free of charge in the Office and around town to interested parties,[21] concerned themselves with the competition between the municipal, "state" project and the private Fire Office of Barbon & Co. The aggressive tone taken by a private company and a private entrepreneur like Barbon to defend his business against the city is absolutely astounding. This tone was also employed by the city and its advocates to write similar attacks against Barbon—clearly, both parties claimed equal footing in the context of these reciprocal argumentative exchanges. Barbon introduces several reasons as to why private insurance is superior to state/city insurance. The vested interest of the private insurers guarantees its careful management: "Mens Credit, Care and industry, are more concerned to preserve *private Interest*, than publick: By this means fires will be better prevented and extinguished. The ill Management and Loss of the Corporation affects no Particular Mans Credit or estate."[22] The administration of the city, it was argued, would also always

21 At the end of the printing address was often the note: "to be had Gratis [...]."

22 Barbon, *An Enquiry*, p. 2.

proceed more slowly, and concerning cases of dispute in particular, it would be much more difficult for customers to demand justice, because the defendant and judge would be united in a single institution:

> The Rule of Justice is not alwayes the same betwixt Private and Publick, as between one private man and another; a man must part with All to satisfie his Debts. But if a very Great Loss should happen to the City, the Rule in paying may not be the same: For the Maxime of Government, is, *That the private concerns should suffer, rather than the Publick be Undone*: because, in the Welfare of the Publick, All are concerned.
>
> Lastly, if the City refuse to pay a Loss; The Remedy is not so easie and certain (by any provision they can make) as it is against Private persons: the Course of the Law is not so easie to Sue a Corporation, as a private person: Besides, all process of Law must come to the Sheriffs, who are Members of the City; if they refuse to serve them; they are Fined, and the City have the Fines: So that at last, The strength of their Security will much Depend on the honour and Justice of the Corporation.[23]

It is significant how often Barbon treats the public sector as a self-serving counterpart, even though he initially emphasizes the positive effect of self-interest, precisely because in his self-interest as a business owner (involving the safeguarding of the goods that he uses as collateral) he is necessarily bound to the interest of his customers. However, he then criticizes the city's interest in the common good as self-interest on behalf of the administrative body, which is not bound to the interests of the customer, but rather runs contrary to their interests.

The city defended itself through a pamphlet written by an anonymous author, in which Barbon is personally attacked as:

> a person whose Chamber-door daily suffers an Assault and Battery, from a whole posse of Bricklayers, Carpenters, Masons, etc. and who, if one may judge by the several Actions of Debt that every Term are brought against him, is in a fair way to Retire for Refuge into one of his own Chambers, which he hath built in the Temple; or by one, who tho he may be thought to be shot up into some considerable height from a low and ordinary Condition […] [Insuring is] too much for one that hath made it his business to run up a multitude of Houses, in the adjacent parts, as if

23 Ibid.

> he were resolved to drain the City of its Inhabitants and Trade. I think I may too truly say, He hath done almost as much prejudice to the City by building Houses, as the Papists by firing them.[24]

Here Barbon is presented as a building speculator: in actuality, as will be shown, there were indeed constant complaints about the low-cost housing he built, which even led to riots at his private residence. Barbon is represented as a dubious businessman whose rampant building speculation destroyed the city almost as fast as the Great Fire itself; accusations of "papistry" were also levelled against him in public discourse. From a continental European perspective, it is indeed strange that the city refused to discipline or bring to justice a citizen who accused city authorities of disreputable behavior and instead engaged with him on the same level of controversy, defamation, and counterdefamation. I know of no comparable case prior to the nineteenth-century within the central European history of insurance. In contrast, in central Europe it was believed that insurance belonged solely in the hands of the state.[25] Even when semigovernmental bodies like the Gotha insurance bank were criticized—in the 1820s in Germany, for example—it was never the state itself that was attacked; rather, blame was placed squarely on the founder of the organization, who made the state complicit. It is remarkable that municipal insurance in London disappeared after only a year and that Barbon's Fire Office emerged victorious from this first competitive dispute. It highlights in an international context how easily early capitalist entrepreneurship in England could freely and strongly establish itself at this point; inversely, of course, it also shows how relatively weak London's municipal administrative bodies were at this time.

1.1.2 Nicholas Barbon: Building Speculator and Growth Theorist

Barbon is known less for his founding of the Fire Office than for his three small general treatises on political economics that secure a place for him in the pantheon of the English political-economic history—a presumably heterodox outsider who enjoyed a strangely selective reception by other outsiders

24 L.R., *To my Honoured Friend Mr. M.T. one of the Committee chosen by the Common Council of London, for the Insuring of Houses from Fire*, London 1682, p. 2f.

25 "Denn diese Anstalt ist überhaupt so beschaffen, daß sie allerdings eine Ordnung der obersten Gewalt seyn muß, und einer Privatgesellschaft nicht wohl anvertrauet werden kann." (Johann Heinrich Gottlob Justi, "Vorschlag von Verbindung der Feuerassecuranzsocietäten mit einer Leihebanco auf die Häuser," in Johann Heinrich Gottlob Justi, *Neue Wahrheiten zum Vortheil der Naturkunde und des gesellschaftlichen Lebens der Menschen*, vol. 1.1, *Vor die Monate Jenner und Hornung 1754*, Leipzig 1754, pp. 561–82, 567).

until reaching the pinnacle of his recognition in Karl Marx's *Capital*.[26] Barbon would have been clearly categorized as a protoliberal thinker of his time on the basis of concepts such as "rent," "interest," "value," and the "balance of trade" had he explicitly opposed the false positions of John Locke;[27] in not doing so, he aligned himself with a group of authors and anonymous publicists, such as Dudley North, Dalby Thomas, Henry Martyn, Francis Gardner, James Hodges, Henry Layton, and John Houghton who, countering the balance of trade theory, had long advanced the theory of economic growth that was finally synthesized by Adam Smith ninety years later. As Joyce Appleby argues, economic theory ran ahead of social development in this case. While the balance of trade classics (by Thomas Mun, Gerard Malynes, and Edward Misselden) maintained that there was an inflexible conception of domestic demand and consumption within the strictures of class society (the aristocracy consumed luxury, the poor adequate nutrition) and the export alone was as growth-enhancing, Appleby demonstrates that demand conditions can cross class and be generally stimulated and that the growth of general prosperity is promoted by consumption.[28]

---

26 Elgin Williams, "Nicholas Barbon, An Early Economic Realist," *Southern Economic Journal* 11 (1944), pp. 45–55, 46. On Barbon, cf. Norman G. Brett-James, "A Speculative London Builder of the 17th Century, Dr. Nicholas Barbon," *Transactions of the London and Middlesex Archaeological Society* 6 (1933), pp. 110–45; Andrea Finkelstein, "Nicholas Barbon and the Quality of Infinity," *History of Political Economy* 32 (2000), pp. 83–102; William C. Baer, "The Institution of Residential Investment in Seventeenth-Century London," *Business History Review* 76 (2002), pp. 515–51; for a more specific and more extensive context of building speculation, see Lisa Margaret Bowers Isaacson, *The Building Society: Speculative Building in the West End of London, 1660–1760*, PhD thesis, Princeton University 1992; Elizabeth McKellar, *The Birth of Modern London. The Development and Design of the City 1660–1720*, Manchester 1999; William C. Baer, "Is Speculative Building Underappreciated?," *Urban History* 34 (2007), pp. 296–316; James H. Ullmer, "The Macroeconomic Thought of Nicholas Barbon," *Journal of the History of Economic Thought* 29 (2007), pp. 101–16; James H. Ullmer, "William Petty, Nicholas Barbon, Multiples, and a New Hypothesis of Inevitable Discovery," *Journal of the History of Economic Thought* 31, 3 (2009), pp. 341–59.

27 Hannah Robie Sewall, *The Theory of Value before Adam Smith*, New York 1901, pp. 58–60.

28 Joyce Appleby, "Ideology and Theory: The Tension between Political and Economic Liberalism in Seventeenth-Century England," *The American Historical Review* 81 (1976), pp. 499–515. Barbon thus is one of the theoreticians who supported the theory that even in the era of mercantilism there was a true conception of overall economic growth. See Steven Pincus, "Rethinking Mercantilism: Political Economy, the British Empire, and the Atlantic World in the Seventeenth and Eighteenth Centuries," *The William and Mary Quarterly* 69, 1 (2012) p. 22 n 66; Barbon also can be included within the interpretation of the "Hartlib" circle and the associated concept of "infinite improvement" (Carl Wennerlind, *Casualties of Credit. The English Financial Revolution, 1620–1720*, Cambridge/Mass. 2011, pp. 117–19); this progressivist part of the mercantilist discourse, made prominent by Pincus, has been now adressed in a more comprehensive way by Paul Slack,

Nicholas Barbon was a physician; as already briefly mentioned, he had studied in Leiden, received his doctorate at the age of 21 in Utrecht in 1661, and since 1664 had been a honorary member of the College of Physicians. Primarily, however, he was engaged as an entrepreneur, and he partnered with John Asgill to found his Land Bank in 1695, which operated similarly to a building society, providing saving and loan facilities. Its main function was to collect savings deposits in order to secure advances on mortgages.[29] He was the owner of large buildings, which he either repurposed or tore down and replaced with a variety of small new buildings—for example, former riverside villas in the West End, York House, and Essex House, which he transformed into an ensemble of taverns, pubs, bakeries, and breweries in 1672–74.[30] He was also engaged in technical invention projects: he secured one patent that is known of, in 1694, for a machine that harnessed the tides in order to pump drinking water out of the Thames and into the city and simultaneously remove sediment deposits from the river. This "water works" was intended to also function as a water pumping station for the purpose of firefighting.[31] Through his construction businesses he often came into contact with the political and intellectual elite (Matthew Hale, Robert Hooke) and several "heterodox" like John Wilkins and John Asgill; he also had a positive opinion of Campanella and Machiavelli. It is believed that his father was a leather salesman and the Baptist-friendly preacher Praisegod Barbon (1595–1679), who sympathized with millennial movements such as the Fifth Monarchy Men. There is, however, certainly no indication that he belonged to any exclusive "standard" scientific network, such as the Royal Society. Like Graunt, who influenced him, he belonged to a rather different breed of thinkers who were "makers" above all and whose publications were essentially used to legitimize their business activities. They were much more than Leibniz men of practice, with whom the history of ideas has a problematic relationship—for example, when one asks for the "influences" of this or that particular small idea unit, e.g. by Giordano Bruno, on their thought. In this sense, one unusual aspect about Barbon is that he seems to diverge at several points from

---

The Invention of Improvement. Information and Material Progress in Seventeenth-Century England, Oxford 2015.

29 John Keith Horsefield, *British Monetary Experiments 1650–1710*, London 1960, pp. 197f., 206–09.

30 Bowers Isaacson, "The Building Society," p. 60; Brett-James, "A Speculative London Builder", p. 321.

31 Finkelstein, "Quality of Infinity," p. 95, cf. for the context van Lieshout, "'The Most Valuable Means'".

all classical mercantilists of his era when it came to ideas of infinite wealth, an unending "natural increase" of economic goods and the inexhaustibility of natural resources—likely because of his experience in economic activity and his fundamental understanding, in contrast to Graunt, that increasing population growth leads to an increase in trade and goods, rather than a decrease.[32] In the literature on economic history and of the history of ideas presented here, there is very little mention of Barbon's insurance project—his growth predictions are referred to in these texts, and thus his activity as a building speculator is mentioned but not related to economic theory.

The history of London's reconstruction after 1666 has been thoroughly examined and is imbued with an almost mythical character—one case in point is Thomas F. Reddaways *The Rebuilding of London after the Great Fire,* which appeared in 1940 during the first year of the German bombing of London and was thus charged with a double historical significance, which unquestionably contributed to the topicality of the title.[33] The concentration in most of these studies is almost always fixed on the large and unrealized new building plans, representative buildings such as Christopher Wren's St. Paul Cathedral, and the involvement of political and intellectual elites.[34]

---

32 Finkelstein, "Quality of Infinity," p. 97. Finkelstein clearly shows how even a very heterodox religious position may have lurked behind Barbon's ideas. Cf. the key passage that is cited from Nicholas Barbon, *A Discourse of Trade*, ed. Jacob H. Hollander (1690; Baltimore 1905), p. 10f., in which the mercantilist *locus classicus* is borrowed from Cato ("oportet patrem familias vendacem esse, non emacem", *De Agri Cultura* 2,7): "The Native Staple of each Country is the Riches of the Country, and is perpetual, and never to be consumed; Beasts of the Earth, Fowls of the Air, and Fishes of the Sea, Naturally Increase: There is Every Year a New Spring and Autumn, which produceth a New Stock of Plants and Fruits. And the Minerals of the Earth are Unexhaustable. [...What] is Infinite, can neither receive Addition by Parsimony, nor suffer Diminution, by Prodigality" is actually quite remarkable; however it seems a bit of a stretch to locate an influence from Giordano Bruno here. The conjunction between *Parsimony* and *Prodigality* seems more like a maxim that Barbon associates elsewhere with Machiavelli. Cf. below n. 56.

33 Thomas F. Reddaway, *The Rebuilding of London after the Great Fire*, London 1940; Cooper, *A More Beautiful City*; Adrian Gilbert, *The New Jerusalem: Rebuilding London: Christopher Wren, the Great Fire and the Royal Society*, London 2002; Baker, *London: Rebuilding*; Field, *London, Londoners*, pp. 59–98. For the wider context of the architects' immersion in intellectual life, learned societies and the public administration (of which the rebuilding of London is just one—however the most exceptional—example), cf. Matthew Walker, *Architects and Intellectual Culture in Post-Restoration England*, Oxford 2017.

34 In place of multiple examples, see instead the integration of the economic significance of fire and reconstruction in Kenneth G. Davies, "Joint-Stock Investment in the Later Seventeenth Century," *The Economic History Review* 4, 3 (1952), pp. 283–301, 288, even though this text concerns itself primarily with longer-term economic developments.

London's post-1666 reconstruction is hardly mentioned in the context of the history of speculative entrepreneurial building—but its climax between 1680 and 1720 is in fact one of the strongest early cases of speculative building—and this is an important and very much neglected context for the emergence of the first fire insurance companies which coincided precisely with this building boom, shortly *after* the reconstruction following the Great Fire. This phase is particularly remarkable: after the resurgence of entrepreneurial speculative urban development in Antwerp during the 1540s (the largest building speculator being Gilbert van Schoonbeke) and in Amsterdam, post-1670 London probably represents the most important example in world history up to that point of a new emergence of entrepreneurial speculative urban development since Antiquity—especially because building speculation was hardly present in the Middle Ages, even, for example, in Renaissance Italy. Nicholas Barbon (though he died in considerable debt) was the largest speculative entrepreneur in this sector and a pioneer in this area; he even owned his own brick factory and strongly pushed building growth in London through his drive for profits. The rebuilding of 9000 houses on the 'ground zero' of the burnt 13,200 houses in a more regulated form is only a part of that building history around since the 1670s.[35] A good study of growth in areas next to the city of London proper, especially in the West End and East End, is a still unpublished 1992 dissertation by Bowers Isaacson.[36] In the absence of other sources, Isaacson evaluates the files of the Westminster Commission of Sewers, which was first established by Henry VIII but functioned regularly only from 1622. The Commission's competency did touch only at its most eastern parts on the region burnt in 1666, but setting aside the Fire court registers, for the burnt area, we do not possess such a source which would help us to understand the character of speculative building of large complexes of houses in that time.[37] The Commission handled disputes and negligence regarding the duty of citizens to ensure that the waste and sewer canals located on their properties that drained into the Thames were clear and operational. In addition to many problems related to the disposal of waste and pollution, the protocols also incidentally yield important information about building history, since they include complaints that addressed issues associated specifically with new buildings

35 Field, *London, Londoners*, p. 66.

36 Bowers Isaacson, "The Building Society," p. 45 for a map of London with the expansion of building development between 1600 and 1750.

37 Cooper's, *A More Beautiful City*. Field, *London, Londoners*, pp. 59–98 (who has analysed the data from the Hearth tax for a reconstruction of the composition of the city after the rebuilding).

constructed by speculators and the housing and pollution problems that went along with them.

According to Isaacson's evaluations of these files, the major construction boom directly followed the 1688 Glorious Revolution with a further small peak from 1720 to 1730.[38] The commission identified a total of 296 building projects between 1661 and 1730: one-third of the projects consisted of fewer than 5 houses, one-half included fewer than 10 houses, and two-thirds of the projects comprised fewer than 50 houses. Only a few projects were as big as Richard Frith's 1679 project, in which 257 houses were built in the Soho fields and space was prepared for another 300. Frith is thus among those individuals who often appear in the files of the commission as a plaintiff or defendant. Frequent appearances generally indicated that the individual was a major player in building speculation; Nicholas Barbon was at the forefront in this regard.[39]

Speculative construction essentially means that buildings are not built for specific individuals, but rather for the nascent housing market itself. Barbon maintained that this would not lead to a housing surplus, because the builders would build first-, second-, and third-class homes (according to the classification of the Rebuilding Act of 1667) in line with the needs of the market. However, Daniel Defoe reported that the activity of the builders in the West End in 1725 clearly did lead to entire city blocks that were empty and ready for sale, as indicated by advertising placards on the front doors.[40] Barbon was quite the profit shark, in spite of claims to the contrary in his verbose *Apology*. There were constant complaints lodged against him by the mason and roofing guilds because of his use of poor materials; disappointed customers often sued him because their houses collapsed, and his streets became slums—in 1684 there was a riot in front of his house protesting one of his buildings in Gray's Inn Fields. Roger North, a contemporary of Barbon, summarized that "all his aim was profit;" North also claimed that Barbon invented the practice of speculative building. The process first involved separating the land into small parcels, which were then individually sold to subcontractors via building leases, who then built houses and assumed the risk associated with actually finding people

38 Cf. the tables in Bowers Isaacson, "The Building Society," p. 66.

39 Anthony Ball (cutler) is named six times; Nicholas Barbon (Doctor of Phys.), Richard Frith (bricklayer), Huntley Bigg (Scrivener), Thomas Ingram (ironmonger) are named five times; John Rossington, Robert Meades, Samuel Aubrey (coachmaker), Charles Peter are named four times; Joseph Robinson, Thomas Lawrence (carpenter), James Pollett, Nathaniel Behoe (bricklayer), Richard Avery, John Meard jr. (carpenter), Benjamin Timbrell (carpenter) are named three times (Bowers Isaacson, "The Building Society," p. 103 n 47).

40 Bowers Isaacson, "The Building Society," p. 89.

to purchase the individual houses.[41] Building speculators were perceived as a plague on the city because their activities were not subject to any regulations or permits: what one contractor built on the one side of a street was not required to have any planned coherence with what was on the other side of the street, which sometimes led to rather absurd effects.[42] Barbon was accused of building houses that lasted precisely as long as the lease contract, after which the roofs would predictably collapse. But Barbon was not representative of all builders: a majority of the over 240 individuals who appeared in the registers of the Westminster Commission of Sewers as builders were respectable craftsmen and small business owners in other trades, who took advantage of the lack of a building trade guild in order to invest a certain sum of their earned assets in the construction of a new home and, while they were at it, to build a couple of neighboring houses that they could then rent.[43]

From the 1670s onwards, there was an active public debate on speculative construction and the resulting rapid growth of London. This was based in part on older legislative discussions about construction and urban growth since the reign of Elizabeth I and was a relatively strong and pointed public discussion: repeated requests for limitation and taxation laws against building speculators often came before both houses of parliament.[44] Barbon used political and economic arguments to defend his practice: in 1678 he published his *Discourse Shewing the Great Advantages that New-Buildings, And the Enlarging of Towns and Cities Do Bring to a Nation*, in which he enumerated the great benefits heralded by the "new-buildings" and the enlargement of the cities of the nation.[45] As major statistical sources he relied on the death lists that had been recorded since 1592, the so-called Bills of Mortality, as well as on their evaluation by John Graunt[46]—that early demographical work that calculated, for the first time, the size of London's population more or less correctly (at 384,000), brought to light the extremely high child mortality rate and the

41 Norman G. Brett-James, *The Growth of Stuart London*, London 1935, pp. 330–42.

42 Two builders on both sides of Newport Street litigated for the entire year about who should lay a new foundation and build according to the standards of the other so that half the houses would not flood every time a rain cloud threatened (Bowers Isaacson, "The Building Society," p. 91).

43 Bowers Isaacson, "The Building Society," pp. 92–94.

44 Carolyn Andervont Edie, "New Buildings, New Taxes, and Old Interests: An Urban Problem of the 1670s," *The Journal of British Studies* 6, 2 (1967), pp. 35–63.

45 Nicholas Barbon, *Discourse shewing the great advantages that New-Buildings, And the Enlarging of Towns and Cities Do bring to a Nation*, London 1678.

46 John Graunt was a friend of William Petty and was inducted into Royal Society on the merit of his *Observations*.

higher mortality rate of London in relation to rural areas, and pointed out the critical fact (for Barbon) that London's population, like the rural population, continually increased.[47] It has been demonstrated in the history of science that Graunt's work, in relation to the knowledge of the time, achieved an incredible degree of exactitude and innovation by combining the general methods of Francis Bacon—whose *History of Life and Death* he cites—with a lay-scientific perspective developed through his experience as a merchant. This facilitated the transfer of mercantile accounting techniques to calculations concerning the balance of life and death in a society.[48] The fact that this mode of thinking was familiar to the philosophical, speculating entrepreneur Barbon, and that Barbon often goes so far as to engage in detailed criticism of Graunt's evaluations of the Bills of Mortality, is hardly surprising. Graunt's basic populationist approach—that population growth is critical for the strength of a nation—corresponds to mercantilist thought[49] and was also accepted by

47 John Graunt, *Natural and Political Observations Mentioned in a Following Index, and Made upon the Bills of Mortality* […] *With Reference to the Government, Religion, Trade, Growth, Ayre, and Diseases of the Said City*, London 1662. Cf. Geoffrey Keynes, *A Bibliography of Sir William Petty, F.R.S., and of Observations on the Bills of Mortality by John Graunt, F.R.S.*, Oxford 1971; Ian Sutherland, "John Graunt: A Tercentenary Tribute," *Journal of the Royal Statistical Society Series A* 126, 4 (1963), pp. 537–56.

48 Philip Kreager, "New Light on Graunt," in *Population Studies* 42 (1988), pp. 129–40; Hald, *A History of Probability*, pp. 82–105; Philip Kreager, "Death and Method: The Rhetorical Space of 17th-Century Vital Measurement," in Eileen Magnello and Anne Hardy (eds.), *The Road to Medical Statistics*, Amsterdam 2002, pp. 1–36; Daston, *Classical Probability*; Hervé Le Bras, *Naissance de la Mortalité: l'origine politique de la statistique et de la démographie*, Paris 2000; Kevin Schurer, *Surveying the People. The Interpretation and Use of Document Sources for the Study of Population in the Later Seventeenth Century*, Oxford 1992; Peter Laslett, "Natural and Political Observations on the Population of Late Seventeenth-Century England: on the Work of Gregory King and John Graunt," in Kevin Schurer and Tom Arkell (eds.), *Surveying the People: The Interpretation and Use of Document Sources for the Study of Population in the Later Seventeenth Century*, Oxford 1992, pp. 6–30.

49 E.g. in Thomas Mun, *England's Treasure by Foreign Trade*, London 1664, p. 12; Samuel Fortrey, *England's Interest and Improvement*, Cambridge 1663, p. 3. Cf. Lars Magnusson, *Mercantilism: The Shaping of an Economic Language*, London 1994. The ongoing discussion on the concept of Mercantilism as such cannot be summarized here *in extenso*, however, as far as I see, since 2011 the new discussion on Barbon and other exceptional mercantilists who recognized *growth* has not paid attention to his role as speculative builder and insurance office founder, but always concentrated on his *Discourse of Trade* cf. P.J. Stern and C. Wennerlind (eds.), *Mercantilism reimagined. Political economy in early modern Britain and its empire*, Oxford 2014; Moritz Isenmann (ed.), *Merkantilismus. Wiederaufnahme einer Debatte*, Stuttgart 2014; Cornel Zwierlein, *Imperial Unknowns. The French and British in the Mediterranean, 1650–1750*, Cambridge 2016, pp. 20–116; Cornel Zwierlein, "Security, Nature and Mercantilism in the Early British Empire," *European Journal of Security Research* 3 (2018), pp. 15–34.

Barbon. In 1685, Barbon released his reworked version of his 1678 *Discourse* to the building industry under another title: *An Apology for the Builder.* While particularly in political theory of early modernity, a medium size is often cited as the wisest and most resilient option for a city republic, Barbon defends a new conception:

> The Citizens, nay the whole Nation is astonished at the flourishing condition of this Metropolis, to see every year a new Town added to the old one [...] The Citizens are afraid that the Building of new Houses will lessen the Rent and Trade of the old ones [...] The Country Gentleman is troubled at the new Buildings for fear they should draw away their Inhabitants, and depopulate the Country, and they want Tenants for their Land. And both agree that the increase of Building is prejudicial to the Government [...] the design of this Discourse is to answer [...] these fears and false conceptions, by confuting these Popular Errors, and shewing that the Builder ought to be encouraged in all Nations as the chief promoter of their Welfare.[50]

As a reason for increasing the size of cities, he offers the constant "natural increase of Mankind" since the "first blessing of the Creation."[51] He argues that "building" is a natural, innate endowment of human beings, akin to foxes or other animals who build their own homes. He defends the consistent population growth since Adam and Eve with proof from a posthumously printed work by the former Lord Chief Justice Matthew Hale. Hale had a definitive role in the special court established by the city in 1666 after the Great Fire, which ruled on ownership and other disputes, and Barbon certainly would have known him at this time.[52] Barbon presents Hale's arguments in the context of Biblical and

50 Nicholas Barbon, *An Apology for the Builder: Or a Discourse Shewing the Cause and Effects of the Increase of Building*, London 1685, p. 2f.

51 Much literature on early modern political arithmetic and population discourse exists, cf. for the French and British contexts Zwierlein, *Imperial Unknowns*, pp. 60–72; for Petty e.g. Ted McCormick, *William Petty And the Ambitions of Political Arithmetic*, Oxford 2009; Rebecca Jean Emigh Dylan Riley and Patricia Ahmed (eds.), *Antecedents of Censuses from Medieval to Nation States. How Societies and States Count*, Heidelberg 2016. For the particularly German early modern population discourse, see Justus Nipperdey, *Die Erfindung der Bevölkerungspolitik. Staat, politische Theorie und Population in der Frühen Neuzeit*, Göttingen 2012.

52 Matthew Hale, *The Primitive Origination of Mankind Considered and Examined According to the Light of Nature*, London 1677; translated and printed in German by Heinrich Schmettau (Cölln an der Spree 1683).

other sources, including the analyses of protostatistical sources for the populations of England, Gloucestershire, and London, such as the *Doomsday Book* (1086) and the Bills of Mortality for London—again, according to Graunt. Barbon considers the most recent evaluations to be the most relevant and uses the *Doomsday Book* to compare recent population numbers with those of the past: according to his calcluations, since William the Conqueror the number of inhabitants in England had increased 30 times.

In addition to drawing support from classical writings and offering comparisons to other empires (such as America's new construction after the arrival of the Europeans), Barbon then begins to formulate a precise calculation, primarily based on the number of marriages in London (which he calculated at 10,000 annually), as well as the number of completed apprenticeships (which he estimated at 9,000), according to which a new requirement of 1,000 houses per year was generated.[53] Barbon also invokes the rise in annual mortality from approximately 13,000–14,000 around 1660–1 to the current mortality numbers of 21,000–22,000 deaths per year. Finally, he refers to the market: if too many houses were really built, as many falsely claimed, builders would not be able to sell their houses because the market would be overstocked; this, he maintains, is not the case. Barbon then develops a kind of "practical economics" doctrine regarding the consequences of construction. Expanding the city limits increases the value of the city as a whole: old houses in the downtown area become more valuable in relation to houses that gradually decrease in value as they near the city's periphery. The expansion of the city would be advantageous because of the increase in business opportunities within the city, but it would also bring benefits for the surrounding rural areas: the expansion of the city would lead to greater consumption of goods produced in these areas[54] and the extra city space could absorb the "supernumerary" and "useless" children of farmers and gentry. More widely, Barbon attempts to show that the growth of the "metropolis" is essential for a nation as a whole. He maintains that prohibitive legislation under Elizabeth I, James I, Charles I ("the martyr"), and a hapless special tax on building contractors under Oliver Cromwell ("the usurper") slowed the growth of London and led to the colonial emigration of many subjects. Growth of the metropolis, in his opinion, was essential for competition with other nations: like the heart within the body, the metropolis

53 Barbon, *An Apology*, pp. 15–18. He thereby disputes Graunt's estimates of the average number of people per house.

54 Paul Slack investigates this aspect in "The Politics of Consumption and Englands's Happiness in the Later Seventeenth Century," *English Historical Review* 122 (2007), pp. 609–31.

transports blood throughout the nation, and when a body has a great heart beating within it, no illness can arise for enemies to take advantage of.[55] Barbon concludes this section with a true Machiavellian judgment: Citizens—in contrast to aristocrats and royalty—make bad advisors; since they are reared through avarice and diligence, they punish vices too severely and are too stingy in rewarding the virtuous. The fear of the petite bourgeoisie toward expansion and progress must be scourged—they are simply incapable of thinking in larger terms.[56] With this argument, Barbon shifts to an imperial tone. Both the Ottoman sultan and the French king practiced the policy of burning hundreds of small villages in wars and depopulating them, in order to create growth in large cities; the Dutch also operated in this way, as evidenced by the example of Amsterdam. Since London had virtually become the "metropolis of Europe," it was necessary to support entrepreneurial growth according to the king's vision.

Behind these numerical arguments and the narrow scope of the economic concerns of a building contractor and founder of an insurance company, we find some far larger guiding principles, frames of thought and perceptions that legitimize aggressive imperial progress, and which link building practices and development of the city with the overall expansion of what was to become the British Empire at large: growth in the metropolis of London was placed within the context of the competitive relationships between the British "nation" and the European powers in colonial contexts. One would hardly expect to find such a perspective in an "apology" for London building contractors. In thought and action, Barbon thereby unifies an imperial model of growth based on business, population, and building stocks, with a plan for maintaining value represented by his insurance project. Nevertheless, he did not explicitly reflect upon the function of what he had founded—insurance companies—within this larger macroeconomic framework, as it has been

55 With this Barbon went against Graunt and his followers, who favored not the metaphor of the heart for the metropolis, but that of the head, and who explained the growth in London that was three times faster than in rural areas the land as the growth of an overly large head in comparison to a too-small body that has been diagnosed as sickly, Kreager, "Graunt," p. 139. The influence of William Harveys medical doctrine of blood circulation was obviously palpable here, which was also integrated as an analogy in the work of other economic thinkers of the time; Cf. Jacob Viner, *International Trade and Economic Theory*, Glencoe 1952, p. 37; Finkelstein, "Quality of Infinity," p. 85.

56 Barbon, *An Apology*, p. 31; probably abstracted vom Machiavelli *Principe*, chap XVI *De liberalitate et parsimonia*, but merged eventually with elements of *Discorsi* II, 19 and 30 (Niccolò Machiavelli, "Discorsi," in *Opere I*, ed. Corrado Vivanti, Turin 1997, pp. 160–162, 380f., 407–410).

recently underlined by James Ullmer. The position these companies came to occupy will be discussed presently.

1.1.3 Insurance Innovation and the "Financial Revolution"

This section will discuss the general and economic importance of fire insurance and its connection to the famous "Financial Revolution".[57] The first defining text devoted to the Financial Revolution (1967) was written by Peter Dickson who had, a few years earlier in 1960, produced the benchmark study of the corporate history of the Sun Insurance Company. Dickson subsumed insurance within the framework of the revolution he presents: insurance, along with banks, formed the "infrastructure of the Industrial Revolution."[58] Banks, like fire insurance companies, quickly proliferated after the founding of the Bank of England in 1695 (Barbon also founded his Land Bank in 1695) and like fire insurance, they also expanded in the provinces beginning in the mid-eighteenth century. In Dickson's second work on the Financial Revolution, he consistently—though typically marginally—mentions insurance agencies as investors in company stocks.[59] The central theory of the Financial Revolution was initially tied in to a wider development: the argument was that the Financial Revolution led the way for the Industrial Revolution and offered it a certain framework—which meant, above all, that state credit was significantly strengthened. First and foremost, a relevant securities/bond market developed, which made debt possible in the first place. The debt was "permanent for the state, but liquid for the individual," i.e., resellable and convertible into cash again on short notice.[60] The state could borrow to a certain extent and, in the face of battles and wars, could thereby access military-political capacities that simply were not possible before the Glorious Revolution of 1688. In 1989, Douglass North and Barry Weingast more clearly described these institutionally and historically significant changes in the history of the financial markets as a "fiscal revolution". This revolution had five key points: beginning in 1688, 1) the archaic fiscal system was replaced with a more efficient one; 2) the legislative and judiciary powers of the crown were restricted; 3) parliament reclaimed its

57 In the following I do not address the common question related to the "revolutionary" concepts in the history of science concerning revolution or evolution. I use the concept in the following only as a referent for the decisive economic and state differences around 1700 and, in particular, as a cumulative cipher for the research related to this topic.

58 Peter G. M. Dickson, *The Financial Revolution in England. A Study in the Development of Public Credit, 1688–1756*, London 1967, p. 7.

59 Cf. Dickson, *The Financial Revolution*, pp. 11f., 41, 94, 145f., and passim.

60 Dickson, *The Financial Revolution*, pp. 457–520, qtd. 457.

prerogative, particularly concerning tax issues, and thus arbitrary actions of the crown were made permanently impossible in this regard; 4) parliament became an increasingly independent actor of resource extraction and the control of expenditures; and 5) a balance of power between parliament and the crown was established. According to this theory, these five points influenced the new stability, and thereby the creditworthiness of the English government, resulting in the enormous growth of state debt from approximately one million pounds in 1688 to seventy-eight million pounds in 1750. This was seen not as a sign of weakness but rather of the strength of the English financial system. This functioning public-credit system, at the very least, would have offered an advantage over France and Spain during the wars of the eighteenth century.[61] The origin and spread of insurance appears in North and Weingast's work only in a footnote, in which they quote Phyllis Deane's work citing insurance companies as one of the actors who benefited from trade in this newly created market of securities and government bonds.[62]

In his book *Sinews of Power*, John Brewer addresses Dickson's theory and emphasizes the expansion of the effective tax extraction apparatus as a vital backbone for confidence in the solvency of the state and thus for the increase of state credit. According to Brewer, insurance companies play no role at all; the same conclusion is reached by Bruce Carruthers[63] and other more recent authors, who increasingly merge the Financial Revolution with the political Glorious Revolution and thereby increasingly write economic history as state financial history. The affinity between insurance and the Financial Revolution established in Dickson's work has been increasingly forgotten. For a time, the revolution was conceived primarily as an element within the historical development of the state.[64] The more recent contributions by Anne Murphy, Peter

61 Douglas C. North and Barry R. Weingast, "Constitutions and Commitment: The Evolution of Institutional Governing Public Choice in Seventeenth-Century England," *The Journal of Economic History* 49 (1989), pp. 803–32. On this, see three points of criticism in Bruce G. Carruthers, "Politics, Popery, and Property: A Comment on North and Weingast," *The Journal of Economic History* 50 (1990) pp. 693–98.

62 "The essence of the financial revolution of the early 18th century was the development of a wide range of securities in which new mercantile and financial companies—the chartered trading companies, the partnership banks, the insurance companies, etc.—could flexibly and safely invest and disinvest" (Phyllis Deane, *The First Industrial Revolution* [Cambridge 1979], p. 185; qtd. in North and Weingast, "Constitutions and Commitment," p. 825).

63 Bruce G. Carruthers, *City of Capital. Politics and Markets in the English Financial Revolution*, Princeton 1996.

64 John Brewer, *The Sinews of Power: War, Money and the English State, 1688–1783*, Cambridge/Mass. 1988, pp. 88–134. The thesis proposed by North and Weingast that the boom in the

Temin and Hans-Joachim Voth, Daniel Carey and Christopher Finlay, as well as Carl Wennerlind have effectively opened new perspectives concerning the Financial Revolution in the history of ideas through the analysis of political economy, above all, through detaching the concept of revolution from the state.

This seems to be a step in the right direction, since especially viewed from the perspective of continental Europe, the unusual ("revolutionary") aspect in the British situation is certainly not that the state becomes a potent debtor in the credit market, but rather that the financial system as a whole—including the credit market—reached such a potential (in spite of "bubbles") and such a level of stability that this secondary, state-supporting effect could also be successful. In more recent studies, the genesis of the private stock market between 1685 and 1695 has been decidedly interpreted as an autonomous development that emerged before state changes. Case studies from the archive of Hoare's Bank have contributed new insights into the decision-making process of an individual financial company during the genesis of capital markets.[65] However, a systematic positioning of the role of insurance companies in the genesis of capital markets fails to emerge from these studies.[66]

Although Robin Pearson used the programmatic title "Insuring the Industrial Revolution" for his comprehensive study on fire insurance, he does not address the preceding Financial Revolution. Instead, he endeavors to demonstrate that insurance is one of the dynamic branches of business in England, rather than simply a "parasitical" element that capitalizes on the actual motor

area of public credit was accompanied on the private business side by a boom in the stock market and that even private individuals could get credit has been recently relativized through the files of a banking house: The main profiteers of the financial revolution were primarily the state and political elites. After 1714 a stagnation actually settled in regarding the development of prices concerning private credit. See Peter Temin and Hans-Joachim Voth, "Private Borrowing during the Financial Revolution: Hoare's Bank and Its Customers, 1702–24," *Economic History Review* 61 (2008), pp. 541–64.

65 Anne L. Murphy, *The Origins of English Financial Markets: Investment and Speculation before the South Sea Bubble*, Cambridge 2009; Daniel Carey and Christopher J. Finlay (eds.), *The Empire of Credit: The Financial Revolution in the British Atlantic World, 1688–1815*, Dublin 2011; Wennerlind, *Casualties of Credit*; Peter Temin and Hans-Joachim Voth (eds.), *Prometheus Shackled: Goldsmith Banks and England's Financial Revolution after 1700*, Oxford 2013.

66 See the brief mentions of insurance in Murphy, *Origins*, p. 29; Wennerlind, *Casualties of Credit*, p. 245 touches on the reference to modernity regarding the system-supporting function of newer financial insurance contracts; in Temin and Voth, *Prometheus Shackled*, p. 111f. insurance (Ram's Insurance) is invoked as a business partner and investment object for Hoare's Bank.

of productivity development.[67] More precisely, however, he includes insurance within the initial question of his study: "If the industrial revolution was, at least partially, insured, did fire insurance ensure the industrial revolution?" Fire insurance would have at least possibly been an "institutional incentive" for the accumulation of material goods for investment and innovation. "Rather than determining economic development and structural change, much of British fire insurance was shaped by it. ... fire insurance might have ensured the general growth of the economy, if not the specific process of industrialization."[68] If Pearson thus more or less negates the suggestion in the title of his study that insurance directly contributed to the Industrial Revolution this could be a result of the internal perspective of his decidedly market-historical approach, which simply does not question the conditions of this market. The trade in increasingly reliable government bonds that arose due to the growth of the state may have been a form of security production that created a framework for the financial market. However, in the context of building speculation and its fundamental significance, the building plots, buildings, and ground rents that served as collateral in the emerging financial market imposed an entirely different significance on insurance.

Although they did not witness a comparable financial revolution, the German cameralists theorized the significance of insurance more explicitly: according to Georg Heinrich Zincke, anyone who insured a house had "in reality *another house in the bank,* or his current house [is] more secure, and this house thereby becomes a more solid component of his assets, it also thereby increases its worth;"[69] in 1754, Johann Jakob Moser added that an insured house was "not a mere ash heap, rather, even if it were to burn down, [it would

67 Robin Pearson, "Ein Wachstumsrätsel: Feuerversicherung und die wirtschaftliche Entwicklung Großbritanniens 1700–1850," *Zeitschrift für Unternehmensgeschichte* 44 (1999), pp. 218–34; Pearson, *Insuring the Industrial Revolution*, p. 364.

68 Pearson, *Insuring the Industrial Revolution*, p. 367f. If one does not stop the historical survey with 1850, as Pearson does in this monograph, then the question becomes not only one of insuring the industrial revolution, but also one of "insuring the empire" in its expansion of global outreach. David Ormrod, in *The Rise of Commercial Empires: England and the Netherlands in the Age of Mercantilism, 1650–1770*, Cambridge 2003, pp. 43–59, positioned the "fiscal revolution" market historically for foreign trade in this sense as an early phase of protectionism politics in the secular competition between England and the Netherlands, while later the massive expansion of the British trade took place. When, beginning in 1786, Phoenix, and beginning in the 1820s countless other English insurers established agencies abroad, the relationship between empire and insurance became a continuation of the relationship between insurance and the industrial revolution.

69 Georg Heinrich Zincke, "Vorrede," in *Leipziger Sammlungen von Wirthschafftlichen, Policey-Cammer- und Finantz-Sachen*, vol. 7, Leipzig 1751, pp. iii-liv, xxvif.

be] worth exactly as much, as what one insured it for, in cash."[70] Countless cameralists from Leibniz to Johann Heinrich von Justi to Philipp Gäng recognized that when individual players in the market accumulated credit, it also created "circulation" as such, thus contributing to economic progress (see below Ch. 5, 2). Despite the striking differences between a protoliberal and a state economic context, these statements by German cameralists can be virtually integrated as interpretative aids for the classification of insurance companies within the fabric of the English Financial Revolution. Without insurance, the rapid speculative house building would have quickly resulted in slums, and without insurance, speculation within the emerging stock market would have had no foundation in secured values. Insurance companies were thus a key component of confidence and the production of security, which was systemically important for the development of the capital market as a whole in conjunction with the state. They were also integral in providing capacity for the system to recover and regain stability through the accrued capital secured within the housing sector in spite of the financial speculation bubble, which was primarily caused by trading in stocks related to the colonial economy.

The "London/Barbon" laboratory of security innovation, and the set of conditions that facilitated the emergence of insurance in this context, was made possible through a number of coincidences:

- The scientific and historical conditions of a publicly acceptable combination of practical empiricism and early mathematics of probability.
- The event of the Great Fire of 1666, which triggered the speculative construction of housing, which created a "boom" in the period following 1680 in the form of the Financial Revolution.
- The emergence of the financial market in the 1680s/1690s, which necessitated a hedge of values; security was dependent almost exclusively on urban real estate, the value of which could be stabilized and expanded through insurance.

1.2 *Hamburg and Leibniz*

It appears to be even more difficult to account for why the municipal General Fire Fund (General-Feuer-Cassa) was created in 1676 in Hamburg at almost the same time as Barbon's first fire insurance company was founded in London. In Hamburg at this point, there was no kind of internationally significant

70 Johann Jakob Moser, *Nachricht von einer freywilligen Feuer-Cassa, Welche zum Besten der Einwohnere des Herzogthums Würtemberg angeordnet und von Hoch-Fürstlicher Landes-Herrschafft gnädigst bestättiget worden ist,* S.l. 1754, p. 3.

historical event comparable to the Financial Revolution or Glorious Revolution, there had been not a megafire close to the dimensions of the London fire, and there was no concentration of scholars that could compete with the London of the time of the founding of the Royal Society. Nevertheless, Hamburg's Fire Fund represents the central germination of German and central European cameralistic security production culture. As such, it will be necessary to reconstruct the development of this laboratory.

1.2.1 The Hamburg General Fire Fund: Innovation without an Inventor

In the case of London, there were named "inventors" associated with even the early insurance projects, and the first successful fire office claimed Nicholas Barbon as its Gutenberg. In contrast, the person who provided the impetus for founding the General Fire Fund in Hamburg can no longer be identified from the archival sources with absolute certitude. These archival sources begin with a senate proposal "to create the General Fire Fund and ordinance that would revoke all other individual ordinances," which was disseminated among the appropriate councils, aldermen, and chambers (September 21, 1676); the establishment of a new parallel fire ordinance, presented to the citizenry (*Bürgerschaft* meaning the large council of Hamburg's constitution) for approval (November 30, 1676), was a bigger topic of discussion than the fire fund itself. After the adoption of this new regulation, in which it was determined "that for each *Compagnie* 12 men [should be designated] annually, going from house to house in case of fire to assist in rescue efforts," the "citizen-captains," i.e. the military principals, were hesitant to conform, therefore it was decided that "certain individuals would be employed and compensated by the Fire Fund" (September 6 and 13, 1677).[71] It was Hieronymus Sillem (1648–1710) who invariably reported to the senate about the progress in the establishment of the Fire Fund, and who had already served in countless administrative functions for the city—Senate meeting minutes refer to him as "H Silm."[72] On November 14, 1676 Sillem announced that "in regards to the Fire Fund, it will now be established as planned," and on November 22 he reports that "the conceived Fire Fund ordinance," has been established as planned. On November 29 Sillem reports that "between them—the alderman and the chamber—the Fire Fund has been established and the mandate on its behalf will be published;" on December 11 he reports "that the Fire Fund ordinance has now been printed, and those who shall form its board will be elected." On December 15 he reports

71 StA Hamburg, Cl. VII Lit. Fe Nr. 1, vol. 1, nr. 2–11.

72 Hans Schröder and Anton H. Kellinghusen, *Lexikon der Hamburgischen Schriftsteller bis zur Gegenwart VII*, Hamburg 1875–79, no. 3747.

> that the deputies of the Fire Fund assembled yesterday and that eight citizens had been elected to that end and that they found reasonable that a mandate should be published decreeing that homeowners should gather in the city hall at a time specified in a previous announcement and that those deputies and citizens present should register their houses and that they should pay the necessary sums, with the provision that after six weeks have passed, the new fire ordinance shall be valid. The question however arose of whether everyone should be obliged to register his house and therefore if the Fire Fund should be altered accordingly. Conclusion: The Fire Fund shall remain unchanged as drafted and as approved by the citizens' council.

A corresponding mandate was supposedly composed by Protonotary Schrödering. Sillem reports that the aldermen agreed to this.[73] One of the first conflicts concerned the General Fire Fund's determination that every homeowner must personally carry "a quarter" of the risk. Immediately, homeowners insisted on the possibility of insuring homes at 100% since "creditors that had money invested in them" would thus also only accept 75% of the home's value as collateral. The senate responded by emphasizing the initial considerations once again: "The new fund is established as partial [insurance] in order to ensure that every property owner personally experiences some risk so that he will better seek to prevent fire outbreaks, if the houses were insured at their total value, one would retain the previous risk [of carelessness due to the fact that the total value is insured]."[74] Furthermore, it was hoped that if the homeowner assumed a quarter of the risk, he would be dissuaded from using the Fire Fund to finance major renovations by means of arson.

Another issue arose after a major fire on the Lehrweet in 1684. Following this fire, some subscribers sought to withdraw and "establish a separate fund."[75]

73 StA Hamburg, Cl. VII Lit. Fe Nr. 1, vol. 4, f. 23r-24r.

74 Ibid., f. 24v-25r.

75 StA Hamburg, Cl. VII Lit. Fe Nr. 1, vol. 1, nr. 2 (February 12, 1685); also the report: "Vnd weil sich unterschiedliche Interessenten der Fewer-Cassa angegeben, und aus derselben ihre Erlassung begehret, in meinung nebenst andern eine particulier Fewer-Cassa auffzurichten, mit dem Erbieten, dz sie zuforderst ihr Contingent vom letzten Fewer-Schaden auff einmahl abtragen wollen, und den E.E.Raht der Meinung ist, daz keiner wieder seinen Willen in solcher Ordnung darin er sich freywillig begeben, zu verbleiben gehalten, auch hiedurch soviel mehr gelder schleünig bey der Cassa würden gebracht werden, als zweifelt E.E.Raht nicht, die löbl. Bürgerschafft werde damit einig sein, das wan jemand noch würd E. Erlegung seines gantzen Contingents dieses letzten Fewer-Schadens auß der Fewr-Cassa scheiden wolte, jhn solches frey stehen möge, und nach

This was, of course, quickly rejected.[76] There is little known about Sillem during this time period—from the meager records of the senate, it is also not clear whether he was actually the "founder" of the fund. In any case, he seems to have advocated for the establishment of the fund and managed the ensuing discussion during the first few months.

The fire of 1684 stimulated a first push in the development of the management and appraisal routines of the Fire Fund. The first detailed damage assessments emerged from this period and the following years, since estimated damages of 1,215,600 Lübisch marks (LM) were at play, resulting in a "project of unpredicted magnitude for the so-called city Fire Fund which should consist in 214 tons of gold … of which 64,200,000 LM" had been distributed. The "income" and "output" is detailed below. Following the establishment of the basic capital in 1676 in the amount of 64,200 LM, annually constant amounts (16,050 LM) were recorded, demonstrating regular payments. The specific damages were further detailed in order to calculate the amount of the special levy that was eventually imposed because of the 1684 fire.

> The fire damages are calculated at 1,215,600 LM
>
> Deducting the amount of cash that, according to the records, should be in Fire Fund: 229,150 LM
>
> There remains: 986,450 LM
>
> Assuming there was no money in the Fire Find, one would have to calculate a capital value of 64,200,000 LM corresponding to the fire damages of 1,215,600 LM, equal to 1 LM, 14 schillings and .05 pence for every 100 LM [of insured value]
>
> If one [rounded this to] 2%, then 1,215,600 LM of the above mentioned capital of 1,284,000 LM would go toward fire damages
>
> 1,994,400 LM therefore remains, which is equivalent to 6 tons of gold
>
> 1,994,400 LM in total.
>
> If no money were present in the fund, the Fire Fund would need 150 tons of gold to pay the damages. Thus, for each 100 LM of insured value [each subscriber would have to pay] 2 marks, 11 schillings, and 2.656 *Pfennige*.[77]

---

dem dato beschehen zahlung auß der Fewer-Cassa weiter nichts zu genießen habe, noch dazu zu contribuiren gehalten sein sollte" (ibid., Lit. Fe Nr.2 vol. 1b, nr. 3—1685).

76 "Herrn Syndici Anderson Rationes warumb die von einigen privatis vorgeschlagene special oder particular-Feuer-Cassa nicht zu autorisiren, oder zu permittieren," ibid., f. 30f.

77 StA Hamburg Cl. VII Lit. Fe Nr. 1 vol. 4, f. 34v-35r.

The special levy imposed on all subscribers was calculated on the basis of exact total estimation of registered homes ("*Erben,*" i.e., inheritances") in the fund. Such status reports appear more commonly during the first years of the fund. For example, the "number and rate of the *Erben* found in the Fire Fund in Hamburg and on the Hamburg mountain in 1708" is given, along with the registered number of *Erben* per "compagnie" (i.e., per hundred city militia members) and their insured value. In total there were 4,226 *Erben* recorded as insured with an insured value of 24,981,400 LM, or 84,840,000 Reichstaler.[78] The early damage estimates are recorded in the Fire Fund main account books from 1684–1713, as shown above.[79]

After the 1720s information about the General Fire Fund becomes rather dense and consistent, but there is no mention of foreign models or fire disasters perceived as threatening—such as the London fire of 1666—found within the files.[80] There is no indication of an earlier transfer of technological or organizatorial knowledge between London and Hamburg or vice versa, which would have been commonly expected in the age of industrialization.[81] The fund can therefore be described as an innovation without an inventor, and this is lack of a single originator is also the reason that so many theories arise as to where this innovation originates and why it proliferated. Earlier cooperative structures (*Genossenschaften*) were often cited as models for the General Fire Fund. Beginning with Otto von Gierke's seminal work (*Das deutsche Genossenschaftsecht*, 4 vols., 1868–1913), these *Genossenschaften* have been consistently presented within German insurance history as "pillars" underlying the emergence of modern insurance.[82] Undoubtedly, the so-called Hamburg fire contracts, emerging after 1591, are key examples of such cooperatives. These contracts assembled approximately 100 homeowners who were obligated to contribute a certain sum toward damages suffered by a member of the cooperative. According to legal historians of the Germanic school, these contracts

78 Ibid., f. 36–54.

79 Cf. 3.1.6. The damages before 1684 (a fire in Stein Street in 1682 and a fire in 1683) are noted in the list just presented. See note 75.

80 It was only in a 1731 project presented by Christian Friedrich Linde that the 1666 and other foreign fires were referred to in the context of a new invention in fire prevention: StA Hamburg, Cl. VII Lit. Fe Nr. 2 vol. 5, f. 1–5.

81 John R. Harris, "Movements of Technology between Britain and Europe in the Eighteenth Century," in David J. Jeremy (ed.), *International Technology Transfer: Europe, Japan, and the USA, 1700–1914*, Aldershot 1991, pp. 9–30.

82 For example, Peter Koch, "Versicherungswesen," in Adalbert Erler (ed.), *Handwörterbuch zur deutschen Rechtsgeschichte*, vol. 5, Berlin 1998, pp. 815–26; Peter Koch, *Geschichte der Versicherungswissenschaft in Deutschland*, Karlsruhe 1998.

persisted into the 1930s and are characterized by early *völkisch* history writing as "ancient, old-Germanic, according to their essence,"[83] as owing their existence to "*völkisch* vigor and cooperative capacity," and as "true German legal cooperatives."[84] In contrast, Wilhelm Ebel attempted to demonstrate that the Hamburg fire contracts were not "guildlike"—like Wilhelm Schäfer, he presumed western (Dutch) influences.[85] The main issue at stake here was determining whether insurance practices that seemed to be especially innovative were to be categorized as "influences méditerranéennes"[86] or as creations of northern European provenance (cf. Gierke's notion of the "Germanischer Assoziationsgeist" [German spirit of association]). This discussion has been taken up again, maintaining to a somewhat surprising extent the main argument of the older narrative.[87]

The fact that the General Fire Fund was, in any case, a hybrid is already made clear by linguistic evidence: in article 2 the word "risk (*Risico*)" is used ("If the homeowner/in the entreaty of the new ordinance/should be obligated/ to assume one quarter of risk of their estate"). The annual advance payment of four schillings per 1000 marks of determined value is also an uncommon mode of premium prepayment for a guild. Of course, the fund does also assume certain traditional elements—for example, as article 15 clearly describes, the location for the fund's safe is the City Hall with "eight locks/the keys for which are constantly and safely kept by the aforementioned eight oldest lords and citizens." This thus connects the fund to the general medieval and early

83 Ludwig Maass, *Die Brandgilden insbesondere in Schleswig-Holstein*, Stuttgart 1910, p. 3.

84 Georg Helmer, *Die Geschichte der privaten Feuerversicherung in den Herzogtümern Schleswig und Holstein, insbesondere die Geschichte der Entstehung und Entwicklung der Brandgenossenschaften oder "Brandgilden" in Schleswig-Holstein*, Berlin 1925, p. 47f., with reference to Gierke, see also pp. 53, 225–227, 242f.; or for a revised version evidently resonating the emerging nationalsocialist agenda, see Georg Helmer, *Die Gegenseitigkeitsversicherung. Eine Schöpfung germanischer Völker*, Vienna 1937.

85 Wilhelm Ebel, *Die Hamburger Feuerkontrakte und die Anfänge des deutschen Feuerversicherungsrechts*, Weimar 1936, p. 40; Wilhelm Schaefer, *Urkundliche Beiträge und Forschungen zur Geschichte der Feuerversicherung in Deutschland*, vol. 1, Hanover 1911, p. 162f.

86 Fernand Braudel, *La Méditerranée et le monde méditerranéen à l'époque de Philippe II* [first ed. 1949], 3 vols., 9th ed. Paris 1990, p. 580, of course without reference to insurance, which is, however, covered by Boiteux, *La fortune de mer*. On the level of the history of ideas, this pattern is also basically followed by Antonio La Torre, *L'assicurazione nella storia delle idee*.

87 Dieter Schewe actually claims that in the first fully developed insurance contracts from Florence of 1385 a direct influence of Danish guild statutes from 1200/1250 can be deciphered: Schewe, *Geschichte*, pp. 166–69, 183f., 194–96, 205–18, 226–34, 244–50, 268–72.

modern tradition of public funds.[88] Furthermore, the insured were described as "interested parties" (art. 3).

In contrast, fire contracts were based on the agreement that an established sum would be paid after a fire occurred (approximately 26 LM from each of 80 participants, in order to pay 2000 LM to the affected parties)—they lacked a consideration of determined values and a prepayment plan. The early fire contracts described themselves explicitly as "contracts" and, from 1591 on, the participants were described as "consorts," which also meant "owners" in early modern cooperative language.[89] Beginning in 1640, they integrated the concept of "interested parties,"[90] which was actually less well-suited to the original medieval guild tradition. Even for these contracts, Ebel has shown that in their first phase until 1640, these contracts had the status of reciprocal contracts, "not dissimilar from civil partnership today." He thereby intended to emphasize that they concerned legal relationships but not "legal subjects," and that they therefore did not have the structure of a legal entity like cooperatives did.[91] While that may be a very abstract and systematic legal definition, one can clearly observe the influence of merchant societies within these organizations.

However, in the end, the administrative process is more important than the pure textual form of contracts and fund edicts—i.e., the accounting of payables and receivables, the assessment of property values and damages, and that which was summarized in simple formulas representing the constant calculations and estimates. This has little to do with either guilds or cooperatives, nor with the hybrid entity of the fire contracts. A more important precursor was definitely the establishment of the Slaves' Fund (*Sklavenkasse*) and the Piece of Eight Fund (*Stück-vom-Achten-Kasse*) in 1622–4 in Hamburg through the mayor and council, into which sailors, sailing toward the Atlantic coast, were obliged to contribute. Theoretically, only those who had contributed had the right to claim ransom money from this fund in order to purchase their freedom in the event of enslavement by Barbary pirates.[92] This

---

88 Ebel, *Hamburger Feuerkontrakte*, pp. 66–77; "Newe Brand-Ordnung Anno 1628," StA Hamburg Cl. VII. Lit. Fe Nr. 2, Vol. 3, nr. 20.

89 Cf. Marie Luise Allemeyer, *"Kein Land ohne Deich …!" Lebenswelten einer Küstengesellschaft in der Frühen Neuzeit*, Göttingen 2006, p. 179.

90 Ebel, *Hamburger Feuerkontrakte*, p. 79; "Fewr Contract […] Hamburg 1654", in ibid., f. BivV; *Copia Dero* […] *Feur- oder Brand-Ordnung Concordia* […],Hamburg 1643, f. BijR; *Copia* […] *Fewr: oder Brand-Ordnung Vigilate* […], Hamburg 1647, f. BiR (StA Hamburg Cl. VII. Lit. Fe Nr. 2 Vol. 3).

91 Ebel, *Hamburger Feuerkontrakte*, p. 18f.

92 Ernst Baasch, *Die Hansestädte und die Barbaresken*, Kassel 1897, pp. 222–38; Robert Bohn, "Von Sklavenkassen und Konvoifahrten. Die arabischen Seeräuber und die deutsche Seefahrt im 17. und 18. Jahrhundert," in Thomas Stamm-Kuhlmann, Jürgen Evert, Birgit

can be interpreted as an early form that combined statism and the principle of premium-based insurance and occurred at the emergence of cameralist insurance. The fact that the Hamburg Slavery Fund is often located in immediate proximity to Fire Fund archival material[93]—a common practice in northern European archives—is yet a further indication that the administrative transfer of the insurance principle proceeded along this path. An examination of Leibniz's reception of the Hamburg sources can also help illuminate the nature of the relationship between "cooperativity" and the "southern European transfer product."

1.2.2 Transformation into an Element of Economic Provisioning: Leibniz

The security product established in 1676 in the practical city administrative milieu of Hamburg needed a mediator to help develop its meaning and significance. It was Leibniz who transformed this singular institution into a model and acted as a mediator across Germany—albeit initially only within limited circles of administrative specialists. His thought and projects should be conceived of as different but highly comparable to the projects and speculations on political economy that, on the other side of the Channel, were framing the "Financial Revolution"—although that term is a bit of a misnomer in the context of the Holy Roman Empire. In comparison with England, the Empire lacked a nationally framed economic and financial system and did not have a "national" asset market that could have paralleled England's during its rise as an imperial power in the eighteenth and nineteenth centuries. Perhaps cameralistic projects—and politics—might be considered a Continental equivalent to these English developments, but only in very well-defined forms of comparison. Beginning in 1678, Leibniz was repeatedly involved in making recommendations to various rulers (Ernst August of Hanover, Duke Johann Friedrich, Margrave and Elector Friedrich Wilhelm and later Friedrich of

Aschmann, and Jens Hohensee (eds.), *Geschichtsbilder. Festschrift für Michael Salewski zum 65. Geburtstag*, Stuttgart 2003, pp. 25–37. Whether this concerns the transfer of the new principle of premium-based insurance or the older principle of the guild into the administrative sphere cannot be determined in an "either/or" framework. It is only in the text, *Vier Revidirte Articuli, Der Boots=Leute Schlaven= Gelder Einnahm betreffend geplacidirt den 27 Januarii Anno 1679*, Hamburg 1679, that the term "Risico" appears in an official document of the Slaves' Fund administration (thanks to my former doctoral student Magnus Ressel for pointing me to the *Articuli*).

93 This is similar in Denmark, Magnus Ressel, *Zwischen Sklavenkassen und Türkenpässen. Nordeuropa und die Barbaresken in der Frühen Neuzeit*, Berlin 2012, p. 312.

Brandenburg-Prussia, Emperor Leopold) concerning the topic of insurance.[94] His discussion of particular issues—the "rebate" (calculations of discounts), mortality or life expectancy calculations in relation to pensions and annuities, and the establishment of pawnbrokers (according to the model of the Italian *monti di pietà*)[95]—could in hindsight be characterized as a contribution to "insurance mathematics." However, it is significant that Leibniz made these considerations in light of actual practice, without insisting on a tight correlation between probability theory and insurance practice, and also without regard for forms relating to life expectancy prognosis (i.e., pension, widow, and orphan funds).[96] In this sense, he differentiates himself from Barbon. Leibniz uses the concept of "assecurance" (*assecuratio*)—common in maritime trade since the late Middle Ages—only in those cases of insurance against accidents (i.e., fire, flood) that are relevant beyond life-expectancy problematics. In contrast, life annuities and pensions were treated by Leibniz as a "purchase" (*emptio*).[97] There is also obviously no relationship in his thought to problems concerning life-expectancy estimations, nor are there any greater sociotheoretical visions or considerations—unlike in the cameralistic insurance projects.

Leibniz's project proposals concerning accident prevention from 1678 until 1688 were directed at insurance-like assistance funds as institutional establishments of the cameralistic state administration.[98] For Leibniz, it is precisely

94 The writings are collected in Gottfried Wilhelm Leibniz, *Hauptschriften zur Versicherungs- und Finanzmathematik*, ed. Eberhard Knobloch und J.-Matthias Graf von der Schulenburg, Berlin 2000, the most important report from 1680 is Gottfried Wilhelm Leibniz, "Öffentliche Assekuranzen," in *Sämtliche Schriften und Briefe* [*Leibniz Akademie Ausgabe = LAA*], series 4, vol. 3, *Politische Schriften (1677–1689)*, ed. Lotte Knabe und Margot Faak, Berlin 1986, pp. 421–32. The other smaller *promemoria* are to be found in LAA I/2, Nr. 10–12, 70, 73–75; LAA I/3, Nr. 17, 21, 27, 28, 40. On Leibniz as an early cameralist (not on his insurance projects) cf. Warren Alexander Dym, *Divining Science. Treasure Hunting and Earth Science in Early Modern Germany*, Leiden—Boston 2011, pp. 158–165; Erwin Stein, "Theoria cum praxi: Leibniz als technischer Erfinder," in Thomas A.C. Reydon, Helmut Heit, Paul Hoyningen-Huene (eds.), *Der universale Leibniz. Denker, Forscher, Erfinder*, Stuttgart 2009, pp. 155–183, 166–179; Andre Wakefield, "Leibniz and the Wind Machines," *Osiris* 25 (2010), pp. 171–188; Justin H. Smith, "Leibniz on Natural History and National History," in *History of Science* 50, 4 (2012), pp. 377–401.

95 Leibniz, *Hauptschriften*, pp. 27–549.

96 Ivo Schneider, "Geschichtlicher Hintergrund und wissenschaftliches Umfeld der Schriften," in Leibniz, *Hauptschriften*, pp. 591–623, 591—a generally opposite tendency is emphasized by Hans Schmitt-Lermann, *Der Versicherungsgedanke im deutschen Geistesleben des Barock und der Aufklärung*, Munich 1954, p. 55f.

97 E.g. Leibniz, *Hauptschriften*, 414.

98 LAA I/2, *Allgemeiner, politischer und historischer Briefwechsel (1676–1679)*, Darmstadt 1927, pp. 74–77: project list "De Republica" for Duke Johann Friedrich von Braunschweig-Lüneburg—in addition to references to a project regarding a widow's fund, there is a short

within the early cameralistic project sketches provided for Duke Johann Friedrich in 1678 that the idea of insurance against natural disasters first emerges: "(7) [An] Insurance fund along with the enactment of fire and water ordinances, through which all subjects get their capitalizable *casus fortuiti* compensated, while they are bound to put a certain [contribution] into the insurance fund according to their means [i.e., their income] annually."[99] Explicit here is the idea that the subjects themselves should generate a fund through premium payments, out of which the damages from catastrophes could be defrayed.

Leibniz's most important insurance project text is a recommendation from about 1860 addressed to Emperor Leopold, however there is no indication of any kind of reaction in Vienna. Although the academic discourse about "insurance" was limited almost entirely to the legal discussion of transport insurance contracts in maritime trade and the classification of such contracts until then, this represents one of the earliest "macrosocietal" insurance theories, which in its almost utopian visionary ductus casts a long shadow onto many cameralistic insurance projects to come, extending even into the eighteenth century. There was no comparable theoretical elaboration of the function and classification of "insurance" in a universal context in England at this point—nor would there be for some time to come.[100] The special quality of this text lies

purely definitional note of a pertinent idea: "Assecurations-Casse samt dazu gehoeriger Feuer= und Waßer=Ordnung, vermittelst deren allen Unterthanen die liqvidable casus fortuiti ersetzt werden, sie aber dagegen jaehrlich ein gewißes nach ihren Mitteln in die Assecurations-Casse zu legen schuldig"; also in the project list for Emperor Leopold from ca. 1680: "Eine Assecurations=Verfaßung; vermittelst deren die armen Leute gegen feuers gefahr gesichert wuerden, ad imitationem deßen so zu Hamburg" (LAA I/3, *Allgemeiner, politischer und historischer Briefwechsel [1680–1683]*, Leipzig 1938); for the main text "Öffentliche Assekuranzen" from 1680 the resumption of the topic of insurance in the form of a template for a *Policey* edict for Emperor Leopold in 1688 is also relevant (Günter Scheel, "Leibniz' Wirken für Kaiser und Reich im Jahre 1688 in Wien nach bisher unbekannten Quellen," in *Leibniz und Europa. VI. Internationaler Leibniz-Kongreß*, part 1: presentations, Hanover 1994, pp. 697–704, 699. Edition presented in Leibniz, *Hauptschriften*, pp. 20–22).

99 Leibniz, "[Gedanken zur Staatsverwaltung] an Herzog Johann Friedrich, Sept. 1678," in LAA I/1, pp. 74–77, 75.

100 Leibniz, "Öffentliche Assekuranzen," pp. 421–32, also reproduced in Leibniz, *Hauptschriften*, pp. 12–19. The research rarely reveals the content summary of a text: Eduard Bodemann, "Leibnizens volkswirthschaftliche Ansichten und Denkschriften," *Preußische Jahrbücher* 53 (1884), pp. 378–404, 392–96; Schaefer, *Urkundliche Beiträge*, vol. 1, pp. 111–14 with a small supplement (p. 114) in the Akademie edition; Schmitt-Lermann, *Versicherungsgedanke*, pp. 46–52 (still dated it to 1697 and thus falsely placed it in the context of the monadology); Traiano Sofonea, "Leibniz und sein Projekt zur Errichtung staatlicher Versicherungsanstalten," *Schweizerische Versicherungszeitschrift* 25 (1957),

in its seamless integration of the premium-based insurance principle within a general state institutional framework that eventually transforms it into an inherent aspect of society as a whole.

1.2.2.1 *Before Leibniz: The Rulership Contract and Disaster Insurance around 1600*

The special status of the Leibniz' text can only be established by comparing it to earlier projects—which will thereby demonstrate how innovative Leibniz's ideas really were. Thus far, we know of two earlier projects involving fire insurance, which conceived of the insurance principle as part of administrative—and thus early cameralistically organized—institutions. The majority of towns and cities would have been subsumed within a territorial area subject to a single putative insurance institution. The first is Wilhelm Stiell's project, presented to the Graf of Oldenburg in 1609 but not actually implemented;[101] the second was published in a *policey* science compendium by George Obrecht of Strasbourg in 1617.[102] A third even earlier project may be even called the earliest state fire insurance project involving the principle of premium-based insurance—heretofore known only in the context of maritime trade. This project is preserved in a subseries of municipal documents in the Strasbourg archives that were collected and combined a posteriori by the archivists themselves.[103] The archival files contain copies and original documents regarding negotiations that took place between 1604 and 1608, foremost from the papers of Michael Daniel

---

pp. 144–48; Eike Christian Hirsch, *Der berühmte Herr Leibniz. Eine Biographie*, Munich 2007, pp. 156–59; Allemeyer, *Fewersnoth*, p. 48.

101 Wilhelm Schäfer, *Urkundliche Beiträge und Forschungen zur Geschichte der Feuerversicherung in Deutschland*, 2 vol., Hannover 1911, vol. 1, p. 116ff; cf. below note 212.

102 "Ordinatio Von einer nothwendigen/vnd hochnutzlichen Fewr-Ordnung", in Georg Obrecht, *Constitutio Von nothwendiger vnd nützlicher Anstellung eines Aerarij Sancti* […], S.l. 1617, pp. 158f.

103 Archives municipales de Strasbourg série III 93/3. The relatively copious bundle is separated into different fascicles, which are not paginated or foliated. A more exact method of citation is not possible. I sincerely thank Gerrit Jasper Schenk for pointing me to this source in 2009. In 2018 and 2020, I worked anew in the Strasbourg and other German archives from Marburg to Schleswig, discovering far more information about this and other projects and their context. There is too much and it is too complex to integrate it into this revised edition of *Prometheus Tamed* and must be reserved for a further investigation. This section therefore has to be kept even shorter than in the German original as, from the point of view of the knowledge gained now, that paragraph in the 2011 *Gezähmter Prometheus* as well as the remarks of Cornel Zwierlein, "Frühe Formen der Institutionalisierung von 'Versicherung' und die Bedeutung der Versicherungsgeschichte für eine allgemeine Sicherheitsgeschichte," in Kampmann, Niggemann (eds.), *Sicherheit in der Frühen Neuzeit*, pp. 441–458, 442–449 need some corrections and important precisions.

von Poland and Adolf Horn, two persons otherwise not known outside the very narrow Strasbourg contexts. Ludwig Pincier and Johann Joungbou[w]en, Isaac Mennius, Joseph Jundt, Georg Obrecht himself and many others were also involved. The Strasbourg inventors obviously tried to sell it as a "secret" project to the people of Strasbourg, as well as to many other German and European princes and cities. In the files the project is described as a "report on how state contributions and revenues could be augmented by the means [i.e., money] of the subjects to create a fund from which they might profit in certain cases of misfortune" and later as a "recovery of fire damages." Its format is similar to earlier cases of scheming, or "project making," which were common in the late seventeenth and eighteenth centuries, and characterized by convoluted trajectories and a high demand for secrecy.[104] It seems that all three early attempts—the two Strasbourg, as well as the Oldenburg projects—were connected to each other, but a full investigation and analysis of that relationship, and of its political context, is still required, and will need more space than can be devoted to it in this context. From the non foliated pages of the Strasbourg manuscript files, I will just quote here one paragraph that illustrates the starting point of the relationship between ruler and subjects, both being threatened by Acts of God:

> It is undeniable that the welfare of each subject and his [material] growth consists foremost in the diligent care for his household and in the trustworthy care for his work and his calling. Beyond this, it is also necessary to prevent that through which an honest subject could be deprived of all that he has gained and achieved during his whole life by his massive efforts, work, care, and industry by way of sudden hazards without him having caused them and without any fault of his own … which would lead to great disadvantage and damage for the ruling class as well … Some of these hazards have the character of resulting from God's violence and thunderous storm—from the power of fire and the ravaging of

104 Cf. on the figure of the 'projector' and the type of enlightenment communication in terms of 'projects' Maximilian E. Novak (ed.), *The Age of Projects*, Toronto 2008; Frédéric Graber, "Pour une histoire des formes projet. Du faiseur de projet au projet régulier dans les Travaux Publics (18e-19e siècles)," *Revue d'histoire moderne et contemporaine* 58, no. 3 (2011), pp. 7–33; Stefan Brakensiek, "Projektemacher. Zum Hintergrund ökonomischen Scheiterns in der Frühen Neuzeit, in *Fiasko—Scheitern in der Frühen Neuzeit*, Bielefeld 2015, pp. 39–58; on Paul Marperger who was importing the concept in Defoe's time into German early cameralist language and practice cf. Kelly J. Whitmer, "Projects and Pedagogical Expectations: Inside P.J. Marperger's 'Golden Clover Leaf' (*Trifolium*), 1700–1730, in: Notes and Records of the Royal Society 72, 2 (2018), pp. 139–158, here 140.

> enemies as a universal ruin for which there is hardly any remedy … But, if a subject should be struck and overwhelmed by conflagration and by the emergency of fire and is deprived, within one hour, of all his temporal belongings—his own and those of his family—the rulers must be sympathetic and redeem those losses … It would be in the highest and most noteworthy interest of the peasant in the country as well as the burghers in town that they were secured against that [those threats] or if the damage has already occurred that the [victims] be compensated.

Insurance was seen as extremely helpful from both the individual perspective as well as from the perspective of the state economy.

The basic idea was that the relevant authorities across the entire country would send estimators to regional administrative seats, where the value of property and possessions would be recorded for each subject. The declarations and determinations of values would be left to the discretion of each individual and there would be no compulsion to participate. Each participant would then be obligated to "annually pay three batzen [i.e. the southwestern German and Swiss coin and currency] in cash for every hundred that he had registered." In other variants of the project, four groats per hundred florins, or two Rtl. per thousand florins were charged. Thus there was a "sum assembled" that served as "capital." One would be deprived of insurance coverage "only through the violence of war and enemies or if the almighty God should destroy entire countries and cities by fire from heaven as with Sodom and Gomorrah. In all other *casibus fortuitis* that involved no moral retribution," the insurance would be recognized. The magistrates recruited by the schemers were enticed by the large sum of money that would be quickly accrued through the consistent annual contract payments (i.e., premiums), and one could furthermore imagine "how many thousands of florins would accumulate, and how many hundreds of thousands within twenty, thirty, or more years, if one laid even a third or a fourth aside in savings." "The ordinary revenues" of state income could thus be substantially increased. One of the schemers compared insurance to a mine, from which treasure in metal could be extracted if one only had the correct technical training; another suggested that since the project was already almost fully implemented in Holland, the Dutch "could, without doubt, find and make use of a little New America in their own provinces."

As rudimentary as such undertakings seem, they are nevertheless integrated into a larger concept of the state and society. The establishment of insurance derives, on the one hand, from the feudal "defend-and-protect" relationship between rulers and subjects. It functions here between rulers and subjects as a "*mutuus respectus*," and a "*reciproce obligatio* between both parties" must be present "if such insurance should be directed toward the utility, growth, and welfare of all involved." The defend-and-protect idea typically involved the defense of one's own territory against external military assaults, and thus was transferred here to a very different field.

There was a single page in the project description entitled "Assurance, Or insurance of merchants over water was discovered approximately one hundred years ago and consists in a doubled or two-sided utility, when it is well and correctly used." This described the Mediterranean transport insurance practice, which apparently could not be supposed to be known by all the members of the magistrate, even in an important trade city like Strasbourg, or in the central and southwestern territorial state administrations: it still needed explanation as I had given above (Ch. 2), and the Hamburg representants had to explain to Speyer advocates and judges of the Imperial Chamber Court around 1630 what an insurance was. In Northern countries, the first insurance policies today preserved in the archives, are usually written in Italian (in the English case all sixteenth century copies with just one exception), showing that it was still a foreign practice mostly in hands of Tuscans or "Lombards" as the name of the center of commercial and financial activities in London (*Lombard street*) still reminds us.[105] Likewise, the first insurance contracts preserved in a German archive (Hamburg) dates to 1588 and the first insurers in that Hanse city were always Dutch.[106] In the Netherlands, the practice of premium insurance had already been introduced during the fifteenth century in Bruges, then Antwerp, as early as 1554 we find "insurance letters"—*polices*—in vernacular Dutch printed and explained to the children of merchants (Gabriel Meurier, 1554).[107] Nevertheless, all this shows that the practice and the textual forms, as well as the theory of insuring, was still far from being part of common everyday knowledge everywhere in Europe. In contrast, it was restricted to some selected places. The fact that in a central European city as far away from the sea as Strasbourg, the principle was being transferred from mercantile practice to a state governmental context and from the risk area of the sea to a future-oriented form of insurance based on the value of building stock is remarkable. It is all the more so because there is surprisingly no evidence of any such

105 Raynes, *History*, p. 26f.; Rossi, *London Code*, pp. 29–34 (the first surviving insurance policy of 1523 in English imitates the Italian style, the next oldest English policies from 1532 onwards are written in Tuscan).

106 Kiesselbach, *Entwickelung*, p. 15.

107 Dave De Ruysscher, "How Normative were Merchant Guidebooks? Of Customs, Practices, and … Good Avice (Antwerp, Sixteenth Century)," in: Heikki Pihlajamäki et al. (ed.), *Understanding the Sources of Early Modern and Modern Commercial Law. Courts, Statues, Contracts, and Legal Scholarship*, Leiden 2018, pp. 144–165.

transfer within the southern European countries at all. It is astonishing that in Italy, where premium-based insurance was invented, institutionalized forms of fire and other types of insurance only begin to appear in the nineteenth century. The development and transfer of the insurance principle that begins in the seventeenth century is clearly specific to Germany and England.

This is why that very important moment around 1600, of something that we might even call the 'birth of cameralism' in Strasbourg, will need further in-depth investigation in the future. In this given context of a long-term investigation, I confine myself to the statement that there were important developments and the first intersections of commercial and administrative thought and practice in Germany before the Thirty Years' War which seem to have been interrupted specifically by that war—and that the story of the invention of cameralist insurance practice begins anew, with a second starting point and take-off, only after that war, in the 1670s.

While the Strasbourg projecters were still conceiving of the state's structure in terms of a late feudal contract between rulers and subjects—a context which one would not suspect of serving as the breeding ground for novel forms of "political economy", Leibniz, was already anticipating a new era, when a new social contract could be conceived of as a kind of general *assecurance* based on natural law and created by a political-economic rationality.

1.2.2.2 *Leibniz and the Territorial Institutionalization of Insurance*

The 1680 seminal project for a general fire insurance for the whole Holy Roman Empire, sent by Leibniz as a manuscript proposal to emperor Leopold I in 1680, was entitled by the editors of Leibniz's Academy edition as "*Öffentliche Assekuranzen*", "Public Insurance." For Leibniz, we must thus assume a rather strict separation between *Assekuranz* and mathematical/probabilistic thought, and that for him it was only *Assekuranz* that denoted *insurance* in our sense of the term.[108] Leibniz was concerned with the idea and concept of (fire) insurance mostly during the years 1680 to 1683, and these projects have nothing to do with his mathematical reasoning about life annuities and probabilistic thought. The 1680 text remains one of the most ambitious and longest ones for nearly a century in Europe.[109]

108 Cf. on that Cornel Zwierlein, "Katastrophe und Prävention—Leibniz, Brandgefahr und Versicherung," in Friedrich Beiderbeck et al. (eds.), *Umwelt und Weltgestaltung. Leibniz' politisches Denken in seiner Zeit*, Göttingen 2015, pp. 433–62, 448–452.

109 "Leibniz did not make any direct contribution to probability theory" (Hald, *A History of Probability*, p. 189).

Leibniz's 1680 text concerning insurance against natural hazards was drawn up from the typical perspective of competition between state powers in Europe. Early cameralist plans for public economic growth did not arise purely from autochthonous political motivations, but instead from an appreciation of the interdependence between external and internal forces in the rivalry between European territories.[110] The project of an insurance fund should contribute to hindering the potential situation where "Germany [is] made completely powerless and subject to robbery by the increasing power of its neighbors." The power of a given territory, in this report, is entirely defined by economic and population theory: "the capital [*Nahrhaftigkeit*]"[111] consists primarily in "the quantity of people"—a numerical value that Leibniz saw Germany as distinctly deficient in following the Thirty Years' War.[112] In such a situation, if further misfortunes, such as storms and especially fire damages, continued to leave their mark upon the population, entire villages and cities would be destroyed in the short or medium term and thus the continued existence of the body politic would be threatened. From the princes' perspective, the main priority was to ensure their military capacity, something only possible with the help of adequate taxation, which, in turn, could only be achieved if the economic well-being of their subjects was not reduced by misfortunes. With an insurance fund in place, victims of fire would receive aid "through real contributions." Society at large necessarily formed a community of solidarity; the state was conceived as the "insurer general," which seems to imply more than the current interpretation of the state as insurer of last resort—instead, the state in society is conceived as an insurance-like system.[113] The insurance institution would serve to make sure that a victim would again become "a productive member of the community." Such an institution would also be in the state's interests, since the state required liquid subjects. Leibniz took

110 For a summary of state competition and rivalry, see Heinz Duchhardt, "Staatenkonkurrenz und Fürstenrivalität. Krieg und Frieden in Europa 1700–1714," in Marcus Junkelmann (ed.), *Theatrum belli. Die Schlacht von Höchstädt 1704 und die Schlösser von Schleißheim und Blenheim,* Herzberg 2004, pp. 3–11; and on the discursive aspect, see Harm Klueting, *Die Lehre von der Macht der Staaten. Das außenpolitische Machtproblem in der "politischen Wissenschaft" und in der praktischen Politik im 18. Jahrhundert*, Berlin 1986.

111 "Nahrhaftigkeit," literally meaning the nutrious power/value of something (i.e., *nähren* = to feed), had a different and more general sense in early modern economic contexts. A "*nahrhaftes Gewerbe*" implied a "prosperous business." Leibniz uses this term in the same context that an English political arithmetician like Petty would use the term "capital."

112 For the relevant German discursive context, see Nipperdey, *Bevölkerungspolitik*.

113 This is not a precise precursor of the current idea of the state as "insurer of last resort"; it implies more the idea that the state *is* in its essential character an insurer.

an almost sociological and social psychological argumentation into account against strict legal thinking: the reason for "rancorousness and misfeasance" (i.e., wanton disregard and defiance) on the part of a ruler's subjects was often that the offender had suffered a misfortune. Such events led to "common desperation," which could lead to "rancorousness" in some and "lethargy" in others. Insurance against misfortunes would apply not to the symptoms, but instead to the roots of a malady. *Before* lethargy and resentment set in, people would be helped out of hardship. Insurance was thus credited with the potential for comfort, stimulus, and the prevention of injustice. In this way, Leibniz integrated insurance more generally with his brand of sociological or socio-psychological reasoning according to its overall social functions into the wider concept of statism and society.[114]

Leibniz made use of at least seven analogical conclusions in order to explain and legitimize the idea of the entire project and indeed to generate it in the first place.

(1) As in "natural societies"—i.e., familiar home communities—"love and loss" must both be endured, and thus "fairness in the republic or in the civil society" is required so that accidents (*casus fortuiti*) affecting one individual are borne by the entire community.

(2) The principle of solidarity on the level of society as a whole level is made plausible through comparison with the regulations in Dig. [*Digest of Justinian*] 14.2 "De lege rhodia de iactu," which explains that in the case of a possible shipwreck, when goods were thrown from the ship in order to facilitate its maneuvering abilities, damages should be "reimbursed from common costs."

(3) The principle of solidarity is also derived from company law: in all "companies" ("*compagnien*") it is typical that "costs and benefits are common," therefore accidents should be shared by all.

(4) Rulers should share the principle of solidarity between all citizens and not only between those who have been victimized, which analogically follows that concerning the "insurers [*Asseurers*] in those trade cities" it was not only the ships that sunk that were insured, because otherwise "the practice of insuring [das Assecuriren] would soon cease," since it would not be profitable.

(5) Leibniz compares the transfer of principles onto rulers as agents with the institution of the *Convoy*, which already existed "in Holland or in

114 Leibniz, "Öffentliche Assekuranzen," p. 426.

Hamburg", where "the ruling class" acted as the "insurer [*asseureur*] general."

(6) The primary "beautiful example" for hazard insurance is that of the "fire company established in Hamburg." In this case, Leibniz understands that while it "originally began privately with the consensus of the ruling authorities," it has now been integrated by the city council "into the commons" for all homeowners.

(7) Just as the ruling authorities assess tariffs in order to maintain roads, bridges, and dams, and charge escort fees, in order to finance escorts to ensure safety against attacks by bandits, insurance funds must also be dedicated to specific purposes. Leibniz therefore also recommends a special management of insurance revenue, separate from city chambers.

The comparison in the first analogy of the nuclear family with the larger "natural society" of "citizens" fully corresponds to the Aristotelian tradition, which conceived of family as the first level of socialization.[115] This first level formed the basis for the next levels of social forms, finally leading to the state at the highest level. Leibniz's identification of a bond of solidarity in society as a whole was not an essential component of this tradition, because the common social and political teachings of the pre-Enlightentment era generally assumed relationships of solidarity among the respective cooperative units, while in super- and subordinate units a principle of subsidiarity rules,[116] or because the organizational principle of society is in itself hierarchical/authoritative, rather than horizontal/reciprocal.[117]

---

115 E.g. Henning Arnisaeus, *De Republica, sev relectionis politicae libri duo* [...], Strasbourg 1636, lib. I, cap. 1, sect. 6, 16 "Definitur autem in hoc sensu Arist. 1. Politc. cap. 1. tex. 10. quod sit societas constituta secundum naturam [...] quotidiani usui gratia"— Arnisaeus would claim that making "bürgerliche societät [*a society of citizens*]" and "Republik" synonymous is erroneous, since the latter is just one type of the former (ibid., lib. I, cap. 1, sect. 1, 3); the ambiguity of the distinction, however, goes back to ancient sources, cf. Werner Suerbaum, *Vom antiken zum frühmittelalterlichen Staatsbegriff. Über Verwendung und Bedeutung von res publica, regnum, imperium und status von Cicero bis Jordanis*, Münster 1976, pp. 66–70. Similarly, in this regard, see also Johannes Althusius, *Politica methodice digesta*, Herborn 1614, lib. 1, n. 24–27, p. 8f. u. lib. III, n. 4, where the concept of consociation is, in any case, more legally semanticized. Cf. Cornel Zwierlein, "Consociatio," in Francesco Ingravalle and Corrado Malandrino (eds.), *Il lessico della "Politica" di Johannes Althusius. L'arte della simbiosi santa, giusta, vantaggiosa e felice*, Florence 2005, pp. 277–93.

116 As articulated by Althusius; Cf. Peter Blickle, Thomas O. Hüglin, Dieter Wyduckel (eds.), *Subsidiarität als rechtliches und politisches Ordnungsprinzip in Kirche, Staat und Gesellschaft. Genese, Geltungsgrundlagen und Perspektiven an der Schwelle des dritten Jahrtausends*, Berlin 2002.

117 In the majority of political theories, Arnisaeus was considered to be the classic example.

Leibniz calls the project as a whole an "assecuratio contra casus fortuitos."[118] The Leibniz' text therefore demonstrates that beyond the history of the specific term, and the word "catastophe"[119] and its early anchoring in the geological discourse of the Royal Society, there was an early modern treatment of this concept that closely paralleled the developmental history of the concept of insurance.

Leibniz obviously takes the "accident" as *Leitbegriff* (guiding concept) and elevates it to a new general level: here, he starts from the semantics preserved by Roman legal language, the *casus fortuiti,* previously also the *vis major*.[120] In Roman law this concept is used in addition to others (*vis maior* [Digestae 19, 2, 25, 6], *casus maior* [Dig. 44, 7, 1, 4], *fatale damnum vel vis magna* [Dig. 18, 6, 2, 1], *vis cui resisti non potest* [Dig. 19, 2, 15, 2], *improvisus casus* [Codex Justinianus 4, 35, 13]) to refer to a "higher power."[121] At no point of development did a unified terminology exist in Roman law.[122] In Roman legal sources "catastrophe lists" often appear in the Digest's fragments and textual environment of such terms, specifying uncontrollable accidents: fire, storms, flooding, severe weather, sinking ships, earthquakes, ambush by bandits, or capture by pirates. *Casus fortuitus* was a newer term than *vis maior* and was not related to the metaphorically loaded conceptualizations of a higher, divine, or heavenly power,[123] but instead framed the problem as a concept of chance.[124] In Roman law, language involving a "higher power" or the "blows of fate or chance" typically appeared as a reason for limiting liability and avoiding responsibility—for example, in the case of property damage or lost rent due to severe weather damage.[125] In the Classical and post-Classical periods of Roman law, relevant terms involving a higher power appeared in relation to specific individuals, whereas *casus fortuitus* was used as a defense and as a limiting factor for the debtor regarding his obligations against his creditor in a contract that was not fulfilled:[126]

118 Leibniz, "Öffentliche Assekuranzen," p. 428, line 19.

119 Briese and Günther, "Katastrophe".

120 Leibniz, "Öffentliche Assekuranzen," p. 424, line 23; p. 425, line 9 and 29; p. 426 line 8, 11, 13; p. 427, line 8, 11, 19; p. 429, line 24; p. 430, line 10.

121 See: Dig. 19, 2, 30, 4; Dig. 16, 3, 1, 35; Cod. Just. 4, 24, 6; 4, 34, 1; 2, 18, 22; Inst. 3, 14, 2; 23, 3.

122 Andreas Doll, *Von der vis maior zur höheren Gewalt. Geschichte und Dogmatik eines haftungsentlastenden Begriffs*, Frankfurt am Main 1989, pp. 91, 118f.

123 Also in Plinius—see Doll, *Von der vis maior*, p. 8.

124 Wolfgang Ernst, "Wandlungen des 'vis maior'-Begriffes in der Entwicklung der römischen Rechtswissenschaft," *Index. Quaderni camerti di studi romanistici* 22 (1994), pp. 293–321, 294.

125 Dig. 39, 2, 24, 4; Dig. 39, 2, 43 pr.; Dig. 19, 2, 25, 6.

126 Ernst, "Wandlungen"—for a critical take on the distinction between object relevance and performance relevance, see Jean-François Gerkens, " 'vis maior' and 'vis cui resisti non potest,' " in Rena van den Bergh (ed.), *Ex iusta causa traditum: Essays in honour of Eric*

anyone who made use of somebody else's property was obliged to take extreme care in how it was used, but in accidents caused by a "higher power," this person was not liable for damages.[127] In such cases, the liability-absolving accident stands in a semantic relationship with diligence and culpability (*diligentia, culpa*).[128] In these contexts especially, where the situation is judged from the perspective of individual possibilities and responsibilities, the concept of *vis maior* is replaced by *casus fortuitus*: here it is not the source of unhappiness that is important, but the fact that the event was uncontrollable.[129] The idea of fairness described by Leibniz above is the basic principle behind these considerations. In Roman law, the "accident" was not associated with a responsibility that could be used to determine whether the affected individual had a right to aid or other compensation, rather, it was simply "an acceptable reason to justify damages"—life and legal considerations were not yet "organized to maximize the suppression of the consequences of bad luck."[130] The *casus fortuitus*, however, had already been extracted from these technical parts of Roman law regarding damage responsibility within purchase contracts, to become the main object of insurance:[131] in Italian policies, of which full copies were even inserted into treatises concerning insurance law, the insurer assumed the risk: "sopra dette robbe d'ogni caso di mare, di fuoco, di getto in mare, di represaglie, o rubarie ... & d'ogn'altro caso portentevole, fortuito, disastro sinistro, impedimento, & caso sinistro [the risk of all manner of shipwrecks, of fire, of damages, of afflictions, and robbery ... and of every other important instance of threat, of dreadful disaster, impediment, and consequence of damage from all those things]."[132] The trade lawyer Benvenuto Straccha dedicated a two-pages-long gloss to the concept in his commentary on the wording of an edited typical insurance contract:[133] By the late Middle Ages, *casus fortuitus* had been transformed from a pure liability-limiting

H. Pool, Unisa 2005, pp. 109–20. For the relationship between *risk* and *liability* in the contexts of contracts of letting and hiring cf. Paul J. du Plessis, *Letting and Hiring in Roman Legal Thought: 27 BCE-284CE*, Leiden 2012, pp. 21–51.

127 The contrast is made clear in Inst. 3, 14, 2.

128 The example from Dig. 2, 11, 2, 3–8 is developed more thoroughly in Ernst, "Wandlungen," pp. 301–06.

129 Ernst, "Wandlungen," p. 310.

130 Gerkens, "vis maior," p. 120.

131 Loris de Nardi, "El caso fortuito: fundamentos culturales y religiosos de una categoría jurídica y de una cosmovisión", *Derecho PUCP* 84 (2020), pp. 337–354, here 348 suggests likewise a generalization of the concept of *casus fortuitus* after the reception of the Digests in medieval Europe, pointing to the 13th c. *Siete Partidas*.

132 See Benvenuto Straccha, *Tractatus duo de assecurationibus ex proxenetis atque proxeneticis* [...], Amsterdam 1569/1668, p. 19 for an example policy.

133 Ibid., p. 60f.

construct to the central object of a contract guaranteeing security—a step that has been recently called the "commodification of risk".[134]

When Leibniz addresses the notion of *casus fortuitus* in the context of his insurance project, he goes beyond the horizon of Roman law and even beyond commercial legal insurance literature: the term that was tailored specifically to individual contractual relationships is used by Leibniz to construct a "state economic" perspective applicable to society as a whole. It is now perceived as unreasonable and, above all, unproductive, that those affected by accidents should be left alone to deal with the damage. In Leibniz' project, it is the question of how to transfer the burden of accidents from the area of private sector risk-transfer regulations to the area of damage redistribution and the equalization of burdens in the collective, especially when there was no legally binding relationship between the affected parties. A new bond of obligation must first be created—one derived from normative quasi-natural law. Leibniz's reflections on the possible objections to his ideas demonstrate the continuing effect of the semantics of Roman law: he questions, for example, if a claimant might seek compensation for self-inflicted damages caused by negligence or spurious intent. The boundary between *casus fortuiti* and culpable dereliction of duty in the context of Roman law is clearly reflected here.

In the context of Roman law, the concept of *vis maior* only serves as an exclusion to claims (*exceptio*) and not as justification for claims (not as *actio*), and so Leibniz seeks and finds a source for justification in another analogically similar legal context; namely, in Rhodian-Byzantinian maritime law. The basic principle here was that when a ship was forced to throw goods overboard to maintain maneuverability during a maritime emergency, the goods were replaced by all, not just by the affected boatmen—a custom established by the principle of fairness (i.e., *aequissimum*). The corresponding fragments for this legal proposition can be found in Dig. 14, 2. Although within the *Digests*, they represent the reception of foreign Greek-Rhodian maritime law, this was the only principle to be adopted: there was no technical legal implementation or true integration of Greek-Rhodian law within the Roman system. The collective liability of boatmen, ship owners, and merchants involved when goods were necessarily jettisoned represents a classic risk community: this is how it was treated in Rhodian law[135] and later in Mediterranean trade, and how

134 Willem H. Van Boom, "Insuring vs. investing in litigation: a comparative legal history of litigation insurance and claim investment", *Comparative Legal History* 8, 1 (2020), pp. 2–26, here p. 9.

135 For the late Byzantine redaction of the *nomos rhodion nautikos* at least, see Walter Ashburner (ed.), *Nomos Rodiōn nautikos = The Rhodian Sea-Law*, Oxford 1909, reprint Aalen 1976, part 3, Ch. 9, p. 27f., 32 and introduction, p. ccxlii; Karl-Friedrich Krieger, "Die

it continued in North Sea and East Sea trade in the notion of the "general average" (*große Havarie*).[136] Roman law integrated the Rhodian principle of loss distribution but *not* through the medium of the community (*communio,* koinonía); rather, Roman law connected this principle to individualized expression within bilateral relationships in the form of labor contracts (*locatio operis faciendi*).[137] This phenomenon of reception without technical integration and the camouflage of "new," poorly fitting legal principles based on traditional contract types dating from late antiquity reappears in the late Middle Ages, and then in modernity where institutions related to customary trade laws are reintegrated into *Ius commune*—not least of all regarding "assecuratio."[138]

Shortly before Leibniz presented his recommendations, however, the details of introducing such compensation claims in an earlier established community were conceptually formulated in the common law literature concerning the *lex rhodia*.[139] When Leibniz integrated this idea of a risk community as an analogy from the *lex rhodia de iactu* and transferred it into society as a whole (because "the entire republic can be taken as a ship as sorts, which is subject to bad weather and accidents"[140] and it would be unfair if "only a few

---

Entwicklung des Seerechts im Mittelmeerraum von der Antike bis zum Consolat de Mar," *Jahrbuch für internationales Recht* 16 (1973), pp. 179–208, 185; Emmanuelle Chevreau, "La 'lex rhodia de iactu'—un exemple de la réception d'une institution étrangère dans le droit romain," *The Legal History Review* 73 (2005), pp. 67–80, 75.

136 Cf. Götz Landwehr, *Die Haverei in den mittelalterlichen deutschen Seerechtsquellen*, Göttingen 1985; Götz Landwehr, "Zur Begriffsgeschichte der Haverei vom 16. bis 18. Jahrhundert," in Erik Jayme et al. (eds.), *Festschrift für Hubert Niederländer zum 70. Geburtstag*, Heidelberg 1991, pp. 57–69; Id., *Das Seerecht der Hanse (1365–1614). Vom Schiffordnungsrecht zum Seehandelsrecht*, Hamburg 2003, pp. 120–30.

137 Chevreau, *La lex rhodia de iactu*, p. 77; less precise Dimitrios G. Letsios, *Nomoi Rhodíon Nautikós. Das Seegesetz der Rhodier. Untersuchungen zu Seerecht und Handelsschiffahrt in Byzanz*, Rhodes 1996, p. 165f.

138 For an overview, see Vidari, "Sulla classificazione".

139 "[…] quippe quo omnes, qui in unam navem merces transvehendas inferunt, quod ea, quae ad navem illam quodammodo communem, conservandam, & a totali interitu ac communi periculo liberandam sunt necessaria, societatem aliquam contrahere videntur; atque sic ab initio, etiam in omnia illa, sine quibus navis illa conservari, & commune periculum averti nequit, consentire creduntur. [… likewise this also unites the] cives in una civitate viventes, in ea quae ad Civitatem conservandam, & eandem a totali interitu, aut notabili periculo liberandam sunt necessaria […]" (Lauterbach, *Q.F.E.F.S. Theseis*, p. 4). In Lauterbach there is also a strong emphasis on equity thought, which is introduced in the *Lex Rhodia* and is at the basis of the development of the notion of legal community. Cf. Coccejus, *De assecuratione*, pp. 32–43, which is, in constrast, quite technically linked to the practice of law concerning damages.

140 The comparison between state and ship is old—even the term *gubernatio* (government) derives from this semantic field: Cassius Dio 52, 16, 4 (Augustus as the captain, who is

meet misfortune, while the others were left unscathed") he went decisively beyond Ancient Roman legal semantics and referred, at least implicitly, to *ius commune* reinterpretations, which did not relate to the Classical basis.

Leibniz's third analogical conclusion originates once more from a branch of law—this time, that of company law (early modern corporative law in its [loose] Roman descendence from types of societas). He refers to the principle of sharing equally in costs and profits (*lucrum* und *damnum*)[141] that also arises in specialized relationships with *vis maior*.[142] To justify why damages from accidents, but not windfalls from good fortune (referring perhaps to the finding of treasure) should be shared by the whole, he explains that such cases of good fortune already indirectly benefit the common good by stimulating consumption and by the fact that the lucky individuals can be presumed to act as charitable benefactors of the poor individually.

In the fourth through sixth analogical conclusions, the topics revolve around the variety of insurance-like institutions in Hamburg: for example, the practice of private maritime insurers, and the "convoy escort" (*Convoyfahrt*) established in Hamburg in 1623, which provided military backup to merchant ships, and finally, the General Fire Fund established there in 1676. Leibniz visited Hamburg in 1678,[143] two years after the founding of the General Fire Fund,[144] and had been in constant contact since the 1670s, with several citizens (such as Vincent Placcius, Heinrich Siver, and Johann Vagetius) who had experienced the great Lehrweet fire of 1684.[145] In the early Hamburg-Hanoverian correspondence of the time, there is no mention of the Hamburg institutions and insurance as such,[146] nevertheless it demonstrates the proximity that Leibniz had to the Hanseatic city, and makes it plausible that he knew of the Hamburg institutions through immediate contacts. In any case, Leibniz's idea for a territorial fire insurance program first appears in a recommendation written to Duke Johann Friedrich Leibniz in September 1678, only a few days after his departure from Hamburg.[147]

---

supposed to save the wreck of the republic, destined for demise), Cicero, *De oratore* 1, 2; 1, 211.

141 Vgl. Inst. 3, 25, 1–3; Dig. 17, 2, 29pr.-1.

142 Dig. 19, 2, 25, 6: in the case of partial lease owners and tenants shall be assessed in respect to the damage caused by force majeure "quasi societatis iure."

143 Cf. Kurt Müller and Gisela Krönert, *Leben und Werk von Gottfried Wilhelm Leibniz. Eine Chronik*, Frankfurt am Main 1969, p. 53; LAA I/2, pp. 56–71.

144 Cf. Zwierlein, *Der gezähmte Prometheus*, pp. 223–31.

145 Ibid., p. 225f.

146 Cf., among others, LAA II/1, no. 38 (p. 126), no. 126 (p. 406), no. 155 (p. 558), no. 169 (p. 607), no. 180 (p. 629); LAA I/2, pp. 55–74; Müller and Krönert, *Leben und Werk*, pp. 53.

147 LAA I/2, p. 75. Strangely enough in the insurance history literature featuring Leibniz as a "pioneer" of insurance, this Hamburg trip is never mentioned as a context.

Leibniz was constantly interested in the risk that fire damage posed for individuals, the state, and society as a whole; as demonstrated by an analysis of our city fire database for his lifetime, the philosopher and scientist lived during the most dangerous period of central European history. Almost all of the largest *nonwartime* German municipal fires occurred during Leibniz's lifetime.[148] Several of the towns and cities in which Leibniz spent a good portion of his life, or that he visited during important periods of his scientific exchange and formation, were cities that were heavily damaged by fire: London's 13,200 houses that burned down in 1666 were in no way all completely rebuilt by 1676 when Leibniz visited the English capital in October 18–29 of that year.[149] Leibniz stayed mainly in Clausthal and Zellerfeld during his journeys through the Harz, where he held discussions about mountain and mining technology;[150] both cities were repeatedly subjected to major fires. Clausthal endured three fires (in 1631, 1634 and 1639) and Zellerfeld was almost entirely destroyed by fire in 1672 (465 houses burned down), only eight years before Leibniz's initial visit.[151]

Leibniz's insurance project, viewed against the backdrop of northern or southern/western European models of the General Fire Fund mentioned above, is thus striking, since the fields of discourse and knowledge from which he draws to create the transfer analogies for the universalization of the insurance principle are, on the one hand, attributable to theoretical law, and on the other hand, to commercial practice—whereby emerges a society-wide "national economic" (*volkswirtschaftliche*) insurance concept, or even that society as a whole is somehow an insurance collective.[152] As previously discussed, even before Leibniz, the early fire insurance projects circulating between 1605–17 never had roots in the cooperative institutions—the *Zünfte*

148 On this cf. the explications and city fire lists in Zwierlein, "Katastrophe und Prävention".

149 Cooper, "Robert Hooke's work as surveyor".

150 On this see, Wakefield, "Leibniz and the Wind Machines". There is no indication in the correspondence on past fires—there is only mention of a single small, recent fire: "Gestern brandte zum Claustal eine Bergschmiede, welche unweit der Zeche Sophia gelegen, und dem Bergwercke bald großen schaden zugefüget hätte." (F.W. Leidenfrost to Leibniz, Zellerfeld, February 7/17, 1679, LAA 1/2, p. 118).

151 Morich, "Die früheren Brandkatastrophen". There is no exact information about the speed of rebuilding in this case.

152 The complex and actual interrelationships between cooperative thinking and southwestern influences is emphasized in more recent Hanseatic historical thought: see, for example, Rolf Sprandel, "Die Interferenz von Gesellschaften und Genossenschaften im hansischen Handel," in Nils Jörn, Detlef Kattinger, and Horst Wernicke (eds.), *Genossenschaftliche Strukturen in der Hanse*, Cologne 1999, pp. 79–100, 99; Albrecht Cordes, *Spätmittelalterlicher Gesellschaftshandel im Hanseraum*, Cologne 1998, pp. 251–60.

(guilds) and *Genossenschaften* (cooperatives) with their traditions of solidarity bonds—but consistently owed their existence to an analogical relationship with maritime premium-based transport insurance. Even later, in the eighteenth century, when cameralists referred to the prehistory of "insurance" in their treatises, they almost always mentioned the mercantile, originally Mediterranean contract practices, and rarely the cooperative precursors.

It is revealing that the genesis of cameralistic insurance discourse employs an entirely counterintuitive interpretation of academic Roman law as a major source of inspiration. The academics of the new scientific field of cameralism align themselves here with one of the most classical forms of academic reasoning, Roman law, or at least with a practice already "coded" in such terms of legal theory. This corresponds to the boundary between theoretical academic law and mercantile common law extending from the late Middle Ages to modernity, which can be examined precisely in the context of northern German insurance disputes. As discussed above, one point of difference between mercantile customs and academic discourse concerned the dispute about the validity of insurance contracts in the case of incomplete premium payments (see Ch. 2, 3). The gap that emerged regarding this dispute between customary law logic in the area of trade and academic law in Hamburg during the 1620s and 1630s, relates to the analysis of Leibniz' texts. The topic of insurance was elevated through legal treatises, such as those of Rutger Ruland, and specialized works on "De assecurationibus," to the sphere of academic debate in Northern Europe—but within the limited scope of Roman law only. It was only through this focus on Roman law that the topic of insurance was relevant for scholars like Leibniz. The *casus fortuiti* discussion in academic law had, in any case, little to do with issues of mercantile law.

An academic discourse about state insurance institutions did not yet exist, however. Leibniz thus addresses different elements of academic legal discourse associatively and analogically, as well as the current but still theory-less practice of municipal mercantilist institutions. In doing so, he offers a new theory: one that tasks the state with provision and prevention in the form of a framework created for stimulating the personal provision of funds. The accident fund, established to offer nothing beyond the accumulation of documents confirming such personal provisions, becomes insufficient.

In contrast to the projects around 1600, it should be noted that Leibniz conceived of his projects primarily from the perspective of the state—to serve the common good—and only secondarily from the perspective of the individual. This no longer has anything to do with the "protect-and-defend" notion, or with the analogy of military defense that a lord owes, by mutual contract, to his subjects. Around 1600, the (as yet unseparated) state and social sectors were essentially conceived of as whole risk communities. A century later, the interplay between self-provision and state provisions provided for the purpose of general welfare—the basic Leibnizian principle underlying individual as well as state activity in this context—and transformed the state into an early organizational form devoted to the goal of securing welfare for its populace. In this sense Leibniz's conception of insurance marks a decisive turning point in the complex development of the theory and practice concerning self-provision and state provision according to northern and southern institutional patterns, which results from an intellectual transfer process instigated by a series of analogical constructs.

1.2.2.3 *"Real-Assecuration": The Founding of the Berlin Academy and Fire Association*

While the character of Leibniz's cameralistic projects has been revealed by highlighting their ingenious use of analogical transfers, in addendum to the development of fire policey discussed above (Ch. 3, 3), the degree to which Leibniz conceived of this insurance project in tandem with scientifically backed security production within the context of a territorial state like Brandenburg-Prussia must also be highlighted.

This issue concerns Leibniz's further technical and practical efforts regarding cameralistic economics. His interests in the invention of the air pump, for instance, and associated problems in physics, such as the vacuum, were not relevant merely in a sense of general theoretical imagination and search for innovation. To a greater degree, he was concerned with the concrete improvement and applicability of fire extinguishing equipment—significantly, this concern dates to 1700–01, precisely the years that he was also closely involved in the establishment of the Berlin Academy.[153] After his return from Berlin in the autumn of 1700, Leibniz met Andreas Schmidt in Helmstedt, where he also presented "De Machinarum Incendiariarium oblatarum conditionibus":[154] Leibniz was interested in

153 Hans-Stephan Brather (ed.), *Leibniz und seine Akademie. Ausgewählte Quellen zur Geschichte der Berliner Sozietät der Wissenschaften, 1697–1716*, Berlin 1993.

154 Cf. the short note of different, important areas of activity, dating as early as to 1700 in the "Feuersprizen mit ander anstalt, *pro re mechanica* aufgeführt werden" (June 1700, LAA I/18, p. 713); also Leibniz to Joh. Andres Schmidt, Wolfenbüttel, September 3, 1700, LAA I/19, p. 139. Vgl. hierzu weiter Schmidt to Leibniz, Helmstedt, February 8, 1701, ibid., p. 387; Leibniz to Schmidt, Wolfenbüttel, February 15, 1701, ibid., p. 423; Schmidt to Leibniz, Helmstedt, January 24, 1701, ibid, p. 458; Leibniz to Schmidt, Hanover March 4, 1701, ibid., p. 469; Schmidt to Leibniz, Helmstedt, March 11, 1701, ibid., p. 489—the picture of the Helmstedt fire hose that was sent with this letter is no longer preserved in the Leibniz papers; Leibniz to Schmidt, Hanover March 15, 1701, ibid., p. 501; Schmidt to

the acquisition of a fire extinguisher for Helmstedt from the specialists of Hamburg and strove to obtain an accurate diagram of such a device.[155] At about the same time, he obtained information from the chief syndic of Maastricht about a craftsman who manufactured "serpentine hoses [*Schlangenfeuerspritzen*] for a reasonable price" in Duisburg.[156] Based on his correspondence with Jablonski, we can see that this interest was actually related to the recently founded Berlin Academy (then called, like the London Royal Society, a *Societät*), regarding which Leibniz was in correspondence with Councilor Chuno. As Leibniz wrote: "I have made various proposals, as Councilor Chuno will attest. Please draw up a relevant agenda *pro memoria* from them. Earlier I wrote to [the city of] Duisburg concerning the hose, [I] ask that one might inquire relating to that on behalf of the Society [i.e., the Academy]."[157] By 1702 it appears that Duisburg hose experts were working in Berlin.[158] Similarly, in his 1703 correspondence we find an interest in asbestos as a fire resistant material that the "Operateur Empirique" A. Janulli had presented to the ducal court in the form of roof tile paint.[159] Leibniz thus acted here as both a scientific-technical expert and as a communication link between the European centers at the forefront of fire extinguishing technology and Brandenburg-Prussia. Leibniz also visited Amsterdam in November 1675, which by then enjoyed nighttime illumination facilitated by the work of the brothers van der Heyden, who had also been central figures in the introduction of the serpentine fire hose. This hose had been tested since 1672 before being advertised and sold throughout Europe (as discussed above in Ch. 3, 3).[160] Van der Heyden's

Leibniz, Helmstedt, March 18, 1701, ibid., p. 508: R. Chr. Wagner was drawing the sketches of the hose.

155 "Mais je vous laisse juger pour les seringues: on en achete une à Helmstedt de l'homme de Hamburg, je croi pour 300 écus. Et Mr l'Abbé Schmidt m'en fera avoir tout le detail, outre que nous en avons ici à Hannover […]" (Leibniz to Johann Jacob Julius Chuno, Hanover, December 31, 1700, LAA I/19, p. 309).

156 Leibniz to Johann Theodor Jablonski, Hanover, January 31, 1701, LAA I/19, p. 377. Johann Theodor is not to be mistaken for his brother Daniel Ernst, cf. Alexander Schunka, *Ein neuer Blick nach Westen. Deutsche Protestanten und Großbritannien (1688–1740)*, Wiesbaden 2019, pp. 81, 125, 152.

157 Leibniz to Johann Theodor Jablonski, Hanover, March 19, 1701, LAA I/19, p. 514.

158 Johann Jacob Julius Chuno to Leibniz, Berlin, April 25, 1702, LAA I/20, p. 194: "Pour les Siringues il est venu un homme de Duisbourg par ordre de la cour, qui en fera pour cette ville icy, Monsr de Fuchs m'a addressé cet homme […]."

159 Leibniz an Lorenz Hertel, Hanover, July 3, 1703, LAA I/21, p. 70: "Cet Empirique monstre une chose curieuse. C'est un morceau de toile faite d'Asbeste, ou d'Amianthe, qui resiste au feu. On n'a gueres vû d'échantillon qui soit si ferme. Il dit de l'avoir des Pirenées: mais je crois qu'un autre le luy a donné tout fait."

160 Müller and Krönert, *Leben und Werk*, p. 46.

Amsterdam original was quickly copied or imitated in other cities—although admittedly, we know very little about the technical history of this process—so it is therefore unsurprising that Leibniz knew of a distributor in Duisburg, a city that traditionally bridged the gap between the Netherlands and the Holy Roman Empire, and more specifically between the Netherlands and Brandenburg-Prussia, which it became part of following the struggle concerning the Jülich dynastic succession. But the widespread acquisition of fire hoses by provincial authorities and villages proceeded very slowly. By the middle of the eighteenth century, the city of Hamburg still possessed more large fire hoses than the entire Kurmark (Electoral March), even though Brandenburg-Prussia and its electors and kings were themselves among the earliest driving forces for such "meliorations," which were later copied by the southwest territories (Ch. 3, 3).

These temporal and technical-administrative issues provide a context for Leibniz's interest in fire *policey*, firefighting technology, and insurance. He was also acquainted with the history of relevant major technical advances and corrected Denis Papin in correspondence about fire hoses in 1704–07, to the effect that the innovation of the air chamber could be traced to the Nuremberg craftsman/inventor Hans Hautsch. Through storing air pressure, this "air chamber" made a continuous stream of water possible instead of spurts interrupted according to the rhythm of the pump handle's operation. It is not documented whether Leibniz had become acquainted with Hautsch's invention (which can be dated precisely to the years 1666–67) on site, when he was staying in Altdorf/Nuremberg, but it is presumed in the literature.[161] Papin was already an important channel of communication through which news of the air pump's invention had reached Robert Boyle and Christiaan Huygens. As Steven Shapin and Simon Schaffer have shown in their now-classic work, knowledge of this invention first remained embedded within complex epistemic disputes between the field of physics, of religion and of political science among well-known circles of natural philosophers; it was only in a second stage that it was distributed as a commercial object within the Netherlands and France during the 1670s.[162] Shapin and Schaffer concentrated on the epistemological controversy between Boyle's experimental approach and Hobbes's deductionism, but they only marginally touch on the diffusion of experimental innovations beyond the circles of natural philosophers (Otto von Guericke and Caspar Schott, Boyle, the Royal Society, Huygens, and Papin). In fact, academies and natural philosophers also occupied themselves with "pumps" in the much more

161 Cf. Knapp, *Wissensgeschichte*, p. 125f.

162 Simon Schaffer, and Steven Shapin, *Leviathan and the Air Pump. Hobbes, Boyle, and the Experimental Life* (Princeton: 1985), pp. 269–76.

practical dimensions discussed here—i.e., water pumps that worked via air pressure. A quick glance at the Proceedings of the Parisian *Académie royale des Sciences* confirms this: in 1676, just as van der Heyden's venture was taking off, in Paris, a certain M. Anaspaize held a *Présentation d'une pompe d'une compression continuelle pour élever les eaux*,[163] which was a discussion of the air chamber issue. In 1716, this problem was examined further by an academy member Philippe de la Hire and one of his sons.[164] In 1725, the year he was admitted to the academy as an important *associé chimiste*, Charles-François de Cisternay du Fay occupied himself with the *Description d'une pompe qui peut servir utilement dans les incendies*,[165] which addressed the question of how pumps could contribute to the general transportation of water in large cities.[166] Other issues concerning fire extinguishing and fireproof or fire-resistant chemical paint for wood were also discussed in 1722 and 1723.[167] Such discussions demonstrate how Leibniz's own method of combining fire *policey*, cameralist institutional and technical innovations, academic reasoning, and the topic of insurance, while exemplary, was not particularly exceptional within the context of the general European engagement with fire danger.

The Brandenburg Electors, who then became the Kings of Prussia, sought to implement massive improvements in fire *policey* during the late seventeenth and early eighteenth centuries. The first attempts to establish a comprehensive fire insurance agency for all cities and rural areas (*plattes Land*) also occurred

163 Archives de l'Académie royale des Sciences Paris, Procès verbaux des séances 13 [1676], f. 79.

164 De la Hire, *Examen d'une pompe présentée à l'Académie par le f. Nicolas Augustin*; De la Hire, *Mémoire sur la construction d'une pompe qui fournit continuellement de l'eau dans le réservoir*, Archives de l'Académie royale des Sciences Paris, Procès verbaux des séances 35 [1716], f. 355, 361; Cf. the series "Pochettes" for the year 1717 (24 July, f. 211: *Pompe nouvelle de la Hire*). On La Hire see Alice Stroup, *A Company of Scientists. Botany, Patronage, and Community at the Seventeenth-Century Parisian Royal Academy of Sciences*, Berkeley 1990, p. 21f. and passim; on the La Hire dynasty see David J. Sturdy, *Science and Social Status: The Members of the Academie des Sciences, 1666–1750*, Woodbridge 1995, pp. 193–213.

165 Archives de l'Académie royale des Sciences Paris, Procès verbaux des séances 44 [1725], f. 63; Cf. the materials on séance: *Pompe à incendie*, p. 35, in the series "Pochettes" 1725 and the examination of a further fire extinguisher of Claesens; on Du Fay cf. Sturdy: *Science*, pp. 286, 362, 370.

166 Donsenbray, *Examen d'une pompe pour donner de l'eau dans plusieurs quartiers de Paris*, Archives de l'Académie royale des Sciences Paris, Procès verbaux des séances 51 [1732], f. 272; Donsenbray, *Examen d'une pompe perfectionnée*, Archives de l'Académie royale des Sciences Paris, Procès verbaux des séances, 54 [1735], f. 125.

167 Cf. Archives de l'Académie royale des Sciences Paris, "Pochettes," 1722 (December 16): *Mesures à propos des reflexions sur moyen d'éteindre le feu*; "Pochettes," 1723 (February 3): *Composition liquide pour éteindre le feu*.

during this time.[168] And here again, when the question arose about which authorities should be responsible for this organization, no firm decisions were made. Leibniz himself apparently saw the new Berlin Academy as the authority for scientific supervision and testing of the best fire extinguishers. Indeed, he saw it even further as "involved in a coordinating role for the plans of large-scale procurement and distribution of fire extinguishers [... as the royal] supervisory authority for their purchase and distribution" in the Prussian provinces.[169]

The context of Leibniz's correspondence with Schmidt and especially with Chuno and Jablonski in 1700–03 is shown in a letter he wrote to King Friedrich I in 1711:

> Furthermore, as Your Majesty has most graciously assigned the *privilegium* of introducing and procuring trustworthy fire extinguishers, of the kind that is not yet in common use in [Kurmark] lands, to the Society which deals with the mathematical arts and mechanics. ... But one is thereto hindered by the fire insurance fund, which has denied the society's *raisonnable* and lawfully offer to second herein, even if that matter cannot be better effected, than by the Society. As such an *obstaculum* has happened now, the Society cannot do otherwise than to request that *Your Majesty* maintain his most gracious concession, in such a way that through a written resolution *Your Majesty* would declare that he upholds the Society [in this power] and gives order, to which everyman thereafter conforms.[170]

It is also apparent that, according to Leibniz's ideas, the recently established Society/Academy was first seen as having an important organizational function in the reorganization of state fire policy. By 1702 there were already public disputes concerning jurisdictions under other state authorities,[171] and later

168 Cf. above p. 211.

169 Cf. Knapp, *Wissensgeschichte*, pp. 55–57.

170 Gottfried Wilhelm Leibniz, *Die Werke. Gemäß seinem handschriftlichen Nachlasse in der Königlichen Bibliothek zu Hannover*, ed. Otto Klopp, series 1, vol. 10, *Historisch-politische und staatswissenschaftliche Schriften*, Hanover 1877, pp. 446–52, here 451f.

171 Chuno to Leibniz, Berlin, April 25, 1702, LAA I/21, p. 194f.: Chuno reports on the invitation of the Duisburg firefighting experts particularly on behalf of Fuchs, and immediately offers assurances that the invitation also happened as "de la part de la Societé," but then recalls that one must be careful at the opportunity, because the establishment of the General Fund as "impots à lever sur les habitants" would make this opportunity "odieuse"; nevertheless it is obviously a societal demand to be responsible in this area, to offer protection, especially at this point in time "Je n'ay pas laissé Monsieur de representer à ces

also concerning the Prussian General Fire Fund (established in 1706 after numerous attempts beginning in 1685). This institution was initially dissolved yet again in 1711, and fire insurance institutions were later renewed only on the level of individual cities (1718 in Berlin) and the provinces of Prussia (1719 in the Kurmark, additional provinces followed).[172] Leibniz's above-cited statement originates precisely from this year of institutional crisis, 1711.

The "Project on Öffentliche Assekuranzen" did not mark the limit of Leibniz's interest in this area; beginning in 1678 at the latest, he demonstrated a far wider interest in and engagement with the problem of fires and security production. At the roots of the royal Prussian Academy were very practical cameralist ideas intended to integrate science into the state for the purposes of fire security. In the buildings of the Berlin-Brandenburg Academy, today a famous and venerable institution at Berlin's *Gendarmenmarkt* again, Noble prize winners and other important university professors and researchers discuss the most recent and most important issues of national politics science and humanities; a historical project of the Academy is even occupied with editing Leibniz's texts, conceiving of him first of all as a philosopher of monads and other very abstract problems. Few would associate this noble institution with Berlin's firefighting cars and their *Feuerwehr* men eventually driving through the neighboring streets with blue sirens, two worlds so very separated from each other nowadays. But in 1700, fire prevention, fire *policey* and fire damage recovery (insurance) were at the core of the issues debated by the founding fathers of that same institution—and Leibniz was very much regretting that the Academy had lost control of those issues to the benefit of the *Generaldirektorium*. The story told in Schaffer and Simon's *Leviathan and the Air-Pump* therefore has a subsequent history that could be titled "Leibniz and the Fire-Hose." The combination of ideas and projects and the momentum of the Leibnizian laboratory as comparable to the British post-1666 laboratory surrounding Nicholas Barbon are already evident in the insurance report of 1680, where Leibniz himself designated all of the practical concerns of fire *policey* as "real-assecuration" and defined the *postfactum* economic compensation for

Ministres, que la societé avoit pensé à ces Siringues et [qu']on avoit obtenu le privilege lors que personne n'y avoit pensé, et qu'on seroit par là en droit de s'en tenir au privilege, mais comme on ne pouvoit pas entreprendre ces Siringues dans la quantité requise dans ce commencement on n'empecheroit point cet etablissement si utile." The (unspecified) great number of fire extinguishers would require a sum of "plusieurs mille ecus," which would lead to substantial "impots."

172 Cf. for more details below Ch. 5, 1.2.

damages as "verbal-assecuration"—a relationship backed by and integrative concept of societal security.

1.3 *Summary: The Power of Analogies*

The laboratories of both London/Barbon and Hamburg/Leibniz prove to be melting pots of projects, practices, and concepts. While there is no evidence of any reciprocal influence in the sources, these "laboratories" simultaneously led to two models of insurance-like security production around 1680: the British model, based on trade and stock exchange, and the German/central European model, based on a cameralistic fusion of the originally "private" construct of premium-based insurance in combination with state institutions. In both cases, the central foundation for the innovation process involved the use of "analogies" on various levels. Just as Graunt had already analogously imposed accounting theory onto populations and thereby linked it to Baconian theory, Barbon similarly used Graunt's findings analogously to generate the first estimates of building stocks. Newbold's projects explicitly align the analogous development of joint-stock insurance companies with colonial trade companies like the East India Company. Characterizing insurers as "adventurers" and combining building speculation with insurance demonstrates how the constant recourse to analogies was not affiliated with any one figure in particular: insurers seem almost like adventure brokers, negotiating the vast sea of buildings, rather than a literal sea, and even the earliest Strasbourg project schemers compared the potential value creation of insurance with profits generated from American colonial expansion.[173]

As his *Apology* indicates, figures like Barbon were of the opinion that activities such as building speculation and insurance were not merely small sectors of a wood- and stone-based economy, but that they were part and parcel of the imperial project of the founding of a British nation complete with its "metropolis." In a similarly large, European-global context, Leibniz also stages his insurance project in a similar fashion: he characterizes accident insurance as a central instrument facilitating the growth of state power

173 Jean-Louis Maunowy, *La genèse des innovations. La création technique dans l'activité de la firme*, Paris 1968, pp. 73–83 described analogous processes under the label of "induction". For a different systematic approach cf. Esa Itkonen, *Analogy as Structure and Process: Approaches in Linguistics, Cognitive Psychology and Philosophy of Science*, Amsterdam 2005 and Adam Biela, *Analogy in Science. From a Psychological Perspective*, Frankfurt am Main 1991.

and the position of the Holy Roman empire within the concert of European powers.

In contrast to the London case, by embracing the transformation of Roman legal discourse, Leibniz constructs a linguistic framework that defines insurance as a valid object in a new kind of academic discourse, which would later be called cameralism. In the case of London/Barbon however, reflection on matters of insurance remained primarily limited to the level of controversial pamphlets that competitively advertised different institutional solutions. Accordingly, the British practice developed further within the London business milieu in forms of societal competition and competing models, while the German practice followed the path of territorial administration. The German model is thus an expression, and, to some extent, also an amplification of differing conceptions of the state/society relationship within which security production emerges. The overall economic function of insurance is, however, quite similar and is primarily based on improving the security of the central asset of property and buildings, which are themselves then offered as collateral for other investments: an element which was obviously of central importance in the English Financial Revolution, but just as important in the still territorially limited cameralistic economy of Germany—even though Germany lacked a comparable period of 'revolution', both as an actual historical phenomenon and a conceptual framework.

The idea of a "culture of innovation" is primarily associated with the technical/scientific systems of the twentieth and twenty-first centuries, in which, in a highly reflexive manner, the production of innovative politically and economically determined conditions were created and specific stimuli were cultivated. Obviously, such a situation did not yet exist in the seventeenth century, however, the described constellation that stimulated the post-fire reconstruction in London, as well as the associated financial revolution, financial speculation, and project schemes (not yet operating on a political/economic metalevel) created an analogous milieu in which new ideas were rewarded and which could be described as an early modern London innovation culture. In the case of the spatially stretched constellation of Hamburg/Leibniz/Prussia, it is more difficult to apply the idea of a simply localizable culture of innovation, since the actual institutional innovation took place in Hamburg without accompanying theoretical reflection. It was only through Leibniz that an epistemic shift—inspired through the observation of practice—occurred, resulting in an associated innovative product: the early cameralistic insurance concept. While it hardly seems fitting to group together the urban mercantile Hamburg milieu and the early Enlightenment environment of Leibnizian influence under the umbrella "German innovation culture around 1700," viewed from the perspective of the final product, the different conceptual and institutional forms of "insurance" were, in fact, both laboratories, functioning as crystallizing chambers for these innovations. In both cases, the two historical laboratories around 1680 anticipated the major aspects of the "construction of the normal secure society" (see Ch. 5). Insurance thus became an important part of economic Enlightenment practice.

2 Religious Culture and "Insurance"

The premium-based insurance principle was established in Italy during the late Middle Ages in relation to the Mediterranean Sea as a commercial space; however, the epochal changes occurring within the laboratories of innovation culture around 1680 took place in northern Europe. The fire insurance institutions emerging in London and Hamburg constitute the first moments—in either a private commercial or state institutional form—when insurance-like security provisions were systematically transferred to a level that potentially included the general public and society as a whole, rather than merely the limited societal sector of merchants and their specific issues. It was also at this moment that "nature" and "culture" became simultaneously shielded from each other through a safety buffer—even when caused by the carelessness of human action, the unleashed fire represented an incursion of uncontrolled nature in the sphere of civilization and society.

The question remains: Why did these epochal thresholds occur in these particular "laboratories" of northern Europe, even though southern Europe had much more experience with insurance and the municipal and mercantile culture of Italy had long been familiar with the principle? In formulating an attempt at an answer, I will invoke the old discussion about Max Weber's "protestant ethic and the spirit of capitalism" of 1905, which due to its 100th anniversary and later debate can still serve as a good starting point to clarify the problem, as, due to that long debate, the critical mass of differentiated arguments in support of Weber's thesis and contradicting his statements have reached a critical mass. 'Insurance' and my specific problem of the South-North relationship regarding the transfer and institutionalization of premium insurance principles in the North have never been adressed within this debate, but the older general debate offers by way of analogy an explanatory framework already well refined. Some indicators suggest, that religious denominational and cultural differences in Europe are surely not the only cause, but at least an important causal factor leading to the emergence of that particular "insurance culture" in northern Europe.[174]

2.1 *Max Weber, the Protestant Ethic, Calvin(ism), and Economics*

Based on the observation that at the end of the nineteenth century the accumulation of capital in mixed-denominational states (Baden, for example) was significantly stronger among Protestants, Weber generated his universal historical theory postulating an elective affinity between the Protestant ethic—in particular, Calvinist/Puritan/Reformed asceticism—and the "spirit of capitalism." As a major motive for the capital-accumulating, and then capital-investing, capitalistic spirit of labor, Weber evokes concerns, or even abject fears, regarding personal salvation—a fear which was intensified through the elimination of all works-based notions of salvation, especially through the Calvinist doctrine of predestination. Accordingly, this led to the fundamental insecurity of the individual concerning the central question of his or her "election," resulting in the "unimaginable inner isolation of the individual."[175] The "decisive problem" was thus for Weber a psychological one: "how could this doctrine be *endured*?"[176] In order to "endure" and survive this insecurity about one's own status as a chosen elect, the *signa salutis* doctrine was crucial. While righteousness based on works was no longer applicable, "relentless vocational work" and good works could be seen as "the technical medium not: to purchase salvation, rather: to eliminate anxiety about salvation."[177] This sublimation of the religious significance of worldly work activity—which is initially practiced just according to the internal logic of profit making and only indirectly as a sign of one's own status of being elected—was the structural mindset that engendered capitalistic productivity before the religious element fell away at some point. Weber thus developed a broad universal historical theory to clarify the apparent superiority

174 Cf. William H. Swatos and Lutz Kaelber (eds.), *The Protestant Ethic Turns 100: Essays on the Centenary of the Weber Thesis*, Boulder 2005; Wolfgang Schluchter and Friedrich Wilhelm Graf (eds.), *Asketischer Protestantismus und der "Geist" des modernen Kapitalismus. Max Weber und Ernst Troeltsch*, Tübingen 2005; Hartmut Lehmann (ed.), *Max Webers Religionssoziologie in interkultureller Perspektive*, Göttingen 2003; Hartmut Lehmann, *Max Webers "Protestantische Ethik." Beiträge aus der Sicht eines Historikers*, Göttingen 1996; Hartmut Lehmann and Guenther Roth (eds.), *Weber's Protestant Ethic: Origins, Evidence, Contexts*, Cambridge 1993. In English, the Weber thesis and the early discussion (Rachfahl, Troeltsch, Sombart) is analyzed in Peter Ghosh, *Max Weber and the Protestant Ethic: Twin Histories*, Oxford 2014.

175 Max Weber, *Die protestantische Ethik I. Eine Aufsatzsammlung*, ed. Johannes Winckelmann, Gütersloh 1991, p. 122.

176 Weber, *Die protestantische Ethik*, p. 127.

177 Weber, *Die protestantische Ethik*, p. 131.

of Occidental, in particular, Anglo-American capitalism, in contrast with areas of different cultural influence, on the basis of the psychological impact of the so-called *syllogismus practicus*—i.e. vocational success is an indication of election, and vice versa, God-given vocational success is a result of election.

Since the appearance of these theoretical essays, they have been subject to countless critiques, which can in no way be fully accounted for here.[178] Recent research has demonstrated that Weber took the theoretical core of his theory almost entirely from the very one-sided 1855 systematical work on the Calvinist theology of Matthias Schneckenburger,[179] who saw the *syllogismus practicus* as a central defining characteristic separating Reformed (i.e., Calvinist) and Lutheran theologies. Schneckenburger, a born Lutheran, was tasked with educating aspiring Reformed pastors in Bern and would have felt thus obliged to thoroughly investigate the differences between the two Protestant schools, especially for the purpose of his didactic lectures.[180] It was argued that the doctrine of *signa praedestinationis* in the Reformation era was actually systematized by the Catholic opponent of Luther, Johann Eck, and that it was integrated into Protestantism by Luther, who then handed it down to Calvin. In this regard, however, Weber was clearly exaggerating: there is no (not even Catholic) theology that recognizes the pure superiority of works over faith for attaining certainty about salvation.[181] It has been biographically ascertained that Weber's Anglophilia[182] and his image of modern Anglo-America was also influenced by the "countermodel" of Catholic Italy, which he experienced first-hand on his trips to sanatoria that occurred at the same time he wrote the

178 A part of the contemporary criticism is assembled in Max Weber, *Die protestantische Ethik II. Kritiken und Antikritiken*, ed. Johannes Winckelmann, 5th ed., Gütersloh 1987.

179 Matthias Schneckenburger, *Vergleichende Darstellung des lutherischen und reformierten Lehrbegriffs*, Stuttgart 1855.

180 Friedrich Wilhelm Graf, "The German Theological Sources and Protestant Church Politics," in Lehmann and Roth (eds.), *Weber's Protestant Ethic*, pp. 27–49, 35f. Cf. Malcolm H. MacKinnon, "Calvinism and the Infallible Assurance of Grace: The Weber Thesis reconsidered," *British Journal of Sociology* 39 (1988), pp. 143–77; Malcolm H. MacKinnon, "The Longevity of the Thesis: A Critique of the Critics," in Lehmann and Roth (eds.), *Weber's Protestant Ethic*, pp. 211–243.

181 Dieter Schellong, "Der 'Geist' des Kapitalismus und der Protestantismus: Eine Max-Weber-Kritik," in Richard Faber (ed.), *Der Protestantismus: Ideologie, Konfession oder Kultur?*, Würzburg 2003, pp. 231–53, 242f., and earlier and more expansively, Dieter Schellong, "Calvinismus und Kapitalismus. Anmerkungen zur Prädestinationslehre Calvins," in Hans Scholl (ed.), *Karl Barth und Johannes Calvin*, Neukirchen-Vluyn 1995, 74–106. Cf. André Biéler, *Calvin's Economic and Social Thought*, ed. Edward Dommen, trans. James Greig (orig. French, 1961; Geneva 2006), pp. 270–76.

182 Guenther Roth, "Weber the Would-Be Englishman: Anglophilia and Family History," in Lehmann and Roth (eds.), *Weber's Protestant Ethic*, pp. 83–121.

"Protestant ethic."[183] Some attention has been devoted to the lack of regard for the precursors of "rationalization" present during the Middle Ages.[184] Fernand Braudel attacked Weber with harsh criticism and pointed out, in particular, that the reality of capitalistic economic activity, but also its conceptual framework—already thoroughly anticipating Franklin's language—is to be found in fourteenth and fifteenth-century Italy (though Weber had already been informed of this last point through Werner Sombart by means of Leon Alberti's *Libri della famiglia*).[185] Relevant here are also Henri Hauser's, Visser't's Hooft's, Conord's and Tawney's impassioned rejections of the elective affinity between Calvinism and capitalism, based on Calvin's famous letter to Claude de Sachin of November 7, 1545, in which he determined that in contrast to the canonical prohibition of usury, charging low rates of interest was permissible-because in Catholic Genoa, there was evidence that charging interest was common.[186] For Braudel the "capitalistic practice" (less so the "capitalistic spirit") would have also thus been an "influence méditérranéenne,"[187] which, like Baroque artistic style, traversed the Alps and then Europe as a "transfert culturel," and thereafter influenced the entire world—though, from 1600 on Italy began to stagnate economically due to the trend toward the global economic redimensioning of the Mediterranean world with the advent of colonial trade.[188]

183 Peter Hersche, "Max Weber, Italien und der Katholizismus," *Quellen und Forschungen aus Italienischen Archiven und Bibliotheken* 76 (1996), pp. 362–82.

184 Lutz Kaelber, "Weber's lacuna: Medieval Religion and the roots of rationalization," *Journal of the History of Ideas* 57, 3 (1996), pp. 465–85.

185 Fernand Braudel, *Civilisations matérielle, économie et capitalisme XVe-XVIIIe siècle. II. Les jeux de l'échange*, Paris 1979, pp. 683–701, 699.

186 Braudel, *Civilisations*, p. 700; on the letter and the historical significance of Calvin for capitalism, see Biéler, *Calvin's Economic and Social Thought*, pp. 403–07 and 420f.: Biéler, in any case, does overestimate the historical significance of his "hero," as far as earlier medieval practices are concerned. For a current summary of the literature on pre-Reformation interest rates and usury in Christian money lending see Rowan Dorin, *Banishing Usury: The Expulsion of Foreign Moneylenders in Medieval Europe, 1200–1450*, PhD thesis Harvard University, 2015, pp. 12–36; as far as Bucer's significance in this regard as a precursor to Calvin, see Cornel Zwierlein, "Reformation als Rechtsreform. Bucers Hermeneutik der lex Dei und sein humanistischer Zugriff auf das römische Recht," in Christoph Strohm (ed.), *Martin Bucer und das Recht. Beiträge zum internationalen Symposium vom 1. bis 3. März 2001 in der Johannes a Lasco Bibliothek Emden*, Geneva 2002, pp. 29–82, 51–53.

187 Braudel, *La Méditerranée*, p. 556.

188 Cornel Zwierlein, "Die Auswirkungen von spatial turn und Kulturtransferheuristiken auf das Epochenkonzept 'Frühe Neuzeit' (exemplifiziert anhand der Transfers des Versicherungsprinzips)," in Michael North (ed.), *Kultureller Austausch. Bilanzen und Perspektiven der Frühneuzeitforschung*, Vienna 2009, pp. 43–67. For recent discussion, criticizing the Eurocentrism of the Weber debate cf. Andre Gunder Frank, *ReOrient: Global Economy in the Asian Age*, Berkeley 1998; or for a more neutral take, see

However, insurance and insuring was nearly never touched upon in these debates, though it was certainly, next to the *cambio*, one of the most important technical parts and inventions of early Mediterranean capitalism and also one of the mechanisms used by Protestant merchants and bankers—is insuring of no importance, because it does not serve first of all the accumulation of capital but "only" its security?

2.2 *A Historiographical Gap: Insuring and Religious Denominations*

In Max Weber's œuvre, the principle, idea, practice, and theory of (premium-based) insurance arises scarcely or not at all. Here and there is talk of obligatory insurance and accident insurance,[189] and at one point the German cameralistic fire societies are characterized as being helpful for the rich, but exploitative for the poor.[190] In his *Economy and Society,* the topic of insurance does not appear at all. This might seem insignificant; however, Weber's theory about the effects of the Protestant ethic focuses precisely on the feeling of anxiety regarding personal salvation that is central to Calvinism. He often invokes the psychological mechanisms of "inner loneliness" elicited by the doctrine of predestination. The doctrine of *signa salutis* serves precisely as a guarantor of security, almost like a type of "insurance." Weber reduces the problematics of faith and theology to a relationship between feelings of insecurity and the production of security: capitalistic activism is not set free through an activist program of its own making, rather it is the unintended consequence of precisely this striving for the security of being elected, for the assurance of salvation. To some extent, Weber uses vocabulary from the field of insurance ("insurance," "premium"—of course in the sense of an award or prize, rather than in the sense of a paid premium); however, he uses these terms in a strictly metaphorical sense.[191] His entire theory is essentially "psychohistorical" in its

---

Kenneth Pomeranz, *The Great Divergence. Europe, China, and the Making of the Modern World Economy*, Princeton 2000. Cf. this analysis that is thoroughly skeptical of capitalism: Braudel, *Civilisation matérielle*, p. 700f., as well as 701–23 of "capitalisme hors d'Europe."

189 Max Weber, *Gesamtausgabe*, ed. Horst Baier, part 1: *Schriften und Reden*, vol. 10: *Zur Russischen Revolution von 1905*, Tübingen 1989, p. 107 n 7a, pp. 206, 219, 497.

190 Weber, *Gesamtausgabe*, part 1: *Schriften und Reden 1892–1899*, vol. 4.1, *Landarbeiterfrage, Nationalstaat und Volkswirtschaftspolitik*, Tübingen 1993, p. 323f.—Max Weber was apparently familiar with the article by Adolph Wagner, "Versicherungswesen," in Gustav Schönberg (ed.), *Handbuch der Politischen Ökonomie*, 3rd ed, vol. 2, Tübingen 1891, pp. 939–1026.

191 Weber even investigates the medieval situation in relation to the psychological motivation of "Versicherung gegen die Ungewißheiten des Zustandes nach dem Tode" (Weber,

approach—punctuated by his usage of highly current Freudian vocabulary.[192] Weber's central concern is the "feeling of the state of grace," the "affective ... security of salvation,"[193] which in a Puritan or Calvinist milieu would be specifically sublimated in the fulfillment derived from work and success—thus arises the image of the "steely Puritanical merchant ... that heroic era of capitalism." The specific "middle-class work ethic," which was based on the "consciousness of standing wholly in God's grace and being visually blessed by him," was essentially a feeling that grew out of the tautologically based formula: *certainty of grace = certainty of success.*[194]

Why Weber himself never engaged with the question of insurance in an economic historical sense remains an enigma. Shortly after World War II the impetus toward a historicization of the sense of security was approached precisely from the other side of the equation: the insurance historian Jean Halpérin,

---

Die protestantische Ethik, p. 63): "Die Leistung der Reformation als solcher war zunächst nur, daß, im Kontrast gegen die katholische Auffassung, der sittliche Akzent und die religiöse Prämie für die innerweltliche, beruflich geordnete Arbeit mächtig schwoll" (ibid., p. 69f.); "Heilsprämien," "psychologische Prämien" (ibid., p. 88f.), and "psychischen Prämien" of "certitudo salutis" (ibid., p. 96).

192 "Vor allem aber war der, wie sich immer wieder zeigen wird, für unsere Betrachtungen fundamentale Bewährungsgedanke als *psychologischer* Ausgangspunkt der methodischen Sittlichkeit gerade an der Gnadenwahllehre und ihrer Bedeutung für das Alltagsleben so sehr in 'Reinkultur' zu studieren" (Weber, *Die protestantische Ethik*, p. 141), "daß die Religiosität direkt *hysterischen* Charakter annahm und daß dann durch jene, aus zahllosen Beispielen bekannte, *neuropathisch* begründete, Abwechslung von halbsinnlichen Zuständen religiöser Verzückung mit Perioden *nervöser Erschlaffung*, die als ‚Gottferne' empfunden wurden [...] gefühlsmäßig—z.B. in der Form des sog. '*Wurmgefühls*' [...] konnte zum Fatalismus werden [...] gefühlsmäßiger Aneignung [...] gefühlsmäßiger Steigerung [...] die Versöhnung und Gemeinschaft mit Gott jetzt (diesseitig) zu fühlen [...] auf eine gegenwärtige innerliche *Gefühlsaffektion* [...] Ein unter Umständen bis zu den fürchterlichsten Ekstasen gesteigerter Bußkampf, in Amerika mit Vorliebe auf der '*Angstbank*' vollzogen [...] emotionelle Religiosität [...]" (ibid., pp. 145, 151, 153—my emphasis). On the popular psychological vocabulary of the *fin de siècle,* see Andreas Steiner, *Das nervöse Zeitalter. Der Begriff der Nervosität bei Laien und Ärzten in Deutschland und Österreich um 1900*, Zurich 1964; Regina Schaps, *Hysterie und Weiblichkeit: Wissenschaftsmythen über die Frau*, Frankfurt am Main 1982.

193 Weber, *Die protestantische Ethik*, p. 154f.

194 Biographical backgrounds are discussed in Arthur Mitzman, *The Iron Cage: A Historical Interpretation of Max Weber*, New York 1970; Klaus Lichtblau, "The Protestant Ethic versus the 'New Ethic,'" in Lehmann and Roth (eds.), *Weber's Protestant Ethic*, pp. 179–93; Nicolaus Sombart, *Nachdenken über Deutschland. Vom Historismus zur Psychoanalyse*, Munich 1987, pp. 22–51; Joachim Radkau, *Max Weber. Die Leidenschaft des Denkens*, Munich 2005.

based on his studies of Swiss insurance history,[195] generalized the historicity of the "concept" of security in broader terms and then Lucien Febvre, the great catalyzer and exploiter of ideas, employed this concept in a religious and anthropological context on the one hand, while on the other hand, kept its relationship to the history of insurance intact, which in turn engendered a number of diverse historical works (Jean Delumeau, Louis Boiteux).[196] In Febvre's contribution, it is almost as if Weber's thesis formed the essential starting point—while he never cites him explicitly, he links the history of insurance with the emergence of the "esprit capitaliste" in the environment of Luther and Calvin. After that moment of alignment, the paths of research on "inner" and "outward" means of security production were separated again: the history of insurance hardly even questioned its relationship to religious and denominational developments,[197] and likewise the religious-historical research into the history of anxiety failed to question its relationship to the notion of insurance.

2.3 *Insurance History and Protestantism: The Search for Evidence*

At first glance, all of this would seem to have little to do with Protestantism: the practice of maritime transport insurance had been largely developed in the pre-reformation era, and little was added to it during the following centuries. Even the *modus* of future planning, which one might judge as weakly linkable with Protestant theology,[198] first emerged in Italy in the form of the political future-planning of the Renaissance era. Thus, this example would also support a Braudelian view of mercantile practices.

As I have already demonstrated, there were developments particular to insurance history that took place only in Protestant countries until these innovations diffused globally during the nineteenth century. When one compares the southern European solutions to the same problems solved by insurance-like

195 Jean Halpérin, *Les Assurances en Suisse et dans le monde. Leur rôle dans l'évolution économique et social*, Neuchâtel 1946; Jean Halpérin, "La Notion de sécurité dans l'histoire économique et sociale," in *Revue d'Histoire Économique et Sociale* 30 (1952), pp. 7–25.

196 Febvre, *Pour l'histoire d'un sentiment*.

197 Peter Koch, "Der Einfluss der Reformation auf die Entwicklung des Versicherungswesens [1983]," in *Beiträge zur Geschichte des deutschen Versicherungswesens*, vol. 1, Karlsruhe 1995, pp. 171–81 summarizes the mostly imprecise mentions in the literature concerning the relationship between insurance and religion.

198 The dissolution of the view of history based on the theology of salvation of the Four Kingdoms doctrine within Calvinism might be interpreted as an indicator for an early general opening of the horizons of history and future, see Arno Seifert, *Der Rückzug der biblischen Prophetie von der Neueren Geschichte. Studien zur Geschichte der Reichstheologie des frühneuzeitlichen deutschen Protestantismus*, Cologne 1990.

solutions in the northern laboratories of insurance culture, a number of distinctions become apparent: in contrast to the Slaves' Funds in Hamburg and in other Lutheran lands, which were created as insurance-like institutions for the purpose of organizing ransom payments, analogous structures in southern Europe were found primarily in the form of charitable donations. In Spain, France, and Italy the former crusade orders of the Trinitarians and Mercedarians were primarily responsible for the provision of necessary ransom funds[199]—here too, they were partially state-based institutions (called "Case di redenzione"), which were managed by religious orders in the different territories of Italy. However, the funds that were collected had not originated from insurance-like practices of self-provision—rather, even in the eighteenth century, certain "old," even "medieval" means of acquiring assets were employed. Thus, for example, the *Casa* for slave ransoming in Lucca was enriched through income raised by the prebend-like collection of the so-called "Lacticines"—dues one was compelled to pay for consuming milk products during Lent. Such dues had been used as early as the Middle Ages to finance the crusades.[200] Church-state institutions were operating under the auspices of donation-type fundraising in the Catholic South. In the Protestant North the same functions were fulfilled by state insurance-like funds—though it should be noted that in Denmark, for example, the Slaves' Fund was, in fact, wholly under episcopal aegis.[201]

The transfer of the insurance principle to sectors other than maritime trade in both northern European "laboratories" and the ensuing diffusion in the provinces of Prussia and the other territorial states further clarifies the trajectory of the development of insurance. The harsh rejection of state participation in the area of insurance in England, as described above, is striking in relation to the cameralistic-state path pursued in the German north where the territories in question were dominantly Lutheran (though the Prussian dynasty, in contrast to its subjects, was Reformed). From 1682 until the nineteenth century, every attempt to establish a state insurance authority in Great Britain was defeated; in contrast, beyond a few, small, unsuccessful attempts during the eighteenth century, private fire insurance companies in Germany did not appear until 1810–20s (Gotha Insurance Bank and Aachen-Munich).

England and Germany were the lone participants in the area of fire insurance during the eighteenth century—beyond insurance for ships and goods,

199 For a review of the literature cf. Ressel and Zwierlein, "The Ransoming".

200 Marco Lenci, "Riscatti di Schiavi Cristiani dal Maghreb. La Compagnia della SS. Pietà di Lucca (Secoli XVII-XIX)," *Società e Storia* 31 (1986), pp. 53–80.

201 F. S. Grove-Stephensen, "Slavekassen og de slesvigske søfartsforhold," *Sønderjysk Månedsskrift* 53 (1977), pp. 410–17 (However, it was a Lutheran bishopric under the king).

no other items were insured to a sustainable and relevant extent in France,[202] Spain, or Italy, and there were only scant examples late in the eighteenth century in the Netherlands and other European countries. Every rule has its exceptions: a Fire Fund was also established in 1752 in a Catholic territory, the Bishopric of Hildesheim, and in 1758 one was established in the Bishopric of Münster.[203] These two named cases are clear exceptions—both territories were situated in the north, bordering Protestant territories in which fire insurance had already long been established. The majority of Catholic territories in the German south, however, were reached by the tide of new fire insurance funds only late in the eighteenth century,[204] in the same way that Catholic Enlightenment was also emulating (in changing forms and patterns) only late Protestant Enlightenment in Germany. It makes no sense to consider here whether the reception of Protestant neologist eudemonism and of Kant by Catholic enlightened bishops such as Johann Michael Sailer around 1800 are phenomena to be treated on the same categorical level as that emulation of institutional innovations, but it is noteworthy that in these cases, at the very end of the eighteenth century, the difference between the denominations had also eroded in so many fields that such a question of origins—whether fire insurance *is* "Protestant," whether Kant *is* "Protestant"—makes no sense any more: both are, to a great extent, transconfessional and *were* neither Lutheran nor Catholic at the moment of their invention, or respectively at the point of writing an extraordinary philosophical œuvre—but there are settings and contextual conditions that made one or the other emerge in a given region and situation, which, themselves, have some denominational cultural roots and preconditions that cannot be denied.

---

202 Michèle Ruffat, "French Insurance from the *ancien régime* to 1946: Shifting Frontiers between State and Market," *Financial History Review* 10 (2003), pp. 185–200. Before 1815 there was no founding of insurance institutions beyond maritime transport; however, there were several unsuccessful attempts to found fire insurance societies in 1756 and 1786: Edgar Blum, "Les assurances terrestres en France sous l'Ancien Régime," in *Revue d'histoire économique et sociale* 8 (1920), pp. 95–104. Cf. the source material in Didier Pouilloux, *Mémoires d'assurances. Recueil de sources françaises sur l'histoire des assurances du XVIème au XIXème siècle*, Paris 2011.

203 Christian M. Zachlod, "Katastrophenhilfe, Kreditwesen und Konjunktur. Zur Wirtschafts- und Sozialgeschichte der staatlichen Feuerversicherung in der Mitte des 18. Jahrhunderts am Beispiel des Hochstifts Hildesheim," in Gerhard Lauer and Thorsten Unger (eds.), *Das Erdbeben von Lissabon und der Katastrophendiskurs im 18. Jahrhundert*, Göttingen 2008, pp. 553–74; Mechthild Siekmann, "Die Brandversicherung im Hochstift Münster 1768–1805. Entstehung, Arbeitsweise, Quellen," *Westfälische Forschungen* 31 (1982), pp. 154–68.

204 Martina Spies, *Feuerversicherung, Waisen-und Kreditkassen bei ostschwäbischen Reichsklöstern vor der Säkularisierung und ihre Auflösung*, Munich 2007.

The globalization of insurance in the nineteenth century was dominantly influenced by the British model—it is an element of capitalism that spread across the entire world (see Ch. 6). A similar developmental structure can be ascertained in the theoretical account of insurance practice. The first academic discourse that directly concerned itself with premium-based insurance was clearly the jurisprudence of the postglossatorial era—beginning in Genoa with the *consilia* of the late fourteenth century and then with the first individual treatise by the Portuguese Pedro de Santarém and lasting until the collection of maritime law treatises by Straccha including his own *De assecurationibus* (1569) in the late sixteenth century.[205] This is a reflection of an older theological-canonical discussion, illuminated by Giovanni Ceccarelli, beginning as early as the twelfth/thirteenth century—assuming that the terminology of "assecuratio" is not strictly adhered to, but rather includes general references to *susceptio periculi* in practice.[206] All in all, the discussion is chiefly concerned with the question of permissibility, and only secondarily, perhaps, with the *iustum pretium*. Cecarelli concerns himself exclusively with medieval and Catholic doctrine up to the threshold era of the seventeenth century—and thus up to the treatises *De justitia et jure* of the late scholastic period—and he was therefore unable to address "the highly relevant question of whether or not there develops specific Protestant thought on the issue of insurance."[207] In this regard, research has focused primarily on the emergence of the so-called "probabilistic revolution" in the context of the life insurance discourse of the seventeenth century, and has supported the theory that this notion could only have arisen within the context of Protestantism because Catholic thought is characterized by qualitative theology and scholastic taboos. Most of the authors in the established history of stochastic-probabilistic thought were indeed Protestants.[208] As discussed above, protoprobabilistic thinking and calculations actually did play a distinct role in the emergence of fire insurance institutions, at least in the London/Barbon "laboratory". There is no mention of the theological permissibility of calculating lifespan averages in either the mathematical-probabilistic treatises or the project schemes for the establishment of life insurance,

205 Petrus Santerna, *Tractatus de assecurationibus et sponsionibus* (facsimile), Lisbon 1961, vol. 2, pp. 8ff, 495ff.

206 Ceccarelli, "Risky Business".

207 Ibid., p. 624.

208 Hacking, *The Emergence of Probability*; Ivo Schneider, "Why Do We Find the Origin of the Calculus of Probabilities in the Seventeenth Century," in J. Hintikka et al. (eds.), *Probabilistic thinking*, Dordrecht 1981, pp. 3–24; Hald, *A History of Probability*.

widow or orphan funds, or pension funds, and it is also true that the relevant authors were all Protestants.

After about 1600, the legal discourse essentially developed only north of the Alps.[209] While the Catholic late-scholastics once again thoroughly exorcise the question concerning the permissibility of the insurance contract on a theological/canonical level around 1600,[210] the northern European Protestant lawyers were as little concerned with this matter as the probabilists.[211] The insurance history literature has long cited *one* singular example for the rejection of insurance as "God averse," which was incorporated into the tradition by the first "technical historian," Johann Beckmann: namely, the Oldenburg project of 1609. Duke Anton Günther of Oldenburg supposedly rejected this project because it "represented a temptation of God."[212] In truth, this formulation of the "temptation of God" was never meant in a profound way to denounce insurance as God-averse, in the sense of an opposition between God's providence and the human right to self-preservation and self-provision eventually neutralizing God's power to punish. None of this was at stake in 1609, as the cliché (*Versuchung Gottes*) appears in relation to the concerns of the country lords that the imposition of the relevant provision might be considered an unjust additional tax, leading to accusations of greed and avarice. It is also telling that the legal opinion of the lawyer Eberhard von Weihe does not invoke any kind of religious issue.[213]

209 In Italy, Casaregis only became significant again in the eighteenth century: see Vidari, "Sulla classificazione."

210 Cf. the source material in Ch. 2, notes 45, 46.

211 See the short treatment of the question in Schragius, *Disputatio*, f. A2r-v.

212 Koch, "Der Einfluss der Reformation," p. 175; Allemeyer, *Fewersnoth*, p. 47f. One reason for this is that in the highly influential entry Johann Beckmann, "Assecuranz," in *Beyträge zur Geschichte der Erfindungen*, vo1. 1.2, Leipzig 1785/86, p. 221 that, which was actually a part of the request for an advisory report by Eberhard von Weihe, was quoted as an excerpt of the duke's decision (Cf. Johann Justus Winkelmann, *Oldenburgische Friedens- und der benachbarten Oerter Kriegshandlungen*, Oldenburg 1671, p. 68. The text of the decision appears as: "jedoch machte der H. Graf den vernünftigen Endschluß/daß Gott dardurch möchte hoch versuchet/seine eigene Unterthanen beschweret/ungleiche urtheil darüber gefället/und ihm der Geitz/unverschuldeter weise/ beygemessen werden; Gott der Almächtige hette Sein uraltes Hauß Oldenburg/so viel hundert Jahren hero/ohne dieses und dergleichen Mittel/erhalten und beglückseligen lassen: wurde auch forthin/ durch seinen kräftigen Segen/demselben beywohnen/und seine Unterthanen für grossen Feuersflammen behüten etc."). More on the original communication as preserved in the Landesarchiv Oldenburg will be published elsewhere.

213 Schäfer, *Urkundliche Beiträge*, vol. 2, pp. 115–29. From the use of language here—certainly not the duke's own words, though he signed the letter, more probably as author is the chancellor Protte—it becomes clear that this writer was a lawyer and well familiar with insurance law terminology ("[...] recompensa pro suscepto periculo"). Von Weihe, in

The authors or respondents of the legal treatises and dissertations that concern themselves with insurance—comprising approximately thirty individual texts[214]—were typically residents of the Hanseatic city of Hamburg (10), Bremen (3), Lübeck, Greifswald, Danzig, or the Netherlands. Only four dissertations originated from universities in southwestern areas (Strasbourg, Heidelberg, Basel); beyond these, the most southern location is Giessen. This reflects the generally accepted fact that until late into the eighteenth century, mercantile insurance law and the insurance principle were uncommon topics within the interior regions of most countries beyond the port and harbor cities.[215] The next step of introducing Grotian natural law into insurance law by Stypmann and his followers (cf. Ch. 2 above) was also being taken in the Protestant north; it was *this* natural law (and not the neoscholastic Spanish-Jesuit school) which was enabling the universalization of the contract types beyond Roman contract typology—natural law was, again, not intrinsically "Protestant", but built on the famous "etiamsi daremus, quod sine summo scelere dari nequit, non esse Deum, aut non curari ab eo negotia humana": even if one admitted that God did not exist, natural law would be binding between pagan and Christian people.[216] As stated above, there was also an absence of theological or moral-theological discourse and arguments within the lawyers' treatises on the insurance context. This type of natural law, however, as well as the post-Grotian natural law rooted in systematic principles (Pufendorf), was emerging only as the offspring of Protestant academic contexts.

In essence, legal theory becomes characterized by a complete distance from explicit references to God or religion. Even in the cameralistic discussion within the solid corpus of eighty treatises and treatise sections concerning fire insurance

contrast, did not refer explicitly to insurance law treatises in this context. The prohibition of insurances by canon law is likewise not discussed by the Protestant authors.

214 Cf. the compilation of the title in appendix 1, and also Scherner, "Die Wissenschaft des Handelsrechts," pp. 829, 841–44, 854–56; on some of the authors, see Heinz Mohnhaupt, "'Jura mercatorum' durch Privilegien. Zur Entwicklung des Handelsrechts bei Johann Marquard (1661–1668)," in Gerhard Köbler (ed.), *Wege europäischer Rechtsgeschichte, Festschrift für Karl Kroeschell*, Frankfurt am Main 1987, pp. 308–23; Karl Otto Scherner, "Zum Sonderrecht der Kaufleute bei Johann Marquard," in Hans-Peter Haferkamp (ed.), *Usus modernus pandectarum: Römisches Recht, Deutsches Recht und Naturrecht in der Frühen Neuzeit*, Cologne 2007, 199–217, and also the literature above in chapter 2.

215 For Germany see Schragius, *Disputatio*, f. A2r: "Cum enim [sc. contractus assecurationis] Iuris romani conditoribus vix cognitus, *ac superioris Germaniae regionibus in hunc usque diem parum usitatus*, eoque paucis interpretibus acriori cura sit inspectus" (my emphasis).

216 Grotius, *De iure belli ac pacis libri tres*, ed. B.J.A. De Kanter et al., Prol. § 11.

written between 1680 and 1810 (see appendix 2), there are only three tiny remarks that one could perhaps use to argue that insurance contradicts divine providence. Such remarks, however, were made only as a rhetorical device: they are posed in order to be discarded as wholly irrelevant. Divine omnipotence was obviously above making itself known to people desperately through simple fire damage.[217]

When one examines the different types of insurance, different European regions, and the different areas of theory and practice, one can thus broadly schematize:

- Until the sixteenth century, premium-based insurance exists almost exclusively in Mediterranean, "Catholic" areas as a common element of maritime trade.
- Beginning in about 1680, the tendencies toward the universalization, diversification and institutionalization of security production through (premium-based) insurance began to become irreversible. This was exclusively apparent in the Protestant/northern European realm.
- Security production proceeds through institutions, which obligated the subjects/citizens to protect themselves through insurance-like means—particularly in *Lutheran* areas in predominantly *state* form. Slaves' Funds were established in Hamburg, Lübeck, and Denmark, while comparable institutions were lacking in Reformed Bremen and the Netherlands; cameralistic Fire Funds appeared first in Hamburg and then in predominantly Lutheran territories, while analogous institutions were absent in Catholic and Calvinist-influenced Europe.
- Security production through large trade societies (joint-stock corporations, mutual societies) was happening initially exclusively in Calvinist-influenced areas.[218]

217 "Gedanken von der Einrichtung und dem Nutzen der in den Fürstenthümern Calenberg, Göttingen und Grubenhagen zu errichtenden Brand-Assecurationssocietät" *Hannoverische Gelehrte Anzeigen vom Jahre 1751*, no. 9 (1751), pp. 291–301, 295f.: "the mighty hand of God has many ways to chastise us poor worms, and nowhere do I find it to be forbidden for one to protect his interests against worrisome misfortunes"; J.V.E., "Vorschlag, wie die Versicherung der Gebäude zu Ersezung des Feuerschadens auf eine leichte Art einzurichten 1756," *Physikalisch-Oekonomische Realzeitung, oder gemeinnüzliche Wochenschrift* 17 (1756), pp. 674–83, 702–16, 708; Ferdinand Friedrich Pfeiffer, *Gedanken über Versicherungs-Anstalten: Hauptsächlich zum Vorteil der Landwirtschaft entworffen, und* [...] *wärend den Akademischen öffentlichen Prüfungen verteidigt*, Stuttgart 1780, p. 103f.

218 The development in the Netherlands was quite slow. The first general insurance societies, which, among other things, also insured against fire, were established in 1720, the first exclusive fire insurance societies were founded beginning in 1771 (*Assurantie Compagnie te Amsterdam*); Cf. Ton Langenhuyzen, "Zekerheid en Brand vóór 1800," in Gerwen, van Leeuwen (eds.), *Studies*, pp. 203–22, 218.

- Contrary to many statements in the literature, the problem of the unification between divine providence and human insurance activities no longer played an active role. In this sense, there is therefore also no actual "Calvinist" or "Lutheran" insurance discourse which was entirely theologically based. The determination that this or that type of security production in this or that denominationally different region was dominant is thus only a correlation or "elective affinity."[219] Which denominational or worldview element would have *produced* or would have *necessitated* the emergence of these realities, and how that elective affinity facilitated the "fit" of the one to the other, cannot be determined. Everything contradicts the idea that this (missing) discursive element could be found specifically in the expression of the predestination theologoumenon: Starting this chapter with Weber was also meant to show that we *cannot* build the argument, as there is no entelechial relationship between a certain type of intrinsic theological reasoning and the idea of insurance in the sense of a relationship between two core idea units in the way that Arthur Lovejoy once conceived of the "History of Ideas", deprived of any contextual elements that did not consist of ideas.

On the contrary, the key to understanding these different realities concerning the institutionalization of premium insurance in Europe probably lies in the distinctively different relationship between the private and public realms and the private and state sectors as following from the differences within the religious and denominational cultures of Europe. The relationship between state, society and the respective concept of "economy" was different, and this was mirrored by the way that the premium insurance principle could or could not be hybridized with state or semi-state institutions.[220]

219 Weber, *Die protestantische Ethik*, p. 77.

220 These questions arise only in a context of European (and than global) comparison such as here, while the question of the specifity and relationship between a complex setting as the Financial Revolution with a given denominational culture played no role or was—because of the sole concentration on Britain—quite clear from the outset in the discussion since Dickson until today. Temin and Voth, *Prometheus shackled,* p. 20 briefly touch on the Weber thesis in relation to the question of work ethics; Seán Moore, "'Vested' Interests and Debt Bondage: Credit as Confessional Coercion in Colonial Ireland," in Carey and Finlay, *Empire of Credit*, pp. 209–28, 210, briefly addresses the Weber thesis revisionistically in relation to the credit potential and affinity with capitalism of the Catholic clergy in Ireland. In Wennerlind, *Casualties of Credit* the main emphasis lies on the epistemological/history of science explanation, so that religion/denomination is not really addressed. One might quote Robert Brenner, *Merchants and revolution.Commercial change, political conflict, and London's overseas traders, 1550–1653*, Princeton 1993—but he was rather trying to explain how a *political revolution* was driven by a core of proto-bourgeois merchants who were bound together by common interests, political convictions and faith—but not

On the other hand, the English (and also Dutch) peculiarity in the development of early capitalistic markets is difficult to account for in comparison with other European countries. Philip S. Gorski has presented models from different perspectives which could help clarify the specificity of the state-building processes, the military revolution in this context, and the spatiotemporal setting of the Financial Revolution in (largely) Calvinist-influenced western Europe.[221] This modelling tends toward a strongly state-centric and a somewhat overwrought representation, suggesting that constitutional control only reigned in Reformed countries, and that simony—contrasted with the separation of person and office—was only prevalent in Catholic countries.

Starting from these reflections on other objects and fields of the relationship between state, religion and economics, linking it with the post-Weberian debate on the relationship between Protestantism and Economics, I deem it useful to conceive of four parameters of comparison here: 1) security; 2) statism (meaning how much a developing actor 'state' is dominating a given sector); 3) economy; and 4) religious denominations. The theories aligning Calvinism with the Financial Revolution always seem to proceed from the observation that a particular type of state—i.e., the idealized constitutional monarchy existing only in the Reform-Calvinist milieu—produces "security": security for the "creditors" of the state, since a state that is not autocratically ruled, but rather controlled by parliament, is a reliable debtor. Security is thereby also ensured for the entire system of capital markets by property guarantees and respect for the division between public and private sectors. In the end, security becomes an object produced through economic transactions, in the form of securities that are exchanged within a market system. However, private insurance practices make it clear that the state is far from being the only institution generating securities and making them available. All those arguments can thereby be condensed into a triadic "clover" scheme that nevertheless clarifies

vice versa, how the *Financial Revolution* was emerging in given contexts and settings of elective affinity to given denominations of belief.

221 Philip S. Gorski, "The Little Divergence: The Protestant Reformation and Economic Hegemony in Early Modern Europe," in Swatos and Kaelber, *The Protestant Ethic Turns 100*, pp. 165–90, 181–87; Philip S. Gorski, "Calvinism and Revolution: The Walzer Thesis Reconsidered," in Richard Madsen et al. (eds.), *Meaning and Modernity*, Berkeley 2001, pp. 78–104; both texts are now in Philip S. Gorski, *The Protestant Ethic Revisited*, Philadelphia 2011, pp. 85–115 and 187–209; Cf. Philip S. Gorski, *The Disciplinary Revolution: Calvinism and the Rise of the State in Early Modern Europe*, Chicago 2003; the statism categories according to Thomas Ertman, *Birth of the Leviathan: Building states and regimes in medieval and early modern Europe*, Cambridge 1997 do not correspond to Weber's use of "patrimonial"/"patriarchal."

a multiplicity as well as the limits of combinations and possibilities. The "clover" illustrates the relationship between denominational cultures, state forms, and forms of security production beyond the realm of transport insurance among the three larger European denominations (Fig. 63):

The scheme reveals its adaptability, however, only after the epochal threshold of 1680–1700, thus only *after* the end of what we usually acknowledge as Europe's "Age of Reformation," when within politicotheological contexts the denominational borders tended to flatten out. This scheme does not, therefore, draw a direct link of causality between, for example, a Presbyterian/post-Dordrechtian Calvinism, or a doctrine of predestination in the precise form of the hypothetical universalism of the Anglican church in the seventeenth century and "insurance agencies" in the form of private business ventures. Inversely, it also does not indicate a direct causal relationship between the Lutheran orthodoxy in Hamburg or the unorthodox elements in the city preceding the Reimarus era that paved the way as a *causa formalis* or *efficiens* for the General Fire Fund and for Leibniz's thinking about insurance. Rather, this scheme can only possibly concern the extratheological conditions of denominational cultures and environments, which in England, as has been seen, created an entirely different milieu facilitating the founding of the Fire Office as a private venture while defeating each state attempt at insurance institutionalization, and which in the German-speaking arena simultaneously engendered the favorable response to a state-governmental form of insurance. Two very opposite

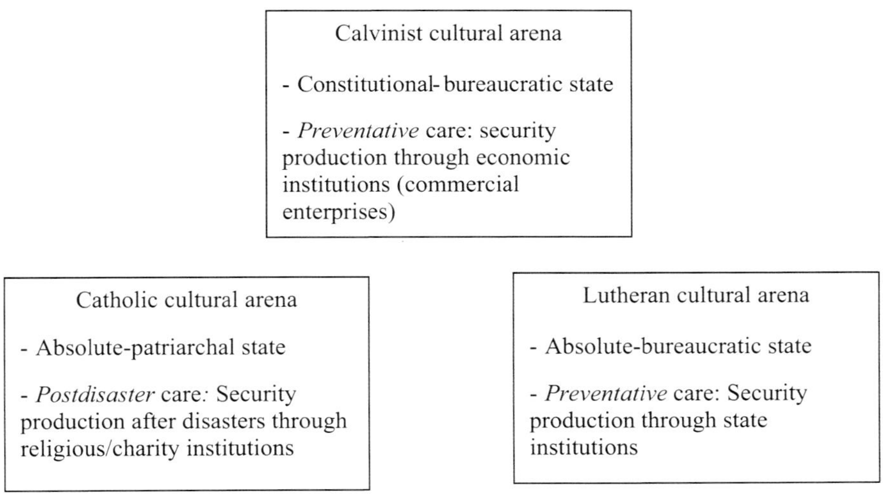

FIGURE 63 "Clover" scheme of the relationship between religious denominations and institutionalized forms of security production

behaviors towards the way how to institutionalize premium insurance, but both happening exactly in the same years around 1680. Before the seventeenth century, diffuse practices and theories of premium-based insurance existed throughout Europe in the area of transportation and also in relation to individual life insurance policies and the like without regard for denominational boundaries—though for other sectors of insurance, such as fire, storm damage, life, etc. there were still no larger institutions or markets. Without a doubt, the above scheme is also limited insofar as it is also only based on the three main Christian denominations named in the Peace of Westphalia. For example, there are indications that Slaves' Funds and other protoinsurance-like forms were also present in Jewish trade communities.[222] In contrast, the Islamic cultural arena displayed a distinct aversion to the principle of premium-based insurance, even though there were functional equivalents corresponding more closely to the Western *foenus nauticum* of antiquity.[223] In early modern times, the merchant diasporas in the Mediterranean continued to operate in the context of partnership contracts and shipping securities, though in Arabic, Sephardic Jewish, or Armenian forms that were not dissimilar to this *foenus nauticum*—it was as if a basic Mediterranean principle from antiquity had been more or less preserved for two millennia resisting the premium insurance technique. I will discuss in the last chapter shortly if and to what extent Asian practices were similar to or different from those forms of European institutionalized insurances, but this is better done with regard to the eighteenth and nineteenth century realities when both techniques were first encountering each other—and when Chinese eventually rejected or started to emulate the European forms.[224]

In the chronological order here, we may stay with the fact that it was only in 1680 that England and Germany alone took the new step of institutionalizing

222 M. Soave, "Malta e gli schiavi ebrei," *Corriere israelitico* 17 (1878–79), pp. 54–57; 81–82; 101–03; 125–27; 147–49; 172–73.

223 This becomes clear with the observation that even in the Mediterranean, maritime transport insurance that had been common since the fifteenth century in Crete obviously disappeared when the Ottomans (re)conquered the island in 1669 and again in 1699, Cf. Molly Greene: *A shared world: Christians and Muslims in the Early Modern Mediterranean*, Princeton 2000, pp. 167, 171f. for the Ottoman transition, although Greene does attempt to locate the end of the practice of this contract type within the (for her argument) important revisionistic perspective of the 1990s as an example of "legal niceties."

224 The contract types are introduced and analyzed, for example, in Francesca Trivellato, *The Familiarity of Strangers: The Sephardic Diaspora, Livorno, and Cross-Cultural Trade in the Early Modern Period,* New Haven 2009; S.D. Aslanian, *From the Indian Ocean to the Mediterranean. The global trade networks of Armenian merchants from New Julfa*, Berkeley 2011.

that principle, of transferring it to other realms, of merging it with different forms of private capital societies or with "the state". It is not a question of praising that as a European "achievement", but it is a historical phenomenon that deserves an explanation for which the clover scheme of denominational differences with regard to the four factors mentioned in Europe might serve as a first attempt.[225]

225 Cf. below (Ch. 6, 4) for a brief discussion of the confrontation between the British model of insurance and early cooperative Chinese hazard communities.

CHAPTER 5

The Emergence of the Normal Secure Society

The early Enlightenment era signifies a decisive drive toward the construction of what I describe as the "normal secure society":[1] the notion of a safe, secure world that existed not only as an ideal, but also as the model and expectation of everyday life—a notion which, incidentally, was entirely counterfactual. Not insignificantly, insurance practices were not only the result of this new order, but also a major formative element. "Uncertainties," "misfortunes," and natural disasters begin to be seen as exceptions to the rule within this secure normal world, and consequently become subject to political and economic regulation.

In this chapter, I will first demonstrate how a multifaceted process of transformation emerges within the practice of pecuniary damage restitution following fire events between the sixteenth and the nineteenth century. This process can also be interpreted as a transformation of community and societal structures (5.1.1). I will then show how Enlightenment social theory is linked to insurance theory (5.1.2) not only in practice, and not only as retrospectively theorized here, but also contemporaneously—albeit primarily in the work of German insurance theorists, who were the only theorists of this time reflecting on such questions in extensive published treatises. One other important element that was indispensable to the entire theory and practice of this "normal secure society" was the increasing conceptualization of the world in the quantified form of numerical values (5.2). On this basis, I will show how a governmentality employing premium insurance schemes clearly reveals structural congruencies with the present-day discourse on sustainability (5.3): the normal secure world is one that attempts to protect itself against the decline and loss of value through insurance-like mechanisms. Great "disasters"—which necessarily have to become larger and larger now to cross the limits of controllability and insurability— appear in this new normal secure world as specters of the unimaginable and incomprehensible, defying the limits of belief.

1 "Sichere Normalgesellschaft" in German: Perhaps "Normally safe or secure society" would be more correct, but a little bit bulky.

 | DOI:10.1163/9789004431225_006

1 Insurance and Social Structures

1.1 *Collective Solidarity: from Risk Communities to Cosmopolitan Aid in the Eighteenth and Nineteenth Centuries*

The concentration on insurance history presented here involves a narrow focus on a form of the monetary compensation for losses specific to certain European areas. In the more recent field of disaster history, there is surprisingly little systematic research on the question of how victims were supported after a disaster, beyond research within the more specific history of insurance.[2] In the broad research on the impact and European perception of the Lisbon earthquake of 1755, for example, I know of no single investigation into the question of how domestic and international aid was transferred after the disaster, beyond occasional brief mentions in passing.[3]

One of the central questions for the purposes of the present study concerns the relationship between social structures and institutions devoted to damage compensation. What relationships did risk and solidarity communities have with each other? How did these relationships change with the advent of insurance? How did they respond to other processes, such as globalization?

In southern and western European countries before the nineteenth century, city and state postdisaster relief was primarily implemented in the form of tax reductions. In a small number of regions—for example, the Dauphiné, Alsace, and Languedoc—institutionalized subsidies were distributed in addition to passive tax reductions. Gilbert Larguier has researched how in Languedoc, beginning in the late seventeenth century and primarily during the climate anomaly of the so-called "Maunder Minimum", the French crown, as well as the estates of Languedoc, pursued special institutionalized strategies with regard to damage management after a series of major crises (harvest failures due to hail, frost, flooding, etc.). Tax relief regulations were carefully balanced, and in 1729 the provincial estates set up their own substantial funds devoted to

2 As important steps in this direction, see Favier and Pfister, *Solidarité et assurance*, or the earlier Favier, "La monarchie"; Favier, "Une aide instrumentalisée"; Matthew Mulcahy, "Urban Catastrophes and Imperial Relief in the Eighteenth-Century British Atlantic World: Three Case Studies," in Geneviève Massard-Guilbaud, Harold L. Platt, Dieter Schott (eds.), *Cities and Catastrophes. Villes et catastrophes*, Frankfurt am Main 2002, pp. 105–22.

3 For a current estimation of damages (in terms of population, buildings, values), of the influence on the gross national product, prices and wages, as well as the state finances in the immediate period following the natural disaster, see Alvaro S. Pereira, "The Opportunity of a Disaster: the Economic Impact of the 1755 Lisbon Earthquake," *The Journal of Economic History* 69, 2 (2009), pp. 466–99, which, however, does not mention any charitable funds received.

damage compensation. Beginning in 1750, reparation payments were also occasionally made for house fires.[4]

Beyond such measures, however, as in the rest of central Europe, damage relief after disasters proceeded here in a rather decentralized manner through different levels of charitable support systems, which also clarifies why a regionally comprehensive investigation of compensation payments in times prior to the introduction of territorial and municipal insurance institutions is hardly feasible: the source material is simply inaccessible or nonexistent. For the Prussian provinces, for example, centralized files for charitable briefs have existed since the sixteenth century—in principle, a permit from the authorities was required in order to collect funds for damage compensation in the territories. Nonetheless, the extant files are only partial accounts of the events; typically, the central administration recorded only the request for permission to collect donations, rather than the pecuniary results of such efforts.[5] Records for fire damage fundraising are given their own category in the archives, which suggests this type of fundraising was quantitatively significant.[6] Of course, files generated from this central governmental perspective are insufficient for accessing the almost inestimable network of requests and acts of assistance occurring amongst different cities, guilds, and private individuals.

The city of Aachen, for example, definitely received assistance from surrounding and possibly distant cities after the great fire of 1656—including, for example, 200 malters of rye that Cologne sent as a gift of aid,[7] offered of help from Maastricht,[8] 1000 pounds of Dutch cheese provided by Liège,[9] Count Ferdinand von Aspremont's offered to house homeless citizens in the city of Rekem in the imperial county of Reckheim,[10] 26 specialized long support beams for the reconstruction of the imperial coronation hall were requested

4 Gilbert Larguier, "La solidarité face aux risques naturels en Languedoc (XVIIe-XVIIIe siècles)," in René Favier and Christian Pfister (eds.), *Solidarité et assurance*, pp. 37–56.

5 Cf. GStAPK Berlin Hauptabt. I, Rep. 9, Nr. G1: 8 Faszikel Bettel-, Armen und Kollektenwesen, Generalia 1565–1806; Nr. G1 Kollektenwesen Spezialia 1621–1808 (more than 30 volumes concerning donations for religious refugees, exiles, and war victims).

6 GStAPK Berlin Hauptabt. I, Rep. 9, Nr. G2: a series of volumes on fire damages, support, and charitable donations beginning in 1573, including files on falsified charity briefs, fire ordinances, donations, timber allocations, decrees of duties, beer taxes, war contributions, etc., all related to fire damages.

7 HStA Köln Rpr. 103, s.d. 15.5.1656. Cf. Hermann Keussen, "Hilfe der Stadt Köln nach dem Aachener Stadtbrande 1656," *Zeitschrift des Aachener Geschichtsvereins* 51 (1929), p. 392.

8 StA Aachen, RA II Allg. Akten 577, f. 4: Claude de la Montagne, Johan Pesters und Antoine Vaes on condolence missions.

9 Ibid.

10 StA Aachen, RA II Allg. Akten 577, f. 14.

from Brussels,[11] 4000 scudi that Pope Alexander VII (who knew the city from his time as a nuncio in Cologne) contributed in response to the city's request for help restoring the church building,[12] and the city of Schweinfurt's exception, allowing Aachen-bound timber to pass duty-free—a concession that was granted even more willingly because the city had "experienced the same misery by God's hand twice over the last century with indescribable woe such that the great damages incurred have not yet been fully rectified nor have they been proportionally shared among the victims."[13] However, a full account of all assistance provided is not available, and there was not yet a central record of donations in a manner reflecting Germany's largest early modern city fire of Hamburg in 1842.[14]

One might compare that to smaller city networks and their communication in moments of emergency[15], however, data for small towns is usually rare. For Wurttemberg, where the central church administration records concerning alms funds have been well preserved, the character of the charitable networks becomes clearer: the so-called "church coffer accounts" (*Kirchenkastenrechnungen*) survived almost completely intact for the period between the great church ordinance of 1559 and the Thirty Years' War.[16] A radius of charitable relationships surrounding Wurttemberg emerges, initially spreading out into the rest of the territory, but later also spanning neighboring regions (Baden, Bavaria, Franconia, Palatinate, Alsace, Hesse, Nassau) and even later including regions located further east (Thuringia, Saxony, Prussia, Braunschweig, Hanover), south (Austria, Switzerland) and west (Montbéliard, France, Netherlands).[17] The references to the more distant regions, however, are sporadic and

11 StA Aachen, RA II Allg. Akten 577, f. 39f.

12 StA Aachen, RA II Allg. Akten 577, f. 30f., 83. The money was then, however, generally used to support the reconstruction of homes and for a memorial in the city hall. Cf. Kunitz, "Aachen," pp. 56–59.

13 StA Aachen, RA II Allg. Akten 577, f. 48f.

14 Kunitz, "Aachen," 54–67; Kraus, "Der Aachener Stadtbrand", pp. 67–70 also illustrate the inconsistent and, even for the time, uncentralized documentation practices.

15 On correspondence exchanged between cities and corporations to stimulate charity actions after fires, see Katrin Keller, "Armut und Tod im alten Handwerk: Formen sozialer Sicherung im sächsischen Zunftwesen des 17. und 18. Jahrhunderts," in Peter Johanek (ed.), *Städtisches Gesundheits- und Fürsorgewesen vor 1800*, Cologne 2000, pp. 199–223, 205f.

16 Gustav Bossert, "Die Liebestätigkeit der evangelischen Kirche Württembergs von der Zeit des Herzogs Christoph bis 1650," in *Württembergische Jahrbücher für Statistik und Landeskunde 1905*, Stuttgart 1906, vol. 1, pp. 1–28; a comparable source would be Franz Konrad, *Der Kirchenkasten zu Wallau an der Lahn. Bericht über die Jahresrechnungen der Kirchenkästen von Wallau/Lahn und Weifenbach von 1588 bis 1832*, Wallau/Lahn 2004.

17 Bossert, "Die Liebestätigkeit," vol. 1, pp. 21–24.

considerably less frequent (Braunschweig/Hannover 3x in 1627–28 only; Alsace 4x, Switzerland 2x, and Netherlands 2x). The locations in "France" are nearly all in the immediate vicinity of the imperial border; Montbéliard, for example, was an enclave of Wurttemberg. To this extent, one can clearly interpret the territorial "neighborhood" (in the, special, early modern sense of the word[18]) as constituting the central basis for a charitable network of this kind. Similarly, Morgan has shown for the 1641 Stratford-upon-Avon fire, that of the many dozens of donation payments collected, nearly all were coming from the local kinship network within a distance of 40 miles, only four came from places farer away—eighty years later, after the establishment of fire insurances, things would have looked different.[19] Permissions for envoys to travel throughout the territory were granted and direct support payments were often made, even when those affected were of different religious denominations—fire events thus occasioned a transdenominational Christian *caritas*. However, it is not possible to establish a clear pattern for such charitable ties: one would assume that less would have been donated in distant regions than in nearby regions, and while denominational differences were not considered absolutely prohibitive, one would imagine that they probably resulted in less generous contributions. Furthermore, one would expect that the size of donations was likely proportional to the magnitude of total damage. In actuality, however, apart from the fact that no large donations were made for very small damages, such regularities are not apparent in the Wurttemberg sources. Culmbach, Bayreuth, Annaberg, and Ilmenau received the maximum amounts of 300, 150, and 100 florins, respectively, in 1554/55, 1601, 1604, and 1605, after which the typical larger amounts fluctuated between 20 and 45 florins. Very often, however, the amounts entailed only a few florins, even in the case of very large fires (4 florins were donated for a 1664 fire in Roßleben/Prussia that destroyed 249 houses).

If one looks at the official, city-wide, mandated collections of donations in Hamburg between 1703 and 1797 as compared to this earlier southwestern data from Wurttemberg, the charitable network profile reveals itself as different, but importantly, another phenomenon is also revealed for the northern city: until about 1730, solidarity networks extended far into other regions. Such networks

18 Beyond the cases addressed by Klaus Nippert, *Nachbarschaft der Obrigkeiten. Zur Bedeutung frühneuzeitlicher Herrschaftsvielfalt am Beispiel des Hannoverschen Wendlands im 16. Und 17. Jahrhundert*, Hanover 2000 the early modern concept of "Nachba(u/w) rschaft" (neighborhood) could include a large range of quasi-obligatory horizontal relationships.

19 Morgan, "The representation", p. 282 for a graph mapping the donations of individuals and communities and the money contributed.

were regularly called upon not only for the benefit of near and neighboring towns and cities, but also for distant municipalities from southwest Germany to northern Europe and beyond (e.g., 1703 Tribsees/Pomerania, 1705 Uppsala/Sweden for the fire of 1702, 1706 Crivitz/Mecklenburg for the fire of 1704, 1711 Nordhausen, 1716 Stralsund, 1719 Frankfurt am Main, 1720 Reichenbach/Vogtland and Solingen, 1722 Greifswald, 1724 for churches in St. Petersburg and New York, 1725 Stockholm and Wolgast, 1728 Goslar, 1729 Copenhagen, Lindau, and Schwäbisch Hall). After 1730, in contrast, collections for such distant locales decreased, while collections increased for the Hamburg poor, the support of orphanages, and the support of the secular municipal almshouse after its founding in 1788.[20] The interpretation of such findings is neither easy nor obvious. There is some evidence that, beginning in the 1730s, the institution of fire insurance was seen as so firmly established in Hamburg that it was deemed no longer necessary to call on outside donations to pay for Hamburg fire damages and, in reverse, the problem of fire damages in many northern German/northern European territories was thought to have been solved by fire insurance, so it was no longer necessary to rely on conventional solidarity relations, even from the perspective of the putative recipient. In contrast to the notion of a constant and uniform process of nationalization, or even beyond that, of transnational network building, the charitable bodies became even more locally framed. Thus, not surprisingly, this process initially appears precisely in those northern European regions where fire insurance was first widely implemented and where older forms of solidarity networks were thereby displaced. In any case, a development can be seen during the course of the eighteenth century that clearly distinguishes the early part of the early modern period from later periods in terms of donation systems and charitable solidarity.

In order to examine the role of fire insurance in the context of community formation and transformation, it will be important to address the resistance of individual large and small communities and other cooperative entities. Such resistance serves to reveal the misunderstandings and cognitive dissonance associated with the establishment of the previously unknown institution and form of administrative communication embodied by the fire insurance funds. For example, in 1768 twelve villages from the Arendsee area stated that for them "in the beginning this establishment [seemed] … useful and beneficial", but now, due to the constant rise in contributions had now become "burdensome and tiresome"; as a result, they requested to go "back"

20 StA Hamburg Bibliothek A 640–77, see attached donation registry. On the almshouse, see Mary Lindemann, *Patriots and Paupers: Hamburg 1712–1830*, Oxford 1990.

to the state of things before the establishment of the fire society. If this were not possible, they requested at least a considerable reduction in their taxation.[21] In the end, these twelve villages and an additional seven requested that they "be allowed to establish a fire society amongst [them]selves," and withdraw from the territorial fire society—a plan repeatedly pursued by villages and even cities: Berlin followed a similar path in 1774.[22] Because of similar sensibilities, the subjects of Pröttlin refused to fulfill their duties in 1815.[23] The village of Alexandersdorf owned by a noble lord in the county of Landsberg also agitated in 1796 for a local substitute regulation—a type of cooperative communal organization agreed upon by the 37 homeowners—for which they wrote a charter. They sent their request to Berlin: "We the mayor and the court districts and the entire community have decided among us and determined that if the threat of fire should arise among us in the community, though everyone prays to be safeguarded from that by the grace of God, but should it happen anyway that God punishes one of us, we have thus decided amongst ourselves, … that every proprietor must pay [the victim] 2 Reichstaler as rebuilding money."[24] The requests from the citizens of Potsdam in 1788 and from Frankfurt/Oder in 1801 were more realistic, proposing the exit from the Kurmark and Neumark fire societies with the goal of founding their own city fire funds:[25] the cities of Hamburg, Berlin, and Magdeburg had demonstrated that once a certain population density was reached, an individual city fire association was indeed feasible.

The main motivation of these separatist movements was always the avoidance of the fees imposed by the fire societies, which were regarded as oppressive and unacceptable. Within these movements, however, a major process of restructuring emerges, newly defining different communities and their relationships. In order to understand this, it is helpful to distinguish between

1. Hazard communities
2. Solidarity communities
3. Risk communities.

Hazard communities implied all the original communities within neighboring proximity that were exposed to natural and manmade hazards—in the context we are dealing with, large fires were a particular problem. The inhabitants of a district, village, town, or city who lived close to each other were determined to

21 BLHA Potsdam Rep 2 D 807.
22 BLHA Potsdam Rep. 2 D 807.
23 BLHA Potsdam Rep. 2 D 815.
24 BLHA Potsdam Rep. 3 Nr. 11435.
25 BLHA Potsdam Rep. 2 D 842, S 4126.

be vulnerable to the same hazards in the same way. These hazard communities originally formed the core of the solidarity communities, although charitable solidarity always encompassed a wider circle than the hazard community itself, including the "neighborhood" of the village, the city, and even the territory. Concerning the state institutionalization of charitable aid, however, as in the case of Wurttemberg, a certain degree of abstraction had already replaced the sense of organic, "felt" closeness. Nevertheless, on a territorial level, there was a growing awareness of the neighborhood of magistrates and governments that was cultivated by the administrative elites and was analogous to the neighborhood of the homeowners at the local level.

The process initially seems to proceed from a transformation of hazard communities into risk communities. Perhaps some of the late medieval cooperative forms of fire guilds can be regarded as results of such transformation, since the fire guilds did identify specific hazards and determined contribution amounts after fire events.[26] However, these amounts were not set according to the size of actual damages, and therefore, the guilds' central concern was not the prediction and understanding of the degree of risk, but rather the regulation of unchanging assistance payments. In addition, membership in the guilds primarily correlated with "natural" neighborhoods in terms of both persons and space. The unusual form of the Hamburg fire contracts already represented a different, intermediate stage, because here, as in a commercial contract, the contract participants were individually listed along with their individual contribution and thus collective agreements were formed solely according to the criteria of spatial proximity and an abstract number of approximately 100 estate (*Erben*) owners. The risk of fire was exclusively specified. Although the damage contribution payment was still not proportional to the estimated value of the estate or to the actual damages incurred, the technical nature of the contract form and the quantification of proportional estate value shares to be paid, damages, and membership contributions already clearly indicate an orientation toward a collective insurance-like format.[27]

The city-wide Hamburg fire fund and, in particular, the territorial fire funds, spreading from the Prussian territories across central Europe, effected something new in this area in terms of social structures: the organically formed, "natural" neighborhood connections were broken apart and the subjects of a province were henceforth defined as single risk community. This had concrete

26 Cf. Schewe, *Geschichte*.

27 Ebel, *Hamburger Feuerkontrakte*.

consequences because residents now were obligated to pay a permanent annual levy in the form of damage assessments. In return, they acquired the right to compensation in the event they were affected by fire damages. This, however, entailed a very abstract logic concerning responsibility and relationships, the positive side of which was hardly noticeable to many at first. Previously, a house, village, or city fire was regarded as a terrible misfortune, and some amount of aid was provided by local solidarity unions and by the authorities. Individual houses were not necessarily built to last forever. When one felt compelled to provide assistance due to some misfortune, it concerned a local incident and the damages involved familiar faces. Damages and ensuing payments were sporadic, and such events were anomalies, exceptions to the rule. Likewise, donations toward distant catastrophes were also perceived as exceptional.

With the institution of the fire funds, everyone was compelled to make constant, regular payments, based on the statistics of mediating damages across the entire territory; individuals were compelled to pay damages for people in far-flung towns and cities, with whom they had no traditional relationship or connection, other than the fact that they both lived in the same territory. In this case, payments were based precisely on the registered value of one's own property and exactly in proportion to the actual damage incurred. The measurability and the concrete assessment of the "risk" formed the basis for these relationships. The traditional solidarity unions were therefore transformed into, or at least supplemented by, risk communities. The solidarity unions, as far as they had traditionally dealt with "everyday exceptions" of larger fires of individual buildings (especially of church buildings) in far-away places, became obsolete. It was now only for very major disasters, which exceeded the capacities of the new risk communities, that new transregional, international, and even global solidarity unions emerged. While this development is certainly worthy of further examination through the lens of a large number of examples, it nevertheless seems to imply that there is something more complex at work than the constant expansion of solidarity unions alone. The media of global networking came into play in this context only after the transformation of local solidarity unions into expansive, abstract risk communities, which also caused older relationships involving large distances to crumble.

When, exactly, the globalization of charitable action and solidarity networks concerning natural and other types of disasters began has been little studied thus far. In the broad field of literature devoted to the 250th anniversary of the Lisbon earthquake in 2005, there is hardly any reference to an international network of charitable aid activity. According to Martin Warnke, Jean-Paul Paurier was the only one to suggest that "the relief efforts for Lisbon probably represent the first case of international disaster relief."[28] However, whether the three or four ships of timber from England and the similarly small Hamburg deliveries merit such a characterization remains questionable—especially since the dynastic national rivalry resulted in the rejection of French and perhaps even Spanish aid, thus casting doubt on the reality of a larger cosmopolitical impulse here.

The Hamburg fire of 1842 serves as an illustrative example for how much had changed between 1755 and 1842: Hamburg was a pulsing trade city of global significance in the early stages of industrialization and was represented by a large number of consuls throughout the world. This led to a globalization of what had heretofore been only a short-term solidarity community emerging after a major fire, which led to the establishment of a public support administration agency (*Öffentliche Unterstützungs-Behörde*) in order to manage and appropriately channel charitable funds and assistance. The reports of this agency offer insight into the subsequent development process of the solidarity network, with the initial report delineating the decisive disaster events that Hamburg had experienced up to that point. In the first half of the nineteenth century, three "calamities" (*Calamitäten*, a term often used in place of "disaster" [*katastrophé*]) haunted Hamburg: the cholera outbreak of 1831–32, the great floods of 1824–25, and the siege of the city during the Napoleonic Wars in 1813. The challenges of the Cholera outbreak, the wars that affected all of Europe, and the flood damages could be met "with [their] own forces," though "generous gifts from abroad, donations by princes and private individuals, which were quickly and gratefully received, provided significant, but only secondary, assistance."[29] After the occupation and siege of 1813–14, the ensuing thirty years of peace resulted in "expanding trade relationships with foreign parts of the world" and a new "state of prosperity and affluence." What is called "globalization" today is characterized here in 1842 as the "great lever of communication between people and nations, and thus of civilization, intelligence, and industry." The great fire, which no one could have predicted, broke out

---

28 Warnke, "Das Erdbeben von Lissabon 1755," p. 18; Jean-Paul Poirier, *Le tremblement de terre de Lisbonne 1755*, Paris 2005, p. 101f. There are many studies on the impact on Lisbon's economy and population and the losses (also of the English and French merchants in the city) caused by the catastrophe but less on the post-disaster relief by foreigners—if there was any. Cf. José Luís Cardoso, "El terremoto de Lisboa de 1755 y la política de regulación económica del Marqués de Pombal", *Historia y Política* 16 (2006), pp. 209–236; Alvaro S. Pereira, "The Opportunity of a Disaster: The Economic Impact of the 1755 Lisbon Earthquake", *The Journal of Economic History* 69, 2 (2009), pp. 466–499.

29 1. Bericht der Oeff. Unterstützungs-Behörde, Hamburg 15 June 1842, HStA Hamburg, Cl. VII, Lit. Fe Nr. 13b Vol. 2.

in the heart of this promising phase of development. Within 79 hours, 4219 homes had been destroyed, 20,000 people had become homeless, and a loss of about 41 million marks (courant) had been incurred.

Immediately, a private aid association, the "Hülfs-Verein," was established, while the public administration agency managing the charitable funds, the "Unterstützungs-Behörde," was created on May 10, and had three members of the Senate (Dammert, Alardus, Büsch) and five members of the Hülfs-Verein (Pastor Plath, A. Abendroth, Blumenthal, Ebeling, Steinthal, Merck) on its board. The Hülfs-Verein was responsible for the administration of gifts in kind and small donations of money, while the *Unterstützungs-Behörde* managed the reception and distribution of larger amounts of both national and international monetary contributions. The donations sketch a pattern of growing concentric circles according to the distance of the donors—a progressively increasing radius of the solidarity relationship between Hamburg and private as well as state and dynastic donors throughout the world (Fig. 64).

In the first report of the Unterstützungs-Behörde on May 31, 1842, donors listed include the king of Denmark, the emperor of Russia, the king of France, as well as most of the German territorial princes (from Prussia to Thurn and Taxis). There were donations recorded from various private individuals and

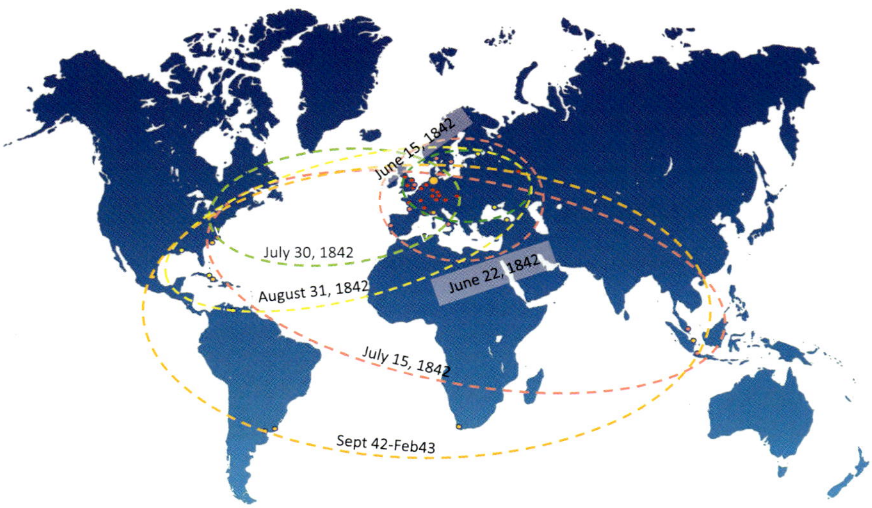

FIGURE 64 The expansion, week by week, of the world-wide solidarity community supporting the post-disaster recovery of Hamburg in 1842: the public support administration, expressly founded for that reason, noted every week the incoming money from every place, individual and collective donators between May 1842 and February 1843 (HStA Hamburg, Cl. VII, Lit. Fe Nr. 13b Vol. 2)

institutions from cities across Germany and Europe: Antwerp, Amsterdam, Carlstadt/Sweden, Copenhagen, Hull, Lyon, London, Leicester, Montpellier, Manchester, Mirow, Marseilles, Nantes, Paris, Prague, Riga, Rotterdam, Tondern, Vienna, and Zurich. A donation of May 20 was also listed from New York, which would certainly indicate that it came from a New York affiliate in Hamburg, since it would have taken a ship well over two weeks to cross the Atlantic. In addition to donations from individuals, the magistrates or councilors of the cities also donated to some extent, and in many cities there were "assistance committees" and "aid societies" formed for the benefit of Hamburg; women's associations are also often named in the records. There were donations from trade societies; charity concerts, theater performances, music festivals, etc. were held in several cities; theaters collected special donations for actors affected by the fire (in Frankfurt am Main, for example); and there were donors from shooting clubs and Jewish communities. In Mannheim, for example, the editorial office of the local newspaper, the *Mannheim Journal*, organized a citywide collection and donated the considerable sum that was collected: this was probably due to the fact that the paper was published by a charitable institution, the Catholic Civic Hospice (*das katholische Bürgerhospital*). Fire insurance companies increasingly appeared as voluntary donors—an impulse likely motivated specifically by a desire to engender public trust.

The radius of donors increased on a weekly basis—on June 15, contributions were received from the king of Sweden and Norway, the duchess of Braganza (the widowed empress of Brazil), from Archangel and Lisbon; on June 22 from the queen of Great Britain and other members of the British royal house, from Bordeaux, Livorno, Naples; on June 30, from New York, where Consul Schmidt had established an assistance committee; from Moscow, where Consul Schilling did the same thing; on July 15, from Batavia, Malaga (via Consul W. Rein), Marseilles (via Consul H.T. Meyer), Montpellier (via Consul Westphal-Castelnau), Nantes (via Consul Collet), and a Philadelphia committee; on July 30 from Algiers, Baltimore; on August 31 from a Havana committee, La Guaíra/Venezuela, New Orleans, Odessa, St. Petersburg, and Puerto Cabello; on September 30 from Trinidad/Cuba and Mazatlán/Mexico; on October 30 from Bahia, Pernambuco, Rio de Janeiro, Tampico/Mexico, Angostura, Buenos Aires, and Montevideo; on December 31 from Cape Town; on February 28, 1843 from Singapore.[30] In total, the *Unterstützungs-Behörde* received the very large sum of 4,296,862 bancomarks—over one-tenth of the calculated total loss. With few

30 1. Bericht der Oeff. Unterstützungs-Behörde, Hamburg 15 June 1842, HStA Hamburg, Cl. VII, Lit. Fe Nr. 13b Vol. 2.

exceptions, this money was all collected from royal families, private individuals and secular aid societies, and assistance leagues and committees. The network presented here aligned itself along political and economic relationships with Hamburg, rather than according to any ecclesiastical ties. These remarkably global contributions, which are clearly direct precursors to today's fundraising campaigns for major, globally recognized disasters, indicate that a complete transformation and differentiation of social and community structures had taken place. At this point, the existing research offers little in the way of comprehensive analyses or developmental histories on this topic.

There is, however, a solid amount of historical research on the history of charitable activity and the history of care for the poor, in particular, as well as on the philanthropy of the eighteenth century in general. Here, a certain *communis opinio* concerning the steps of the process of transformation has been established: Wilbur Jordan has shown, for example, that in the post-Reformation transformation of the sixteenth and seventeenth centuries, charitable activity became focused mainly on donations managed by foundations and trusts. The money was mainly used for four types of support services: 1) the prevention of acute poverty in individual families; 2) the financing of poorhouses; 3) the unspecified use by communities and city officials for the care of the needy; and 4) the support of elderly people incapable of working.[31] The particularity of England—in contrast to the rest of Europe, but also to the rest of Great Britain—was that there were Poor Laws in place in England starting in 1572 and, beginning in the late seventeenth century, there was regular taxation assessed on property for the benefit of the poor: the state was introduced as an institution engaged in fighting poverty. At most, this operated only as an indication of Enlightenment politics regarding poverty. The Enlightenment tenor of *policey* reforms during the eighteenth century favored work over charity, fought begging practices, supported "charitable gifts," and urged citizens to volunteer for fundraising and offering assistance to the poor.[32] Particularly

31 Classic for England would be W.K. Jordan, *Philanthropy in England, 1480–1660: A Study of the Changing Patterns of English Social Aspirations*, London 1959; W.K. Jordan, *The Charities of London, 1480–1660*, London 1961; for the state of research in this field: Nigel Goose, "The Rise and Decline of Philanthropy in Early Modern Colchester: The Unacceptable Face of Mercantilism?" *Social History* 31 (2006), pp. 469–487.

32 Joanna Innes, "The State and the Poor: Eighteenth-Century England in European Perspectives," in John Brewer and Eckhart Hellmuth (eds.), *Rethinking Leviathan: The Eighteenth-Century State in Britain and Germany*, London 1999, pp. 225–281; Owen Hufton, *The Poor of Eighteenth-Century France 1750–1789*, Oxford 1974; Martin J. Daunton (ed.), *Charity, Self-Interest and Welfare in the English Past*, London 1996; H. Cunningham and Joanna Innes (eds.), *Charity, Philanthropy and Social Reform in Europe and North America*

during the eighteenth century, analogous private "friendly societies" proliferated, which contributed to individual self-provision efforts (*Eigenvorsorge*) on the one hand, and to neighborhood-based assistance on the other. In 1803 in England there were 9,000 such associations comprising over 700,000 members.[33] Such associations certainly operated as a model for organizations such as the Hamburg's 1842 "Hülfs-Verein." According to Gertrude Himmelfarb, the eighteenth century generated an entirely new definition of "poverty" as a problem to be solved, representing an exceptional situation rather than a normal state of affairs.[34]

The secularization of charitable activity is recognized as a significant, comprehensive step taken during the eighteenth century in the context of private and state welfare institutions. Classic studies on this development include Michel Vovelle's work on the de-Christianization that began in the eighteenth century in regions like the Provence, indicated by the decline of religious movements and church endowments and donations. He argues that it was not only the Revolution that facilitated the sudden secularization of France, but rather a development emerging from slow incremental changes in the practice of everyday life.[35] For the Anglo-American world, it is worth noting that in England, beginning in the mid-eighteenth century, and in the United States beginning in the 1780s (starting in Philadelphia), the notion of "philanthropy/love of mankind" conceptually replaced the older "caritas/love thy neighbor," precisely where modern institutional forms of organization devoted to the common good were sought and found. These included "friendly societies," provided that they were organized according to the structure of a joint-stock company and were devoted to managing capital for charitable causes. Also included were the mutual savings banks that proliferated beginning in the 1820s on the East Coast of the United States, and institutions such as the Massachusetts Hospital

---

1690–1850, London 1998; Paul Slack, *From Reformation to Improvement: Public Welfare in Early Modern England*, New York 1999 [until about 1730]; Ole Peter Grell and Andrew Cunningham (eds.), *Health Care and Poor Relief in Protestant Europe 1500–1700*, London 1997; Ole Peter Grell, Andrew Cunningham, and Robert Jütte (eds.), *Health Care and Poor Relief in Eighteenth and Nineteenth Century Protestant Europe*, Aldershot 2002.

33 Peter H.J.H. Gosden, *Self-Help: Voluntary Associations in Nineteenth-Century Britain*, New York 1974, p. 12.

34 Gertrude Himmelfarb, *The Idea of Poverty: England in the Early Industrial Age*, London 1985.

35 Michel Vovelle, Piété *baroque et déchristianisation en Provence au XVIIIe siècle. Les attitudes devant la mort d'après les clauses des testaments*, Paris 1973; Rudolf Schlögl, *Glaube und Religion in der Säkularisierung. Religiosität in der katholischen Stadt, Köln, Aachen, Münster 1700–1840*, Munich 1995 transferred the approach to German comparative urban history and complemented it especially through an analysis of book ownership.

Life Insurance Company, which was organized as a life insurance company but generated its primary profits through the management of nonprofit endowments and foundations. Such organizations paved the way for the development of American "civil society," characterized by charitable donations, foundations, and the nonprofit sector.[36]

With all these references to welfare and donations, the parallel development of insurance is hardly even recognized as an institutionalization of individual means of self-provision (*Eigenvorsorge*); however, there is little information on how the transformation of charitable and welfare institutions and practices operated both as an indicator as well as a formative element for the emergence of new stable and sporadic communities. At most, the patriotism of these new institutions of philanthropy has been linked to the patriotism of the nation. Recently, Amanda Bowie Moniz has stressed that the literal human relationship implied by "philanthropy" should be taken seriously and that there was a thoroughly cosmopolitical aspect to such welfare institutions: the Humane Society of Philadelphia assured their supporters in 1795 that they had organized "the same plan in Europe, India, and the Caribbean" after so many thousands had once again found themselves back in the clutches of the "general family of mankind." Their model, the Royal Humane Society, assured its supporters that it was based on liberating "the whole human race from sudden calamities." The form of those "human societies" was based on an initiative of wealthy men in Amsterdam, who founded a company in 1767 to resuscitate drowning victims, and whose methods were imitated beginning in the 1770s in Hamburg, Milan, Padua, Paris, Venice, Vienna and elsewhere, as well as in 1774 in London. Several of these societies created networks among themselves by receiving honorary members from other societies.[37] Although such societies always oriented their operations toward the immediate vicinity, the focus was aimed at "mankind" and the "calamities" that threaten it: this language was even chosen by the Hamburg *Unterstützungs-Behörde* in 1842 ("Calamitäten). But despite promising attempts at investigating such questions in recent times, we still do not

---

36 Barbara L. Bellows, *Benevolence Among Slaveholders: Assisting the Poor in Charleston, 1670–1860*, Baton Rouge 1993; Lawrence J. Friedman and Mark D. McGarvie (eds.), *Charity, Philanthropy and Civility in American History*, Cambridge 2003; Kathleen D. McCarthy, *American Creed: Philanthropy and the Rise of Civil Society, 1700–1865*, Chicago 2003; Conrad Edick Wright, *The Transformation of Charity in Postrevolutionary New England*, Boston 1992.

37 Amanda Bowie Moniz, "Cosmopolitanism and Philanthropy in the Early American Republic," *GHI Bulletin Supplement* 5 (2008), pp. 9–22. Ead., *From Empire to Humanity: The American Revolution and the Origins of Humanitarianism* (Oxford: 2016), pp. 25f., 89–96, 165, 171 on "disaster relief networks".

know much about the developmental steps taken by the cosmopolitical relief movement during the long nineteenth century—though the donation records of the 1842 Hamburg fire disaster make a strong case for the existence of an already relatively well-established practice.[38]

At what point did such transnational donation networks develop in the wake of natural disasters and, later, technological disasters?[39] And how can these scattered networks that clustered together as a global network after major disasters be conceptualized? One option is to interpret such networks as an unusual type of "cosmopolitical community." Ferdinand Tönnies compared and contrasted two social types: the "community" (*Gemeinschaft*) based on mutual bonds, versus "society" (*Gesellschaft*) based on the individual interests and aims. "Community" in this context is associated with pre-modernity due to the prevalence of common values, small populations, close ties, familiarity, tradition, frequent interactions, and emotional bonds, while "society" correlates with modernity and its loose ties, different and pluralistic beliefs and values, distance, infrequent interactions, temporary arrangements, and regulated competition.[40] Since Tönnies' time, manifold categories and subdivisions of these broad and general types of societal structures have been differentiated and have moved beyond the dichotomy between the "premodern" community on the one hand and "modern" society on the other—obviously communities still do arise within modernity at given moments and in certain contexts. In the currently revitalized discussion about "community," the basic conditions for community-building (spatial proximity or active choice) are generalized and typified according to the modes and frequency of interaction among

---

38 Anne O'Brian, *Philanthropy and Settler Colonialism*, New York 2015, pp. 10–86 also points to the domestic charitable activities of Australian offshoots of philanthropic societies in about the same time period, but does not address potential externally oriented charitable donation campaigns.

39 On the parallel phenomenon of the expansion of the radius of the respective news coverage from the local to at least a national level, see Jürgen Wilke, "Von der Lotterie zur patriotischen Opfergabe. Vorstufen und Erscheinungsformen von Spendenkampagnen in der älteren Presse," in *Massenmedien und Spendenkampagnen. Vom 17. Jahrhundert bis in die Gegenwart*, Cologne 2008, pp. 15–72; on the contemporary phenomenon see Gabriele Lingelbach, *Spenden und Sammeln: Der westdeutsche Spendenmarkt bis in die 1980er Jahre*, Göttingen 2009; see Julia F. Irwin, "Raging Rivers and Propaganda Weevils: Transnational Disaster Relief, Cold War Politics, and the 1954 Danube and Elbe Floods," *Diplomatic History* 40, 5 (2016), pp. 893–921 for a contemporary case study while there is still comparatively little of research on the long-term history of the phenomenon.

40 Ferdinand Tönnies, *Gemeinschaft und Gesellschaft. Abhandlung des Communismus und des Socialismus als empirischer Culturformen* (orig. 1887), Darmstadt 2005.

members, how the members are situated, and the amount of face-to-face communication.[41] At the core of the definition of "community" are shared activities or beliefs and bonds created by common community values. By imposing these criteria on the analysis of the developments occurring at the intersection between insurance-type and charitable donation activities and campaigns, the following overview emerges from the examples discussed above:

1) The introduction of regional forms of fire and life insurance, primarily in the eighteenth century, contributes significantly to the dissolution of local communities in the central and northern European arena. As common trade "neighborhoods," these communities were based purely on geography. Early institutional forms that compensated for damages and losses, such as guilds or even fire contracts, could be classified as "elective communities" of increasing degrees of abstraction. They were hazard communities, in so far as they were justified by the common concern that when disaster struck, the same danger existed for all members. Resistance against these forms of loss prevention, involving the shifting and reestablishment of resource streams away from the relatively narrow circle of the face-to-face community, was perceived as a process of "alienation" (*Entfremdung*).
2) Risk communities could no longer be considered "communities" at all, because they were either created "from above" by the state or because the voluntary entry into an insurance group was based purely on economic interest. Risk "communities" thus actually became risk "societies" in line with the definitions above. State and private insurance organizations thus continued to maintain a de facto hybrid form of language and self-reflection that continued to correspond

41 The discussion surrounding "communities" is once again active in sociology, particularly in the context of disaster analysis (following 9/11), since post-disaster communities, their behaviors and their potential for a remodeling of civil society, have been subject to analysis. Journal foundings like *City & Community* serve as further indications of this. Stephen Brint, "*Gemeinschaft* Revisited: A Critique and Reconstruction of the Community Concept," *Sociological Theory* 19 (2001), pp. 1–23, 8: "I will define communities as aggregates of people who share common activities and/or beliefs and who are bound together principally by relations of affect, loyalty, common values, and/or personal concern (i.e., interest in the personalities and life events of one another)." The typological sketch on p. 10 in Jochen Gläser, *Neue Begriffe, alte Schwächen. Virtuelle Gemeinschaft*, in Michael Jäckel and Manfred Mai (eds.), *Online-Vergesellschaftung? Mediensoziologische Perspektiven auf neue Kommunikationstechnologien*, Wiesbaden 2005, pp. 51–72 can also be seen as a critique of Brint's, in his view, overly broad concept.

to the central criterion of the "community" as a societal form: namely, a shared sense of responsibility regarding the security of values to be protected by the community. It is precisely this aspect which is highlighted in terms of the concept of community generalized above.

3) The comparison between England/Great Britain and Germany reveals that there was even less of a community-type substrate existing in England during the transition to insurance-like forms of loss prevention—at least concerning the joint-stock companies, in which a sense of cooperative social responsibility for the maintenance of the common good was less apparent as compared with the state-managed German projects. In the mutual societies on the other hand, hybrid elements were indeed present. Private insurance companies founded in Germany around 1820 that were based on the English model had to contend with what could be called an expectation to organize security in patterns of charity and community forms; this clarifies why the Gotha Insurance Bank, for example, established itself as a "completely nonprofit" organization based on reciprocity, while the Aachen (-Munich) joint-stock company found it necessary to devote half its profits to the financing of a charitable "Association for the Promotion of Employment."
4) The existence of a large-scale solidarity community before 1730 is suggested by the fact that solidarity contributions for major fires, especially for the purpose of church reconstruction, were directed into quite distant (though not typically global) regions with great regularity up to that point. However, this community was only made visible in the context of sporadic "exceptional cases" and through individual relationships; there was no consistent contact between all potential distant locations within the network. This solidarity community is thus only accessible on a sporadic basis and is visible only at these moments. The commitment induced by notions of Christian charity is a binding element, which entails a love for needy fellow Christians, even those in distant places. The community could be thought of as "imagined," although only on a sporadic basis. Compensatory organizations, such as fire insurance institutions, apparently contributed to the dismantling of these imagined communities during the course of the eighteenth century with regard to small and medium-sized disasters, since such disasters began to be handled locally, or in cooperative/state community-like hybrids. However, this change seemed to correlate with an increase in damage-value thresholds that could no

longer be met locally, demanding solidarity and voluntary humanitarian assistance.

The transition here occurs between the eighteenth and nineteenth centuries. A sporadic, spontaneous, and surprisingly effective short-term community arises after major disasters, effecting massive donations and functioning as a hybrid between virtual and imagined communities: virtual, in that in the majority of cases involving contributions, a one-time donation decision accounts for the emergence of this community and there does not remain a real reiterated bond or continuity of this community; imagined in that it rests on a cosmopolitical ethic. While a variety of friendships and other networks contribute to its emergence, the community of emergency donors as a whole can never be reduced to such networks. In its momentary quality and its singularity, this community reveals the element described above (4), even if only implicitly, which is often adopted in addition to a notion of community—namely, the element of a certain temporal extension. But perhaps this element is not really necessary, because in any case, the charitable intentions—whether purely altruistic or with mixed intentions—constitute an impulse that unites all donors within a community, especially from the perspective of the receiver. If the cosmopolitical element is accentuated, and if the proof is brought to the fore, that there is a global multitude of human societies, each created with cosmopolitical intent, behind every donor community emerging from a major disaster—then one could argue that it is precisely through a series of major global disasters where a large intersection of the same donor groups offering assistance becomes visible as active again and again, that a usually latent imagined community of consistent cosmopolites is revealed over time.

1.2 *Insurance as Emerging from Notions of Social Contract and Moral Duty*

To what extent was the insurance theory of the eighteenth century integrated into already generalized societal concepts and, in this respect, how did it contribute to the prehistory of sociological thinking? As noted at the outset, if we do not take theoretical contributions to probability theory and life insurance into account but rely solely on concepts related to insurance as an institution and principle, the source material for the English region before the nineteenth century is rather scant. This is due to the privatized business roots of the local insurance bodies, in contrast to state-administrative integration within German-language areas. While shorter articles and smaller treatises did not always explicitly express a societal vision or philosophy, when they did, the earliest such cases belong to a eudemonistic tradition propagating the notion of securing happiness as the main goal of rulers and the state.[42]

> The state is a society, in which many unite to preserve and promote their common well-being: Thus, all members must have access to the common happiness, which is their common purpose, according to a convenient proportionality. ... A ruler is bound by the essential duties of his profession to make the general happiness of his people the main purpose of his cares and occupations, and to maintain such happiness through the easiest, most efficient, and safest means possible. ... Furthermore, since all members of a state are connected by virtue of their life in society to one

---

42 The eudemonistic current of state thinking in Europe began with Leibniz, but was popularized by the work of Christian Wolff, Cf. Clemens Schwaiger, *Das Problem des Glücks im Denken Christian Wolffs. Eine quellen-, begriffs- und entwicklungsgeschichtliche Studie zu Schlüsselbegriffen seiner Ethik*, Stuttgart-Bad Cannstatt 1995; and also Ulrich Scheuner, "Die Staatszwecke und die Entwicklung der Verwaltung im deutschen Staat des 18. Jahrhunderts," in Gerd Kleinheyer and Paul Mikat (eds.), *Beiträge zur Rechtsgeschichte. Gedächtnisschrift für Hermann Conrad*, Paderborn 1979, pp. 467–89; Ulrich Engelhardt, "Zum Begriff der Glückseligkeit in der kameralistischen Staatslehre des 18. Jahrhunderts (J. H. G. v. Justi)", *Zeitschrift für Historische Forschung* 8 (1981), pp. 37–79; Jörg Cortekar, *Glückskonzepte des Kameralismus und Utilitarismus: Implikationen für die moderne Umweltökonomik und Umweltpolitik*, Göttingen 2007. The other standard works concentrate less on the "happiness" of the state: Robert Mauzi, *L'idée du bonheur dans la littérature et la pensée françaises au XVIIIe siècle,* Paris 1960 (1994); Carminella Biondi et al. (eds.), *La Quête du bonheur et l'expression de la douleur dans la littérature et la pensée françaises*, Geneva 1995; Massimo Mori, "Glück und Autonomie. Die deutsche Debatte über den Eudämonismus zwischen Aufklärung und Idealismus" *Studia Leibnitiana* 25 (1993), pp. 27–42; Frank Grunert, "Die Objektivität des Glücks. Aspekte der Eudämonismusdiskussion in der deutschen Aufklärung" in Frank Grunert and Friedrich Vollhardt (eds.), *Aufklärung als praktische Philosophie*, Tübingen 1998, pp. 351–68; Cornel Zwierlein, "Das Glück des Bürgers. Eine begriffsgeschichtliche Skizze zur optimistischen Grundmentalität der Aufklärung," in Hans-Edwin Friedrich, Fotis Jannidis, and Marianne Willems (eds.), *Bürgerlichkeit im 18. Jahrhundert*, Tübingen 2006, pp. 71–113; for the parallels in English ethical, political and economic discourse, albeit quite different in its setting, see Jeffrey Barnouw, "The Pursuit of Happiness in Jefferson and its Background in Bacon and Hobbes" *Interpretation: A Journal of Political Philosophy* 11 (1983), pp. 225–48; James F. Jones, "Prolegomena to a History of Happiness in the Eighteenth Century," *The French American Review* 6 (1982), pp. 283–95; Robert Darnton, "The Pursuit of Happiness," *Wilson Quarterly* 19 (1995), pp. 42–52; Carli N. Conklin, *The Pursuit of Happiness in the founding Era: An Intellectual History*, Columbia 2019 stresses common law influences (like the legal thought of Blackstone, albeit usually opposed by Jefferson) next to Anglican theology and philosophical concepts of improvement on the formula in the US Declaration of Independence (1776), and stresses less the tradition of sentimentalism (Shaftesbury circle) of the early eighteenth century.

> purpose, namely to the preservation of general happiness, they are also bound, by virtue of divine and human rights, in line with their Christian and civic obligations, to carry another's burdens, even when there are such burdens that would threaten the utter destruction of some members, of the state as a whole, or at least of some main portions.[43]

Johann Heinrich Gottlob von Justi, who is known as perhaps the most eudemonistically focused cameralist,[44] treated the contents of insurance in his *Staatswirtschaft* (State economy) under the main heading: "Of the Means and Measures of the Rulers to Preserve and Multiply the Assets of the State, and thereby to Further the Happiness of the Subjects."[45] Besides the fact that the focus on happiness in general was the universal topos of cameralist thought during the 1750s, this was especially central in this context because insurance companies were conceived of as "institutions against *misfortunes*"[46]: *Un-Glücke* (misfortunes) such as major fires and other conflagrations constituted the exact opposite of *Glück* (happiness) that the state claimed as its goal. Additionally, the old concept of *casus fortuitus* came into play in all its metamorphoses—from the narrow sense of the legal treatment of performance defaults in Roman law up to its application beyond the law of obligations (see above Ch. 4, 1.2). *Unglücke* such as floods, fires, and epidemics constituted minuses on the balance sheet of happiness and were therefore to be avoided.

More strongly generalized remarks on the importance of insurance in the Enlightenment social and political model can be found in later insurance treatises. Ferdinand Friedrich Pfeiffer (1759–1806) was Schiller's classmate at the military academy *Karlsschule* in Stuttgart, and was later a teacher, then

43 [Anonymous], "Vorschlag, wie die Versicherung der Gebäude zu Ersezung des Feuerschadens auf eine leichte Art einzurichten," *Physikalisch-Oekonomische Realzeitung, oder gemeinnützliche Wochenschrift* 17–19 (1756), pp. 673–84, 701–16; 675, 702f.

44 Cf. Engelhardt, "Zum Begriff der Glückseligkeit" and Ulrich Adam, *The Political Economy of J.H.G. Justi*, Oxford 2006.

45 Johann Heinrich Gottlob von Justi, *Staatswirthschaft oder Systematische Abhandlung aller Oekonomischen und Cameral-Wissenschaften*, part 1 (1755; Leipzig 1758), pp. xlv, 284–89.

46 This is a general rubric, under which insurance in general introductory texts about cameralism was handled: Johann Friedrich v. Pfeiffer, *Natürliche aus dem Endzweck der Gesellschaft entstehende Allgemeine Policeiwissenschaft*, 2 vols., Frankfurt am Main 1779, reprint Aalen 1970, Ch. 3: "Wider die Unglücksfälle, sind vernünftige Maßregeln zu nehmen," (pp. 332–340); Johann Friedrich v. Pfeiffer, *Grundsätze der Universal-Cameral-Wissenschaft oder deren vier wichtigsten Säulen* [...], Frankfurt am Main 1783, Ch. 32: "Von den Verwahrungsmitteln wider die Unglücksfälle,", pp. 567–680; similarly Johann Heinrich Jung-Stilling, *Lehrbuch der Staats-Polizey-Wissenschaft*, Leipzig 1788, reprint 1970, 4th main part: "Unsicherheit des Eigenthums durch Unglücksfälle," pp. 360–90.

secretary of the pension fund (*Rentkammer*), and finally a war council member (*Oberkriegsrat*) in Munich.[47] In 1780, the same year that Schiller received his doctorate in medicine from the Karlsschule, Pfeiffer defended his dissertation, published unter the title *Gedanken über Versicherungs-Anstalten* (Thoughts on insurance institutions), in the presence of the duke. He published this remarkable discourse with part-neo-Aristotelian, part-Rousseauian roots, locating the origins of human society in the union of many pursuing a common goal and by that fulfilling the destiny of mankind that lies in the drive toward self-preservation and the necessity to unite forces for making human progress possible.

> Should the security of a nation be threatened from any side, then the whole society must be united to remove the danger where possible, or at least to share the inevitable misfortune that affects every individual, because nobody lives for himself alone any longer. And this is indisputably the reason for the organizations for the insurance of property and, to some extent, also of revenue, *in the nature of the social contract*—because they are based on creating a greater security of life by ensuring the basic maintenance of it.[48]

In formulating the notion of the social contract, Pfeiffer then cited not Rousseau's *Du contrat social*, but his *Emile*:

> L'homme naturel est tout pour lui [*Pfeiffer* adds: même] [*Pfeiffer* departs here from the original: il est l'unité numérique, l'entier absolu qui n'a de rapport qu'à lui-même ou à son semblable]. L'homme civil n'est qu'une unité fractionnaire qui tient au dénominateur, et dont la valeur est dans son rapport avec l'entier, qui est le corps social. Les bonnes institutions sociales sont celles qui savent le mieux dénaturer l'homme, lui ôter son existence absolue pour lui en donner une relative, et transporter le *moi* dans l'unité commune; en sorte que chaque particulier ne se croy plus un, mais partie de l'unité, et ne soit plus sensible que dans le tout.
>
> [The natural man is all for himself. (He is the numerical unit, the absolute whole, connected with nothing other than himself). The bourgeois man is nothing but a broken unit that depends on the denominator, and

47 Matthias Luserke-Jaqui, *Schiller-Studien. Der ganze Mensch und die Ästhetik der Freiheit*, Tübingen 2018, p. 27; Matt Erlin, *Necessary Luxuries. Books, Literature, and the Culture of Consumption in Germany, 1770–1815*, Ithaca/NY: 2014, p. 28.

48 Ferdinand Friedrich Pfeiffer, *Gedanken über Versicherungs-Anstalten hauptsächlich zum Vorteil der Landwirtschaft entworffen*, Stuttgart 1780, p. 2, my emphasis.

> its value lies in its relation to the whole, which is the social body. The good social institutions are the ones that know best how to denaturalize man, deny him his absolute existence, to make it relative and to transform the "I" into a common unit, so that nobody feels alone any longer, but feels himself as part of a unit, having feeling only in context of the whole.][49]

Rousseau expresses the difference between the people in the state of nature and those anchored in the social contract within the mathematical language of his time, as is often the case in his political theory. While this seems astonishing for some scholars today,[50] such mathematical societal definitions likely resonated with Pfeiffer in the context of insurance. If one considers how fire insurance funds worked in the German territories, Rousseau's image of the citizen as a subset of the whole corresponds to the way in which the citizens were listed as owners of houses in insurance registers in terms of the value of the houses and the sum insured, which, in turn, formed the basis for proportional assumption of the damage of others (see Ch. 5, 2 below): in this context, citizens were, in fact, nothing more than a fraction of a whole, and the whole was, in terms of value, precisely determined by the sum of all registered values. Fire cadasters and fire insurance corresponded to an administrative perspective that took a quantified approach to "citizens" and "built capital," expressing them as measurable units. Concerning this classification, Schiller's friend Pfeiffer can insert insurance into general societal functionality and the greater whole of the state. Insurance is thus not a mindless apparatus of a bloodless territorial administrative machinery; rather, it belongs to those "social institutions" that are necessary for the "denaturalization" of man. It makes men into citizens united within the social contract.

Other scientific treatises construct their general social theories on the basis of English and French models and are typically eclectic in the introductory parts: Different approaches are combined, and one approach is not exclusively followed. A lawyer and propagator of the insurance concept from the imperial city of Reutlingen, Johann Jacob Fetzer (1760–1844), used the Scottish

---

49 Jean-Jacques Rousseau, *Émile*, in *Œuvres completes*, vol. 4 (Paris: 1969), 245–868, 249, qtd. Pfeiffer, *Gedanken über Versicherungs-Anstalten*, note 2, my trans.

50 For an analysis of Ch. 3, 1 of the *Contrat social*, where the mathematical language of the eighteenth century is similarly employed—inscrutable to many modern Rousseau interpreters—see Marcel Françon, "Le Langage mathématique de J.-J. Rousseau," *Isis* 40 (1949), pp. 341–44 and Marcel Françon, "Sur le langage algébrique de Rousseau," *Annales de la Société Jean-Jacques Rousseau* 33 (1953–55), pp. 243–46.

Enlightenment philosopher Adam Ferguson in his writings from 1783: "all external aspirations of the people ... [have] as a final purpose, to acquire the means of livelihood, security, convenience, or the ornamentation." Therefore, within man the "tendency toward sociability is embedded, thus they are well-suited toward life in a community and according to certain instinctual tenets, guided by reason to understand how to share their activities. ... When now, following the instinct of sociability, several people agree to achieve a specific purpose through the employment of common powers, thus emerges a society."[51] Fetzer aligns himself with Montesquieu and against Justi and Joseph von Sonnenfels to argue against the power of self-preservation as decisive for the state and—in accordance with Gottfried Achenwall (*Grundsäze der Staatsklugheit*) and Anton Büsching (*Vorberaitung zur Erdbeschreibung*)—defines the state as "a society of families ... , united under a ruler for the purpose of ensuring external security and inner welfare. The achievement of the "common best" was important, necessitating the adjustment of "private use." Like Fetzer, Johann Brückner (1751–1836) also leans on English political and social doctrine in his prize writing[52] on a cantonal fire insurance fund; he uses the "common best" in his introduction as the principle on which the fire fund was based and identified with the English "public spirit," likely in the tradition of David Hume and Adam Smith (*Theory of Moral Sentiments*).[53] Fetzer further clarified that the common best is "given both from outside and from inside." For the outside, politics and the wisdom of the state intelligence were responsible. For "inner security," the science of *policey* took over: Fetzer divided this into "public" and "private security," the latter implying primarily a "security of goods" that was threatened by "accidents" (*Zufälle*). Fetzer then mentioned a few projects in totally different areas, a "health insurance company" of

51 [Johann Jakob] Fetzer, *Ueber Brandschadens-Versicherungsanstalten, ein patriotischer Vorschlag eines Reichsstädters*, Reutlingen 1783, p. 7.

52 The Economic Society in Bern announced this prize question in 1787, at the instigation of the Canton government: "Is a fire insurance institution in the Canton of Bern advisable, and on what grounds. And what would be, according to the needs and situation of the Canton, the best and most efficient arrangement for such a fire insurance institution?" Fifteen answers were submitted, and five of them were published (by Council Secretary Wild, Commission Secretary Wyttenbach, and by Pastor Gruner, Brückner received the prize) (*Allgemeine Literatur-Zeitung* 252 (September 22, 1792), p. 641f.

53 J[ohann] A[lbrecht] Brückner, *Abhandlung über Errichtung einer Brand-Assekuranz-Casse im Canton Bern*, Zürich 1790, p. 8. Brückner was a learned theologian, served 1799–1811 as a private tutor in Russia in the family of Duke Alexis of Kurakin and in 1811 became a censor in Leipzig. On the development of the "public-spirit" idea particularly at the end of the eighteenth century, see Roberto Romani, *National Character and Public Spirit in Britain and France, 1750–1914*, Cambridge 2002, esp. p. 175 (Adam Smith).

(Laurent[?]) D'Houry, the insurance societies devoted to Prussian cattle disease, but also lightning rods and some of the fire *policey* inventions common in Enlightenment discourse (see above Ch. 3, 3.2). Fetzer's remarks in relation to the (in his opinion) three existing types of fire insurance organizations and his solution for a fourth type needed not be detailed here; once again, the important point is that insurance was incorporated into general conceptions of the state and society.[54]

Pfeiffer similarly classifies insurance within a matrix of external and internal security politics according to a historical perspective on the insurance principle and contributes a reflection as to why fire insurance in antiquity and the Middle Ages was unknown. In the form of a political-economic *Querelle des anciens et des moderns,*[55] he emphasizes with pride that now, in more enlightened times: "internal security is no longer [sought] from the outside, but the external from the inside, then one noticed that those states in which one found the best possible *policey* facilities commonly also had the most effective means of creating security from the outside, and that this security was durable."[56] According to Pfeiffer, therefore, insurance is part of an enlightened governmentality that replaced the primacy of external politics with a primacy of internal politics. The greatest perfection of such institutions would be an insurance of "revenues":

> But there is still one step to take. ... I mean the insurance of revenue, that part at least, which would secure the necessary support of life for the poor and a comparable income for the rich.—Insurance itself, against accidents, which one never or rarely can control.—Security, especially against the rage of the elements.—Imagine a country, in which the most violent storms could make nobody poor, the most ferocious current, could create no hunger.—Taken from this perspective at least, who wouldn't consider himself lucky![57]

According to this social and security theory, the concept of the "normal secure society" is formulated as an ideal. At this moment, the theory of insurance against risks approximates a notion of social security, in that both are oriented

---

54 Fetzer, *Ueber Brandschadens-Versicherungsanstalten*, pp. 7–12.

55 Cf. Ch. 2, 4 above.

56 "[...] daher wundere man sich nicht, wenn man weder in ältern noch mittlern Zeiten Polizei-Anstalten der Art antrifft [...] Durch Länge der Zeit, zur Aufklärung, durch Schaden selbst wurde man klüger." (Pfeiffer, *Gedanken über Versicherungs-Anstalten*, p. 3).

57 Pfeiffer, *Gedanken über Versicherungs-Anstalten*, p. 4.

toward the maintenance of stability and balance in the area of prosperity and the production of security by the state for the good of its citizens.

Embedding insurance theory into such a general social theory also implied integration into associated conceptions of duty and morality. In this respect too, it can be described as symptomatic that Pfeiffer connected Rousseau's conception of the social contract with *Emile,* which is based far more on the education of individual men and women in a society. While the economic concepts had, in part, a very utilitarian connotation, the implied relationship to general social theories and other related topoi also carried a strong moral and ethical dimension. In the words of Brückner, writing in Chardonnay/Pays de Vaud, the ultimate goal of humanity should be "moral perfection". Insurance was a means of attaining this goal, since it could prevent the fall into deplorable ruin for many individuals. In Brückner's work, there is clear evidence of the cosmopolitical philanthropy discourse that formed the background for the nascent cosmopolitical practice of philanthropy in the Anglo-American arena. It is true, he argues:

> that mankind, taken as a whole, moves toward moral perfection in a measure proportionate to the development of their wondrous intellectual abilities of spirit. ... The philanthropist is indifferent to nothing that affects humanity: his happiness and unhappiness is shared with his brothers, he feels pleasure throughout his being at the sight of the first of them, and mourns with compassion at the sad scenes surrounding the last of them, believing that it is happening to himself, and in his sympathy, he often goes beyond the reality of the misfortune which he witnesses; how then could he not rejoice when he realizes, that in his age discoveries are made, the immediate objects of which are general utility and the well-being of either the individual or society. ... He often feels stimulated by his wishes for the general interest in common activity.[58]

For Brückner, at the origins of "insurance" as a social institution lies less a sophisticated concept of social contract than a highly elaborate and emotionally charged concept of the human community, strongly influenced by the Scottish Enlightenment.[59] The words "scene" and "compassion" gesture toward the gradual transfer of theoretical elements and the structure of appeal from

58 Brückner, *Abhandlung*, p. 10.

59 Annette Meyer, *Von der Wahrheit zur Wahrscheinlichkeit. Die Wissenschaft vom Menschen in der schottischen und deutschen Aufklärung*, Tübingen 2008.

drama theory onto the human-nature relationship, in particular the *eleos* and *phobos* discussion associated with Lessing's reading of Aristotle: the "compassion" (*Mitleid*) that can be felt here is related to a natural "catastrophe." Cameralistic insurance theoreticians were obviously also participants in the Enlightenment aesthetics discourse of the eighteenth century.[60] Brückner formulated a kind of conception, if not of globalization, then at least of the general circulation of inventions and principles devoted to the common good: "A major part of the world like Europe simultaneously engages itself with the same perspectives, same occupations, and same interests. From its extreme ends, it shares knowledge and enlightenment." A 1792 treatise emphasizes that the tradition of charitable donations emerging from "feelings of philanthropy" and practiced by subjects of the Electorate of Trier was transferred onto the establishment of fire insurance. Each individual was encouraged to remember "that he is a member of a community of charity, that his small contributions are part of a kind, sympathetic support system devoted to easing of the misfortunes of his common man, his neighbor."[61] In the same year, Philip Gäng of Salzburg questioned what "such institutions" were based on, in answer stated that they were based "on the sense of need." "Whoever [was] ... once an eyewitness to a horrific fire, or who [had] ... only once read the dreadful reports and accounts of such a mournful scene" has hardly escaped

> without being gripped with powerful compassion toward those victimized, without feeling deep empathy about the unnamed misery that a single accident of this kind often spreads over whole families, entire towns and cities, without at the moment of this intense compassion feeling the most ardent wish in himself to share in the hardship of the unfortunate, or at least to know that it has been somewhat relieved.[62]

Even if the "perfect equality" of rich and poor cannot be attained in a given state, insurance is one institution that, in relation to the value of each property, in accordance with its requirements, treats everyone the same. Since "in this way an entire country can be [united] in a society,"[63] insurance techniques

60 Cf. above p. 169.

61 J.L.Schaeffer, *Nachricht über die Vortheile des Kurtrierischen Brand-Versicherungs-Institut*, s.l. [Koblenz] 1791, f. *4v.

62 Philipp Gäng, *Von Versicherungsanstalten wider Feuerschäden und ihrem Nutzen im Allgemeinen* [...], Salzburg 1792, p. 5f.

63 Pfeiffer, *Gedanken über Versicherungs*-Anstalten, vol. 2, p. 20.

promised equality.[64] Everyone, he argued, should be a member of this institution: philanthropic feelings of compassion were aligned with pre-revolutionary social contract notions and with ideas of equality.[65] While the English insurance business could hardly appeal to similar ideas about the common good, it nevertheless could and did appeal to notions concerning individual advantage. However, institutions that were sanctioned, supported, introduced and sometimes forced upon communities by the state, such as German insurance institutions, were themselves represented as embodiments of beneficence and philanthropy: philanthropic "language" was indeed partly imported from the social contract discourse of western Europe, where it was typically applied to other fields. In German-speaking countries, however, in addition to the basic mindset, emotional and moral language constituted crucial elements of "insurance." This could lead to the point that the amalgamation of the raw fragments of reality into what should become the "normal secure society" could only be expressed in the literary form of the political dream: the anonymous author of a 1792 miscell in the journal *Der Anzeiger* "read in the shadow of a hundred-year-old linden" about the history of the Hanseatic, the Rhenish and Swabian Leagues, which he admired as "monuments of the spirit and power of our fathers." In the dream, the author sees emperor "Francis the Second sitting on a throne." A deputy of the imperial cities presents Francis with a petition proposing an everlasting covenant between the imperial cities, not as before for the purpose of military defense, but for the mutual "promotion of the well-being of the citizens and community support and assistance in case of need." The petition proposed to establish:

1. A joint fire-flood-hail-cattle insurance institution
2. A common credit system, mutually guaranteed by government stock
3. Community citizen rights in all imperial cities
4. Non-profit societies for the promotion of agriculture, arts, and trades [modeled after Hamburg, Bremen and Nuremberg]

This partial imperial reform project[66] in the form of a dream included an empire-wide development on the level of the imperial city and was written just prior to the meeting of the Mainz archchancellor at the conference of princes

64 "Es ist ganz richtig, daß man eine völlige Gleichheit unmöglich heraus bringen kann." However, insurance effects equality more strongly than other institutions of the Enlightened state: ([anonymous], "Vorschlag, wie die Versicherung der Gebäude," p. 710).

65 In this regard, Pfeiffer is also a remarkable author, who already in 1780 separated society (going against a categorization according to "estates") into a "class" of producers, a "class of refiners" of raw materials, and a "class" of transporters/traders (p. 9f.) This occurs already earlier in English economic writings (James Whiston for instance prioritizes the function of labor and offices over belonging to rank and estate as early as 1693), but it is quite rare for the German context..

66 On the topic of imperial reform projects in the eighteenth century, see Wolfgang Burgdorf, *Reichskonstitution und Nation. Verfassungsreformprojekte für das Heilige Römische Reich Deutscher Nation im politischen Schrifttum von 1648 bis 1806*, Mainz 1998.

(Fürstenkonvent), which intended to organize a monarchical military response to the French Revolution, shortly before the revolutionary armies crossed the Rhine in September 1792. The first topic of "imperial matters" was, probably wisely, the dissemination of insurance across the empire, and was, according to the dream, welcomed by Francis II with his "imperial word in favor [of it]. About that my heart yearns so strongly, that I [i.e. the narrator of that piece] with a loud *Vivat Francis the Second!* awakened and became aware—that I had been dreaming."[67] The curiosity of a utopia that imagines the imperial federation of cities as an insurance company should not encourage the dismissal of the emotional appeals of the insurance discourse as basically utopian. Rather, it demonstrates the extent to which Enlightenment thinkers saw insurance as a highly important medium of reform, and the even utopian excitement associated with the far-reaching "dreams" of this very practical, viable, and easily deployable management tool.

Overall, the examplary analysis of some of the cameralistic insurance treatises demonstrates how strongly "insurance" was abstracted as a principle within the German Enlightenment, and how, in the context of a social and state model founded in natural law, it was installed as a means of security production particularly oriented toward "accidents" caused by nature. In this sense, something survives on an institutional and social level from a process that began in 1400, when insurance was used as a mechanism for decoupling the natural world from the world of values (cf. Ch. 2, 2). Its integration within a social theory at the intersection of cameralism, social contract doctrine, and philanthropy leads to the fact that insurance companies also employed considerable emotional and appellative language, framing insurance as the "moral" duty of citizens and subjects. Thus, as illustrated above (Ch. 5, 1.1), insurance institutions and corporations in Germany were even more strongly tied to elements of "community" at the end of the eighteenth century than they were in the Anglo-American context.

In the sociological debate regarding the relationship between "morality" and "risk/insurance" there are some authors who, like Anthony Giddens, make the argument that with the emergence of insurance as a design tool of the social state, "morality" is banished from human relationships: "Moral principles run counter to the concept of risk and to the mobilizing of dynamics of control. Morality is extrinsic so far as the colonizing of the future is concerned."[68] It may

67 [Anonymous], "Patriotischer Traum eines deutschen Reichsstädters," *Der Anzeiger* 11/12 (July 16/17, 1792), pp. 81–83.

68 Giddens, *Modernity and Self-Identity*, p. 145. Similarly, David Garland, "'Governmentality' and the Problem of Crime: Foucault, Criminology, Sociology," in *Theoretical Criminology* 1 (1997), pp. 173–214, 182; for a current, critical perspective, see Mary Douglas, "Risk as a

well be true that in the individual examination of an incurred loss, the insurance principle makes the concept of forensic guiltiness meaningless—at least in terms of the recompensation of damages—but thinkers like Giddens disregard the complexity of the overall development through concentrating on a single aspect of the equation. The notion of insurance does not entirely replace morality; rather, it is shifted to the prior commitment to establish insurance, to the participation in insurance, and to the features embedded within the logic of insurance. This is nowhere more evident than in historical discussions presented here concerning the introduction of insurance into the social fabric.

2 State and Society in the Code of Numbers

Previous chapters have discussed the diffusion of a "quantifying spirit," associated particularly with the widespread diffusion of insurance across the territories. This quantifying spirit was undoubtedly closely related to Pfeiffer's reception of Rousseau's mathematical/speculative conception of the social contract, along with the associated representations of power, the state, and society in numerical form. Alfred Crosby has argued that in occidental society this quantifying tendency begins in the thirteenth century, intensifies until the sixteenth century, and then becomes an essential aspect of European global imperialism. Somewhat less pointedly, the gradual transition from "statistics" as a more or less narrative description of the state toward "statistics"—or political arithmetic—as a mathematical discipline assisting a variety of governmental institutions has already been frequently mentioned. The eighteenth century in particular was elevated by a decisive boost of this "quantifying spirit;"[69] at this point, however, this general process primarily discussed in the history of ideas becomes less important, while concrete questions concerning how a "quantification" and a recoding of things into numbers emerges from the introduction

---

Forensic Resource," *Daedalus* 119 (1990), pp. 1–16, 10 and Ericson, Doyle, Barry, *Insurance as Governance*, pp. 66–98.

69 Alfred W. Crosby, *The Measure of Reality: Quantification and Western Society, 1250–1600*, Cambridge 1997; Mohammed Rassem, *Statistik und Staatsbeschreibung in der Neuzeit. vornehmlich im 16.-18. Jahrhundert*, Paderborn 1980; Mohammed Rassem, Justin Stagl, and Wolfgang Rose (eds.), *Geschichte der Staatsbeschreibung: ausgewählte Quellentexte 1456–1813*, Berlin 1994; Philippe Desan, *L'imaginaire économique de la Renaissance*, Fasano 2002; Tore Frängsmyr et al. (eds.), *The Quantifying Spirit in the Eighteenth Century*, Berkeley 1990; John L. Heilbron, *Weighing Imponderables and Other Quantitative Science around 1800*, Berkeley 1993; Theodore M. Porter, *Trust in Numbers: The Pursuit of Objectivity in Science and Public Life*, Princeton 1995.

of insurance become more prominent. In particular, when the fire funds are looked at from this perspective, their strongly innovative character is revealed and their capacities to reshape administrative activities are highlighted—qualities that are occluded from the eyes of economic historians who focus on the British "model" of the nineteenth century alone. Through being founded within the territorial state, fire funds had at least one advantage over the early English fire insurance agencies—including joint-stock companies such as the *Sun*—namely, the impetus of a (tendential, at any rate) conception based on the totality of government/society from the outset. Private fire insurance agencies were not interested in the "totality" of the state and society—rather, they simply worked successively to expand their markets.

The inaccuracy of risk and premium estimates in England when premium-based fire insurance was being "invented" can be reconstructed as follows. Initially, value appraisals of houses and land according to specific criteria (location, proximity to water, incorporation of the existing prescriptions, in addition to estimates of reconstruction costs and expenses and how they should be distributed among landlords and tenants) was a very concrete matter. Particularly after the *Great Fire* of 1666, so-called "pattern books," were created on these topics in London, which probably constituted the earliest relevant practical instruction manuals and included corresponding tables and numerical examples for landlords and tenants along with instructions guiding the modalities of reconstruction.[70] Stephen Primatt, for example, elucidates the calculation of property values after the fire of 1666 and highlights the not entirely surprising fact that the location of the land is decisive concerning its overall value and the ratio of the relationship between basic ground rent and the rental or leasing rates for the building.[71] Such texts thus articulate in written form the practical knowledge concerning the market-determined value of houses and land—also and especially concerning dynamic situations such as the rapid speculative reconstruction following the fire of 1666 in one of the largest cities in Europe, which cultivated the strongest early capitalist environment in Europe other than Amsterdam. The early fire insurance projects in

70 S.P. [Stephen Primatt], *The City and Countrey Purchaser and Builder* […], London 1668, pp. 1–50 for the discussion of how to calculate the values of building stock, p. 51–167 for practical advise to the builder for drawing the groundplot and what building materials to use etc. Cf. similarly, William Leybourn, *A Platform for Purchasers, Guide for Builders, Mate for Measurers*, London 1668.

71 A house that is worth 100 pounds in annual rental revenue and that could be sold for 1500 pounds would cost about 500 pounds to rebuild: in one area Cheapside in London, one could get a 20-year ground rent for 50 pounds per year—thus 1000 pounds in total (ibid., 29. On this textual genre, Cf. Baer, "The Institution of Residential Investment".

London built on this practical knowledge, for example, when they determined the average value of a house at 250 pounds and thus calculated the amount of 50,000 pounds of premium income as the basic capital to establish insurance for 4000 homes at 5% premium payments.[72] At this point, these were highly imprecise estimates, perhaps justifying the self-conception of English insurance founders as "adventurers."

In contrast to the situation on the other side of the channel, in England, no further data exist, for example, concerning average fire and damage totals—with the exception of the data collected by Barbon. Furthermore, the value of the buildings was determined in each individual case, but the value of the entire building stock was only calculated as a rough estimate only. There is no evidence that any more complex calculations were attempted regarding fire insurance until the end of the eighteenth century. As a point of contrast, it will thus be useful to now review the implementation of fire insurance in Prussia, based on the evidence generated by the early Prussian fire firms around 1700, who attempted the earliest widespread territorial application of a governmental fire insurance program.

The "Great Elector" himself—inspired, perhaps, by Leibniz—initially undertook the project to transfer the Hamburg General Fire Fund model to Berlin in 1685. Following Wilhelm Schaefer's studies of 1911, there has been no additional archival research on the topic of the early history of fire insurance in this region.[73] In texts summarizing the topic, the "failure" of the first territorial fire fund, the Prussia-wide General Fire Fund of 1706, is always invoked first, followed by the establishment of the Berlin Fire Society of 1718 and then the further formation of city fire societies in Prussia. Not much is known about how the earliest forms of these institutions really worked.[74] However, the discussion between Elector Frederick III, who took up his father's project in the 1690s, and various other groups—in particular those in Berlin, the city of royal residence—adds insight. On the suggestion of the electoral prince, thoroughly

72 Augustine Newbold, *Londons Jmprovement and the Builder's Security asserted, by the Apparent Advantages that will Attend their Easie Charge, in Raising such a Joint-Stock, as may Assure a Re-building of those Houses, which shall hereafter be Destroyed by the Casualties of Fire* (presented January 1, 1679; London 1680), p. 2.

73 Schaefer, *Urkundliche Beiträge*, vol. 2, pp. 137–40, 169–80; Oskar Simon, "Die Entwicklung des Immobiliar-Feuerversicherungswesens in Preußen bis zum Jahre 1866," *Annalen des Deutschen Reiches* 21 (1888), pp. 62–95, 69; Peter Koch, "Berlins Bedeutung für die Versicherungswirtschaft," *Versicherungswirtschaft* 46 (1991), pp. 236–50, 238.

74 Schaefer, *Urkundliche Beiträge*, vol. 2, pp. 131–238. Reinhold A. Dorwart, "The Earliest Fire Insurance Company in Berlin and Brandenburg, 1705–1711," *The Business History Review* 32, no. 2 (1958), pp. 192–203 paraphrases the edicts.

practical operational guidelines were created for the city of Berlin. The first of nine points outlining such measures led to the proposal for appraisals to be made by masons and carpenters. They prescribed that the value of the buildings (apart from the inflammable land) should be "appraised, so that nobody could register or declare his house in the fund accounts for more than it was worth." Everyone had the right to "declare half or three-quarters of the value, assuming the rest of the risk himself," but nobody would be allowed to insure less than 50 percent of the value. Each homeowner was then to appear before the authorities at a specific location to declare the registered value of the home, "so that an orderly process going from the residents of one street to another" could ensue. Everyone would receive an insurance document "with a special stamp" and a number which would correspond to in an entry in the "fund account book from the appropriate accountant." According to the initial concept, every insured member would pay an annual fixed amount in proportion to the insured value—thus fully in accordance with the concept of a premium-based insurance. A system of proportional allocation would be applied only if damages exceeded the assets of the fund. With every fire, the masons and carpenters would again operate as specialists in damage appraisal, and the compensation would occur within six weeks of the event, with the amount paid corresponding to the proportion of the insured value of the building.[75]

Initially, an anonymous, apparently mercantilist thinker from Berlin wrote a manuscript advisory report in favor of the project and agreed with the electoral board of commerce: he introduced the idea "that there would never be trade or commerce in a city, in which the residents lacked credit." Flourishing trade cities featuring ample credit should be taken as models, and fire insurance was seen to play a role in credit circulation: the anonymous writer concluded that "in the city of Hamburg, among other good establishments and ordinances, the Fire Fund Ordinance is almost the most important piece, and has placed this city in its flourishing state." Several points were offered to explain how fire insurance had increased credit and monetary circulation in Hamburg: 1) the investment necessary for building a house would not necessarily proportionally reduce mercantile capital; 2) those affected by fire do not automatically lose

75 "[W]an zum exempel ein haus 1000 Rth werth, er aber dasselbe nur zur helffte versichern lassen und der schade würde nachgehen auf 700 Rthaler taxiret, so kan er nicht mehr alß 350 Rth auß der cassen praetendirn daß übrige aber mußß er selbst zu durchfirung solches baues hergeben." (*Unvorgreiffliche erinnerung welcher gestalt diese feuer Cassen ordnung eingerichtet und observiret werden müste* [ca. 1690 Berlin], LA Berlin A Rep. 005-05 Nr. 8, old shelfmark Stadtarchiv 27423, no pag.).

their entire creditworthiness as before; 3) he reasoned, in particular, about the unfavorable capital outflow arising when those with capital deposited their money in Hamburg or Danzig because they knew it is safer there, as it was supported by mortgages. In the 1690s, one could deposit money in Hamburg for a relatively low interest rate of 2–3% plus an additional tax to the municipal finance chamber of "¼ or perhaps ½ pr. Cent," with the result that money flowed out of Prussia and secured liquidity for Hamburg's merchants. With the introduction of a fire fund, "wealthy people [could] draw better benefits or interest from their capital in local places where 6 or at least 5 percent would be given. … Wealthy foreigners would also be encouraged to leave their money there, where their capital would be safely accommodated and their pensions would thus survive." Even private investment deposits for the maintenance of widows and orphans would be easier to procure from Berlin if securing credit were supported by insurance-like institutions. The anonymous expert also believed that insurance would help to ensure that fire-damaged houses would increasingly be rebuilt in stone because damage compensation would be immediate.[76]

However, all these delicate economic-political arguments came to nothing in terms of the largely conservative Berlin guilds (*Gewerke*). They argued that there was "a major difference between these [royal] residence cities and the city of Hamburg," specifically concerning the constitution of the cities: they credited Hamburg with greater agility and ability to reform in comparison to Berlin, since there was only one authority in Hamburg: the city council, which necessarily answered to both large and small groups of citizens. In Berlin, in contrast, "the council does not have so much to say, and its citizens are subject to many different jurisdictions, with the result that much good that eventually could be established, has to be put aside [i.e., for future concerns]." Geopolitically and economically "there is a residence [city] here and no Hanseatic city." If, in so-called residence cities, fire funds were "deemed practical," then "Dresden" would have long established one. It was not known in Berlin that at almost the same time in precisely that city of Dresden, the introduction of both a large, and later a small fire fund (1700/1705) were under consideration. In terms of structural economics, Hamburg was also a city with citizens "of large capital," while Berlin consisted only of "shopkeepers and poor craftsmen," and "50 of them would still not compare to one wealthy Hanseatic merchant." The very wealthy citizens in particular were not "subject to civil jurisdiction." The possibility even existed that in case of a destructive fire, residents of more

76 Gutachten, LA Berlin A Rep. 005-05 Nr. 8, old shelfmark Stadtarchiv 27423, no pag.

moderate means would actually have to subsidize the wealthy for an expensive reconstruction—as exempted individuals, some wealthy individuals did not contribute, but would have nevertheless have hoped to profit from a fire funds if the need arose. Supposedly countless "corrupt" people had profited through the contributions of others by negligently allowing their houses to burn during the 1684 Hamburg fire. Many residents of Hamburg were even of the opinion that "if they had not been subject to the fire fund, the last fire would have been sooner extinguished and had not spread so widely, since many dishonest people may have thought: 'let the old house burn, since a new one is ensured by such an ordinance.'" The poor residents of Berlin should not be burdened so harshly; they should be spared "until the assets of their city [could be] compared with those of Hamburg."

The authorities of the council and the four *Gewerke* (the guilds or industry corporations of Berlin) thus decided on July 15, 1696 that they would maintain the old discarded position and communicated to the "high ministry" that they were "not in agreement concerning that issue."[77] As the future would show, the whole argument was quite unfounded and based only on initial appearances. However, it can be taken as yet another (see 5.1.1) sign of resistance from old cooperative social structures against the transformation of society into a new risk collective that also tended to cross hierarchical boundaries relating to the estates (i.e., the first, second, and third estates). The mercantilistic appraisals of municipalities and city residents served only to determine the conditions influencing cash flows and viewed the fire fund as a credit- and value-stabilizing mechanism, in which the contributors and recipients were recognized as numerical quantities, subject to the same dangers and contributions. In contrast to this view, the council and guilds of Berlin argued with concrete entities—the position of the cities, the conditions of the jurisdictions, differences according to estate hierarchies, etc. It was a question of numbers versus things.

This was the situation in the city of Berlin, which did not establish its own fire society until 1718 and resisted incorporation into larger territorial projects.[78] Thus, strangely enough, tiny Kurmark and Neumark hamlets like Drossen (today Osno, Poland), Wriezen, Strasburg, and many others served as the compliant municipalities of central bureaucratic control and as hubs

77 Ibid., underwritten by Martin Reichardt, Andreas von Echern, Georgen Heinrich, Jacob Schmidt, Hannß Caspar Vollungen, Hans Hamilton, Johann Vegelten.

78 Cf. now Simone Ladwig-Winters, *300 Jahre Feuersozietät Berlin-Brandenburg*, Berlin 2018, pp. 20–35.

No	Names of citizens	Building appraisal	Amount registered	Contribution to Fire Fund	Salary contribution	Invoice tax
1	Christoph Maachs, shoemaker's widow, an old home	50 Rtl.	50	0 Rtl., 1 *Groschen*, 6 *Pfennige*.	3 pf.	3 pf.

FIGURE 65 Sample of an entry for the citizens' fire fund of Drossen (BLHA Potsdam Rep. 8 Drossen 646)

for major steps of development. These municipalities implemented what was conceived of and initially rejected in Berlin—namely, the establishment of a comprehensive regional fire fund inclusive of both cities and rural areas. On October 7, 1706, it was recorded that the "specification of residents in and around Drossen, their houses and other buildings will be assessed as such and registered in the fire fund."[79] 369 houses were appraised in the town and recorded as here above (Fig. 65).

In the list, the total amount to be paid by Drossen to the Neumark Chambers of War and Domains was then calculated; the fire fund register proceeded similarly in the following years, resulting in what was probably the first estimate of the total value of the building stock for the town (ca. 58,000 Reichstaler, insured in the amount of 36,000 Reichstaler).[80] Given that after 1700 insurance coverage in England was no longer widespread, and was especially patchy in the slowly developing provinces outside London, it is no insignificant feat that a solid 60% rate of insurance coverage was reached in this region, especially considering how many tiny hamlets and villages were included in the Kurmark and Neumark regions. In Wriezen, approximately 350 houses were recorded in Wriezen (with a total value of ca. 63,090 Reichstaler and 43,000 Reichstaler (Rtl.) insured, thus also representing a comparable value of 65%).[81] In Strasburg/Uckermark, all 207 houses were recorded in 1706 and were assessed at 28,610 Rtl., of which 14,800 was insured—thus

79 BLHA Potsdam Rep. 8 Drossen 646.

80 Cf. BLHA Potsdam Rep. 8 Drossen 646, 647, 648 (1706–08).

81 "Designatio der Häuser somit anndern Gebäuden mit dem Taxen und den Assurance und Casse-Geldern, wie solche zur Königl. Preußischen General-Land- und Stadt-Feuer-Casse

making an insurance rate of about 50%. This proportion, however, grew quickly: in 1707, 210 houses were similarly assessed (28,640 Rtl.); at this point 21,750 Rtl. had been insured, or about 76%. This value remains fairly constant, with slight increases, until the end of the first project (1711); for Strasburg, the registers from 1706 to 1711 also reflect a similar trend.[82] According to the rules of the administrative procedures, the annual payments—the actual insurance contribution, the proportional contributions for the salaries of officials and an invoice tax —were paid for the entire city. The form created to document this end result calculation was filled in at the short-lived "Collegium of General Fire Fund" in Berlin.[83] The Neumark also submitted an auxiliary register documenting the active contribution of at least forty-four cities and administrative regions.[84] According to their contributions between 1706 and 1711, the largest municipalities were Züllichow (1,605 Rtl.), Cottbus (1,315 Rtl.), Crossen (887 Rtl.), Küstrin (694 Rtl.), Drossen (534 Rtl.), and the city of Sommerfeld (496 Rtl.). Among the 23 "immediate" cities were the thirteen offices of the Neumark and some special municipalities and co-operatives, such as the French colony in Cottbus. The total assessment of 12,101 Rtl. was almost completely paid. In the absence of other summary values, extrapolating from the existing Wriezen, Drossen, and Strasburg data, total values likely amounted to about 8,000–9,000 houses, with an appraised value of approximately 1,300,000 Rtl., and a likely insurance rate of 60–70% of that. Considering the effort that would have been invested in operating this fund, updating the assessments annually in each city, and collecting the fees, the efforts required in a province like the Neumark were naturally much higher than those required in Hamburg, which with its approximately 4,200 houses around 1700 contained about half the number of houses in all the cities and administrative regions included in the Neumark. The Prussian General Fire Fund of 1706–11 quite explicitly functioned as an anticipation of premium-based insurance.

---

ein zu bringen, verfertiget bey der Stadt Wrietzen an der Oder in 1706" (BLHA Potsdam Rep. 8 Wriezen 630; the values were still almost identical in 1707: BLHA Potsdam Rep. 8, Wriezen 631).

82 BLHA Potsdam Rep. 8 Strasburg 836, 840: The number of houses rose from 210 to 216, the tax value rose to 29230 Rtl., the insured value rose to 22,350 Rtl.

83 Receipt of the General-Feuer-Cassa-Kollegium for the payment of 70 Taler taxes by the city of Wriezen for the year 1706, BLHA Potsdam Rep. 8 Wriezen 630.

84 BLHA Potsdam, Rep. 3, Nr. 11855, table after p. 54: approximately 17,702 Rtl. were contributed. For the more important Kurmark there is no comparable document for the general fire society 1706–1711.

The king and central bureaucracy insisted for some time on their resistance against "rebelliousness and opposition" by the estates and guilds. In 1708 they attempted to better ground the institution in empirical reality, with Friedrich II issuing the following orders to cities and administrative regions across the land:

> to establish an extract and report, based on inquiries, from which one can determine how much common church collections or other tax reductions have been granted and paid out during the last 15 years to the victims of fire; likewise, what kind of fire extinction equipment, fire hoses and buckets, and other fire instruments existing at your location? How much is usually spent yearly to buy and to repair [the equipment]? How much money do you have to withdraw and to deposit for that each year? What provisions and ordinances are undertaken and made against the threat of fire hazard? How can it be improved?[85]

While the supervision and inspection of fire extinguishing equipment was consistently carried out in this way, the retrospective survey of all fire damages collected since 1693 served the purpose of obtaining a better statistical overview of the average occurrence of fire damages at a time when there simply were no data about this. Comprehensive answers seem not to have been received, or at least there is no remaining archival evidence of them. An increasing knowledge base of the average fire damages in a region emerged only slowly with the establishment of the fire funds themselves and with their increasingly widespread proliferation.

In 1711 this General Fire Fund ceased operations due to protests by the estates and the city guilds. Instead, as already mentioned, the Berlin Fire Society was established in 1718,[86] and one year later the fire societies of the Kurmark and Neumark cities was established,[87] which from 1720 until the early nineteenth century assessed the buildings in each city every five years ("Quinquennien"), insured them for a specific amount, and handled damages according to a system of repartition, in which damages were proportionally distributed. The *Domänenämter* (domain administrations), and thus the rural areas, were initially excluded from this system until Friedrich II pushed for the formation of separate fire societies in the rural parts of the Prussian provinces (the so-called

85 Erlass Friedrich I., Cölln, 21.3.1708, Kopie in BLHA Potsdam Rep. 8 Strasburg 829.

86 Feuersozietät (Berlin, West) (ed.), *260 Jahre Feuersozietät Berlin*, Berlin 1978; Karl Ernst Rimbach, *Feuersozietät Berlin 1718–1968*, Berlin 1968; Ladewig-Winters, *300 Jahre*.

87 Cf. Instruction of the GD to Steuerrat Lütkens, Berlin, 4.12.1719 (Copy sent to city of Strasburg), BLHA Potsdam, Rep. 8 Strasburg 829, no pag.

plattes Land, beginning with the Kurmark, the duchy of Silesia, and the county of Glatz).

The fire societies of the Kurmark and Neumark cities were based more or less seamlessly on the accumulated experience the cities had with the earlier General Fire Fund, which, for the time, boasted an impressive form and systematicity. They operated according to the principle of subsequent allocation of funds, rather than as a form anticipating premium-based insurance: 118 small Kurmark and Neumark towns were included, regardless of whether they were "mediate" or "immediate." In each city, the magistrates set up a cadaster, in which the homeowners declared one part of value of their homes, with the other part determined by new appraisals made by the carpenters and masons.[88] The total insurance value of the city was entered into a corresponding registry of district cities by the *Steuerrat* (i.e., administrator of city affairs in a district or region) of one of the ten Kurmark and Neumark districts.[89] The ten *Steuerräte* then sent these registries to the Chambers of War and Domains, and finally to the General Directorate, where each district was entered with the insurance values for each of its cities into a general cadaster.[90] For all three levels—city,[91] *Steuerrat*,[92] Chambers of War and Domains (CWD)[93]—so-called repartition tables (i.e., proportional calculation/conversion tables) were established at the beginning of each quinquennium. The General Directorate submitted the general cadaster, including tables and instructions for the *Steuerräte*, to the king for approval, after which copies were sent to both Chambers of War and Domains.[94] When fire damages were suffered in a city, the degree of damage was reported to the CWD by the *Steuerrat*. The tables then determined how much each district contributed; each *Steuerrat* determined the amount that each city contributed, and every city magistrate determined the amount contributed by each individual homeowner.[95] This meant that

88 Taxing (Estimation) records for the city of Strasburg for the six *Quinquennia* 1720–1749, BLHA Potsdam, Rep. 8 Strasburg Nr. 836.

89 Example: Table in BLHA Potsdam Rep. 8 Strasburg 829 no pag.

90 Example: "Haupt-Catastrum 1725–1729," GStAPK Berlin, HA II, Abt. 14 Kurmark, Materien, tit. CCXVII, sect. a, Nr. 3, Bd. 1, f. 15r-23v.

91 Example: "Feuer-Societäts-Rechnungen 1731 bis 1739," BLHA Potsdam Rep. 8 Strasburg 840.

92 Example: see note 89.

93 Example: GStAPK Berlin, HA II, Abt. 14 Kurmark, Materien, tit. CCXVII, sect. a, Nr. 3, Bd. 1, f. 24r-v (2nd Quinquennium); something similar, also in the city magistrate files: BLHA Potsdam Rep. 8 Strasburg 829, no pag.

94 Example: "GD an den König, 8.9.1725," GStAPK Berlin, HA II, Abt. 14 Kurmark, Materien, tit. CCXVII, sect. a, Nr. 3, Bd. 1, f. 5v-12v.

95 Example: July-Sept. 1746 in BLHA Potsdam Rep. 8 Strasburg 830, no pag.

every year the amounts that individual cities contributed toward fire damages in the other cities were recorded. In November 1721, it was noted that residents of Drossen paid 3 Reichstaler (Rtl.), 22 *Groschen* (gr.), and 5 *Pfennige* (pf.) for damages in Salzwedel; 17 gr., 8 pf. For Beelitz's damages; 3 Rtl., 18 gr., 8 pf. For Beeskow; 4 Rtl., 6 gr., 6 pf. for Straußberg; 4 Rtl., 5 gr., 6 pf. for Berlinchen; 8 gr., 9 pf. for Zehden; 2 Rtl., 5 gr. 10 pf. for Prenzlau; and 2 Rtl., 2 gr. 9 pf. for subsequent damages in Straußberg.[96] This had the advantage of providing concreteness and relevance—money could be collected for specific damages. The collections thus seemed structurally congruent with parish-based charitable collections, which at this time were banned or subject to strict licensing procedures.[97] Consequently, the new risk group became concretely tangible, not too far removed from the realm of experience of city residents, and all the while, the entire discussion was oriented toward value estimates, proportional value partitioning, and income streams. Beginning in 1747, the whole repartition process no longer took place with each individual fire; rather, the Chambers of War and Domains projected the amounts, collected the incidental damage amounts, and then extended them to the *Steuerräte* on a quarterly basis.

The fire society recorded an increase of insurance sums from 5.44 million Reichstaler to 7.14 million Reichstaler between 1720 and 1744. In the absence of any cadaster at the city level, one can extrapolate here only by means of existing sampling records in relation to the total amounts—leading to the rough estimate of about 40,000 homes insured within the 118 cities in 1744. The fire damages requiring compensation ranged from 24,568 Reichstaler (in the 2nd quinquennium) to 82,612 Reichstaler (in the 5th).[98] Over the years, damages were distributed accordingly (as in Fig. 66).

The fire fund functioned quite impressively once the royal administration yielded to the resistance of the estates and guilds and initially settled on the compromise of city associations. The process was very clear and specific regarding each situation, but also quite complicated. The completion of the general cadaster and the special computation tables for each level at the beginning of each quinquennium alone lasted six whole months—a special excise

96 BLHA Potsdam Rep. 8 Drossen 650.

97 *Patent Worinnen alles eigenmächtige Collectiren Bey den Bürgerschaften Und Communen Oder Dorffschaften verboten wird*, Berlin, March 4, 1728, BLHA Potsdam Rep. 8 Strasburg 829.

98 In Peter Borscheid and Annette Drees (eds.), *Versicherungsstatistik Deutschlands 1750–1985*, St. Katharinen 1988 the insurance statistics serve as no indication of Kurmark and Neumark society. Data according to GStAPK Berlin, HA II, Abt. 14 Kurmark, Materien, tit. CCXVII, sect. a, Nr. 3, Bd. 1, f. 14r-24r; ebd., Bd. 2, f. 21r-29r, 47r-55r, 90r-96r und BLHA Potsdam Rep. 8 Strasburg 829, no pag.

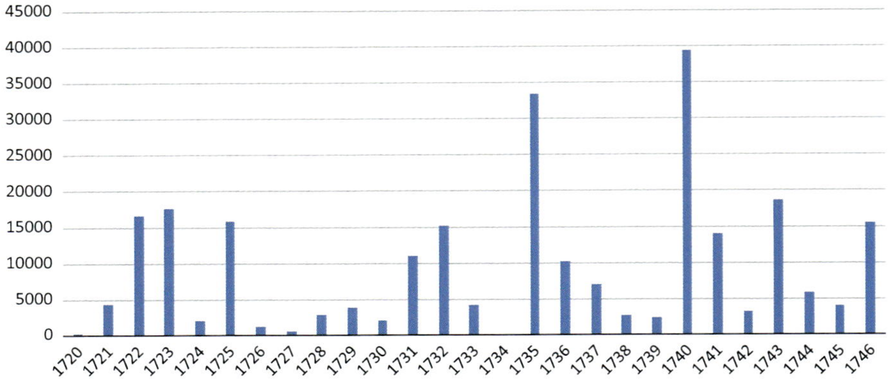

FIGURE 66 Fire damage in the Kurmark cities' Fire Society 1720–46 in *Reichstaler* per year (BLHA Potsdam Rep. 8 Strasburg 829)

estimator was brought in for this purpose under the supervision of the war council; his salary was paid every five years according time invested in the project. His salary increases were actually quite remarkable, demonstrating how much his work was valued (Fig. 67).[99]

In reality, this calculation of allocations on the provincial, district, and city levels constituted the only "mathematics" that were applied in these reciprocal forms of insurance and also accounted for the arts of risk assessment and premium calculation. However, the importance of premium calculation for the eighteenth and much of the nineteenth century should not be overestimated, nor should the importance of risk assessment for repartition among risk community members be underestimated: these two arts differed only slightly in form. Even in the realm of life insurance, the calculation of premiums was made according to preformed tables—assuming, of course, that true "calculation" and risk classification were actually at play, which occurred primarily only from the last third of the eighteenth century on.

Within this framework, the calculations of Johann Heinrich Waser of Zurich, a former pastor and official member of the Physical Society established in 1778, are extraordinary:[100] he based his argument on surveys of the Physical

99 Cf. "GD an den König zur Approbation des Generalkatasters des zweiten Quinquenniums, 8.9.1725," GStAPK Berlin, HA II, Abt. 14 Kurmark, Materien, tit. CCXVII, sect. a, Nr. 3, Bd. 1, f. 5v-12v. The salary of estimators rose from originally 20 to 50 Rtl. from 1725 to 1745, Cf. BLHA Potsdam, Rep. 3, Nr. 11855, tables after p. 54.

100 Johann Heinrich Waser, *Betrachtungen über die Zürcherischen Wohnhäuser* [...], Zurich 1778—The work somewhat loses itself in many tangents.

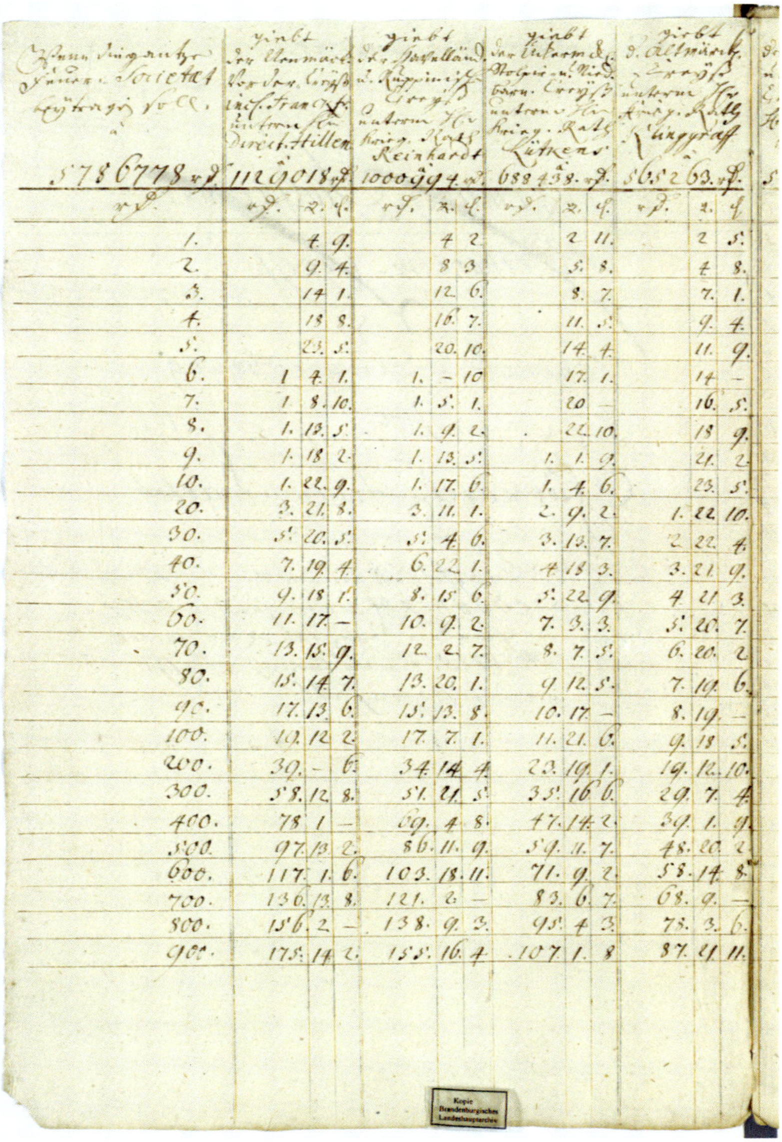

Wenn die gantze Feuer-Societaet beytragen solte	giebt der Neumärck. … Hillen			giebt d. Havelländ. u. Ruppinische Creyß … Reinhardt			giebt der Uckermärck. Stolp. u. Niederbarn. Creyß … Lütkens			giebt d. Altmärck. Creyße … Langgraf		
5786778 rt.	1129018 rt.			1000994 rt.			688438 rt.			565263 rt.		
rt.	rt.	g.	pf.	rt.	g.	pf.	rt.	g.	pf.	rt.	g.	pf.
1.		4	9.		4	2		2	11.		2	5.
2.		9.	4.		8	3		5.	8.		4	8.
3.		14	1.		12	6.		8.	7.		7.	1.
4.		18	8.		16.	7.		11.	5.		9.	4.
5.		23.	5.		20.	10.		14.	4.		11.	9.
6.	1	4.	1.	1.	–	10		17.	1.		14	–
7.	1	8.	10.	1.	5.	1.		20	–		16.	5.
8.	1.	13.	5.	1.	9.	2.		22.	10.		18	9.
9.	1.	18	2.	1.	13.	5.	1.	1.	9.		21.	2.
10.	1.	22.	9.	1.	17.	6.	1.	4.	6.		23.	5.
20.	3.	21.	8.	3.	11.	1.	2.	9.	2.	1.	22.	10.
30.	5.	20.	5.	5.	4.	6.	3.	13.	7.	2.	22.	4.
40.	7.	19.	4.	6.	22.	1.	4.	18.	3.	3.	21.	9.
50.	9.	18.	1.	8.	15.	6.	5.	22.	9.	4.	21.	3.
60.	11.	17.	–	10.	9.	2.	7.	3.	3.	5.	20.	7.
70.	13.	15.	9.	12.	2.	7.	8.	7.	5.	6.	20.	2.
80.	15.	14.	7.	13.	20.	1.	9.	12.	5.	7.	19.	6.
90.	17.	13.	6.	15.	13.	8.	10.	17.	–	8.	19.	–
100.	19.	12.	2.	17.	7.	1.	11.	21.	6.	9.	18.	5.
200.	39.	–	6.	34.	14.	4.	23.	19.	1.	19.	12.	10.
300.	58.	12.	8.	51.	21.	5.	35.	16.	6.	29.	7.	4.
400.	78	1	–	69.	4.	8.	47.	14.	2.	39.	1.	9.
500.	97.	13.	2.	86.	11.	9.	59.	11.	7.	48.	20.	2.
600.	117.	1.	6.	103.	18.	11.	71.	9.	2.	58.	14.	8.
700.	136.	13.	8.	121.	2.	–	83.	6.	7.	68.	9.	–
800.	156.	2.	–	138.	9.	3.	95.	4.	3.	78.	3.	6.
900.	175.	14.	2.	155.	16.	4.	107.	1.	8.	87.	21.	11.

FIGURE 67 First page of the repartition table of the Kurmark fire society for the second quinquennium 1725–1729: top left, the total sum insured (5,786,778 Rtl.) and the column: "If the whole fire society were to contribute" (1, 2, 3, … 900 Rtl.), then further right the columns with the pro rata estimate of the contributions of the Neumark anterior districts (*Vordere Kreise*), of Havelland, of the Ruppin district, of the Uckermark, Stolper, and Niederbarn district, and the Altmark district along with the names of their four war councils and *Steuerräte* (Copy from BLHA Potsdam Rep. 8 Strasburg 829)

Society, which stated that there were 1189 houses in Zurich. To account for an accurate determination of value, he draws from archival sources to compile a historical list documenting the purchase prices of 181 houses between 1221 and 1700. He then determines the average house price for each time period and proportionally converts actual historical purchase prices into contemporary values, taking inflation into account.[101] Thereafter, he divides all buildings in the city into nine value classes, arriving at a total value for the building stock of 6,162,350 guilders, and prescribes staggered potential contributions for each of these nine classes for a putative fire fund. These very complex calculations replace the work of masons and carpenters who were otherwise primarily employed as appraisers. Waser reinforces the common argument of insurance proponents that the interest rate of mortgages on insured homes would fall by 0.5–1%, and thus those insured would recapture the costs paid into the fire fund by twenty to forty times, bringing "several million more guilders than before ... into circulation."[102] Finally, Waser calculated that the interest-bearing premium income would rapidly grow into a substantial amount, which would then be available for investment. This result, however, was only valid under the assumption that a house fire would occur only once every ten years. This was an estimate based on past experience,[103] but as most premodern cities did not possess any kind of fire-resistant building stock, the corresponding calculation would have been much more risky. In London, ninety-seven houses with an estimated value 24,000 pounds had suffered fires in the fifteen years following the Great Fire.[104] According to the criteria of early insurance schemers, the probability of loss in London was thus more than three times higher than in Zurich. These calculations of home values in Zurich and the corresponding conditions for insurance seem almost picturesque in that they are based on replacing one house every ten years. The legitimacy of the corresponding contributions, which would lead to ever-increasing funds, could thus certainly be called into question.

---

101 He proceeds from the dubious starting point that the absolutely different epochal average prices of each period are completely equal in terms of purchasing power "der Preis jeder Sache [hänge] von der vorhandnen Menge des Gelds und von der concurrenz der Käufer ab [], folglich bey geringerer Bevölkerung und kleinerm Geldvorrath der Preis eines Wohnhauses in Zürich zu 432 ½ fl. verhältnißmäßig eben so hoch und dem Käufer eben so beschwerlich gewesen seyn muß, als jetzt bey geänderten Zeiten und Umständen der Preis von 5000 fl." (Waser, *Betrachtungen*, p. 8f.).

102 Waser, *Betrachtungen*, p. 12.

103 Ibid., p. 16.

104 Barbon, *An Enquiry*, p. 3.

The anonymous author of one the first important articles on the subject, written in 1751, offered a more realistic picture. In his impressive report on the principalities of Calenberg and Göttingen and the electorate of Hanover, he provided precise calculations: He scoured the archives for fire statistics—which were not easy to find *before* fire insurance was introduced in an area—and found that 700 buildings in smaller towns and 351 buildings in the villages of the region had been destroyed by fire during the twenty-one years between 1729 and 1749, amounting to a financial loss of 96,000 Reichstaler, or an average of 4,600 Rtl. per year. He estimated the insured sum total of all insured buildings at 10,000,000 Rtl. In order to cover the annual average loss of 4,600 Rtl., a homeowner insured for 3,000 Rtl. would have to pay 1 Rtl., 19 *Mariengroschen*, and 13/5 of a *Pfennige*. However, if, due to the greater security provided by insurance, the homeowner paid 0.5% less mortgage interest, he would save 5 Reichstaler for each 1000 Reichstaler of debt, which would, in turn, be available for investment and would therefore enter circulation. The savings captured by a reduction in interest, available on mortgages of insured homes, would thus be higher than the cost of insurance.[105]

Waser's treatment of the topic, however, demonstrates something significant for the time: instead of simply assuming a rough estimate of average home values, the process of establishing insurance offered the opportunity for complex historical-economic investigations concerning wages, prices, developments in housing values, comparative housing conditions, and many other topics that unfortunately cannot be explored in further detail here.

In the neighboring canton of Bern, Brückner had also cited the two main conditions for the necessary calculations to establish fire insurance: the determination of "the sum total of the actual value of all homes in the entire canton, in addition to the amount of any damages to be awarded." Regarding the "total sum," however, he had to admit that the "first of those requirements ... entirely [depends] on something, which a private individual is not in the position to prove," and that also concerning the estimation of damages, one could determine this "at the most, only through probability, through induction ... through news reports ... in the registers of state administration." On the basis of such issues, it was determined that fire insurance should necessarily be established by the state.[106]

As these examples demonstrate, the establishment of fire insurance in the premodern era, whether through private companies or state institutions,

105 [Anonymous], "Gedanken von der Einrichtung," p. 299.

106 Brückner, *Abhandlung*, p. 40.

led to calculations, estimates, at the very least cumulative information, concerning the value of large housing stocks, even regarding an entire city (Brückner's "total sum"). This phenomenon seems to be especially associated with forms of insurance—even in medieval tax registers or early cadasters, which as is well known had been recorded in Florence as early as the fifteenth century, households are listed individually and in great detail. When aggregations had arisen in the context of political theory and "statistics" in its older sense, they related primarily to individual households and their inhabitants. The value of the entire building stock of a city was not really subject to calculation—there was simply no need. This cumulative aggrege form of valuing in financial terms what urban residents of the premodern era unconsciously experienced materially as family homes lived in by generations, loaded with individual memories, shows a shift of perception. A process of doubling occurs: home values begin to become abstracted and detached from their half-timbered or brick substrates as a separate reality and another balance sheet emerges, in which values are again fixed and bound by means of insurance. The one abstraction, that of the aggregation and assessment of value, becomes linked to another abstraction, namely, the potentiality and dimensions of future loss.

As mentioned briefly above (Ch. 4, 1.1.3), it should be emphasized that these rather practical economic considerations concerning the impact of widespread insurance coverage on the liberation of investment capital is much more substantially discussed in the cameralist literature beginning in the mid-eighteenth century than it is in the pamphlets, business, and advertising texts of the three different models of insurance companies in London during the 1680s (joint-stock company, mutual company, public fund of the city of London).[107] There, the focus almost exclusively concerned the possible profit the company or the institution could realize in comparison to the security provided for the insured—this functioned in a far more differentiated manner than was initially the case around 1700 in German-speaking areas. The impact of the introduction of insurance on the capital market, however,

107 In addition to Newbold and the *Enquiry* (see Ch. 4, 1.1.1 above), see *An Advertisement from the Insurance-Office*; *A Letter to a Gentleman of the Insurance Office, Concerning the Cities Insuring Houses*, London 1681; R., *To my Honoured Friend Mr.* 1682; Nicholas Barbon, *A Letter to A Gentleman in the Country, Giving an Account of the Two Insurance-Offices; the Fire-Office & Friendly Society*, London 1684; *The Friendly Society, Or, A Proposal of a New way or Method for Securing Houses from any Considerable Loss by Fire, by way of Subscription, and Mutuall Contribution*, s.l. 1684.

remained rather unnoticed—it had nothing to do with the interests of the founders of the insurance companies and would have been only of interest to a social or economic theorist who had the reciprocal effects as a whole in view.[108] The German cameralists, on the other hand, offer an imaginative reflection about the putative macroeconomic effects related to the introduction of insurance: as early as 1680, Leibniz understood that with the potential of excess capital generated by the premiums, one could form a type of investment fund for "start-up" ideas:

> Just as insurers are highly necessary in cities of commerce, because they are able to offer work to the poor, advance the middle class, and give support to many beautiful ideas that would otherwise usually perish in their first bloom, thus here [i.e., through establishing an empire-wide natural hazard insurance] capital would be available simply for the improvement of the country, and by means of this new manufacturing and commerce, that which so many have [only] theorized before could be put into practice to the inexpressible benefit of the prince and his subjects.[109]

Along the same lines, Justi suggested on a number of occasions from 1754 onwards that inexpensive loans from the surplus premium capital could be distributed by connecting an insurance institution with a "lending bank," which could use the same insured houses as collateral for the loans. In this way, the idea of increased monetary circulation and the growth of credit—the "doubling or at least the increase of value"—could be accomplished in a circular

---

108 In economically oriented pamphlets that reflect the "Financial Revolution," there is hardly any mention of insurance. For example, *Angliae tutamen: or, the Safety of England. Being an Account of the Banks, Lotteries, Mines, Diving, Draining, Lifting, and other Engines, and many pernicious Projects now on foot; tending to the Destruction of Trade and Commerce, and the Impoverishing this Realm* [...], London 1695; *The Importance of Publick Credit* (London: 1699); Nicholas Barbon, *A Proposal for Raising the Publick Credit, By Setting up an Office for Transferring and Discounting Tallies* [1697]; *Publick Credit, Under consideration, how to be Retreiv'd* [1700]; J. Whiston, *The causes of our present calamities in reference to the Trade of the Nation fully discovered* London 1696; *A Letter to a member of Parliament, occasioned, by the growing Poverty of the Nation, from the want and Decay of Trade, and wrong Management thereof*, Edinburgh 1700. Also, in a practical dialogue, such as *Bank-Credit or the usefulness and Security of the Bank of Credit Examined; in a Dialogue between a Country Gentleman and a London Merchant*, London 1683 there is still no reference to insurance. In the famous chapter in insurance in Daniel Defoe, *An Essay on Projects*, London 1697 fire insurance is only briefly mentioned (p. 115f.), its effects on the economy as a whole have not yet been recognized.

109 Leibniz, "Öffentliche Assekuranzen," p. 432.

manner within a *single* institution. The constant question of how to create money for manufacturing and entrepreneurial endeavors found an answer in insurance.[110]

In Justi's political economy (*Staatswirthschaft*), these ideas were explicitly connected to the concept of credit: "The second chapter, which encourages the circulation of money, is the perfect credit of the country." Justi subdivided this into three types of credit: "1) the credit of the prince and his coffers ...; 2) the public credit of a country," which included a) credit with foreign nations, b) the credit of the assemblies of the estates (*Landstände*), or c) the credit "of a large general trading company, imagined as an entire nation, as it were"; 3) "special credit," which implied security and ease for private persons to borrow money for their own purposes.[111] Despite similar terminology, Justi's concept of "public credit" (*öffentliche Credit*) is not identical to the English idea of "public credit" as it existed around 1700. For Justi, public credit only related to the external economy—it did not concern the security of state bonds or securities. As a result, the most important thing was the "special credit" with which individuals of means "could loan out their money" with "convenience and security." Justi thus reflects on how this "convenience and security" of potential investors and creditors could be improved, and it was in this sense that institutions offering protection from accidents—especially fire insurance companies—were relevant, since they helped prevent the situation where "even the most honest debtors [are] against their will unable to pay their debts."[112] In this arrangement, fire insurance was not conceived in a simple cameralistic sense as a way

110 Johann Heinrich Gottlieb von Justi, "Vorschlag von Verbindung der Feuerassecuranzsocietäten mit einer Leihebanco auf die Häuser," in *Neue Wahrheiten zum Vortheil der Naturkunde und des gesellschaftlichen Lebens der Menschen*, vol. 1.1: *Vor die Monate Jenner und Hornung 1754*, Leipzig 1754, pp. 561–582. Cf. p. 254, note 25. One could, in any case, use half of the registered insurance sum as credit capital, in comparison to other lending banks and tontines, such a bank would have the advantage that personnel were already employed for the administration of fire insurance and therefore the interest rates could be set at 5% instead of 8% because of lower administrative costs. Through linking it with fire insurance, which could serve as funding capital, so that "mittelmäßige[] Lande, so wie die altfürstlichen Häuser in Deutschland gemeiniglich besitzen" could also become bank owners. Much of this recommendation is repeated in Johann Gottlob Justi, "Vorschlag, wie durch die Feuer-Assecuranz-Anstalten eine Leihebank errichtet werden könne, um Nahrung und Gewerbe zu befördern," *Göttingische Policey=Amts Nachrichten auf das Jahr 1755. oder vermischte Abhandlungen zum Vortheil des Nahrungsstandes* [...] 48 (December 15, 1755), pp. 189–191.

111 Ibid., § 260f., 276f.

112 Ibid., §§ 269–273, 284–289.

for the state to keep taxpayers solvent,[113] but instead as an institution that could offer a stabilizing cushion to the credit and capital markets. Justi thereby employed the perspective of the entire (state) economy to formulate that which, in most other texts, is expressed from the perspective of the insured, who are able to secure lower interest rates to borrow money by using an insured house as collateral.[114] This idea was still present in 1792 in Philipp Gäng's work, which promoted fire insurance in Salzburg, where the concept was late to arrive.[115]

This results in the paradoxical image of German cameralists pondering the practical economic uses of insurance, the effects of the establishment of insurance on the capital market, and the relationship between "credit" and "insurance" in a much more general form and manner than the English insurers. Paradoxically, this was because the cameralists always had the territorial framework in mind—i.e., a tightly limited capital market—while the English practice of the capital market was already much more advanced. Insurance companies were only one group of players on the market among other, often more potent, trade companies, such as the colonial merchants, from which the insurers took their structural model for establishing new firms.[116] English insurance focused mostly on the acquisition of profit and dividends, while the cameralists, although they dreamed of such profit for their organizations, in practice implemented either retroactive assessment, which compensated only the actual damages, or models for which the premium income was not large enough to build any significant capital.

---

113 More clearly, according to Leibniz: "Daraus folgt nun, daß in einer wohlbestelten Republick man dem jenigen so ohne seine schuld durch unglücksfälle, vim majorem und casus fortuitos in schaden geräth, nicht nur durch nachlaß einiger onerum wie ins gemein zu geschehen pflegt, sondern auch durch würckliche beysteuer zu hülffe kommen solle, damit er ein düchtiges glied der gemeine seyn und bey Nahrung bleiben möge," ("Öffentliche Assekuranzen," p. 425).

114 As above in Waser, *Betrachtungen*, note 102; a corresponding calculation can be found in [anonymous], "Gedanken von der Einrichtung," pp. 293–301, 299. Cf. Wilhelm Hagena, *Die Ansichten der deutschen Kameralisten des 18. Jahrhunderts über das Versicherungswesen*, PhD thesis University of Erlangen, 1910, pp. 23–26.

115 Philipp Gäng, *Von Versicherungsanstalten wider Feuerschäden und ihrem Nutzen im Allgemeinen* [...], Salzburg 1792, p. 20: the advantages of insurance supposedly consisted in "a) Vermehrung oder Erhöhung des Werths aller Gebäude, und folglich auch Erhöhung oder Vermehrung des gesammten Staatsvermögens; b) Beförderung des Credits, und Vervielfältigung der Gelegenheit zu sicheren Darlehen; c) Beförderung des Nahrungs- und Gewerbsstandes in mehrfacher Rücksicht."—similar statements follow regarding the high interest rates in the capital markets and the stance of the "careful capitalist" when using uninsured houses as security.

116 Newbold, *Londons Jmprovement*, p. 4: fire insurance is compared to the East India Company, Guinea Company, and other corporations.

It is true that mathematical theories of probability were not normally applied in the field of fire insurance until the mid-nineteenth century, even if authors repeatedly invoked the idea of "probability."[117] However, of greater importance is the precision with which the total value of all buildings was determined for the purpose of insuring them. Political arithmetic led to an appreciation of entire countries and cities based not on their characteristics, history, and traditions, but as a number of buildings with a certain value. The insured sum constructed the "capital" of fire insurance; sometimes the values were also based only on premium income as "capital." The whole feedback loop of insurance and rises in mortgage value and credit was also seen as a generation of "capital."[118]

3 Time Instead of Space: Sustainability and Insurance

The present study began with the correction that—in spite of some assumptions purported in historiography—premodern insurance practices in the maritime sector in the late Middle Ages were essentially spatially based, functioning with little temporal or future orientation (Ch. 2). Following the epochal threshold that occurred around 1700 and the transfer of the insurance principle to the areas of "fire" and "life," however, an impetus arises by which we might characterize insurance as a phenomenon to be integrated into the contemporary analytical concept of the "colonization of the future." It is no

117 Especially Brückner, *Abhandlung*, pp. 12, 14, 21s., 27, 34, 41 repeatedly uses the word "Wahrscheinlichkeitsrechnung" (probability calculations) without invoking any complex mathematical methods. Johann Georg Büsch, *Allgemeine Uebersicht des Assecuranz-Wesens*, Hamburg 1795, p. 9 makes the general remark that "Es muß eine Rechnung über die Wahrscheinlichkeit Statt haben, wie hoch der Verlust im Verhältniß der Kräfte dererjenigen ausfallen könne, aus deren Beiträgen der Ersaz geleistet warden soll"; Lorraine J. Daston, "The Domestication of Risk: Mathematical Probability and Insurance 1650–1830," in Lorenz Krüger, Lorraine J. Daston, Michael Heidelberger (eds.), *The Probabilistic Revolution*, vol. 1: *Ideas in History*, Cambridge/Mass. 1987, pp. 237–60, 240; Pearson, *Insuring the Industrial Revolution*, p. 4.

118 There were often arguments against this, that it would be better to pay the proportionally owed contribution to resolving fire damages after the fact, rather than to pay in advance, so that the money could still circulate and it would not become "todte [...] Capitalien" (dead capital) (Georg Heinrich Zincke, "Vorrede zum Siebenden Bande der Sammlungen. Von den nöthigen Policey-Anstalten in Städten, wegen derer Folgen und Wirkungen der Unglücks-Fälle, die man nicht verhüten können; insonderheit aber denen Feuer- und Brand-Cassen," in *Leipziger Sammlungen von Wirthschafftlichen, Policey- Cammer- und Finantz-Sachen*, vol. 7, Leipzig 1751, p. xxv.

coincidence that Torsten Hägerstrand originally employed this conceptual notion in an ethnological context. In doing so, he focused on the process that people engage in with the construction of houses and shelters, which not only have a certain physical position, but since they have a longer lifespan than their builders, also project the latter's existence or at least memory somehow temporally into a future that they can "colonize" beyond their own death. In a certain sense, the architect uses buildings to determine, mold, and "colonize" a future that cannot otherwise be experienced.[119] Therefore, in principle, the act of building already presumes a certain aspect of future domination. It has become clear that for London, the phase following the major fire of 1666 was influenced precisely by the extraordinary coincidence of the construction boom, building speculation, and the Financial Revolution. As never before in the history of Europe, early capitalist thinking and practice collided with excessive building and insurance thought. In Hamburg, a not dissimilar climate was present at this time—though in this case the catalyst for innovation was the existence or at least memory city administration. The security production enabled by home insurance was anthropologically linked to the most important exosomatic "instrument" of mankind, with which durability and a future could be guaranteed for one's self and one's descendants. That houses were valued as "*immobiliar*" ("immobiles," i.e., real estate), and created for long duration is, however, not self-evident, because it was precisely due to the risk of fire that houses in villages were, for a long time, typically not built "for eternity," but rather out of wood, so that they could be easily rebuilt. If fire insurance is both the effect and engine of such thinking, with the preservation of homes—at least in terms of value—at its core, then it should be recognized as an early form of sustainability thinking. The historiography of the concept of sustainability typically focuses—in terms of its early modern roots—on discourses like cameralist forestry science.[120] I would argue that early modern insurance thinking should be added as another structural comparison for the concept's core.

In twentieth and twenty-first century discourses, sustainability and insurance appear to have little in common. The famous *Brundtland Report* (1987)

---

119 Torsten Hägerstrand, *Time and Culture*, Berlin 1985, p. 40f.

120 Martin Stuber, *Wälder für Generationen. Konzeptionen der Nachhaltigkeit im Kanton Bern (1850–1880)*, Vienna 2008; Jeremy L. Caradonna, *Sustainability. A History*, Oxford 2014; a recent overview concentrates again on the forestry discourse: Paul Warde, "*Sustainability*, Resources and the Destiny of States in German Cameralist Thought," in Katrina Forrester and Sophie Smith (eds.), *Nature, Action and the Future. Political Thought and the Environment*, Cambridge 2018, pp. 43–69.

defines sustainable development as "any development which meets the needs of the present generation without damaging future generations."[121] Emerging from the UN conference "Environment and Development" in Rio de Janeiro in 1992, it claimed worldwide applicability. More detailed definitions followed, conceiving of sustainable development as

> actions accomplished by humans that show respect to a variety of living beings, which ensure the preservation of life, nature and culture for future generations within the ability of the environment, as well as establishing relationships in order to build a better society in the future and seeking a higher level of happy life for maximum population independent of time and space.[122]

Insurance institutions were indeed early participants in the environmental sector—for example, through their contribution to environmental pollution policies. Incidentally, it would seem that sustainability thinking affected insurance only indirectly, since insurance operated in the area of damage liquidation *after* something an event that was only possibly caused by nonsustainable action.[123] Some attempts have been made to unite insurance thinking with sustainability-like thinking; for example, the proposition to create a leasing institution through which the current, nature-exploiting generation leases the environment and—similar to a planetary life insurance policy—makes payments to future generations.[124] All in all, the global insurance industry has addressed the sustainability debate with hesitation,[125] presumably under the assumption that the financial sector is relatively "clean" and environmentally

121 Gro Harlem Brundtland (ed.), *Our Common Future: The World Commission on Environment and Development*, Oxford 1987.

122 Vladimir F. Krapivin and Costas A. Varotsos, *Globalization and Sustainable Development. Environmental Agendas*, Berlin 2007, p. 20f.

123 Cf. Howard Kunreuther, "The Failure of EIL Coverage: Gridlock in Environmental Insurance," *Environment* 29, 1 (1987), pp. 18–35; and the discussion with Robert Mason, *Environment* 29, 4 (1987), p. 1.

124 Brendan J. Quirke, "Leasing the Environment," *Sustainable Development* 4 (1996), pp. 98–102.

125 There is no mention of insurance theory and practice given in the 72 historically influential programmatic articles on sustainability collected in Michael Redclift (ed.), *Sustainability: Critical Concepts in the Social Sciences*, 4 vols., London 2005. The risk of an unsustainable reputation for companies was analyzed by Peter Forstmoser and Nikodemus Herger, "Managing Reputational Risk: A Reinsurer's View," *The Geneva Papers on Risk & Insurance* 31 (2006), pp. 409–24.

neutral in comparison to the heavy industry and technology.[126] But how then can the historical understanding of the relationship between "sustainability development" and "insurance" thinking be further explored?

While "sustainability" is currently a key concept dominating thousands of government reports, meetings, and policies worldwide beginning in the 1960s,[127] works by Rolf-Peter Sieferle and others have demonstrated that the corresponding German term "Nachhaltigkeit" (sustainability) was not coined in the nineteenth or twentieth century—even in its explicit programmatic sense, the term dates as far back as at least the eighteenth century. Hans Carl von Carlowitz's *Sylvicultura oeconomica* of 1713, an early work in scientific forestry, is considered the earliest reference to date.[128] Other early synonyms could likely be found in other languages. This concept, with its implicit idea of the finitude of natural resources, emerges at more or less the same time and as a counterpoint to the general Enlightenment belief of the universal progress of civilization. Research has determined that in the wider context of the German cameralistic administrative sciences, German forestry economics seems to have formulated a concept of *Nachhaltigkeit* that was not primarily ecological, but rather economically oriented. The forest should be well managed in order to ensure that resource extraction remained constant and continuous.[129] Based on source material from the Swiss canton of Bern, Martin Stuber has clearly traced the development of sustainability thinking from 1750–1880,

---

126 It was only in 1997 that 92 leading insurers, reinsurers, and pension funds signed a statement of opinion: *Environmental Commitment by the Insurer Industry*, UNEP Insurance Industry Initiative steering committee, Geneva, http://www.unepfi.org/signatories/statements/ii/). This project was then expanded: *The inaugural report of the Insurance Working Group of the United Nations Environment Programme Finance Initiative - Insuring for Sustainability. Why and how the leaders are doing it.* (2007), 28, https://www.unepfi.org/fileadmin/documents/insuring_for_sustainability.pdf). Cf. Herwig Peeters, "Sustainable Development and the Role of the Financial World," *Environment, Development and Sustainability* 5 (2003), pp. 197–230, 207 here.

127 For the more recent history of the concept with a decidedly global approach, see W.M. Adams, "The Origins of Sustainable Development," in Redclift, *Sustainability*, vol. 2, pp. 69–100 (first published 1990).

128 Cf. Rolf-Peter Sieferle, *Der unterirdische Wald: Energiekrise und industrielle Revolution*, Munich 1982; Joachim Radkau, "'Nachhaltigkeit' als Wort der Macht: Reflexionen zum methodischen Wert eines umweltpolitischen Schlüsselbegriffs," in François Duceppe-Lamarre and Jens Ivo Engels (eds.), *Umwelt und Herrschaft in der Geschichte*, Munich 2008, pp. 131–36.

129 Cf. Henry E. Lowood, "The Calculating Forester: Quantification, Cameral Science, and the Emergence of Scientific Forestry Management in Germany," in Tore Frängsmyr et al. (eds.), *The Quantifying Spirit*, pp. 315–342; Scott, *Seeing Like a State*, pp. 11–22; Radkau, "Nachhaltigkeit," p. 133.

distinguishing three types: the cameralist type beginning in 1760, which had forestry economics as its aim and which guaranteed a *reasonable rate of return* for the *necessary demand*; the liberal type beginning in 1800, which had an *optimized profitability* in a purely economic sense as its goal; and finally, the first ecological type beginning in the 1850s, aimed at the *sustainable protection of resources*.[130] Accordingly, it seems that until 1850, concepts of "sustainability/ *Nachhaltigkeit*" in the field of forestry and also mining were motivated solely by the quest for a constant yield. The differences lay only in the conceptual relationship between government and the economy respective to cameralism and liberalism. It is much more likely that it was this Enlightenment economic perspective that gave birth to sustainability thinking, rather than an early Romantic view of nature as a threatened resource.[131] Günter Bayerl has employed the perhaps too radical metaphor of nature as a warehouse to characterize the tendencies of Enlightenment thought,[132] which certainly contradicts a more ethically charged notion of sustainability thinking.

In combining the statements made in the previous chapters and sections about the first future-oriented attempts at insurance during the seventeenth and eighteenth centuries, remarkable congruities with the concept of sustainability are revealed. Both the first published draft of a municipal fire insurance program in London by Augustine Newbold (1680) as well as Barbon's thinking contained clear concepts of the progress, growth, and "immortalization" of the value of the building stock: the basic capital to be applied "might *for ever perpetuate the Re-building of all the Casualties by Fire, happening to those Adventurers.*"[133] The future-orientation of the project and, to a certain extent, a break with the traditional medieval and guild-oriented perception of the "city" can be read in the words "the immortalization of reconstruction": major fires thus generated an accelerated and dynamic perception of the community as a whole, which simultaneously offered a counter element facilitating the "immortalization of reconstruction." Similarly, the community of owners was conceived of as analogous to early global and colonial players—i.e., the merchant adventurer companies (explicitly mentioned by Newbold is the *East-India Company*). Just as the efforts of these trade companies were oriented

130 Stuber, *Wälder für Generationen.*

131 Like Adams, "The Origins" and others suggest.

132 Günter Bayerl, "Prolegomenon der 'Großen Industrie'. Der technisch-ökonomische Blick auf die Natur im 18. Jahrhundert," in Werner Abelshauser (ed.), *Umweltgeschichte. Umweltverträgliches Wirtschaften in historischer Perspektive*, Göttingen 1994, pp. 29–56, 31–50.

133 Newbold, *Londons Jmprovement*, pp. 1, 2, 4.

toward the stabilization of their streams of trade, leading to the establishment of maritime insurance mechanisms, so too should the whole city insure itself in order to stabilize the value of its buildings. This objective of value preservation can be identified as the germination of a kind of economically sustainable thinking inherent to all insurance projects and practices, which is, in this case, not bound to "natural resources" (such as forests) as the carrier of value, but to the buildings of the city themselves. It is important also to mention that the idea of "conservation" and the long-term survival of London's houses, was, from the beginning, associated with the idea of progress/development and the growth of economic prosperity, also in terms of insurance: "And as it may be employed, in few Years they may receive Dividents out of the Improvement." It was no accident that the title of Newbold's insurance proposal was called "London's *Improvement*":[134] as in the case of forest science, the optimization of resource utilization coinciding with concepts of conservation, improvement, and the preservation of value were also inextricably bound to the first conceptual moments of fire insurance.

For the cameralists, it was of great importance that natural disasters reduced the prosperity and thus the credit of individuals. Insurance helped not only to balance the plus and minus columns; rather, the dream of Enlightenment thinkers was the capacity to more or less double individual loans through the higher mortgage values of insured houses. These considerations were typically introduced with the notion that uninsured homes were essentially understood as future ashes, while insured homes were conceived of as including a second, "virtual" house that could be built in the future and could thus be added to the owner's capital. Today one might formulate it as the idea that insured homes facilitate an intertemporal transfer of ownership and can be recognized as an asset.[135]

---

134 Ibid., p. 2.—the memorandum analyzed above from Leibniz to Emperor Leopold is written much more from the perspective of the dukes and the state, and primarily questions how Germany can be strengthened in competition with the European powers. The state-owned insurance project was only seen as one type of growth in the well-being of the state. With exception of the goal "damit er [der Untertan] ein düchtiges glied der gemeine seyn und bey Nahrung bleiben möge" (Leibniz, "Öffentliche Assekuranzen," pp. 421–32, 425), there is no precise mention of the concept of "conservation"/"sustainability" that lies at the foundation of this concept of progressive thinking, even if it is silently implied.

135 Someone who insures his house has: "also in der That noch ein Haus in der Casse, oder sein ietziges Haus sicherer, und dieses wird dadurch ein solideres Stück seines Vermögens, dieses aber dem Werthe nach eben darum erhöhet." Uninsured houses can be considered "als ein[] Brandhaufen im voraus." One could take credit out with greater security on insured houses, insurance thus serves as an instrument for the expansion of a city's

> It is a truth confirmed by experience that the security with which we can enjoy an object, immensely increases its value, and in reverse, that the value of a thing is reduced to the extent of how great and manifold the threat we are exposed to in that respect … It is therefore quite natural that buildings must be accorded a much higher value if they are not subject to fire hazards, or, amounting to the same thing, if in the event that they are destroyed by fire, that nevertheless secure compensation is provided for … Therefore, many people of means can decide more easily to purchase or build a house if they can be certain that the capital invested in it can no longer be lost through a fire. As the value of buildings rises, the assets of the proprietors of such buildings thus also increases, and consequently the total assets of the state as well, of which buildings comprise no inconsiderable part. With such assets, furthermore, the credit in a country also gradually grows.[136]

The calculations introduced in the previous chapter—as far as they (like Waser's) lay deep in city history, or (like the anonymous author in 1751) reviewed at least the previous 30 years—always had the impetus of stabilization and reinforcement, but also growth in time as goals. In 1783, the Reutlingen lawyer Fetzer more or less predicted how the history of his own city would have been altered, if in 1502 when a major fire destroyed 150 buildings, there had been fire insurance in place, patterned after his model of using an interest-bearing fund. If at that time there had been only 10,000 guilders as capital in the fund, "and it had been [left to] its own growth by trustworthy administration," "then, assuming a continuous rate of interest, and that it would have not been otherwise used until the year of 1726, it would have multiplied by so many millions, that it would have hardly been impossible, not only to generously restore the full value of all houses, but the city could have been gilded according to the latest architectural trends." The major fire disaster resulting in the almost total destruction of Reutlingen in 1726, from which the imperial city had never quite recovered, is described as a problem of non-forward-thinking, unsustainable

---

wealth (Zincke, "Vorrede," pp. iii-liv, xxvi.). On the topic of uninsured houses as potential or future ash piles, see also auch [anonymous], "Gedanken von der Einrichtung," pp. 293–301, 294; [anonymous], "Vorschlag, wie die Versicherung der Gebäude," pp. 673–684, 675; Johann Jakob Moser, *Nachricht von einer freywilligen Feuer-Cassa, Welche zum Besten der Einwohner des Herzogthums Würtemberg angeordnet* […], s.l. 1754, p. 3; Schaeffer, *Nachricht*, p. *1v; Waser, *Betrachtungen*, p. 5.

136 Philip Gäng, *Von Versicherungsanstalten wider Feuerschäden und ihrem Nutzen im Allgemeinen*, Salzburg 1792, 20.

security policies from a time before advent of fire insurance. Thus the pressure on the imperial city to either establish its own insurance institution or to join Württemberg's increased.[137]

In light of these examples and quotes, if one attempts to compare historical and contemporary sustainability and insurance thinking, a few semantic elements can be abstractly defined:

1. "Sustainable development" includes a (more or less) strong future-orientation of thought.
2. "Sustainable development" comprises a strong moral aspect of individual and collective responsibility. The moral appeal even attempts to form a cosmopolitical collective (i.e., humanity) that actually (still) does not exists as an operating actor.[138]
3. "Sustainable development" includes an element of conservation—since the object of conservation is "nature" as a value, which is sometimes expressed by the economic metaphor of "natural capital."[139]
4. "Sustainable development" also entails, sometimes in contradiction to the first point here, an element of development as a goal of state and individual actions that nevertheless implies a form of "progress"/growth thinking[140]—the objects of development are usually economics, politics, living conditions.

Because of the central factors 3) and 4) "sustainable development" assumes an element of "predictability/calculability of nature." It owes a large part of its conceptual frame of reference, at least in terms of metaphors, to economics, and seeks methods and criteria to measure nature and sustainability itself.[141]

137 Fetzer, *Ueber Brandschadens-Versicherungsanstalten*, 33f.

138 Cf., for example, Jonas, *Das Prinzip Verantwortung*; Jean-Pierre Dupuy, *Pour un catastrophisme éclairé*, Paris 2002.

139 Cf. Salah El Serafy, "The Environment as Capital," in Redclift, *Sustainability*, vol. 3, pp. 49–57; Peter A. Victor, *Indicators of Sustainable Development: Some Lessons from Capital Theory*, in Redclift, *Sustainability*, vol. 3, pp. 58–79 (both first published in 1991); Begüm Özkaynak, Pat Devine, and Dan Rigby, "Operationalising Strong Sustainability: Definitions, Methodologies and Outcomes," in Redclift, *Sustainability*, vol. 2, pp. 369–394 (first published 2004); David W. Pearce and Jeremy J. Warford, *World without End: Economics, Environment, and Sustainable Development*, Oxford 1993, pp. 41–62.

140 Whether the often employed concept of "co-evolution" can actually eliminate the final residues of "progressive thinking," is more than questionable, on this point see Richard B. Norgaard, "Sustainable Development: A Co-Evolutionary View," in Redclift, *Sustainabilty*, vol. 2, pp. 131–48 (first published 1988).

141 Cf., for example, part C only: "Measuring environmental quality perceptions," in Gert de Roo and Donald Miller (eds.), *Compact Cities and Sustainable Urban Development. A Critical Assessment of Policies and Plans from an International Perspective*, Aldershot 2000, pp. 127–78; Pearce and Warford, *World without End*, pp. 83–95; Kent E. Portney,

In comparing the abstracted elements of Enlightenment insurance practice and conception, we find that:

1. Post-1700 "Insurance" includes a (more or less) strong future-oriented mode of thinking.[142]
2. "Insurance" of the Enlightenment era entails a strong moral aspect of individual and collective responsibility, assisting in preserving the property of others. Communities were remodeled according to this objective (cf. 1.1).
3. "Insurance" includes an element of conservation as an action goal—the object of conservation is the value of property (particularly of buildings).
4. "Insurance" also includes an element of development that implies a type of progress/growth thinking as an action goal: because housing values are magnified by insurance, the welfare of the nation grows to the extent that its credit increases.

Accordingly, especially when taking the history of thought regarding sustainability insurance thinking in the eighteenth century into consideration, both form and thought are strikingly similar in structure; in fact, they are almost homologous. Other details that support this similarity can unfortunately not be discussed here.[143]

The major difference between the current sustainability discourse and the topic of insurance seems to center on the object to be protected: on the one

---

Taking sustainable Cities seriously. Economic Development, the Environment, and Quality of Life in American Cities, Cambridge/Mass. 2003, pp. 31–75.

142 In contemporary sociology, insurance is commonly used as an example for a modern manner of thinking of the future as an open, "colonizable" area [field], capable of being operationalized, see Barbara Adam, *Time and Social Theory*, Philadelphia 1990, pp. 138–142; Giddens, *Modernity and Self-Identity*, pp. 109–143 and Gerda Reith, "Uncertain Times: The Notion of 'Risk' and the Development of Modernity," *Time & Society* 13 (2004), pp. 383–402.

143 The ideological-critical analysis of the concepts would also be comparable: The staging of sustainability by public and private actors sometimes invokes the suspicion, that "it is merely a public relations tool by which the signatories [of the UNEP Insurance Industry Initiative] can be seen to be doing something without having to commit to change" (Peeters, "Sustainable Development," p. 208); the sustainability discourse could allow us to move towards a type of scientific-progressive thinking without a fundamental change. Early modern insurance thinkers and practitioners would be analogously confronted with the suspicion that fire insurance might lead to an increase in the number of major fires and burned houses, since owners of insured houses might allow their own property to burn, in order to finance a new and better building with the insurance proceeds (Moser, *Nachricht*, p. 5; [anonymous], "Gedanken von der Einrichtung," p. 682; Fetzer, *Ueber Brandschadens-Versicherungsanstalten*, p. 80f.) —which is a standard problem of assymetrical information in the insurance business.

hand, nature should to be preserved in its biodiversity and its condition within the biosphere; on the other hand, economic capital should be extracted. However, as has been seen, early sustainability thinking also focused on nature as an economic resource.[144] Taking this into account, we can question whether today's sustainable development programs supported by the UN and other governmental and nonstate actors can escape this perception of nature—in the event of a crisis, economic criteria for resource generation and prosperity production still tend to take priority.

If, at first glance, it seems that insurance thinking follows a *reverse* logic from sustainability thinking (i.e., insurance protects against nature's threat to humanity; sustainability practices protect nature from humanity's threat to nature), a closer look at the history of these concepts in order to understand their structural proximity will be helpful. Against this background, it is no longer any wonder why and how insurance and sustainability thinking were linked so late in the discussion. The concept of "insurability" was introduced as a heuristic instrument and transferred from insurance mathematics and later the sociology of risk and insurance to the discussion of sustainable governance. The existing relationship between technological innovation, risk definition by the insurer, and sustainable development meant that the expertise of the insurance business could be used to define insurable risks. By assuming that uninsurable technology was also unsustainable, this expertise could be used as an indicator for the boundary between public and private businesses, as well as the border between sustainable and unsustainable technology.[145] This idea illuminates a final compound of the concepts, which lies in the specificity of their respective access to the future: both sustainability thinking and insurance identify controllable risks within a huge number of uncontrollable threats. Only actions where the risk of environmental degradation is measurable and, when necessary, controllable or preventable, can claim legitimacy from the perspective of sustainability. In the same way, an insurer will only identify those risks that are measurable and that are payable without causing the company to fall into ruin.

---

144 Cornel Zwierlein, Historicizing Environmental Security, *European Journal of Security Research* 3 (2018), pp. 1–13, 3–5.

145 Walter R. Stahel, "The Role of Insurability and Insurance," *The Geneva Papers on Risk and Insurance* 28, 3 (2003), pp. 374–81; Kristina Dahlström, Jim Skea, and Walter R. Stahel, "Innovation, Insurability and Sustainable Development: Sharing Risk Management between Insurers and the State," *The Geneva Papers on Risk and Insurance* 28, 3 (2003), pp. 394–412.

These final considerations suggest that the temporal orientation, and specifically, the future-orientation, of the insurance practices and theories established during the eighteenth century has unexpected structural correlations with what is currently one of the most definitive global political concepts of future-planning. The near simultaneous emergence of these concepts in the early cameralist milieu around 1700 reveals a basic temporal foundational structure of cameralism as a whole.[146] The development of a counterfactual expectation of a permanent "normal secure society"—meaning the expectation that an internally and externally secure state is the norm, while disasters represent the exceptions—and the fact that such security must continually be produced, is based on a specific form of targeting the future. A clear time index is implied: this normal secure society is constantly designing itself as such, projecting itself and its assets forward in time and space, simultaneously attempting to improve and to preserve what has already been achieved. Consequently, such a society is in structural affinity with a sustainably operating economy, at least in its early embryonic forms conceived within cameralist thought.

146 For a different approach, underlining the importance of spatiality to the concept of cameralism cf. Marcus Sandl, *Ökonomie des Raumes. Der kameralwissenschaftliche Entwurf der Staatswirtschaft im 18. Jahrhundert*, Cologne 1999.

CHAPTER 6

The Globalization of Safety Regimes: The Return of Space

This chapter illustrates the diffusion of a unique security regime wherein the principle and practice of "premium insurance," was distinguished by the fact that it was clearly identifiable as a new and different system resting on a purely cultural substrate. It is not a technical tool or machine, but is based solely on ideas—in particular, on ideas about the "future," a specific writing system, and the application of a few basic political and social accounting "rules" such as the law of "large numbers." In some respects, premium-based insurance is therefore a product that is easily transferable across cultures; in other ways, however, its cultural character makes it more difficult to transfer than a simple technical instrument. If one controls the means and requirements of production, a mechanical tool as an artifact is only loosely bound to a cultural context. As long as there are pens, paper, and corresponding writing systems, the abstract and "impalpable" idea of insurance is, in principle, quite easily transferable; but it is also much more tightly integrated within a system of cultural hermeneutics and, therefore, is also burdened with substantial invisible barriers. Historically, these barriers were bound up with the transition to modern forms of security regimes, in ways that will be further specified—after all, even the early capitalist premium-based insurance of the Renaissance marked a difference from older economic styles—but they were also intertwined with cultural differences. Through the process of globalization, these temporal/epochal and cultural/spatial movements and resistances to transfer became increasingly intermingled.

It would be tempting to begin with a section on "Europeanization": Even when different cultures were encountered in the early modern era there were obstacles to the transfer of insurance and even cases of backward development: I have noted above instances of the European-Ottoman encounter referred to in the literature, with effects on the import and decline of premium insurance business on islands occupied by either Venetians or by Ottomans. Similarly, I noted that the institutionalized (fire) insurance model in no way simply proceeded unobstructed through space and time, as for instance, it advanced not at all in France.[1] The British insurance model thus began to make strides on the European continent only relatively lately, in the 1810s and 1820s, when private

1 Cf. above Ch. 4, 2.3, notes 202, 221.

insurance companies had begun to pattern themselves after the English model in France, Germany, and then rapidly in other countries as well.[2]

The diffusion of such new capitalist *techniques* somehow implied a clash of cultures or even of periods: this was the case in Germany where at least the cameralist form of institutionalized premium insurance was already well in practice, albeit within the forms of early modern territorial cameralist state administration. How forms of the modern capitalist society were received in such an institutional-administrative environment of the Old Reich's *structure* is beautifully illustrated by the following case record. A Hamburg citizen, Johann Augustin Level, had founded a paper mill in Walksfelde, an aristocratic estate located in Holstein, which he had insured with the British insurer Phoenix twice for three-year terms at the rate of 13,000 Courant Marks. Phoenix was the first insurer to expand outside the British islands, and had been represented in Hamburg by Hanbury and Lantz since 1786. On March 22, 1793, the mill burned down, along with a quantity of insured "raw materials" for making paper—rags. The insurance company did not want to pay because it suspected insurance fraud; Level resorted to "confiscating" the private assets of Hanbury and Lantz—a response that had been legally deemed unjustified in several previous instances. Phoenix insisted that Level could, at most, demand justice and his money at the central headquarters in London, because "the London Phoenix-Company [had] never established a fund in Hamburg or designated from where the putative damages should be paid." Level appealed to the Imperial Chamber Court (*Reichskammergericht*) and cluttered the proceedings with the *acta priora* of the previous Danish and Hamburg instances summarized in two voluminous fascicles, but this became one of those court cases that remained undecided when the Imperial Court disbanded with the end of the Old Reich.[3]

The same problem essentially reappeared later, after the 1906 San Francisco earthquake: in this case too, legal disputes typically revolved around the location of the court; the US thus subsequently increasingly placed restrictions on foreign companies, obliging them to establish capital funding on-site in US locations, rather than relying on funding at distant company headquarters (e.g. London).[4] The conjuncture between different eras, this "achrony"[5] that pitted

2 *Festschrift 225 Jahre Braunschweigische Landes-Brandversicherungsanstalt 1754–1979*, s.l., s.d. [1979], p. 20.

3 StAH RKG L 35 Teil 1 u. 2.

4 Cf. Tilman J. Röder, *Rechtsbildung im wirtschaftlichen "Weltverkehr." Das Erdbeben von San Francisco und die internationale Standardisierung von Vertragsbedingungen (1871–1914)*, Frankfurt am Main 2006.

5 On the concept of achrony, see Zwierlein: "Return to Premodern Times?".

the leading insurance company of globalization in the British Empire against the backdrop of the beleaguered highest court of the thousand-year-old Reich, offers an almost concave reflection of the problem that British insurers encountered regarding societies outside of Britain. While this problem obviously concerns "space," there is also an element of "time" invoked by the fact that these companies often teetered on the temporal edges of "modernity," or even beyond the framework that defines "modernity." Insurance agents were sometimes very aware of this, framing the "unmodern other" they encountered in Orientalist modes, but at other times they seemed unconscious of the invisible distinction between modernity and premodernity, which was constructed and reified through the cultural transfer of a hitherto unknown European cultural practice to non-European countries. Great Britain's domination of the market on the other side of the Channel was remarkably quick—it began in about 1810 and was accomplished by mid-century. Simultaneously, and just as quickly, competing forms of insurance were also established across the Continent. These developments are clearly illustrated for the Rhine regions in the form of the insurance supervisory permits that were required in Prussian territories beginning in 1837 (the regulations were modified in 1853).[6] Through these it is known that the following fire insurance companies were active between 1837 and 1848 in the district of Düsseldorf:[7]

1. Fatherland Fire Insurance Society of Elberfeld
2. Aachen-Munich Fire Insurance Society
3. Fifth Hamburger Assecurance Company
4. Second Hamburg Sea and Land Fire Assecurance
5. Leipzig Fire Insurance Society
6. London Phoenix Fire Assecurance Society
7. Fire Insurance Bank of Gotha
8. English Sun Fire Office
9. English Royal Exchange
10. West of Scotland Fire Insurance Company
11. Mecklenburg Fire Assecuration Society of Neubrandenburg
12. Hamburg Patriotic Assecurance Company
13. Cologne Fire Insurance Company
14. Adriatic Security Society of Trieste [*Reunione adriatica di sicurtà a Trieste*]

---

6 Michael Tigges, *Geschichte und Entwicklung der Versicherungsaufsicht*, Karlsruhe 1985, pp. 20–53.

7 "Verzeichniss der höhern Orth concessionirten Versicherungs-Gesellschaften," in Stadtarchiv Solingen S-3143.

15. Rhine Prussian Fire Insurance Society of Dusseldorf
16. General Assurance Company of Paris [*Compagnie d'assurances générales à Paris*]
17. Leipzig Chattels Fire Insurance Bank [*Leipziger Mobiliar-Brand-Versicherungs-Bank*]
18. Fire Insurance Institute Borussia in Königsberg
19. Berlin Fire Insurance Society
20. Magdeburg Fire Insurance Society
21. National Insurance Society in Szczecin
22. Silesian Fire Insurance Society of Breslau.

While the history of each of these individual insurance companies would certainly be of great economic-historical interest, this topic cannot be pursued in detail here; it is sufficient to point out that the list demonstrates that within a relatively short period of time, up to the middle of the century, a large number of new domestic "start-ups" had emerged in addition to the territorial fire societies, and also that more than a quarter of these were foreign companies, from Britain in particular.[8] Many of the early companies listed above no longer exist, and in most cases, even if they do still exist in a merged or otherwise transformed form, the company archives do not go all the way back to the companies' beginnings. If they do go back this far, they often contain only a few pieces of information such as founding contracts, meeting notes, and general assembly protocols, and such pieces of information do not lend themselves to extensive cultural-historical analysis. The four major French fire insurance companies that were founded between 1818 and 1829 (*Compagnie des Assurances générales, Le Phénix, L'Union, Le Soleil*) all rapidly took hold in Europe, but were then quickly stymied in most parts of the German-speaking areas—although admittedly this is based on scant research relayed through archival transmission. Consequently, it is difficult to track the European expansion of, for example, the French *Compagnie d'assurances générales* based on its own archival materials.[9] Newspaper advertisements and advertising brochures

8 For the parallel case of German life insurance, see Heiss, *Die Institutionalisierung*, for the fifth long-lasting German establishment, see Alexander Müssener, *Die Entwicklung der Aachener Feuer-Versicherungs-Gesellschaft im 19. Jahrhundert unter besonderer Berücksichtigung ihrer Allgemeinen Versicherungsbedingungen*, Hamburg 2008.

9 According to *La Compagnie d'Assurances générales contre l'incendie 1819–1919. Cent ans d'Assurance et de Sagesse Financière*, Paris 1920, p. 93 the company gave up its foreign business "bientôt," following an initial expansion, and reentered the global market only after World War I. The existing archives do, however, indicate a continued presence from 1820 to 1848 (in Stuttgart, Munich, and Koblenz as the main agency locations along the Rhine). The significance of this branch of the business, is, in any case, difficult to determine, since the

help establish the position in the German market (for example, concerning the *Phénix* founded in 1819, *Le Soleil*, founded in 1829, and the *Union Paris*). The French Phoenix had paid 25 million francs for fire damage between 1819 and 1837, of which 2 million was accounted for abroad.[10] Some fires reveal a larger spread: in 1834 Greußen, a rural town of 2,315 inhabitants in the principality of Schwarzburg-Sondershausen near Erfurt, was afflicted by fire—of the 226 destroyed houses almost all (219) were insured with private companies, since the principality had no fire fund of its own. The inhabitants relied on German companies in only 88 cases; 131 homes, in contrast, were insured with the *Compagnie royale* and the *Compagnie du Soleil*. The radius of effect associated with these companies was thus in no way limited to French border areas;[11] foreign insurance companies, it would seem, filled precisely those gaps not covered by territorial fire funds. German insurers always regarded foreign companies as strong rivals, first learning from them, but then trying to suppress them, using state revenues when necessary.[12] Beginning

accounts in the surviving files are not separated into regions; see Archives nationales du monde du travail, Roubaix, 46 AQ 1ff. (reviewed up to vol. 10): Procès-verbaux du conseil d'administration 1819–1846. The historical research into the early French history of private insurance is not terribly dense, summary syntheses such as Serge Magnan, "L'assurance incendie. Du risque industriel et la concurrence de 1820 à nos jours," *Risques* 18 (1994), pp. 133–49, are more strongly concentrated on later time periods. More research exists on social insurance, cf. Michel Dreyfus, Danièle Voldman, Michèle Ruffat, Vincent Viet (eds.), *Se protéger, être protégé: Une histoire des assurances sociales en France*, Rennes 2015.

10 Cf. Baldur Graf, *Versicherungskaufmann* 3 (1987), p. 122 in Aachen AMB Generali Archiv Kasten 1825 as well as the newspaper clips collected by Graf from the 1830 Aachen and Cologne city newspaper: In the 1830 issues of the Cologne newspaper there were approximately 18 published advertisements—of those 8 were foreign (from Union Brussels and Union Paris). See also Rudolf Bergmann, *Geschichte des rheinischen Versicherungswesens*, Essen 1928, where the appearance of the French Phoenix is given for 1810—according to StA Düsseldorf, Reg. Aachen, Abt. I Nr. 31 however, the French Phoenix was not founded until 1819.

11 Paul Lürmann, *Der große Brand der Stadt Greußen in der Nacht vom 16. zum 17. Oktober 1834*, Greußen/Thuringia 1934, pp. 11–16.

12 Cf. the agent of the Lower Rhine insurance of goods (*Niederrheinische Güterassekuranz*), Wesel's comments to the Düsseldorf government on October 22, 1840: "Von den französischen Gesellschaften ist namentlich die Pariser Compagnie d'assurance générale die gefährlichste und drückendste Konkurrentin, die überall und auf jede Weise uns entgegentritt und jedes Jahr bedeutende Summen dem Vaterlande entzieht. Blicken wir vollends auf die politischen Verhältnisse zwischen Frankreich und Preußen hin, wodurch die Sicherheit der bei französischen Gesellschaften Versichernden möglicherweise sehr gefährdet werden kann, so dürfen wir wohl das feste Vertrauen haben, daß uns der nötige Schutz gegen das Ausland zuteil wird" (qtd. Rudolf Bergmann, *Geschichte des rheinischen Versicherungswesens*, Essen 1928, p. 132).

in 1837, the permits of some foreign insurers in Prussia went unapproved or were even revoked.[13] Local insurers quickly profited from this: Aachen (later Aachen-Munich) Insurance used the Treaty of 1840 to assume the insurance obligations of *Compagnie d'assurances générales*, which amounted to 100 million francs of insured value; Colonia Insurance used the German stock of Union Paris to establish itself in 1837.[14]

Even a brief look at this still largely unexplored topic of European expansion and competition in these markets during the first half of the nineteenth century demonstrates that in the initial phase, German insurance companies were strongly fixated on the domestic market and tried to dominate it. Throughout the nineteenth century, very few German insurance operations conducted business outside of Germany—with the exception of the Hamburg-Bremen Fire Insurance Company, founded in 1854, which entered the global market very aggressively.[15] German-owned insurance companies operating from 1884

13 Tigges, *Geschichte und Entwicklung*, pp. 32–37. When this law with its permit conditions was enacted, the fire insurance institutions were already involved. In 1830 the founder of the Aachen society, David Hanseman, had already created a recommendation about the law's decree regarding fire insurance bodies for the Westphalian provincial estates, which was printed in Aachen in 1834; the restrictive permit requirement for foreign companies was firmly established there in §§ 17–23 (Aachen AMB Generali Archiv, Fach Bücher).

14 Aachen AMB Generali Archiv, Kasten 1839: Excerpt from the general meeting minutes of March 30, 1840, Director Seyffardt made the Aachen business possible. Colonia acquired capital at the Rothschild Bank in order to buy out the French holdings; Klara van Eyll, *"... genannt Colonia." 150 Jahre Kölnische Feuer-Versicherungs-Gesellschaft AG 1839–1989*, Cologne 1989, p. 16.

15 The first agencies were: 1855 in St. Thomas (Caribbean); Honolulu, San Francisco (1857); Maracaibo, Hong Kong (1858); Santiago de Cuba, Havana, Haiphong, Tonkin (1859); Valparaiso, Mexico City, Veracruz (1865); Caracas, La Guairá (1866); Port Elizabeth (South Africa), Mazatlán (1868). In 1871 the Chicago fire disaster led to a substantial expansion of the business in the USA: in 1879 Hamburg-Bremen Insurance sent over 250 agents to the USA; in 1871, agencies were also founded in Yokohama, Humacao (Puerto Rico); beginning in 1883 there was activity in India (1886), Siam (1887), The Straits Settlements and the Philippines (1888), Turkey (1888), Egypt (1891), Curaçao (1888), Costa Rica (1911), Trinidad (1897), Nicaragua (1891), Colombia (1889), Ecuador (1897), Peru (1889), Uruguay (1900), Togo (1896), Russia, Switzerland (1886), Malaga (1911), Ceylon, Cameroon, German South West Africa (1896), German East Africa (1896), Cape Colony (1896), Transvaal (1912), Natal (1896), Nigeria (1912), Guatemala (1912), Algeria, Tunis, Morocco (1912), Shanghai, The Congo (1896), Spanish Guinea (1896), Manchuria. The entire foreign business, however, collapsed during World War I in 1914. The archival holdings are currently combined in the ERGO archive; there is unfortunately very little data for earlier periods, the general meeting minutes are only preserved beginning 23 May 1856. The 1954 commemorative publication *100 Jahre Hamburg-Bremer Feuer-Versicherungs-Gesellschaft Hamburg 1854–1954*, pp. 53–84 does not offer much additional information (I thank the archivist Markus Holmer for helpful information). The holdings of Magdeburg Insurance, which also had

onwards in the German colonies, where the applicable legal standards were disputed anyway, were more a curiosity than a statistically relevant phenomenon.[16] In any case, the cameralistic insurance model was not an institutional format that was adapted to globalization, and none of the territorial fire funds experienced any kind of true "expansion." A remarkable and still scarcely researched exception can be found in the attempts by German governmental advisors from 1868 onwards to transfer the state insurance model to Japan during the Meiji period.[17] However, the Japanese insurance founders proceeded to emulate the private companies that existed in Germany and were modelled on British companies such as the Gotha Insurance Bank.[18] From the perspective of German insurers, it was not until the advent of World War I in 1914 that "the bridges with other countries" finally collapsed.[19]

For the British insurers, Germany was a market among many in the world, albeit always recognized as the first—the long history of cameralistic insurance

---

a global presence, can currently be found in the Allianz archive in Munich; however, there is not yet an inventory of these holdings. For smaller German insurers, who were active globally, there are hardly any surviving individual archival materials; however, they can often be found to a small extent in the files of British insurers for many cities—for example, in Japan 1877, the Hamburg-Bremen, Berlin-Cologne, Lübeck, and the Hamburg Hanseatic fire insurance companies (CLC/B/192/019/31522/176, pp. 60, 107, 124).

16 In German East Africa, where the Hamburg-Bremen and the Aachen-Munich fire insurance companies underwrote a few policies, there was a discussion concerning the applicability of legislation from the *Reich*: since only the criminal law was directly transferable in the protectorates, insurance found itself in a more or less lawless arena—for example, the Prussian law from May 8, 1837 regulating the fire insurance bodies was not applicable. The governor's police authorities tended to check the contracts for overinsurance without explicit authorization to do so—which was, of course, met with objection by the insurers. Through this process, insurance operations in the German colonies are thus archivally documented, more direct sources from the company holdings are not available (Bundesarchiv Berlin-Lichterfelde R 1001/826, December 1902-August 1903).

17 Paul Mayet, *Landwirthschaftliche Versicherung* [...] *Vorschläge zur Besserung der Lage des Japanischen Landmanns, im Auftrag des Kais. Jap. Ministeriums des Innern*, Tokyo 1888, pp. v-vii, as well as Paul Mayet, "[Mitschrift eines Vortrags gehalten vor der Deutsch-Asiatischen Gesellschaft in Tokyo]," LMA CLC/B/192/019/31522/175, p. 281. There were no fire insurance-like institutions in Tokugawa Shogunat until 1868, Cf. Sand and Wills, "Governance, Arson, and Firefighting in Edo". On Mayet cf. Richard J. Smetshurst, *Agricultural Development and Tenancy Disputes in Japan, 1870–1940*, Princeton 1986, pp. 58–60.

18 Takau Yoneyama, "Tsuneta Yano, Founder of the First Mutual Company in Japan," in Robin Pearson and Takau Yoneyama (eds.), *Corporate Forms and Organizational Choice in International Insurance*, Oxford 2015, pp. 29–46.

19 Borscheid, Vertrauensgewinn und Vertrauensverlust, p. 327.

institutions had evidently resulted in a general familiarity with the principle of insurance, so that a strong demand quickly emerged. British insurance companies were the only ones who were extensively present outside their home country, using the knowledge and infrastructure developed within the official and unofficial economic structures of the British Empire.

Among these companies, the archives of the Phoenix (investigated by Trebilcock) and those of the Sun Fire Office offer the material best suited for going beyond a pure balance sheet analysis toward a cultural-historical analysis of globalization from the perspective of an expanding company and its independent agents. Pursuant to an executive committee decision, the Sun Fire Office began to quickly expand outside of Great Britain in Europe and beyond. Beginning in the 1850s and within the space of a few years, a massive wave of agency founding occurred throughout the world, after which the Sun began to be a truly globally operating actor in the field. In so doing, the Sun was following a trend, started by Phoenix in the 1830s and actualized through a variety of English insurance companies: globalization, in this case, appears as the "self-spreading" of a branch of an industry and a market, expanding from central England/London into every corner of the "empire." This development also reveals a certain continuity with the early movements of diffusion of the insurance principle and insurance industry in the Renaissance, in that this expansion always took place through—and almost exclusively in—port cities. Insurance was implemented in Japan not in Tokyo, but in Yokohama; in Chile, it appeared in Valparaiso rather than in Santiago; in India it emerged in Calcutta, Bombay, and Madras, but not in Delhi. The territorialization of insurance and its expansion into rural areas did not occur for a long time—with the exception of Hamburg, where a network of subagencies expanded into a Germany that was already predisposed toward accepting the concept of "insurance." The "globalization of the insurance industry" (like the globalization of economics as such) can be conceptualized not as a widespread expansion, but rather as the construction of a network of port-city nodes.

The fact that this movement proceeded from the center of London outward should, however, not be dismissed. In the nineteenth century, global insurance management was an almost exclusively British endeavor. On-site operations in Latin America, the USA, and Asian cities typically used local insurance companies, which were in competition with the English, but such companies did not typically operate outside of their country or even outside their native city or region—with a few, mostly short-lived exceptions, such as a few Dutch companies and the Hamburg-Magdeburg Fire Insurance, which had a presence in Bombay.

Following the previous chapters that concentrated on the Enlightenment era threshold of 1680/1700 and the lines of development traced here for the eighteenth century, this chapter will present a cross-sectional study on globalization based on the files of the Sun Fire Office. Unfortunately, a symmetrical comparison with German files is no longer possible, since there is no comparable German archival material for this period: Germany had not yet begun to substantially participate in the global market at this point. The asymmetry of the comparison is thus unavoidable; however, I will continue to address systematic questions concerning the consequences, functions, and patterns associated with the diffusion of insurance, as well as the accompanying perceptual modes. I will show that despite important epochal differences between the internal crucial European moment of 1700, with the institutionalization of the insurance principle, and the global European/Non-European interaction around 1830, there are phenomena and characteristics that are quite comparable on a structural level.

The Sun Fire Office preserved an impressive archival source body in the form of its so-called "Memorandum Books." This set of sources is unlikely equaled by any other insurance archive and will facilitate this study of the company's globalization. In addition to balance sheets, financial statements, management decision protocols, and correspondence, the Memorandum Books—of which there are approximately 300, each about 300–500 pages long—contain detailed reports from foreign branches of the company, which provide information about the local fire insurance markets, fire prevention measures, and the number and intensity of recent fires. Some parts of these volumes were collected pursuant to inspections of the agencies by members of the Sun's Foreign Department (often by Francis Boyer Relton, who served as Foreign Department superintendent from 1868, and George Saward Manvell, clerk in the Foreign Department from 1864 on). Material was often collected retrospectively, going a few decades back. In addition to handwritten analysis by on-site representatives and the members of the Foreign Department in London, the volumes contain drawings, maps, early photographs, newspaper clippings,[20] printed premium agreements, agent-association protocols for the respective locations, extracts and other parts of original correspondence, forms, and statistics. These volumes thus functioned as a kind of ever-growing encyclopedia and fund of knowledge for the London head office and its global network of agencies and insurance markets. The volumes are a collection of strategic operational knowledge about hundreds of locations all over the world where

20 Unfortunately, the newspaper excerpts and clippings do not always include all the requisite information (newspaper name, date, page); tracking down the exact page numbers in original would be like looking for needles in a haystack and hardly be worth it. When in doubt, therefore, I use the Memorandum Books to cite instead.

insurance had been implemented, and thus, in some way, represent global economic memory.[21] Peter Dickson first made extensive use of these Memorandum Books in his monograph on the Sun in 1960;[22] in the past couple of years they have attracted some further interest from historians, but the volumes still contain a wealth of heretofore unaddressed material.[23] The company made the volumes accessible to the public in 1994 by entrusting them to the Guildhall Library in London, and they were subsequently transferred to the London Metropolitan Archives.

Assuming that the general question concerning the relationship between security and modernity is relevant to these materials in the context of the globalization process, there are a number of follow-up questions that take us beyond a strictly economic perspective: How was "security" sold and produced in new regions? How did insurance orient itself in new environments? How did "empiricism" relate to previous normative-type beliefs and the frameworks that structured perception? How was "modern" separated from "premodern"? How structurally similar or dissimilar were the cities in which insurance companies attempted to implement a notion of insurance based on the European model

21 As far as I know—outside of perhaps the Phoenix material in the Cambridge University library and that of the Norwich Union—there is no comparable body of insurance history in the nineteenth century covering the global outreach of an agency. The natural disasters branch of Munich Re would be a comparable well-furnished archive for a specialized research question from the 1970s or later, going back retrospectively to the 1950s; the Swiss Re manages a similarly dense archive in Zurich, which, however, does not cover to the same extent the agency network of a primary insurer.

22 Peter Dickson, *The Sun Insurance Office 1710–1960. The History of Two and a Half Centuries of British Insurance*, London 1960, pp. 162–233.

23 Since the publication of the German language edition of this book, a few additional works have appeared that use the files of the Sun Fire Office for studies of insurance globalization, including Lee Kam Hing, *A Matter of Risk. Insurance in Malaysia, 1826–1990*, Singapore 2012, pp. 11–96; Hüseyin Akbulut, *Chumhuriyet Dönemine Kadar Osmanli Imparatorluğunda Sigortacilik*, PhD thesis Istanbul Üniversitesi 2014; Gabriel Tortella Casares et al. (eds.), *Historia del seguro en España*, Madrid 2014, p. 114; Monica Keneley and Grietjie Verhoef, "Establishing Insurance Markets in Settler Economies: A Comparison of Australian and South Africa Insurance Markets, 1820–1910," *African Historical Review* 47, 1 (2015), pp. 76–105; Monica Keneley, "Business Strategies and Conditions of Uncertainty: The Rise of Mutual Life Insurers in Colonial Australia," in Pearson and Yoneyama (eds.), *Corporate Forms*, pp. 169–92; Kristen Alff, "Levantine Joint-Stock Companies, Trans-Mediterranean Partnerships, and Nineteenth-Century Capitalist Development," *Comparative Studies in Society and History* 60, 1 (2018), pp. 150–177. In addition to the Sun and Phoenix holdings, comparably large records can be found with the Norwich Union Fire & Life holdings for questions concerning globalization. The material for European areas (Portugal, Frankreich, Deutschland) seems somewhat scant, but after 1864 (Calcutta), a not insignificant global expansion began; see Ryan, *A History of the Norwich Union*, pp. 614–741.

of cities and markets? In a global context, can the fire danger/fire safety distinction still be superimposed onto the premodern/modern distinction?

For achieving the most meaningful comparison possible (eventually resulting in showing more contrasts than similarities), based on cities with very different and special characteristics, a choice of data from the Sun business balance sheets offers a very good first overview—and reveals also immediately some surprising differences that need explanation—and that, as we will see, the insurance agents could not explain or were not realizing immediately themselves. Evidently, some of the characteristics were learned only the hard way in the moment when the agents began to engage in new cities. In looking at the most significant cities simply in terms of premium payments, remarkable differences can be seen in the relationship between premium payments and damage payments over a decade or two ("loss ratio" with and without agency expenses, Table 4). This comparison was made by the insurers themselves, sometimes for the purpose of analysis of one or the other individual agency, but the records do not contain any global comprehensive narrative or analysis of a loss ratio comparison that might have been effectuated in the Sun Fire's central agency in London around 1880 or 1900, as one might expect. To some extent, what follows is therefore, in this way, only a retrospective historical analysis apparently not accessible to, or conceivable to, the nineteenth century businessmen themselves, though the data was produced and gathered by them.[24] Indeed, the different cities engender very different values in a global comparison, which shows that even from the perspective of urban environmental history there were very different elements at work in these places.

The difference between Asian, American, and European cities is striking. As shall be seen, even the Sun was, in part, quite surprised by these facts. For the following discussion, I have chosen Hamburg, Constantinople (Istanbul), the Indian metropolises of Bombay (Mumbai) and Calcutta (Kolkata), "China" (Hong Kong and Shanghai), and New York (representing the USA), for the purpose of comparative analysis. It would certainly be preferable to be able to work with local archival material for each of the cities mentioned above in order to cross it with the Sun records, especially for the "non-Western" cities; this would provide insight into how insurance was perceived from the perspective of the non-European inhabitants of the countries included within the insurance market. A separate research agenda would involve

24 This so-called "underwriting" or "loss ratio" with the requisite business expenses not yet deducted, represented one of the most common n "measurements used by contemporary underwriters to ascertain the performance of a company's business over time" (Pearson, *Insuring*, p. 36 n 37).

TABLE 4 Comparison of premium income/loss payments/expenses of the Sun Fire Office for agency maintenance in select cities.[a]

City, Years	Premium Income	Damages (loss ratio in %)	Expenditures (loss ratio plus agency costs %)
Batavia (1858–1888)	33,729	0	5,006 (14.84%)
Bombay (1852–1895)	54,772	4,840 (8.84%)	9,705 (26.56%)
Calcutta (1852–1895)	51,944	8,615 (16.59%)	4,925 (26.07%)
East Indies, China, Japan (1852–1888)	39,9538	101,698 (25.45%)	52,999 (38.72%)
Shanghai (1852–1895)	66,806	17,902 (26.79%)	11,318 (43.73%)
Hong Kong (1852–1895)	33,398	13,977 (41.80%)	5,002 (56.80%)
Smyrna (1863–1895)	166,585	77,595 (46.58%)	32,784 (66.26%)
Yokohama (1864–1895)	97,676	68,628 (70.26%)	13,104 (83.68%)
Valparaiso (1857–1895)	128,312	101,159 (78.80%)	21,028 (95.00%)
Constantinople (1865–1895)	158,214	137,087 (87.00%)	20,892 (99.85%)
Hamburg (1837–1895)	383,930	313,545 (81.67%)	106,854 (109.50%)
New York	(average of 22 foreign (US) insurance agencies: 60% loss ratio)		

[a] Sources: LMA CLC/B/192/019/38852/1 u. 2; LMA CLC/B/192/019/31522/275, p. 184f., LMA CLC/B/192/019/11935K.

examining the transfer processes in more precisely differentiated sociohistorical contexts, extending to a linguistic study of the translations of insurance terms into different languages where the phenomenon was new. The data preserved within the advertising materials included in the Memorandum Books provide a wealth of information for such an endeavor in Arabic, Chinese, Bengali and many other languages; perhaps this information could even be compared with existing local court and administrative records. Unfortunately, this topic cannot be further addressed here, nor has it yet been tackled in any work of economic history. Nevertheless, attempting what is primarily a qualitative investigation into the perceptions of the field agents and the head office rather than looking at market trends should be a step forward in that direction. The possibility of integrating a more balanced, "non-Eurocentric" investigation into the process of the European expansion of insurance companies would therefore have to be separately and carefully evaluated for each city and region; until now, existing research does not suggest that there is a wealth of archival material that would be easily accessible to more knowledgeable scholars such as the present author for such an investigation, though such findings would certainly be very desirable.[25]

1 Hamburg[26]

The Sun began to conduct business outside of Great Britain in 1836 in Hamburg under the agent William Elliott. Initially the company expanded rapidly—it established 81 agencies throughout Germany that were linked to the head office in London via the main German office in Hamburg. The value of insured properties in Hamburg rose in the six years from 1835 to 1842 from approximately £857,000 to £6,422,000, and premium income rose more than tenfold from about £2,045 to £25,000—no other agency outside England brought in more revenue.[27] But just six years after the Sun entered the Hamburg market, the Hamburg fire of 1842 broke out, leading to claims amounting to £117,710[28]—more than double the revenue that had been accrued to date—at which point the Sun withdrew from the German market (which seemed to be plagued by intense competition anyway, as well as a certain hostility toward foreign insurers), only to return again, just as quickly, focusing exclusively on Hamburg, Lübeck, and Bremen until World War I.[29] Between 1845 and 1871 annual premium income stabilized at the moderate rate of about £4000, slowly rising by 1855 to about £6000 and once again yielding the highest revenues outside of England.[30] As seen above concerning the Imperial Chamber Court case, the Phoenix, represented since 1786 by its agent, William Hanbury, had maintained a presence in Hamburg for a much longer period. Over the course of

25 In any case, the recent literature quoted above in n. 23 does not differ substantially in terms of sources used and there are not many hints there to the wealth of local sources available that offer the possibility of an easy de-Europeanization of research approaches in this field. The first step must be a close analysis of the European perceptional frames of thought of these.

26 Some aspects of this paragraph have been further developed in Cornel Zwierlein, "Perceiving Urban Fire Regimes in Europe and China (1830 to 1870): British Fire Insurance Businesses and the Sudden Challenge of Globalisation," in Gerrit J. Schenk (ed.), *Historical Disaster Experiences*, Heidelberg 2017, pp. 327–52.

27 According to LMA CLC/B/192/019/38852/1.

28 "Losses 1842," LMA CLC/B/192/019/38853/1.

29 Dickson, *The Sun Insurance Office*, pp. 164–71.

30 LMA CLC/B/192/019/31522/111, p. 12.

seventy years, the second headquarters on the Elbe contributed more policies representing far greater revenues than the rest of the Phoenix foreign-agency network combined until the company entered the American market.[31] With the Hamburg agency, Hanbury created "a full-blown satellite company of notable independence."[32] Although Phoenix ended up paying more than twice as much as the Sun in losses following the 1842 fire in Hamburg, Phoenix did not respond with the same panic and did not completely withdraw from the German market.[33] Taking this situation into account, one might conclude that the Sun files on Hamburg and Germany would be of little interest. But the opposite is actually true: particularly in light of its rapid entrance into the market and the rash, almost panicked exit after 1842, it is possible to reconstruct the image the Sun had previously had of Germany, and Hamburg in particular.

While this image is reflected in the letters written by Elliott to the headquarters, it is more strongly presented in the reports written for the Sun by Richard Atkins, the head of the newly established Foreign Department, about his travels to Hamburg and Germany in 1837, 1839, and 1842. Atkins examined Elliott's accounting and the informational basis for all the foreign cities where agents had been established, and reported being very impressed.[34] Elliott, who had lived in Hamburg for over 50 years, pointed out the 150 sugar mills as particularly lucrative targets of insurance; although they obviously carried a degree of unavoidable risk, he could not recall more than one or two fire incidents in the past 50 years. At one point there were more than 500 refineries in Hamburg, but the larger ones had pushed the smaller ones out of the market over time.[35] Atkins went on-site to inspect the various "risks," i.e., the main

31 Clive Trebilcock, *The Phoenix Assurance and the Development of British Insurance*, vol. 1, *1782–1870*, Cambridge 1985, p. 189.

32 Trebilcock, *The Phoenix Assurance*, p. 193.

33 Trebilcock, *The Phoenix Assurance*, pp. 284–92. On Hanbury and his descendants, who followed the agency for decades, see HStA Hamburg, Best. 621-1, Hanbury posthumous work: Main Account Books of the Company, 1815–1864.

34 "In addition to the Office System, which he carries out completely, he has a book of streets for each of the considerable towns wherein are entered all the insurances in that street, so that at any moment of time he can tell the sum insured in each street by the numbers of the houses & references to maps & thus effectually control & supervise the extent of his risks & the conduct of his subagents" (LMA CLC/B/192/019/31522/106, p. 2).

35 Ibid., p. 3, 23. According to Petersson, *Zuckersiedergewerbe*, pp. 58, 229, the year 1807 was the highpoint with approximately 428 refineries. The decline of the Hamburg sugar industry was, above all, due to the progress of modern competitors in Hanover and Prussia (not in the still insignificant influence of beet sugar). The highpoint occurred within the Continental Blockade era; cf. Burghart Schmidt, *Hamburg im Zeitalter der Französischen Revolution und Napoleons 1789–1813*, 2 vols., Hamburg 1998 und Silvia Marzagalli, *Les*

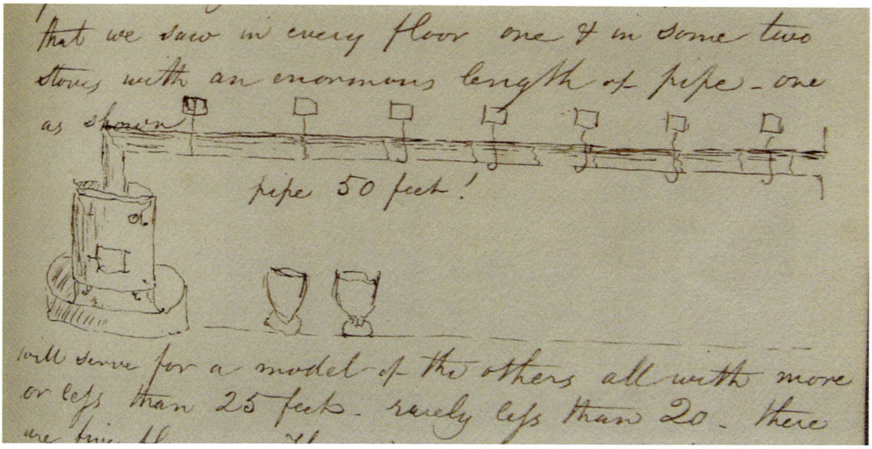

FIGURE 68 Drawing of a risk-elevating heating pipe by Atkins (LMA CLC/B/192/019/31522/106, p. 23)

potential properties to be insured, and drew a number of small sketches to illustrate the report. The steep gabled roofs of the reservoirs, for example, were considered to be a major risk in the event of a fire.[36] In one traditional sugar refinery, which, from an English perspective, was built in the "simplest & most primitive form," Atkins asked the longtime owner about the fire risk. At first, this appeared highly amusing to owner, since he claimed never to have heard of a fire in a sugarhouse: "though I brought him to confess that the insides of stores were sometimes burned."[37] Atkins examined the design of the ovens—for example, in the Dannenfeld sugarhouse on Altewall Street, which had been operated by the same owner since 1799—and found the fire risks associated with the equipment to be shocking: "This forms the most extraordinary risk I ever saw & it is to me quite unaccountable how the Hamburg refineries escape accident." Each floor of the building was heated by ovens equipped with old, unsafe, six to seven-meter-long furnace pipes extending over the entire length of the room. Atkins drew a picture to illustrate the setup (Fig. 68) and expressed his astonishment at the owner's claim that his rate with a local insurance company was nevertheless still far too high (1/5%).

---

boulevards de la fraude. Le négoce maritime et le blocus continental, 1806–1813, Villeneuve d'Ascq 1999.

36 LMA CLC/B/192/019/31522/106, p. 5.

37 Ibid., p. 4.

Later, Atkins encountered a factory in Hanover with almost twenty meters of unprotected heating pipes.[38] The Andlys couch factory, in contrast, made a good impression because "everything [was] conducted on the English mode"[39]—although he noticed that unlike in England, "preindustrial" work took place in the homes of daily laborers instead of in factory buildings.[40] He deemed the water supply, which was operated by three water companies, to be very good; importantly, it was independent of the Elbe canals.[41] Nevertheless, Atkins was not quite certain about the actual degree of fire risk and determined that it would be "desirable in the highest degree … to limit the sums insured in quarters" and that Elliott should be "extremely cautious". After touring northern and eastern Germany, his final judgment about the 'motors of early German industrialization" remains somewhat ambiguous:

> From my survey of the German manufactories of many kinds—especially Sugar Refineries, Cotton, Woollen & Flax Mills, I am decidedly of opinion that they present generally very bad risks—The Stoves with very long iron pipes in universal use cannot be considered but as greatly increasing the hazard. It is important however, to take into consideration the following facts—1st In respect of Sugar Refineries after a very large experience in Hamburg the insurance of them has proved a source of great profit both to the London & Local Offices. Whatever causes may be in operation, they have proved much better risks than the English refineries. In Hamburg the chief seat of them, the losses have been very few & in proportion to the number, present a remarkable contrast in the number of accidents to those which have happened in the London Refineries.[42]

Also, regarding the cotton mills, he judged that in spite of the strange and dangerous pipe furnaces, "The systems of management in regard to the discipline

38 Ibid., p. 29.

39 Ibid., p. 12.

40 "The marked difference which subsists between all such manufactories in England & those in Hamburg, consists in the fact, that all the work that can be possibly done away from home, is sent out to be wrought in small workshops, or in the dwellings of the workmen—This I was told was the case with all branches of manufacture, Carpenters, Cabinet makers in short every trade, which in England is often brought into large buildings is here given out in piece work to journeymen who find a place for themselves to work in. There is not for example a large cabinet makers shop known in Hamburg" (ibid., p. 13).

41 Ibid., p. 15f.

42 Ibid., p. 102f.

of the workpeople & the internal arrangements are remarkable for strictness & as regard to security against fire."[43] The miraculous fire safety of the antiquated German refineries, which he apparently believed to be credible, is quite inexplicable to Atkins, and the only compelling explication seems to be sociocultural—namely, there was a particular discipline and culture of safety that was at work within these factories. The meticulousness that he perceives as a positive attribute in this regard attracts his derision in a different context, when he encounters the cumbersome and obsessively detailed German business practices during his second visit to Germany in 1839:

> The amount of correspondance is absolutely incredible. The lengthy proposals of the agents requiring for every policy however simple a careful perusal & dissection—the habit of carrying every thing out into the most minute & inconsequential details & raising an interminable correspondance upon them, present an amount of counting house labor which appears to me in contrasting it with English business absolutely frightful.[44]

During an inspection, he offers a thoroughly positive evaluation of the 700 regular and 1800 replacement firefighters—mostly boatmen and other trustworthy individuals having access to water—however notes that their uniforms were inconceivably dysfunctional.[45] The Sun became increasingly engaged with insuring the existing steam-powered factories.[46] After a further inspection of the situation in Germany by Atkins in 1839, Elliott's agency headquarters were transferred from Hamburg to Berlin, as they increasingly focused on Germany as a whole, or at least on the northern areas. On May 6, 1842 Atkins headed to Germany for a third time—for a routine inspection and the evaluation and appointment of a couple of agents—and arrived in Cuxhaven on May 8. The news of the disastrous Hamburg fire that raged in the city on May 4–8 reached him here: first, he returned briefly to London and was given instructions at the headquarters to go back to Hamburg on May 20, where he was to support Elliott in handling payments to the 287 victims who were insured by the Sun.

The massive scale and strong growth commitment of the Sun during the first six years shows, first and foremost, that according to the qualitative inspection analyses of their foreign specialists, Atkins had hardly anticipated

43 Both quotes, ibid., p. 102f.

44 LMA CLC/B/192/019/31522/106, p. 112.

45 Ibid., p. 119f.

46 E.g., the machine factory Gleichman & Büsse: ibid., p. 131f. with drawing.

such a large fire. From the outside it may seem surprising how little the "expertise" of the analyzing "experts" matched the reality. The files do not indicate any kind of awareness of an increase in the risk of fire, even though there were certainly indications: in Ch. 3, 1.6 I have used the Hamburg fire statistics by Bieber to explain Hamburg's "fire ecology", and in the files of the Sun, there is a version of Bieber's work that extends to 1834 and one extending to 1840.[47] The findings should have been alarming for insurance companies: following the relative paucity of fires throughout much of the eighteenth century, both the average number of firefighting incidents (over 100 per year), as well as the number of destroyed buildings significantly increased, especially between 1830 and 1841, in spite of substantial population stagnation. These findings curiously did not make their way into the merely quantitative statements included within the files of the Sun Fire Office, even though, as Atkin's report shows, the Sun was particularly interested in the manufacturing systems of Germany. As clarified above in Chapter 3, 1.6, a large number of the approximately 575 buildings that burned between 1772 and 1841 were protoindustrial commercial sites, including 48 sugar factories and 17 distilleries/breweries, and there were many indications at this point that such buildings carried a particularly high fire risk. An evaluation of Bieber's statistics could certainly have raised awareness for the Sun Fire Office, but the signs simply were not recognized. The analytical judgments present in Atkins's qualitative inspections were thus somewhat indecisive and ambiguous, and in the end, insurance coverage was offered less discriminately here than was probably wise.

2 Istanbul

Istanbul—or Constantinople, as it is typically called in the English sources—is not one of the first cities where the Sun started business in outside of Europe,[48] but the Tanzimat reform movement beginning in 1839, as well as the construction of the Orient Express railway, attracted new attention to the Ottoman Empire within London's business community. Initially the Sun acquired an agent in Smyrna in 1863, and from there gained access to the most important cities in the

47 LMA CLC/B/192/019/31522/107, p. 2–8; LMA CLC/B/192/019/31522/109, p. 83f.

48 Dickson, *The Sun Insurance Office*, p. 188f. starts his chapter on the non-European engagement of the Sun with Smyrna. The balance sheets, however, betray its earlier activity in the Indian and West Indian outposts in Valparaiso, Singapore, Batavia, Hong Kong, Shanghai and Cape Town: LMA CLC/B/192/019/38852/1 und /2.

Middle East. Foremost among them was Istanbul/Constantinople, to which the Sun obtained access in 1864.

This city of approximately 500,000 residents was one of the most dangerous cities in the world at the time in terms of fire danger due to its heavy reliance on wood construction. As with other Middle Eastern cities, research into the history of fire in Istanbul is not terribly substantial; nevertheless, in her monograph on Istanbul in the nineteenth century, Zeynep Çelik includes an examination of the extent to which serial fires contributed an important element in the process of urban development —and sometimes stagnation—in contrast to, for example, the situation in Indian cities.[49] Between 1633 and 1839 there were 109 major fires recorded in Istanbul, and 229 fire in the period from 1853 to 1906 significantly influenced the contours of the modern cityscape.[50] The Sun sent one of its employees, Mr. Woods, from Smyrna (Izmir) to Constantinople in June 1864 to create a situational analysis evaluating where and how they could establish themselves in the city, and he was able to gain access to historical and current statistical material concerning the frequency of fires—albeit from unknown sources. Here, for example, is a list of unknown origin recording the major fires of the eighteenth century. This list would have certainly made any fire insurer queasy (Table 5).

In the period from 1841 to 1848, 13,750 destroyed houses were recorded in the city district of Pera;[51] for the last four years, from 1859 to 1862, Woods was able to get exact statistics from the city administration, stating that for the Sixth district (Galata and Pera), which was of great interest to insurers, "only" 337 houses burned down. These statistics included a detailed breakdown of construction types and identified destroyed and rebuilt plots of land, which suggested that that this district had ultimately grown.[52] "From 1865 to 1869" the

---

49 There is not much research on the fire history of medieval and early modern Middle Eastern cities, cf. Anna Akasoy, "The Man-Made Disaster: Fire in Cities in the Medieval Middle East," *Historical Social Research* 32 (2007), pp. 75–87; Syrinx von Hees, "The Great Fire in Cairo in 1321: Interactions Between Nature and Society," Schenk (ed.), *Historical Disaster Experience*, pp. 307–326; Baer, "The Great Fire of 1660"; Doris Behrens-Abouseif, "The Fire of 884/1479 at the Umayyad Mosque in Damascus and an Account of Its Restoration," *Mamlūk Studies Review* 8 (2004), pp. 279–97.

50 Zeynep Çelik, *The Remaking of Istanbul: Portrait of an Ottoman City in the Nineteenth Century*, Berkeley 1993, p. 52f. It is not true that no fires occurred in the first 180 years of Ottoman rule; cf. H. İnalcik, "Istanbul," *The Encyclopaedia of Islam* 4 (1978), pp. 224–48, 237 and especially Alfons Maria Schneider, "Brände in Konstantinopel," *Byzantinische Zeitschrift* 41 (1941), pp. 382–403.

51 LMA CLC/B/192/019/31522/257, p. 36.

52 "[…] le 6me Cercle s'est augmenté, en quatre années de 424 bâtiments en pierre, et 206 bâtiments en bois, comprenant ensemble une surface de 45.127 piks carrés" (LMA

TABLE 5 List of major fires in Constantinople of unknown origin, recorded by Agent Woods (LMA CLC/B/192/019/31522/257, p. 89).[a]

1729	12,000 houses
1745	Fire: 5 days
1749–50	12,000 houses
1750	10,000 houses
1751	4,000 houses
1756	500 houses
1756 July	15,000 houses
1761	large fire
1765	large fire
1767	large fire
1769	large fire
1771	large fire
1778	2,000 houses
1782 February	600 houses
1782 June	7,000 houses
1782 August	10,000 houses 50 mosques 100 corn mills
1784	10,000 houses
1791 March and July	32,000 houses
1792	7,000 houses
1795	7,000 houses
1799 Pera	1,800 houses and other buildings

[a] It reads on LMA CLC/B/192/019/31522/257, p. 32 "No records of number of Fires have been kept for any length of time." This list does not match the list of fires given in Schneider "Brände in Konstantinopel" and is not reliable; in terms of the dimensions, however, it would have given the insurance company a more or less accurate image of the high fire danger. This is also indicated by the evaluations found in the files of the Levant Company through Matterson in 1936, who gives numbers of destroyed houses for several years: in April 1696 there was a major fire, destroying almost all the houses of the English nation; in 1729, 20,000 houses were destroyed; in 1731, 75% of Galata was destroyed; in 1751, 11,000 houses were destroyed—in contrast, the major fires in Aleppo und Smyrna were no major issue for the Levant Company, so that in reality this all specifically concerns Istanbul. See Clarence H. Matterson, *English Trade in the Levant, 1693–1753*, PhD diss. Harvard University 1936 (Harvard University Archives HU 90.3052), pp. 31–32 drawing on The National Archives Kew, PRO SP 97/20; 97/24; 105/110; 110/72; 105/118.

city experienced "the most active phase of urban planning in [the] nineteenth-century."[53] At the same time, the city had been relentlessly plagued by uncontrolled fires. Even in the short time that Woods was in Constantinople, he witnessed a fire in the old town that "within three hours burnt out 500 families, about 300 houses, 80 of these families were Greeks."[54] Woods made observations about the existing fire-extinguishing technology in the city, which he judged as essentially ineffective.[55] Constantinople was thus a city of fire, a city whose reform-era urban construction dynamics were strongly influenced by medium-sized and large fires. At times, urban planners took advantage of the fire-cleared areas as a development opportunity; however, the sheer misery of the fires often curtailed such an impetus.

The Sixth district had been selected in 1858 by the city council as a testing ground for modernization:[56] the external pressure on the Ottoman Empire had become so great that in the atmosphere of the Tanzimat that it appeared necessary to approach Western standards and emulate the urban development of cities like London, Paris, or Vienna.[57] The establishment of insurance at this point indicates that those in London hoped for a profitable outcome.[58] Accordingly, on a map of the predominantly European-populated Sixth district (Pera, Galata,

---

CLC/B/192/019/31522/257, p. 34; Pik Hâlebi = "Ell of Aleppo" = 0.686m ; cf. *Meyers Konversationslexikon*, 4th ed., Leipzig 1885–92, vol. 15, p. 922). A similar statistic for the years 1859–1864 served the calculation of a *Times* journalist that among 4113 destroyed buildings in 160 fires Constantinople during these six years results in a very high average number of 26 houses burned per fire (*Times*, September 8, 1865, in LMA CLC/B/192/019/31522/260, p. 38).

53 Çelik, *The Remaking*, p. 63.

54 LMA CLC/B/192/019/31522/258, unpag. [p. 42].

55 Ibid. The "dry- or destruction fighting techniques" of the Turks, above all through the tearing down of buildings and only secondarily through the relatively inefficient use of small portable water sprayers is emphasized; cf. Lionel Frost, "Coping in Their Own Way: Asian Cities and the Problem of Fires," *Urban history* 24, 1 (1997), pp. 5–16, for a different take on these "Asian" extinguishing techniques, see Jordan Sand and Steven Wills, "Governance, Arson, and Fire Fighting in Edo, 1600–1868," in Greg Bankoff, Uwe Lübken, and Jordan Sand (eds.), *Flammable Cities: Urban Fire and the Making of the Modern World*, Madison, Wisconsin 2011, pp. 44–62.

56 Cf. Christoph K. Neumann, "Modernitäten im Konflikt. Der Sechste Munizipal-Bezirk von Istanbul, 1857–1912," in *Istanbul: vom imperialen Herrschersitz zur Megapolis. Historiographische Betrachtungen zu Gesellschaft, Institutionen und Räumen*, Munich 2006, pp. 351–75 (see p. 369f. on the fire of 1870).

57 Emel Ardaman, "Perspective and Istanbul: The Capital of the Ottoman Empire," *Journal of Design History* 20 (2007), pp. 109–30, 110–12.

58 False, according to İnalcik, "Istanbul," p. 237: "fire-insurance began only in 1890." At most what might be meant here is that on August 2–14, 1892, the first Ottoman fire insurance, the *Société Générale d'Assurances Ottomane* began its work and even until 1906 collected

FIGURE 69 Map produced by C. Stolpe, with the allegedly better insurable risks colored in red by the Sun Office's agent Wood—Plan der zum 6.ten Communalbezirk vereinigten Vorstädte Galata, Pera und Pancaldi von Constantinopel [...] ergänzt [...] bis ins Jahr 1861 durch C. Stolpe LMA CLC/B/192/019/31522/257, Annex

Pancaldi),[59] which included the original distinct delineations between "Muslim" and "Christian" districts and neighborhoods, Woods once again marked in red those blocks that in his opinion were good enough to be insured: without exception, only "Christian" housing blocks were deemed insurable (Figure 69).[60]

---

by far the most premiums among the 41 competing insurance companies in Istanbul, cf. LMA CLC/B/192/019/31522/266, p. 285, 349. In 1906 the insurance law was also enacted as an annex to the book of commercial law, see ibid., after p. 350. The implementation of national insurance companies only beginning in 1890 refers to the protectionist tendency in the global market at this time—described as a globalizing "backlash," cf. Peter Borscheid and Kai Umbach, "Systemwettbewerb, Institutionenexport und Homogenisierung. Der Internationalisierungsprozess der Versicherungswirtschaft im 19. Jahrhundert," *Jahrbuch für Wirtschaftsgeschichte* (2008), pp. 207–26, 217f.

59 A graphic presentation of the mix of cultures in Galata/Pera, especially from the Greek perspective can be found in Akylas Mellas (ed.), *Pera: The Crossroads of Constantinople*, Athens s.d. (ca. 2003).

60 For a later version of Stolpe's map, see http://www.lib.uchicago.edu/e/su/maps/asian-cities/G7434-I8-1882-S86.html. Announcement of the Routh & Son agency for the Sun in French, English, Turkish, and Greek: LMA CLC/B/192/019/31522/259, p. 14–17.

The heuristics of insurability extracts and delineates the supposedly secure European, stone-built, "modernity" from the closely spaced plurality of "nations," cultures, and, in some respects, eras. The sizes of the eighteenth-century fires in the table above demonstrate that if one adopted the simplistic material and structural standpoint that modernity equates with fire-security (not accepted in this study, but always helpful as a quick reminder of an association and yardstick shared by many in Western societies then and now), Constantinople was still properly located in a phase of "premodernity": In the eighteenth century, Constantinople suffered fires equal to the size of London's famous fire of 1666 (ca. 12,000 homes destroyed) exactly ten times—a fact of which the European historical memory is hardly even aware. In 1865, just one year after Wood's report, there were multiple serious fires, among them a fire on August 10, in which the old Serail palace was largely destroyed. On September 5–6, the largest ever fire broke out in Istanbul's historic district. Early high estimates indicated that 8000 buildings were destroyed; the number was later revised down to 2744 buildings, including 1,879 houses and 751 shops.[61] English newspapers took great pleasure in reporting on this fire, comparing it to the London double-disaster of 1665/66: just as in London, the Great Fire followed a major outbreak of the plague, in Constantinople, fire followed a major cholera epidemic. The comparisons clearly suggested that the state of Constantinople's firefighting technology and administration also seemed to date from about "two hundred years" ago. People made jokes about the little hand fire extinguishers operated by running soldiers, typically imputed more to Oriental folklore, which were comparable to the English firefighting techniques of the early modern period (Fig. 70).[62]

The catastrophe encouraged European newspapers to publish great visions of urban planning à la Baron Haussmann for the burned areas of Istanbul. Regrets were even voiced that a larger part of the city had not succumbed to the flames; other voices, however, took a moralistic tone against the cynicism that allowed speculators "to fondle future landscapes from imaginary

---

61 LMA CLC/B/192/019/31522/260, p. 44f. On the reconstruction after the "Hocapaşa" fire, see Çelik, *The Remaking*, pp. 55–59—and for a more precise map of the fire damaged area than in Çelik, *The Remaking*, p. 56, see LMA CLC/B/192/019/31522/260, p. 33. Cf. Ardaman, "Perspective," p. 115.

62 LMA CLC/B/192/019/31522/260, p. 37f. (Times of September 8). "[…] but to form any idea of a Turkish fire-engine we must call to mind those prodigious leathern squirts which are still preserved, with their attendant buckets, in a few City churches [sc. in England], and which, during the seventeenth century, were worked at London fires by the Bridewell boys. […] Moreover, in old Stamboul the term ‚house' is eminently a title of courtesy. A Turkish house is, in plain language, a very dirty wooden box, with a big hole in it called a door, and a few little holes assumed to be windows." (ibid., p. 39—Telegraph).

FIGURE 70 Late Ottoman fire fighters with hand-held fire extinguishers (engraving from a British newspaper, 1870, in: LMA CLC/B/192/019/31522/261, p. 11)

minarets"[63] above the heads of the suffering residents. In any case, the appointed reconstruction commission worked relatively successfully until 1869. The Western disdain for the situation in the Muslim and Armenian districts was intensified by the fact that the fire broke out in "one of the most crowded and dirty districts of the old city" with a large number of wood buildings,[64] while the British newspaper author imagined here the "professedly better-off quarter of Pera" as highly superior in comparison: although three to four large insurance companies (Royal, Imperial, Sun) had operated in Pera and Galata for two years, they would "naturally shrink from having anything to do with wooden-built Stamboul."[65]

Almost five years later, however, they were proven wrong by a major fire on June 5, 1870, when about 8000 houses burned down in Pera[66] and about 1,300 people died—and this, by tragical irony, in a quarter which had been

63 Ibid., p. 41.

64 Ibid., p. 39.

65 Ibid., p. 35.

66 3,000 houses, according to Çelik, *The Remaking*, p. 64.

FIGURE 71 Plan of the part of the 6th district of Constantinople destroyed by the 1870 fire (LMA London CLC/B/192/019/31522/260, p. 149)

designated as "insurable" by Wood and where stone construction was already dominant (cf. the damage area plotted on Figure 71).

In the five years since the establishment of insurance, only a few houses had been insured, so the initial estimates of the damage amounts of the insurers (£30,000 Royal, £25,000 Imperial, £26,000 Sun) were minimal in comparison with the total estimated damages of £6 to 7 million. In fact, only 20 houses insured by the Sun were affected, amounting to a damage value of £17,174.[67]

Once again, visions of planning were sparked by this burned tabula rasa—for the most part, however, they remained largely unrealized and the special administration of the Sixth district was dissolved again within the same year.[68] Pera's reconstruction proceeded sluggishly, in a similar fashion to many

67 *The Levant Times*, June 13, 1870, p. 661 in LMA CLC/B/192/019/31522/260, p. 113; for the risks actually encountered, cf. ibid., p. 160. In 1869 the total insured amount in Constantinople was £146,176 and the premium amount collected was £905 (cf. the general balance of the Foreign Department 1869/70 and "paid losses" 1870 in LMA CLC/B/192/019/38852/2).

68 Çelik, *The Remaking*, pp. 64f., 73 (with an imprecise map of the fire area featuring a damage zone that is much too small: this is also made clear by the photographs in in LMA CLC/B/192/019/31522/260, p. 171ff. See Ardaman, "Perspective," p. 114 for a retrospective map of 1917).

European cities in the early modern period. The fire of 1870 was often compared to the Chicago fire one year later: while in Chicago, 18,000 houses had been rebuilt thirty months after the fire ("1 house rebuilt every hour"), in Pera 600 houses had been rebuilt after forty-seven months (one house every 56.5 hours).[69] While Western observers saw a huge side-effect of Chicago's Fire in the massive growth of a hitherto unprecedented modern urban form,[70] Pera's fire was, for European observers, a significant indicator of the limited ability of the city to modernize.

Such an assessment is also reflected in the hard insurance numbers: the agents in Smyrna, Henry & Rose, had continued to judge this major fire as "quite an exceptional occurrence" and had determined that "the field of insurance at Constantinople" was "favorable for operations."[71] In fact, insurance did initially seem to be profitable. Income soared from £146,176 in 1869 to more than 10 times that number (£1,554,247) within four years. After another year of high damage payments (1874), insurance income settled at about half that number until about 1879. After 15 years of activity in Constantinople, the sum of premium income between 1865 and 1879 was only slightly higher than the total amount of paid losses (about £76,750 versus £74,434).[72] The high fire danger within the city was thus reflected in some initial attempts at defensive policy writing practices for only a few "European" "good risks" which thus yielded lower returns; this was not an expansively growing, modern market in the same way as the North American cities, at least not until the Young Turk Revolution of 1908. Even as late as 1918, a large fire destroyed more than 7,000 houses.

Fires seemed to be an endemic, "progress"-halting problem in the city. It was no coincidence—or at most, a coincidence in the microarena of collective action patterns of ethnic groups and religious communities—that the endemic nature of this situation became an almost constant factor in the calculus of its building and residential patterns. While Baer has shown that in the reconstruction after the largest fire of Constantinople in 1660 (280,000 houses destroyed, 40,000 deaths), previously Christian and Jewish properties inside the city gates were Islamicized,[73] Woods notices two hundred years later that the opposite

69 LMA CLC/B/192/019/31522/262, p. 31.

70 "No other city in America had ever grown so large so quickly" (Cronon, *Nature's Metropolis*, p. 9; "In the 1880s and early 1890s, Chicago was probably the fastest growing metropolis in the Western hemisphere" (Sawislak, *Smoldering City*, p. 283 n 21).

71 LMA CLC/B/192/019/31522/260, p. 205.

72 Based on the balance sheets and damage amounts in LMA CLC/B/192/019/31522/2.

73 Baer, "The Great Fire of 1660".

tendency was true in Istanbul for the many "small" fires that affected several hundred houses in areas home to a plurality of ethnic and religious groups. The enforcer in such cases was not the state, which with its Western-oriented urban concepts had only minimal success; rather, it was the collectives of "nations" or "millets":

> The Fires appear to act as the Pioneers of other Nationalities & amongst the Turks. The Turks allow the Jews to settle amongst them, & the Jews do not object to Greeks amongst themselves, so by degrees the Quarter becomes a mixed Jewish & Greek quarter & these two nations side by side help one another to shoulder out the Turk. Other nations eventually inhabit the ceded Quarter, & so the whole of Stamboul seems likely to undergo in time a total change.[74]

If even in such catastrophic moments, divisions between "nations" remained intact, then the knowledge of local differences, customs, and even different jurisdictions within the city is significant. At the London headquarters, in addition to pure statistical material and information about the housing stock, there was also relevant cultural information about the ethnic/religious and national pluralism existing in Constantinople,[75] albeit at a rudimentary level that hardly helped to overcome a strong Western "Orientalist" perspective.[76] "Nationality" in Woods's report refers on the one hand to "millets," officially recognized as separate cooperative groups of non-Muslim subjects of the Ottoman Empire,[77] and on the other hand, to citizens of European countries. Here it seems that in spite of concurrent processes of nationalization, the premodern "natio" continues to be entangled with modernity until the mid- to late nineteenth century.[78]

74 LMA CLC/B/192/019/31522/258, unpag. [p. 43, 45].

75 E.g. LMA CLC/B/192/019/31522/259, p. 89–108.

76 The fact that long into the twentieth century, even in the field of urban history, such an "oriental" view of "the" Muslim city as being a depraved form vis-à-vis the cities of antiquity, is demonstrated by: André Raymond, "Islamic City, Arab City: Orientalist Myth and Recent Views," *British Journal of Middle Eastern Studies* 21 (1994), pp. 3–18.

77 Benjamin Braude and Bernard Lewis (eds.), *Christians and Jews in the Ottoman Empire: The Functioning of a Plural Society*, 2 vols., New York 1982 and Michael Ursinus, "Zur Diskussion um 'millet' im Osmanischen Reich," *Südost-Forschungen* 48 (1989), pp. 195–207.

78 For example, an effect of the Greek War of Independence of 1821 is explicitly indicated: the houses, business, and lands of Greek settlers (primarily immigrating under Selim III, 1762–1808, ruled 1789–1807)—who quickly began to dominate a large part of the Mediterranean trade, and more or less took over the position of the Levant Company that was founded in 1581 and gradually lost significance in the mid-eighteenth century

"Safety" and "modernity" had not yet established a balanced relationship with each other in Constantinople: the available forms of modern design, modern introspection, and modern belief in progress did not lead to sustainable fire safety. Elements of "premodernity" and "modernity" were intertwined here and the litmus test of fire safety did not function. The Sun's perception of Constantinople was strongly influenced by the impression that if spatial planning was exact enough and areas where construction and (partial) fire safety were relatively "modern" were clearly identified, it should be possible to run an insurance business according to "modern conditions" in the Christian district. The map created by Woods served as the most important orienting instrument for precisely distinguishing "premodern" from "modern" areas. But this was more of a suggestion of predictability and empirically grounded truth, which, of course, failed in the end.

3 Bombay/Calcutta

The Sun Fire Office established itself in India in 1852, earlier than in the Middle East.[79] Although further away, the relevant cities were part of the British Empire, and therefore "felt" closer than the Ottoman Empire cities that were, at most, economic players in the sphere of colonial powers. In India, Leckie & Co. was appointed as the managing agency in Bombay, Parry & Co. in Madras, and Cullen, Muir & Co. in Calcutta.[80] In 1842 the local Bombay Fire Office had already terminated its activities and offered its remaining inventory and agents to the Sun; however, instead of the Sun, Imperial Insurance took over and the Sun was, in the end, only involved with the reinsurance of some outstanding risks. It was only in June 1852 that the

until it was dissolved in 1825—were seized in the course of pogroms during the Greek uprisings in Constantinople. In response, the majority of the wealthy merchants assumed a new nationality (French, Austrian, English, or Russian), in order be endowed with the corresponding protections: this meant moving from the old corporative nationality of the "millet" system to a new nationality within the modern system of the state. See LMA CLC/B/192/019/31522/259, p. 94f. For the context cf. Cornel Zwierlein, "Mediterranean Transformations: From the security of mercantilist trading empires to a modern security regime," *Pedralbes. Revista d'Història Moderna* 40 (2020) [forthcoming].

79 LMA CLC/B/192/019/18249, vol. 13, f. 1, 2, 13, 14 (letters exchanges on establishing contact with the agencies in the three cities; cf. also the first balance sheets with Indian revenues) LMA CLC/B/192/019/38853, vol. 1 (1852).

80 Dickson, *The Sun Insurance Office*, p. 192f.: Dickson treats the Indian engagement of the Sun only in these two pages.

Sun commissioned its own agents. In Calcutta at this point, in addition to Imperial Insurance, Alliance, Phoenix, and Royal Insurance were already active players in the market. By 1852, Royal Exchange had entered the market, and they were quickly joined by a succession of companies: Liverpool & London Insurance (1853); Northern (1856); and Globe, London and Phoenix (1859). In the 1860s and '70s an additional twenty fire insurance companies from England, as well as one German company (Magdeburg-Hamburg) were also established.[81] There was a similar progression in Madras.[82] By 1827/28, Phoenix had agents in Calcutta and Madras,[83] but beyond this example, the expansion of the insurance market in India began primarily in the 1850s—thus, the Sun was one of the first to be established there. The major trade cities of the Indian subcontinent, had, of course, encountered the concept of premium-based insurance earlier, but only in the form of maritime transport insurance and essentially only within the internal trade circuit with Europeans: there is no documentation indicating an earlier integration of premium-based insurance in the Indian hinterlands or in other insurance arenas, such as "fire" or "life."[84] In the seventeenth and eighteenth centuries, there were already insurance-like practices in India, which apparently resembled the Mediterranean *foenus nauticum*, i.e. the "insurance purchase" (bīmā) associated with maritime transport. But this was not a principle that had been generally applied to other areas.[85]

In practical terms, the European insurance of the nineteenth century was initially directed toward the three major British economic centers, which were also India's largest cities at the time. The East India Company was established

---

81 LMA CLC/B/192/019/31522/151, p. 21–76. In Brian Supple, *The Royal Exchange Assurance: A History of British Insurance, 1720–1970*, Cambridge 1970, p. 214.

82 LMA CLC/B/192/019/31522/164, p. 37.

83 Trebilcock, vol. 1, p. 233f.

84 There is still no comprehensive work on early Indian insurance history. Cf. G.R. Desai, *Life Insurance Business in India*, Delhi 1973, pp. 1–6; Benoy Kurmar Sarkar, *Social insurance Legislation and Statistics*, Calcutta 1937; Tirthankar Roy, *The Economic History of India 1857–1947*, 2nd ed., Oxford 2006, p. 288f.; Adrian Jitschin: *Über die Versicherung der Inder. Indische Lebensversicherung 1874 bis 1988*, PhD thesis, University of Marburg 2011, pp. 3–8; G. Balachandran, "India: From a Colonial Outpost to a Leading Market," in Peter Borscheid and Niels Viggo Haueter (eds.), *World insurance: the evolution of a global risk network*, Oxford 2012, pp. 444–71, 444–47. Often, the insurance sector is omitted in general economic histories. Binay Bhushan Chaudhuri (ed.), *Economic History of India from Eighteenth to Twentieth Century*, New Delhi 2005; Sabyasachi Bhattacharya (ed.), *The Cambridge Economic History of India*, New Delhi 2005; Dietmar Rothermund, *Indiens wirtschaftliche Entwicklung. Von der Kolonialherrschaft bis zur Gegenwart*, Paderborn 1985.

85 A. B. Leonard, "Underwriting British Trade to India and China, 1780–1835," *The Historical Journal* 55, 4 (2012), pp. 983–1006, 987–90.

in Madras and Calcutta in 1639 and 1690 respectively, and beginning in 1668 the British Crown backed Bombay through the East India Company. Following the adoption of British rule after the uprising—i.e., the First War of 1857—these port cities were the undisputed economic centers of the country. At the beginning of the nineteenth century, Calcutta had about 350,000 inhabitants (in the year 1820 CE), Madras about 300,000 (1802), and Bombay 200,000 (1825). Of the remaining non-British-occupied cities, only Lucknow, Lahore, and Hyderabad likely had populations topping 200,000—Delhi, in contrast, had only 100,000 inhabitants.[86] Bombay, in particular, encountered strong growth beginning in the 1830s.[87] According to the 1865 census, there was a population of 41,713 within the old, increasingly dilapidated city ramparts (i.e., within the fort) and 783,980 (or 816,562 including the "floating population") in the city as a whole.[88] For Bombay, Calcutta, and Madras the rates of population growth were roughly comparable to the major American cities. Between 1845 and 1875, Bombay experienced an intense period of urban planning and modernization.[89] The London headquarters were initially very cautious and, in the beginning, limited policy amounts to £5000 and set relatively high premiums due to the degree of perceived risk. The main means of risk containment were issuing policies in the "European city" only, while prohibiting policies in the "native city."[90]

86 Christopher A. Bayly, *Indian Society and the Making of the British Empire*, Cambridge 1988, p. 68.

87 1826: 185000, 1835: 250.000, 1844: 524121, 1864: 816562, 1872 decrease to 644405: S.R. Kapse, *Police Administration in Bombay, 1600–1865*, Bombay/Delhi/Nagpur 1987, p. 181; Mariam Dossal, *Imperial Designs and Indian Realities: The Planning of Bombay City, 1845–1875*, Oxford 1991, p. 2. Robert Home, "British colonial civic improvement in the early twentieth century: E. P. Richards in Madras, Calcutta, and Singapore," in: *Planning Perspectives* 31, 4 (2016), pp. 635–644.

88 Jim Masselos, "Changing Definitions of Bombay: City State to Capital City," in Indu Banga (ed.), *Ports and Their Hinterlands in India (1700–1950)*, New Delhi 1992, pp. 273–316, 293.

89 Dossal, *Imperial Designs*, p. 3.

90 In the following, I am entering into an analysis of a specific form of colonial segregation within a field of economic interaction. It is not the place to render a full bibliography and discussion of segregation studies. One might note that systematic approaches define segregation as "the unevenness of the distribution of [a society's] members across places or categories" (O. D. and B. Duncan, "A methodological analysis of segregation indices," in: *American Sociological Review* 20 (1955), 210–217; C. J. Dawkins, "Measuring the Spatial Pattern of Residential Segregation," in: *Urban Studies* 41, 4 (2004), 833–851, 833; P. Hennerdal/M. Meinild Nielsen, "A Multiscalar Approach for Identifying Clusters and Segregation Patterns That Avoids the Modifiable Areal Unit Problem," in: *Annals of the American Association of Geographers* 107, 3 (2017),

Compared to other cities, by 1852, as insurance began to be established, there was still no really reliable knowledge about the history and risk of fire in Indian cities: the only event that was repeatedly mentioned was the 1803 fire in the Fort district of Bombay.[91] For Madras, it was reported as late as 1852 that "the destruction of an European house by Fire is unknown and no record exists thereof." There were only reports of small fires in warehouses ("godowns") in the European part of the city. On the other hand, statements about the risk of fire in the "native town" were uncertain and contradictory: "8 or 9 years since there was an extensive Fire in the native Town ... difficult to judge of risk—Surrounded by and attached to small Native dwellings—There was danger on every side and such risk never could be insured.—It was extraordinary, few fires even there."[92] The information about this fire in 1843 or 1844 is very imprecise: while it suggests that under no circumstances should insurance be established there, this quote also reports that there were relatively few fires in the native town. After commissioning the respective agencies with the business of insurance, fires were documented from 1852 to about 1872 in the Memorandum Books: reports always focused on individual buildings, warehouses, jute-presses; the insured or uninsured damages were always noted. The fact that there were also other types of fires that attracted little attention from either the authorities or the insurance companies is occasionally alluded to. On February 28, 1866 there is a report concerning the eastern suburbs of Calcutta: "This fire one of the most serious of its kind known at Calcutta for many

555–574), and tend to historicize the phenomenon by applying either an ahistorical or an unquestioned definition of race found in the sources, asking for the exact distribution, distinction, legal treatment etc. Another approach, followed here, is to concentrate more on the perception of the observers themselves while they are (consciously or unconsciously) (re-)producing borders. For an example of an approach taking the subjective representations of segregational patterns as the object of historical investigation itself cf. S. L. Smiley, "Mental maps, segregation, and everyday life in Dar es Salaam, Tanzania," in: *Journal of Cultural Geography* 30, 2 (2013), 215–244 for a very different period and place.

91 The mention of an 1803 fire that destroyed 1000 houses in the Madras Memorandum Books must refer to Bombay, LMA CLC/B/192/019/31522/164, p. 46; cf. LMA CLC/B/192/019/31522/154, p. 1; the fire of 17 Feb. 1803 is also the only one repeatedly invoked in: *The Gazetteer of Bombay City and Island*, 3 vols., Bombay 1909, vol. 1, 362, 413, vol. 2, pp. 24 n 1, 130–132 (damage: 471 houses, 6 sacred buildings, and 5 barracks destroyed), vol. 2, pp. 326, 342f., vol. 3, p. 53—exception: one mention of the current fire statistics in 1908/09 (vol. 3, p. 51). On the 1803 fire, see Mariam Dossal, "Knowledge for Power: The Significance of the Bombay Revenue Survey, 1811–1827," in Banga (ed.), *Ports and Their Hinterlands*, pp. 227–43, 232–35.

92 LMA CLC/B/192/019/31522/164, p. 46.

years. Hundreds of native huts, straw, haystacks and such like burnt. ... Loss 8000 to 10,000 Rps [rupees]."[93] This is the only such entry in the Sun documents concerning any of the three cities—the economic damage of such a fire of several hundred "huts" of "natives" was equivalent to only a tenth to a twentieth of the monetary loss represented by a single burned jute warehouse. Therefore, very little can be established about the frequency and intensity of fires in the native towns of the three cities during this time. The Sun requested a "fire record" in 1871 from the police commissioner of Calcutta, Stuart Hogg, and received a list of 53 fires that had taken place from 1865 to 1870. While fourteen fires in "filed huts" were listed, there was no further description; all in all these fire statistics were not terribly meaningful since they contained no damage estimates.[94] In the very detailed fire documentation for Bombay,[95] there is a separate—but rather unspectacular—section devoted to fires in the native town; one of the larger fires in 1868 apparently affected barracks of Ethiopian migrant workers, but the actual scope and magnitude of the fire is not given:[96] relatively speaking, fire events appear in these sources quite rarely.

Beyond the Sun Fire Office sources there is comparatively little mention of major fires in the literature on the urban history of Madras, Calcutta, and Bombay—let alone any kind of systematic investigation of fire trends in these cities. When the topic of fire did attract attention, it would seem that contemporaries, at least during the first half of the nineteenth century, were primarily interested in fire events in the "native" sections of Indian cities. Two fires were reported in Calcutta in 1780—in one 700 huts burned down, in the other 15,000.[97] Travelers to India often clearly articulated their perception of a dual city, divided into a "black town" and a "white town," which were, respectively, highly prone to fires and almost fireproof. Thus, Walter Hamilton reported frequent fires in the native town in 1820, while the European district seemed almost "incombustible."[98] Wallace lists nine major fires in the native town

93 LMA CLC/B/192/019/31522/158, p. 257.

94 LMA CLC/B/192/019/31522/158, p. 253–276; vol. 160, p. 167; vol. 161, p. 47 (fire with great damages in the amount of 300,000 rupees on May 15, 1882); vol. 164, p. 46–48.

95 LMA CLC/B/192/019/31522/154, p. 1–136—the Post Office fire of 1869 is described as one of the largest fires.

96 Ibid., pp. 71ff.

97 James Long and J.H. Stocqueler, *British Social Life in Ancient Calcutta (1750 to 1850)*, ed. with notes P. Thankappan Nair, Calcutta 1983, pp. 19, 31, 76f.

98 Walter Hamilton, "A Geographical, Statistical, and Historical Description of Hindostan and The Adjacent Countries [1820]," in *Calcutta in the 19th Century (Company's Days)*, Calcutta 1989, pp. 220–44, 224.

between February and May of 1813; in one of them 400 huts were destroyed, in another 300, and in a third fire 1,000 huts were burnt. He summarizes the situation thus:

> Throughout India, for some time before the commencement of the monsoon, the atmosphere becomes heated to an extreme degree, and very high winds prevail. During the months of May and June, this agitated state of the air is severely felt in Calcutta; and destructive conflagrations from fire often desolate the native quarters of the city, where the houses are constructed of such inflammable materials, that a spark bursts forth into a blaze in a moment, and is carried like lightning along by the wind. The houses of the Europeans and rich natives, being built of brick and terraced, escape with little damage on these frightful occasions; for the course of the destructive element is so rapid that such buildings are passed before any serious impression is made on their external parts. It is melancholy to contemplate the effects of these fires. The little all of the poor natives, their furniture, cows, goats, horses, are consumed; and frequently the lives of the owners lost in vain endeavours to save their property. Yet the evil is of such a nature as not to be easily remedied; for were the huts to be removed from the town, the same destruction would attend them, as is seen in the suburbs, and all over India where the *ghurs* are made of reeds.[99]

Arson and negligence in the handling of fire were often at work, but the somber spectacle of these fires was reported as "so common all over India, that the eye becomes familiarized by habit, and feels a shock every time of less painful disgust."[100] The incidence of fires in the huts of the "natives" is mentioned until the mid-nineteenth century in travelogues,[101] while, with the exception

99 Excerpt from Lt. R.G. Wallace, *Fifteen Years in India: Or Sketches of A Soldier's Life. Being an Attempt to Describe Persons and Things in Various Parts of Hindostan* [1822, 2nd ed., 1823], in *Calcutta in the 19th Century*, pp. 301–34, 318f.

100 Ibid., p. 320.

101 William Huggins, "Sketches in India, Treating on Subjects connected with the Government; Civil and Military Establishments; Characters of the European, and Customs of the Native Inhabitants [1824]," in *Calcutta in the 19th Century*, pp. 411–26, 415 and in Emily Eden, "Letters from India [1836/37, 1840/42, pub. 1872]," ibid., pp. 650–756, 710: "80.000 homeless people last week" due to a fire; Leopold von Orlich, *Travels in India, Including Sinde and the Punjab* [1845, trans. from German, here a letter from Alexander von Humboldt 1843], in ibid., pp. 860–906, 867: referring to the division into one city of Christians and one of natives, the native town "as in all Oriental cities" is closely built. The palm-roofed bamboo

of the 1866 incident above, such fires were not mentioned in the Sun Fire Office files originating from the second half of the nineteenth century. Were they simply not recorded, even though every individual house fire in the European section of the three cities was precisely described? While this is certainly conceivable, it is nevertheless strange, since there was certainly a discussion about whether insurance should play a role in the native towns, which would have made such information significant and necessary.

One gets the impression that travelers to India and merchants were endowed with a "European eye:" they imagined Indian cities as divided between "European" and "native" sections and on site they simply reproduced and reified this notion.[102] While the information in the insurance files initially appears to be collected and communicated using seemingly unbiased and impartial inductive-empirical data collection techniques, on closer inspection it becomes clear that certain thresholds of prejudice were maintained. The insurance business began in 1852 under the assumption that only "European" buildings (i.e., not those of "natives") would be insured,[103] and for this purpose it became necessary to cognitively support the idea in the London headquarters of a divided and distinct "white town" and "black town" that could be precisely delineated cartographically and topographically on a map (Fig. 72).[104]

---

huts are "excessively liable to take fire, and during my stay here, no less than 700 of them were burnt to the ground in three days."

102 Swati Chattopadhyay, *Representing Calcutta. Modernity, nationalism, and the Colonial Uncanny*, London 2005, pp. 21–28 and passim. Dossal, *Imperial Designs*, p. 4 emphasizes, in contrast, that the Indians also describe themselves as "natives": "Nineteenth century Bombay was thus inhabited not by 'Indians' and 'Englishmen,' but by 'Natives' and 'Europeans.'" Cf. Cornel Zwierlein, "Insurances as Part of *Human Security*, their Timescapes, and Spatiality," *Historical Social Research* 35, 4 (2010), pp. 253–74, 264–66.

103 Sun Fire Office London to the Leckie & Co. Agency, Bombay, August 12, 1852: "[…] but it is considered a point of great importance, to limit our Insurances strictly to those for European Firms & of the best Class of Buildings only. Our rule of business must, for the present at least, altogether exclude insurances for native Firms & be confined as to locality to the European part of the Port & Colaba on the same terms as the Imperial—say 18 annas per Cent for the former & 14 annas per Cent for the latter" (LMA CLC/B/192/019/18249/13, f. 14v).

104 LMA CLC/B/192/019/31522/158, after p. 241—earlier, p. 238–240 the copy of a letter from Henderson to the London headquarters from February 23, 1872, in which, however, reference is made to a different map. The placement of the boundaries for the "white town" correspond quite precisely to those that were described by R.G. Wallace in 1822, cf. Chattopadhyay, *Representing Calcutta*, p. 78 (Fig. 2.1—there the author had actually marked the areas himself, while here a contemporary cartographic drawing followed). On the "divided town," see also Dossal, *Imperial Designs*, pp. 16–20.

FIGURE 72 Plan of Calcutta 1872: Sun Fire Office's agent Henderson colored brown the circumference of the City and in red the part recognized as "European town"

There is evidence that Agent Peel offered to procure a high-resolution map of the "European part of the City (late Fort)" of Bombay; however, there is no such map to be found in the files.[105] Today, there is only a commonly available map from 1866 delineating the districts.[106] Nevertheless, there are many photos, as well as precise measurements of the length and width of the sixteen main streets in the Fort district and the three main roads in the native town.[107] In 1869, Agent Scott, a Sun employee, described the southern section of the Fort district, which in the meantime had become primarily an office and business area, since even Europeans lived in the suburbs at this point: "It could be expected as is the case, that houses in this Quarter should be substantially built, the streets are tolerably wide, & the whole aspect & appearance is European. … All Buildings throughout Bombay, native & European, are roofed with tiles: & water pipes with a plentiful supply of water are laid through the principal streets."[108] Here, and particularly for the prestigious buildings erected in 1864 through Scott and McClelland's Elphinstone Circle project, insurance initially was established because Scott and McClelland offered certain assurances:

105 A letter or note written to the headquarters on November 12, 1870, LMA CLC/B/192/019/31522/151, p. 161.

106 LMA CLC/B/192/019/31522/152, p. 127.

107 Ibid., pp. 178–99.

108 LMA CLC/B/192/019/31522/151, p. 201 ('European part of City').

> It is our decided opinion that there is less risk of fire with the Elphinstone Circle buildings than with those of the best class in England. We believe we are right in stating that English fires usually originate from fireplaces, Chimneys, gas or lucifer matches.
>
> None of the buildings under consideration have fire places or Chimneys in any room except the Cook room (Kitchen) such not being required in the hot Bombay Climate.[109]

Plans and photos of the magnificent new buildings were sent to the headquarters in order to convince the Board of the Company of the trustworthiness of their wares.[110] In 1869 Scott had, however, used much more negative undertones when writing about the "Native part of Fort" (Fig. 73):

> The native quarter of the Fort 600 Yards long, by 300 yards broad, is of an entirely different character. The streets are narrow and torturous, the houses are divided by narrow cross streets and passages.
>
> A Native house is constructed of Wood, simply and entirely. Upright posts are first put up, upon which the floor beams are fixed, it is thus carried on from floor to floor (often five stories in height) till the roof is fixed. Then brick walls are built from bottom to top leaving the posts to project slightly beyond the plaster either on the inside or outside of the house: often the posts and walls of the upper floors, supported upon curiously carved wood blocks—project beyond the posts of the ground floor and this feature taken together with the wood projecting Verandah which invariably goes with it, gives the peculiar aspect to Eastern Houses. The brick walls being intended merely as a shelter from the sun in the hot season, and from the rain in the monsoon are built as thin as possible. ... There is at least one Cook room on each floor, and Chimneys are rather the exception than the rule. The floor of the Cookroom will be in most cases of plaster and brick, and the fire place is invariably of the same materials and well built, by the side of it is the sink with a store of water: so that really there is not much need for a Chimney as wood is used, to be lighted only when a meal is being cooked.[111]

109 Ibid., p. 212.

110 Map of *Elphinstone Circle* between the "native" and "European" sections of the Fort, ibid., 214f.

111 Ibid., p. 232.

FIGURE 73 View of a street in the "native town" of Bombay with characteristic wooden buildings and verandas, 1875, LMA CLC/B/192/019/31522/156, p. 18f. (*Bombay Gazette*?)

Photographs and prints in newspapers like the *Bombay Gazette*, which were cut out and collected at the headquarters, gave the impression of a high fire-risk situation based on the sole use of wood in construction—apart from the reassuring news, of course, that there was hardly any danger from heating and ovens.

Despite the very stereotypical characterizations of the opposition between the European and native towns, there were agents in Bombay, who were unsure how to draw the boundary lines for the Sun headquarters and also for the other insurers active there, since they had not yet been precisely determined by 1852. Since some insurance companies offered policies on certain streets of the native town at the lower rates offered to the European town, the lack of clarity concerning the boundary between the "white" and "black" towns led to excessive competition, where rates of a united, "weakly collusive" market of companies otherwise prevailed.[112] In 1855 the Sun gave agents permission

112 LMA CLC/B/192/019/31522/151, p. 228. Cf. Robin Pearson and Mikael Lönnborg, "Regulatory Regimes and Multinational Insurers before 1914," *Business History Review* 82 (2008), pp. 59–86, 69; comparable with Monica Keneley, "The Origins of Formal Collusion in Australian Fire Insurance, 1870–1920," *Australian Economic History* 42 (2002), pp. 54–76.

to sell insurance in the native town as well "if Agents [were] satisfied as to their [i.e., the natives'] respectability."[113] Following that, the agents of Leckie & Co. also wrote in 1861, when the question came up again, that they thought "it would be desirable to extend the business of the Office to Risks in the Native Town and other places outside the Fort within the Island of Bombay."[114] At this point it became a serious struggle to define "native town".[115] The headquarters, probably under pressure due to competition with other companies, decided very carefully that Leckie & Co. should be allowed to issue policies in the native town and across the whole city ("Bombay Island") in strict accordance with six criteria of exclusion: in particular, insurable buildings had to be made of brick, have tile roofs, not be adjacent to wooden structures, and not shelter "hazardous" businesses—timber trade, fire-intensive manufacturing, etc. Such policies were offered with a ceiling of 30,000 rupees of insured value.[116] When the Sun later lost out on a lucrative deal in 1863 because of this limitation, the maximum limit was raised to 50,000 rupees.[117] It was only in 1893 that the Bombay Association of Fire Insurance Agents came to an agreement among the agents of all locally operating insurance companies, which finally delineated the borders of the native town and established common rates (Fig. 74).[118] To educate and orient local insurers, a special map of the native town boundaries was created.

The disputed boundary between the European and native towns, or even—as a visitor in 1876 formulated it—between "Modern Town and Native Town,"[119] and the tendency of agents in Bombay to issue policies in the native

---

113 LMA CLC/B/192/019/31522/152, p. 108.

114 LMA CLC/B/192/019/31522/151, p. 258f.

115 "First, as to the definition of the term 'Native Town,' we think that this term should be applied only to that part of the island inhabited exclusively by natives and which you will find marked 'Native Town' in the small map accompanying the Times of India Calendar forwarded to you by last mail [i.e., the map in LMA CLC/B/192/019/31522/152, p. 127]. We do not as a rule, recommend the acceptance of such risks, but your Managers might leave us a discretionary power in the matter, in that case we should use the greatest caution, and with regard to the rates we submit that they should be the same as those charged by the Alliance & Phoenix, in preference to that given by the North British" (ibid., p. 258).

116 Decision of 8 Oct. 1861 (LMA CLC/B/192/019/31522/152, p. 108).

117 Ibid., p. 109.

118 LMA CLC/B/192/019/31522/156, p. 307 (Circular of Clement Poole, secretary of the Bombay Association of Fire Insurance Agents, 4 Oct. 1893).

119 Mrs. Guthrie, "Modern Town and Native Town," in *Charm of Bombay: An Anthology of Writing in Praise of the First City in India* (ed.), R.P. Karkaria, Bombay 1915, pp. 314–316, qtd. in Prashant Kidambi, *The Making of an Indian Metropolis. Colonial Governance and Public Culture in Bombay, 1890–1920*, Aldershot 2007, p. 35.

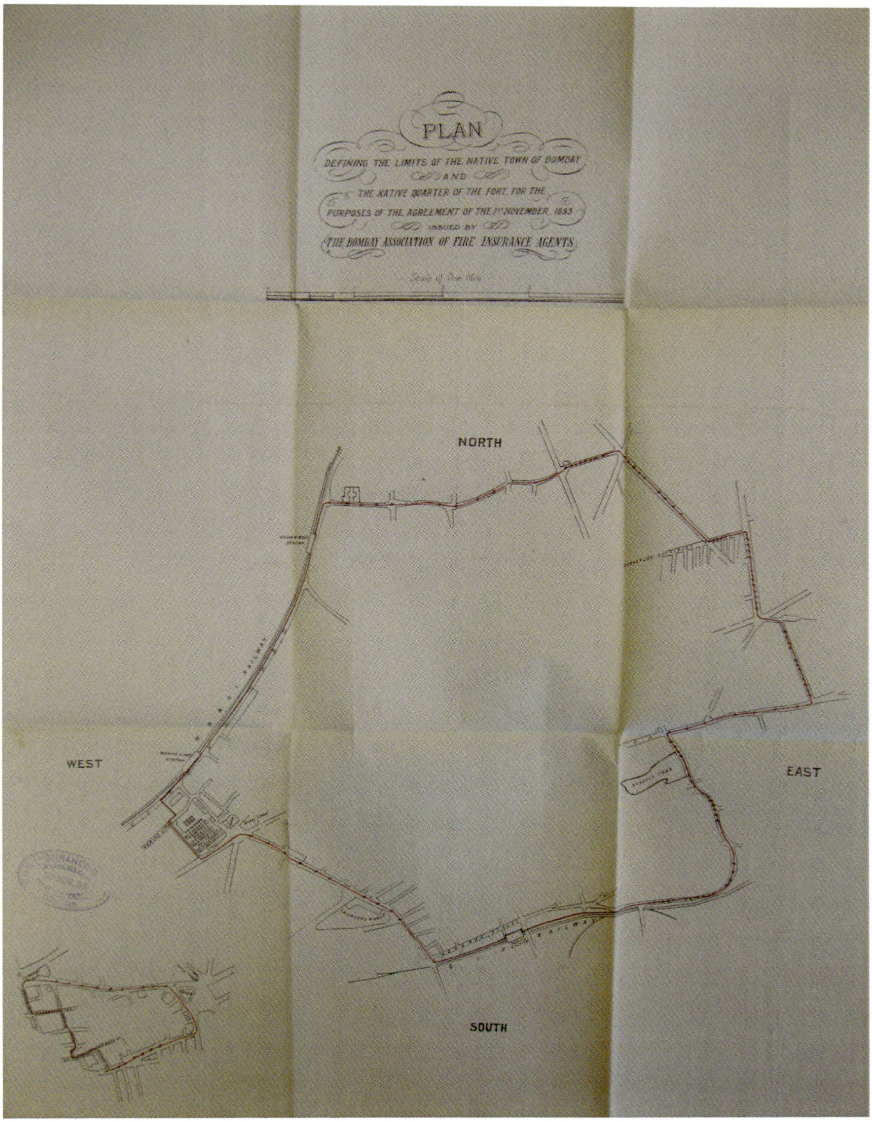

FIGURE 74 *Plan Defining the Limits of the Native Town of Bombay and the Native Quarter of the Fort, for the Purposes of the Agreement of the 1st November, 1893, issued by The Bombay Association of Fire Insurance Agents*, LMA CLC/B/192/019/31522/156, p. 307

town in spite of the borders, ultimately indicates how tentatively this spatial division was anchored in the mind, while also demonstrating the prevalent mixing of societies and habitats, particularly in economic contexts, as pointed out by Swati Chattopadhyay for Calcutta: "the black and white towns were far from being autonomous entities; the economic, political, and social conditions of colonial culture penetrated the insularity of both towns, although at different levels and to varying degrees."[120] In fact, taken together, the fire trends of both the European and native town—in comparison with other contemporary, similarly rapidly growing cities of the nineteenth century—reveal a highly unusual character. Insurance companies recognized the quite remarkable numbers that emerged from India only after several years: in Bombay, according to Chamber of Commerce data, the ratio of losses to premium income of all insurers was 10.05% for the period of 1847–69;[121] for the Sun, the figure was 7.81%;[122] in Calcutta, the ratio was only 1.07% for the period between 1852 and 1871; in Madras no damage payments at all were made between 1852 and 1878 (Table 6).

The Indian agencies were thus consistently some of the most profitable foreign agencies, because while agencies in cities such as Hamburg and Yokohama always drew significantly higher premium incomes (two to three times higher), they also had to invest a much larger proportion of that income into damage payments—the major Hamburg fire of 1842 and the Yokohama fire of 1867 alone constituted two high loss values that persisted on the balance sheets for a long time.

The Chamber of Commerce of Bombay stressed that "in 22 years' time (from 1847 to the beginning of 1869) only eight fires of any importance have occurred within the limits of the Port, *Including Both the Native and European parts.*"[123] These figures thus run counter to the "Orientalist" perception that the native town contained low-quality buildings and high fire frequency and possessed a completely different, premodern character. Perhaps the fire extinguishing system that was established and repeatedly extended by Henry Conybeare, the superintendent for repairs from 1852, contributed to the relative fire immunity of Bombay.[124] However, such an explanation is clearly not entirely sufficient,

---

120 Chattopadhyay, *Representing Calcutta*, p. 76.

121 Report of the Chambre of Commerce 1869, LMA CLC/B/192/019/31522/153, p. 66, 81.

122 LMA CLC/B/192/019/31522/156, p. 63ff.

123 Report of the Chambre of Commerce 1869, LMA CLC/B/192/019/31522/153, p. 66, 81—my emphasis.

124 Kidambi, *The Making*, p. 39; Dossal, *Imperial Designs*, pp. 95–124: Dossal offers an in-depth treatment of the water supply system, but never touches on the possible connection to questions about firefighting. As such, the link to the fire question remains very hypothetical.

TABLE 6 Business figures of the Sun for the Indian agencies (LMA CLC/B/192/019/31522/164, p. 40)

	Premium revenue	Regular expenses	Fire damage
Bombay 1852–1878	£38334	£5289	£2995
Calcutta 1852–1871	£18365	£2646	£198.9
Madras 1852–1878	£4829	£663	-

especially since Calcutta also boasted similar exceptional fire characteristics, even though there was still no comparable fire extinguishing system in place by 1870. There were many theories that attempted to explain this exceptionally low fire rate, which almost seemed like a reversal of the descriptions in the above-cited reports and travelogues about Calcutta from the late eighteenth century to about 1840. When the news of the Chicago fire broke in India in 1872, astounded writers comparing the situation in America arrived at the shocking conclusion that Indians lived in much safer cities in terms of fire danger.[125] Among other things, this was obviously indicated by "natives, who, encouraged by the immunity from Fire which Bombay has so long enjoyed, prefer to take their chance rather than pay the heavy rates asked for."[126] There was thus a persistent debate at the headquarters in London about whether or not insurance rates should be reduced in India to make insurance purchases more popular. Sun agents prepared a report that initially aligned with the popular perception of European superiority, asserting that "there are no buildings whatever in Bombay constructed according to scientific principles to guard against fire." They maintained that the houses were all built "with a very heavy proportion of wood-work, in a large number of cases the frame work, the floors and roofs are finished before a wall is commenced either external or internal"—a fact that would significantly raise the fire risk in a typical English house. But, as they also conceded: "it is surprising how few fires there are either in the Fort or thickly populated native town, and still more surprising when a fire does occur,

125 "The non-inflammable nature of the woods principally used in the construction of houses here in addition to the manner of constructing them, [...] would render almost impossible a great conflagration like that in Chicago" (W. Nicol & Co., Agents of the North British and Mercantile Insurance Company and the other active insurance agencies in Bombay, Bombay, 23 Aug. 1872, LMA CLC/B/192/019/31522/153, p. 3).

126 Ibid.

how very rarely it is known to spread to the adjoining houses."[127] The miracle of fire immunity also explicitly applied to the native town. As D. Gostling similarly noted in the *Bombay Gazette* ten years later on May 18, 1883: "We have never been, in a large city, so exempt from serious fires as Bombay. In one year a greater destruction of property from this cause takes place in New York than we have known in 35 years in Bombay. We confess that we are unable to explain the comparative immunity of Bombay in this particular." Gostling then attempted to come up with some underlying reasons for this, which coincided with the main points of Scott and McLelland's explanations from 1872: In a tropical city like Bombay, fireplaces are not needed for heating; furthermore, after cooking, fires were always immediately extinguished, since with the absence of chimneys, cooking produced a lot of smoke. Fireplaces for cooking were made of fireproof *chunam*, a mortar made of sand and mussel glue. On the wooden planks of the foundational flooring, six inches of *chunam* or cement was applied, and wood covered with *chunam* does not burn. Like other tropical hardwoods, Indian teak does not burn as quickly as wood grown in temperate climates. "Hence, practically, native houses are, if not fireproof, at least fire-preventive." Thus, over the course of more than twenty years, fires only affected individual houses. "Where is there an instance of a fire burning down half a street? It is not to be found." English fire insurance companies shied away from insuring "native" houses due to a fear over ritual arson, but it turned out to be a groundless concern: "I believe human nature is much the same all the world over, and that there is just as large, or small, a proportion of rogues in Europe and America as in India."[128]

The phenomenon of overly densely populated areas, though typically interpreted as a sign of inadequate housing conditions,[129] is cited as another reason that major fires rarely take hold: fires were always immediately detected. While it was suggested in the travelogues mentioned above that monsoon winds caused fires to tear through the huts, the emphasis now shifted to the high humidity of the season, which was thought to literally dampen any such proclivity[130]—the insurance companies used Bombay's rainfall statistics between 1825 and 1870 to vividly establish the state of permanent moisture reigning in the city between June and September or October.[131] It is astounding to take note of the reversal of the Eurocentric/Orientalist sense of superiority: suddenly, Indian

127 Extract from report by Messrs. Scott & McLelland [1872], LMA CLC/B/192/019/31522/153, p. 112f.

128 LMA CLC/B/192/019/31522/156, p. 164.

129 In 1849, sometimes a very high number such as 43 inhabitants per house were registered, cf. Kidambi, *The Making*, 33.

130 *Star of India*, 30 Nov. 1872, LMA CLC/B/192/019/31522/151, p. 182f.

131 LMA CLC/B/192/019/31522/153, p. 183.

houses were deemed exemplary in terms of their fire-resistant construction in contrast to the far inferior, average English house, which burnt quickly and easily.[132] While the Chicago fire of 1871 was invoked in Constantinople as an example of the high modernism of the American city, at least in terms of the speed of reconstruction contrasted with the slow "Oriental/premodern" pace of reconstruction in the Ottoman capital, in India the Chicago fire was used to show that Indian cities should be understood as firesafe areas. It was, of course, a type of safety that was not really "modern" from an outside perspective, but through a strange reversal of standards, it was a phenomenon that emerged from "native" practices and the general climatic conditions of India. The result was a kind of "premodern safety," whereby Indian cities experienced population growth as stunning as that of American cities and some European metropoles, but did not experience the typical side effects of large fires, such as those that occurred in the United States. The strange discrepancy between European travelers' reports of frequent slum fires in the early nineteenth century and their utter disappearance in the sources beginning in the second half of the century is difficult to reconcile. Most likely it was because they were reporting fires in bamboo and palm leaf huts, and while these probably also occurred in the later period, as is still often the case in the megacity slums of the twenty-first century, they did not attract the attention of insurers, who were only concerned with solid-built sections of the city, be they part of the native or the European towns.

However, the observations of contemporaries concerning the fire immunity of Indian cities was clearly limited to handwritten recommendations, evaluations, and a few newspaper articles—there is no echo of them in the recently burgeoning field of Indian urban history, though this would be one means of approaching an important element of a specific "colonial modernity—*not* a colonizer's modernity."[133] On the level of global historical comparison, this

132 "On the other hand what is the case with an ordinary English house. The dons, windows, floors and roofs are constructed of deal a wood that cracks splits, expands and explodes on being fired. This wood is very soft and takes fire with the greatest ease. The roofs are covered with slates, a light material that flakes and turns to powder under fire. Houses have been known to take fire in consequence of heavy flaming embers falling on the slade roofs. Party walls on the roofs are an absolute necessity. The chimney flues also are a constant source of danger, many fires having arisen from the ends of beams and joists being built into the walls too near the flues." (LMA CLC/B/192/019/31522/153, p. 113f.—*report* by Scott and McLelland). One might be surprised that this move from ignorance toward gathering data and experience with regard to such essentials as the monsoon were noted only now or now anew a bit more systematically by British agents after nearly 300 years of experience of European travelers and colonists through and in India. However, it is known that this climatic specifity of India, though basically known to ship captains, was "étrangement ... peu évoquee dans les correspondances entre Lorient et Paris" around 1700 (Marie Ménard-Jacob, *La Première Compagnie des Indes. Apprentissages, échecs et héritage 1664-1704*, Rennes 2016, pp. 79-81).

133 Chattopadhyay, *Representing Calcutta*, p. 6.

phenomenon is not addressed at all—in fact, Lionel Frost seems to impose a common "premodern" typology of fire characteristics, fire prevention, and firefighting upon all "Asian cities" from Constantinople to Hankou to Tokyo, which effectively obscures the entire urbanity of India with its special monsoon climate. While Frost's analysis became a popular theoretical position, it nevertheless completely elides the reality of the obvious diversity and complexity of the problematics concerning the dichotomies of "modern/premodern," exemplified by findings such as those invoked by Indian cities between 1850 and 1900.[134] The correlation between "security" and "modernity" is disrupted here, since Indian "premodernity" appears to be at least as safe as colonial modernity. Nevertheless, the type of fire "safety" observed since the 1850s is of a different kind—it is at this point in history that the modern and Western schema of threat predictions, risk assessment, and security production takes hold, and the relative lack of danger first begins to be noticed. Through the successful reduction of threats in premodernity, combined with colonial horizons of expectations and concepts of security, a kind of "other" modernity is revealed.

4 China

In 1842, following the Opium War, the Treaty of Nanjing forced the Qing government to accept the installation of five "Treaty Ports" in Shanghai, Canton,[135] Ximen, Fuzhou, and Ningbo, which were opened to British trade.[136] The Chinese were required to pay an indemnity of $21 million over four years. "Consuls were to be permitted at these five places. ... The island of Hong Kong was to pass to Her Majesty in perpetuity," and the British fleet would not leave Nanking until it had loaded the first six million of the total indemnity sum.[137] Before the Treaty of Nanjing, Shanghai had been a medium-sized port serving the more important cities of Suzhou and Hangzhou (Fig. 75), but now the roles changed

134 Lionel Frost, "Coping," pp. 5–16.

135 On fires in Canton, see the recent work, John M. Carroll, "Slow Burn in China: Factories, Fear, and Fire in Canton," in *Empires of Panic: Epidemics and Colonial Anxieties*, Hong Kong 2015, pp. 35–55.

136 See John King Fairbank, *Trade and Diplomacy on the China Coast: The Opening of the Treaty Ports, 1842–1854*, Cambridge, Mass. 1964; for Shanghai, see Linda Cooke Johnson, *Shanghai: From Market Town to Treaty Port, 1074–1858*, Stanford 1995; the contributions in Robert Bickers and Christian Henriot (eds.), *New Frontiers: Imperialism's New Communities in East Asia, 1842–1953*, Manchester 2000 generally address the years after the Opium War.

137 Peter Ward Fay, *The Opium War, 1840–1842*, 2nd ed., Chapel Hill 1997, p. 362; Brian Hook, *Shanghai and the Yangtze Delta: A City Reborn*, Oxford 1998, pp. 8–12; William T. Rowe, *China's Last Empire: The Great Qing*, Cambridge 2009, p. 172.

and Shanghai became a far more important city than Hangzhou. Its population exploded from 400,000 to one million by about 1900. The British arrived in Shanghai in November 1843, and the Treaty of Nanjing allowed them to reside in the city. In 1845, the Qing intendant, Gong, published the Land Regulations, which precisely defined the Chinese and foreign districts.[138] The British established their district north of the walled Chinese city near the Huangpu River. The Americans merged their settlements with those of the British in 1863, while the French established a separate settlement between the walled city and the British sector. The jurisdiction and government of the international settlement was in the hands of a municipal council which, from the very first days, applied the principle of extraterritoriality. Europeans were judged by their own national laws rather than by Chinese law. The economic importance of Shanghai rose quickly: its internal trade in silk, cotton, tea, silver, and opium comprised close to 50% of domestic trade by the 1850s. Therefore, on contemporary maps,[139] Shanghai was represented as a conglomerate of spatially distinct clusters of supposedly pure ethnic identity, plotted next to each other with clear-cut boundaries. In general, this represented reality accurately, but examining the situation in more detail revealed many hybrid elements within the city.

Hong Kong had been ceded to the crown after a two-year-long occupation in 1843 (cf. for a view on the port Fig. 77). Even more quickly than Shanghai, the city grew and took on the role of one of the world's most important trade centers and port cities, a development which the British diligently observed and proudly noted. In a note on "the rise and progress of Hong Kong," they showed the change in relevant indicators—the rise in the local population from about 7,500 in 1841 to 121,970 in 1869; the rise in receipts (revenues and parliamentary grants) from £38,813 in 1841 to £192,428 in 1869; the rise in the number of ships (1845: 334; 1869: 4426) and the tonnage of shipping in the harbor (1845: 117,210; 1869: 2,256,049).[140] A contemporary analysis of 1876 represents the function of the city as a center of trade and finance, stressing that it was much more important than, for example, Cape Town:

> It has been well observed that one of the most profitable and useful purposes to which the surplus means of an old country can be applied is the prudent development of the resources of newer settlements where

138 See Marie-Claire Bergère, *Histoire de Shanghai*, Paris 2002, pp. 35–44.

139 See Catherine Vance Yeh, "Representing the City: Shanghai and Its Maps," in David Faure and Tao Tao Liu (eds.), *Town and Country in China: Identity and Perception*, Basingstoke 2002, pp. 166–202.

140 The Rise and Progress of Hong Kong, LMA CLC/B/192/019/31522/62, p. 13.

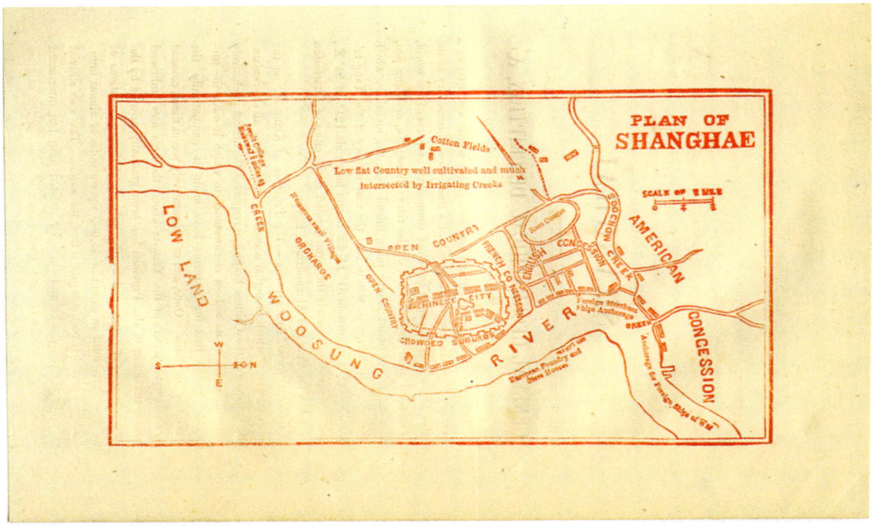

FIGURE 75 Nicholas B. Denny's map of Shanghai, 1867 (Original size 12x20cm), Nicholas Belfield Dennys (ed.), *The Treaty Ports of China and Japan 1867*, digitized in the Digital Library "Visual Cultures in East Asia," accessed September 15, 2019, http://www.vcea.net/Digital_Library/Maps_en.php?ID=401

> money is scarce. In this way the investing class in England has within its reach the higher scale of interest constantly ruling in the East. In China for a long time past money has been exceptionally dear. Banks of undoubted standing and credit have been paying at the rate of 12 and 14 per cent. per annum on deposits in Hong Kong, and elsewhere much higher figures have prevailed. These loans have at the same time proved highly advantageous to the borrowers. Amongst the permanent causes of high rates may be pointed out the active employment of all accumulated capital; the large business profits which are generally made by native traders; and their instinctive fear of any display or acknowledgement of wealth. … The rich retired native merchants generally like to own property in Hong Kong, as it is beyond the reach of the Mandarins' *squeezes*.[141]

Business in China was extremely good for the Sun.[142] All in all, the company earned £66,806 in Shanghai between 1852 and 1895, while the expenses for the

141 Granville Sharp, Bank Buildings, Statistical Account of Hong Kong, Nov. 1876, LMA CLC/B/192/019/31522/62, p. 79.

142 I am using the foreign Memorandum Books China 1–4 of the Sun Fire Office here, LMA CLC/B/192/019/31522/60 to 63.

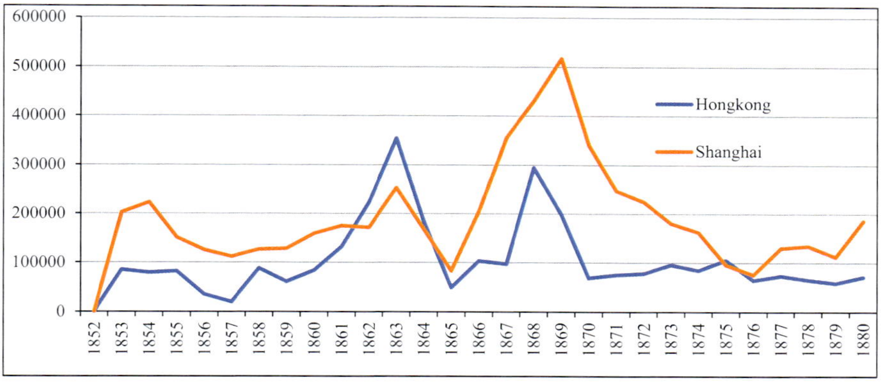

FIGURE 76 Total insurance sums Sun Fire Office (Hongkong, Shanghai, in pounds insured), LMA CLC/B/192/019/38553 vols. 1 and 2

agency were only £11,318, and only £17,902 had to be paid out in losses. In Hong Kong during the same time, the Sun earned a £33,398 premium income, had to pay only £5002 in overheads, and only £13,977 in losses. In comparison with many of the Sun's other agencies around the world, the Chinese agencies were important for business volume as well as for net gain (Fig. 76).[143] In comparison with other European agencies, the Sun became established early in the Chinese market. The Phoenix, otherwise the pioneer in the globalizing insurance market alongside the old sugar trading networks, entered Shanghai only in 1858, and Hong Kong in 1862, roughly 10 years after the Sun.[144] In 1960, Peter Dickson devoted two pages to the history of Sun agencies in the Chinese treaty ports: according to this analysis, the prosperous early years, inferred from the numbers above, lasted until around 1870. At first, the Sun competed only with other British insurers, the only early German exception being the Hamburg-Bremen Insurance Company.[145] During those years, several attempts by European merchants to set up local fire insurance businesses (the first China Fire Insurance Company of Shanghai in 1864, the Victoria of Hong Kong, 1870) soon went bankrupt. Only a merger of the Hong Kong Fire Insurance Company of

143 See Zwierlein, "Grenzen der Versicherbarkeit," p. 438.

144 Trebilcock, *The Phoenix Assurance*, p. 313.

145 In Hong Kong, the Imperial, London, Alliance, Royal Liverpool, Phoenix, the Liverpool & London & Globe, the North British, the Northern, Queen, Commercial Union, Guardian, the Manchester, London & Lancashire, and Norwich Union are mentioned; among the non-British insurance companies are the China Fire Insurance, the Australian, and the Hamburg Bremen. In Shanghai, all the same offices are represented, with the exception of the Guardian, the Norwich Union, and the Australian, LMA CLC/B/192/019/31522/60, p. 55f.

FIGURE 77 General view of Victoria, Hong Kong, from a newspaper clipping cut out for the Sun Fire Office headquarters in London (from *The Illustrated London News*, 5 May 1866, lma clc/ B/ 192/ 019/ 31522/ 60, fol. 141)

1866 and the second China Fire Insurance Company of 1870 was competitive. Anticipating changes, the Sun slowed its engagement in China. In 1887, German insurance companies entered the market, and in 1895 the Sun transferred its representational duties to the German mercantile firm Siemssen & Co., where they remained until World War I.[146]

Looking at the development of the insurance sum, it can be seen that although the market was favorable, the Sun faced some unforeseen difficulties, mainly due to changing agents, many of whom declared bankruptcy. By contrast, the payments of losses were not of great importance, even for years with large fires, such as 1867 and 1878 in Hong Kong. The only effect was the increase in insurance sums in Hong Kong after 1867, as was usual after great fires: demand for insurance always rose quickly in the aftermath of such disasters before normalizing again after a time. However, the other peaks and valleys of the graph cannot be explained by the influence of major fires. The part of the city that was particularly interesting and important for the Sun was Victoria Harbor.

Two great fires struck Hong Kong, one in 1867 and one in 1878. On the evening of October 30, 1867, a conflagration broke out at about 7:00 pm. This destroyed 200 houses, warehouses, and harbor storehouses known as "godowns (Fig. 78)."

Most Chinese houses and property were uninsured; the total losses of the insured properties ($453,505) were divided among the insurance companies in Hong Kong.[147] As was usual after every fire of that size, the nineteen active

146 Dickson, *The Sun Insurance Office*, 196–197.

147 The losses were divided as follows: China Insurance $110,000, Royal $110,000, Hong Kong $100,000, Imperial $50,000, Sun $24,005, Guardian $17,000, Northern $16,000, Northern

FIGURE 78 Great Fire of Hong Kong, 1867, *The Illustrated London News*, newspaper clipping in LMA CLC/B/192/019/60, 75

insurance companies met and discussed possible improvements, but there were no decisive changes such as the introduction of an organized volunteer or even a professional fire department. However, it should be emphasized that large fires were rare in China in comparison to other East Asian markets such as Japan, where the Yokohama fire of 1867, among others, testified to the hazardous situation there.

The second great fire in Hong Kong struck on Christmas in 1878. It started at Endicott Lane at the corner of Queen's Street and destroyed 350 to 400 houses. A sketch of the portion of Victoria Harbor destroyed by the fire was published and the agents of the Sun meticulously marked the fortunately small number of destroyed houses they had insured (Fig. 79).

The distribution of the damages, estimated at $1 million, was different from the fire eleven years earlier.[148] The Sun was even luckier than in 1867 and only

British & Mercantile $12,000, Liverpool & London $5000, London $5000, Phoenix $4500, LMA CLC/B/192/019/31522/60, 79.

148 The Hong Kong Fire: $150,000; China Fire: $70,000; the first real Chinese Company, the On Tai: $20,000; Phoenix, Liverpool, London and Globe: $80,000; Transatlantic

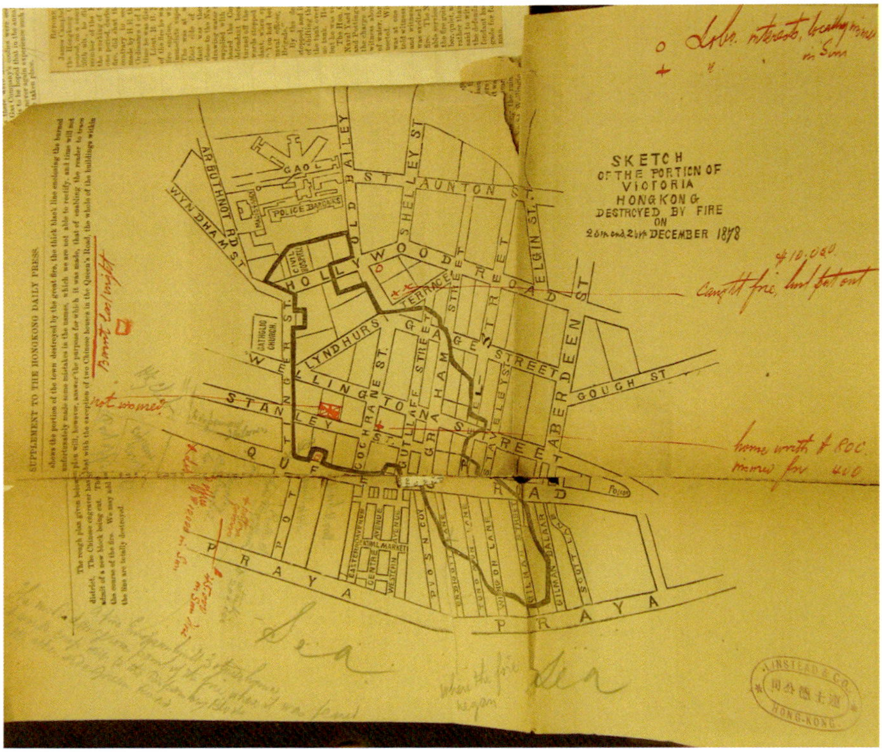

FIGURE 79 Sketch of the affected district in Victoria Harbor, Hong Kong, with the red markers placed by the Sun's agent indicating the houses insured by the Sun (LMA CLC/B/192/019/31522/62, p. 129)

had to pay $500 (£91) in losses.[149] For the first time, an insurance company exclusively operated by the Chinese was afflicted: On Tai Insurance, which had been founded just one year previously and insured maritime transport as well as fire (Fig. 80). The Hong Kong and the first and second China Fire Insurance Companies were run largely by Europeans, even though they were based locally. Around 1876–77, three Chinese companies were founded, run exclusively by Chinese citizens but based on European principles. With the advent of On Tai Insurance (and the brief existence of the Sheong Insurance Company), advertisements and policy forms in Chinese were sent to London by Sun agents for the first time.

Fire Insurance Company of Hamburg: $50,000; Imperial: $30,000; Lancashire $17,000; Royal: $10,000 (LMA CLC/B/192/019/31522/62, p. 128. These are rough estimates).

149 Ibid., p. 135.

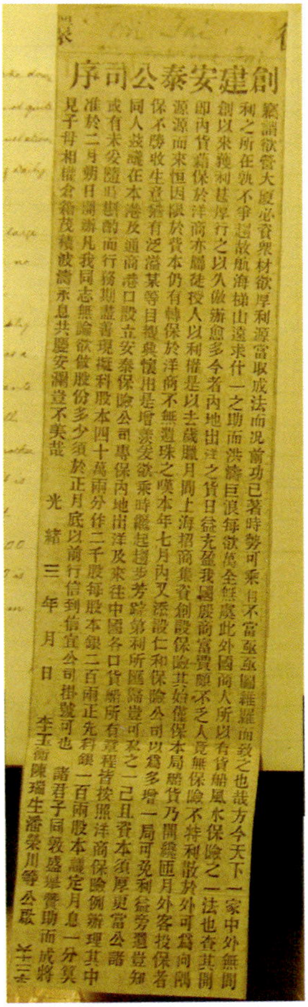

FIGURE 80
Newspaper advertisement by the On Tai company foundation to attract new customers, 1877 (Figure 80a, top), and its police forms (Figure 80b, bottom) (LMA CLC/B/192/019/31522/62, PP. 110-112); The translation by the British contemporaries of the text on the left reads "Prospectus—Mr. Santos asked me to take down the comprador's translation as he could not quite make it out of [! without any punctuation] the following is a liberal translation [!]." The first half is printed in the Hong Kong Daily News. It goes on to say: "Foreign Insurance Companies pay well; a large business is done by Natiues and they have no Insurance Companies. There are plenty of rich Chinamen and if they wish they can easily make a Company on same principles as a Foreign one. Last year the China Merchants started a Company which paid well. This year number seven moon another was started also doing very well. It is now proposed to start a third called 'On Tai' on Foreign principles. 2000 shares in Taels 200 of which Taels 100 is paid up. Guaranteed 1% p. mensem from time of starting." Letter to Linstead 16 Oct. 1877. In fact, this renders only the last, most technical sentences of the text. It starts with some interesting general remarks on economic competition ("The whole world is one family and there is no difference between China and the West. As for the opportunity for enrichment, who doesn't fight about it?") and goes on to remark that "Our country [China] has many merchants, but surprisingly there is no insurance" and finally espouses the national-mercantilist vision that the profits of the insurance business should not be left to foreign merchants. It gives an account of the prior foundations of the first two Chinese insurance companies. The document only refers to maritime cargo insurance at that moment, but On Tai Company was active later on in the fire insurance market. Thanks to Florin-Stefan Morar for a translation of that clipping. Cf. a brief remark on the On Tai Insurance Company in David Faure and Elisabeth Köll, "China: The Indigenization of Insurance," in Borscheid and Haueter, *World Insurance*, pp. 472–494, 475 and n 10.

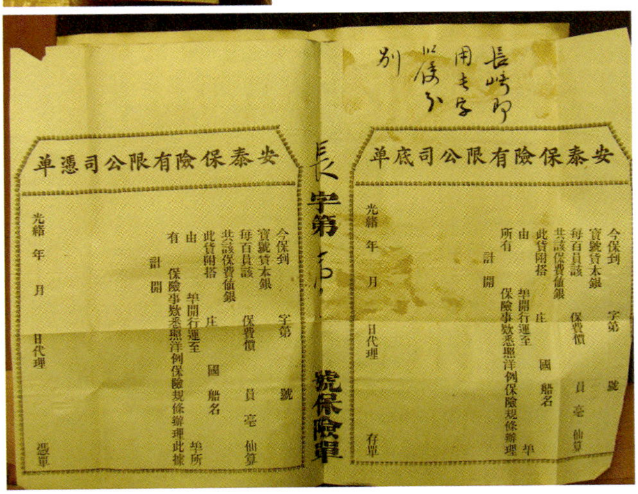

There are glimpses of a "cultural transfer" of the insurance principle within the melting pot of European/Asian business in Hong Kong.[150] It is typical that this transfer happened in the last third of the nineteenth century: at that time, there was a certain national, sometimes nationalist, "backlash" in many regions of the world, and local companies were being founded in competition with the British.[151] Also, apparently as early as 1847, there was an unsuccessful project to construct bankruptcy insurance as a hybrid between the existing form of guild associations (*gongsuo*) and the newly introduced principle of European premium-based insurance. However, evidence of this is scarce and it would be necessary to consult the archival sources for the further precise (and also linguistic) descriptions of that transfer process.[152] Consequently, this analysis will focus on how Europeans managed themselves and their insurance businesses in Hong Kong and Shanghai.

---

150 The directors of the On Tai Insurance Company, as first founded for Marine Risks on Goods, 1 June 1877, were Lee Sing, of the Lai Hing Firm; Chan Shung Lai, of the Lai Yuen Firm; Wong Yik Pun, of the Chun Cheong Wing Hong; Loo Yee, of the Yee On Firm; Fong Soey Fung, of the Tung Sang We Hong; Wong Pak Cheong, of the San Tye Lee Hong; and Pun Pong, of the Wy Sing Firm. The Manager was Ho Amei, LMA CLC/B/192/019/31522/62, 109.

151 The founding of the first local fire insurance institution in Singapore represents a typical global parallel for this point in time in the last quarter of the nineteenth century, Hing, *A Matter of Risk*, pp. 49–96.

152 It is the account of Linda Cooke Johnson that illuminates this phenomenon; she refers here to the "Reports from the Consulate at Shanghai," Public Record Office, Kew, Foreign Office 223/76, f. 287r. Johnson, *Shanghai*, pp. 143–44, 167, 219. However, we cannot judge from her text whether it relies on contemporary English accounts or on Chinese sources handed over to the Consul; the nineteenth-century English description may already have influenced the account by identifying some guild funding operation as "(premium) insurance." I would hesitate to conflate the *gongsuo* plan, as Johnson does, with "insurance" (in the sense of premium insurance) because there may persist some misunderstanding about the character, history, and functioning of European insurance (cf. the statement "Insurance—a newly introduced practice in nineteenth-century Europe and America [!...]" on p. 322). In medieval and early modern insurance history, the relationship between guild practices of security production and the Mediterranean premium insurance system, and later on, the British joint stock company system, is much discussed: some give the guilds priority in a history of the invention of *the* insurance; others rather distinguish both practices from one another; some third voices stress the hybridisation of both types, see the rather one-sided Schewe, *Geschichte*. Faure and Köll, "China," do not touch upon the *gongsuo* problem. For an overview on the state of research on the *gongsuo* or *huiguan* guilds or guild and corporation halls cf. Christine Moll-Murata, "Merchant and Craft Guilds", in: Ead., *State and Crafts in the Qing Dynasty*

Fire insurance companies approach new cities through the lenses of "rates" and "risk" classification. In China, the typical difference between first-class risks (brick or stone-built buildings) and second-class risks (wood or half-timbered construction) was complicated by the location of the house within the city. Insurance companies exclusively insured buildings in foreign-settled sections of the city. They noted in their lists of rates whether the dwellings and warehouses to be insured were "surrounded by Chinese houses," whether the "Native Town [was] separated by a Street 20 feet wide," whether the buildings were "in risk of Native houses" or whether they were "near" or "surrounded by Chinese houses." In the latter case, in some cities, they added an extra fee to the premium.[153] The Memorandum Books illustrate the evolution of this crucial question from 1852 onwards. At first, there were no Chinese houses in these settlements, but in 1853, the city was "taken by the Rebels" and many Chinese refugees began to settle within their borders until 1865.[154] In 1854, the company's directors asked their agent in Shanghai if the company should insure 1) buildings on foreign property, erected in the European manner and in all parts resembling those occupied by Europeans, if belonging to and occupied by Chinese; 2) the same buildings, belonging to Europeans, but occupied by respectable Chinese merchants; 3) property in those houses. At that point, the London headquarters decided to insure only European houses occupied by Chinese (the second case) and signed a £304,000 insured sum at 1.5%. From 1868 onward, the Sun also insured Chinese houses in Hong Kong, but doubled the premium rates (50 shillings per £100—i.e., 2.5%). A photograph in the *memorandum books* shows three typical buildings, marked on a cadastre map that has not survived as Lot. 190 (Chinese), 196, 197 (European) which

(*1644–1911*), Amsterdam 2018, pp. 321–347, here 330, 334: Moll-Murata reminds that some guilds (specially in Port cities) had taken over (for some time) rather public offices as fire-fighting, but this seems not to have been linked with post-fire damage recompensation. For the bankrupt insurance-like scheme mentioned by Johnson one would have to check for the language and concepts used or if there was a conflation of ideas originating from the descriptive language of the English consuls.

153 China. Rates for Foreign Settlements (Tariff 13 April 1864 to 9 June 1866), LMA CLC/B/192/019/31522/60, p. 36.

154 See Bergère, *Histoire de Shanghai*, pp. 46–57. This explains why the situation of European and Chinese housing was far more complicated than in the 1820s and 1830s. For Canton Carroll, "Slow Burn in China," p. 41 states that "[f]or the most part, the foreign factories and their contents were less vulnerable to fire than the rest of Canton and its suburbs"—but the very distinction between the 'European' and the Chinese became difficult, at least in Hong Kong in our later period when European fire insurances had settled.

FIGURE 81 A Chinese and two European Houses in Shanghai ca. 1877, LMA CLC/B/192/019/31522/62, p. 242f

illustrates what the agents had in mind when they distinguished roughly the type of building (Fig. 81).

By 1895, the description and differentiation of risks for Hong Kong had developed into ten classes of "European occupation and construction" (residential homes, retail shops, godowns, saltpeter, coal, kerosene oil godowns, … explosives). Moreover, buildings that were not "purely" European were now listed under the second subcategory of "European risks" as "mixed occupation:" There was a distinction between:

1. Houses of Semi-European or Chinese Construction, occupied entirely by Europeans and/or others, except Chinese [a. (sic, subcategory like this in the source) If partly occupied by Chinese]
2. Buildings of European Construction, occupied partly by Europeans and/or others, and partly by Chinese and the Rates were formed for Europeans and/or others except Chinese.[155]

155 LMA CLC/B/192/019/31522/63, p. 101.

Under “Europeans risks,” agents included “Europeans, Americans, Armenians, Portuguese, Parsees, Indians and Japanese.” It is telling that Japanese and Indians were treated as equivalent to Europeans, and that Portuguese had to be included explicitly as Europeans.[156] In the second major category, “Chinese Risks,” the agents had to distinguish between 1) those Chinese buildings occupied by Europeans, and 2) those Chinese buildings occupied by Chinese, the latter category having ten different subcategories, including “4) Houses of European construction, occupied exclusively or *partly by Chinese and partly by Europeans* (including Compradors' effects in European Offices),[157] 5) Houses of Chinese or Semi-European construction used as Shops and/or Dwellings,” and 10) an additional 25% if the property was “Suburban (outside City Boundaries).”[158] In Shanghai, also in 1895, rates were similarly differentiated according to European or Chinese occupation, but with fewer subcategories.[159]

The many categories and distinctions were the result of long discussions and communications, which are mirrored in the Memorandum Books. As early as 1854, we read:

> The Imperial and the Alliance take risks on buildings owned by Europeans and occupied by Chinese as godowns only at the same rate as on buildings occupied and owned by Europeans. … Risks on European built buildings owned by Europeans occupied by Chinese, as before remarked we consider in every way as desirable as if occupied by Europeans, indeed even more so, for as in the one case the property within the godown being utterly uninsured (Insurance being unknown to the Chinese) it is but reasonable to expect even greater precautions against Fire than where the property as in the case of Foreign occupation, is covered against fire.[160]

156 Japanese Immigration in “Treaty Port China” started in the 1870s but grew quickly, especially after 1904; for Shanghai, see Christian Henriot: “ ‘Little Japan’ in Shanghai: an insulated community, 1875–1945,” in Bickers and Henriot, *New Frontiers*, p. 148.

157 “Comprador” was the usual term in colonial trade for Cantonese mediators, translators, and trading contact men who worked for the British and who “themselves did business mostly with other Cantonese” (for the importance of those compradors to Shanghai see Johnson, *Shanghai*, p. 197; Bergère, *Histoire de Shanghai*, pp. 80–84).

158 LMA CLC/B/192/019/31522/63, pp. 101–102.

159 Ibid., p. 138.

160 LMA CLC/B/192/019/31522/60, p. 156f. (August 1854).

The statement that insurance was completely unknown to Chinese is only partly true if we think of the *gongsuo* system.[161] Normally, in the Treaty Port of Shanghai, the Chinese were not allowed to own houses, but they managed to buy them through the mediation of Europeans:

> The French side contains a very great number of Native houses and as a rule we would not accept risks there unless on property situated on the Bund or entirely separated by a large open space from Native buildings, nearly all the respectable people live on the Bund. The American side contains also a good many Native houses, but they are clustered together and any risk we would be likely to be offered is quite away from them.[162]

In 1866/67, the Sun's agent Chapman counted 11,774 Chinese houses and only 354 "purely" European houses in the English settlement. However, most Chinese houses were in English hands: "Chinamen cannot own property in the settlement; if they buy houses, it is in the name of some European."[163] Such practices undermined the clear native/European border, and although insurance companies reacted as flexibly as possible, they were unable to remove the distinction between European and Chinese. The steering committee of the Sun in its London headquarters was likely concerned about the problem of "Chinese houses" and issued an official definition in 1869:

> It may be useful to remark that the term "Chinese houses" has the definite meaning of "Houses occupied by Chinese, whether owned by them or by Europeans." The evidence of persons who have resided in the Settlements and Treaty Ports seems to establish the fact that at such places the Natives only dwell in houses of a kind which Europeans never occupy. Hence the occupation sufficiently and accurately defines the construction of a house.[164]

161 Johnson, *Shanghai*, pp. 143–144, 167; but see above note 152. Faure and Köll, *China*, pp. 472–73 point to the first introduction of (maritime) insurance in Guangzhou (Canton) as early as 1805.

162 Agents Chapman/King & Co. to Sun Fire Office, April 8, 1869, London, LMA CLC/B/192/019/31522/60, p. 166f.

163 LMA CLC/B/192/019/31522/60, p. 173.

164 LMA CLC/B/192/019/31522/60, p. 112.

The committee was concerned but could not give any clear advice: "The Committee, however, do not feel able to submit any definite recommendation, as the intermixture of European and Chinese risks is so general and complicated that the question can only be satisfactorily dealt with on the spot." Agents were tasked with dividing the town into districts corresponding to that intermixture of Chinese and European premises.[165] All in all, the definition was strange because it drew conclusions from habits and manners and imposed them onto the material substance of buildings.

Given that the intermixing between ethnicities became more and more complex, the efforts made to distinguish between "European" and "Chinese" seem to have followed an increasingly ethnic or even racial logic, rather than an economic one. For the case of India, the even stricter separation between the European and native districts of Bombay and Calcutta was economically dysfunctional, as has been seen. For Shanghai and Hong Kong, the ethnically hybrid situation of both cities does not lend a good legitimation, from our ex post perspective and from a point of view of simple economic calculation, for all those subtle categories and subcategories centered on the Chinese/European distinction. The reason for this ethnic mixing lies in the difficult political situation in the surrounding Chinese Empire and the attractiveness of open trade ports. European settlements grew massively: the London headquarters even noted that they did "not apprehend so much danger from Chinese buildings in the English settlement] as you [the local agents] seem to" have done.[166] Given the short distances between them, a fire starting at a Chinese house would have immediately affected nearby European houses; however, Chinese houses were as safe as those of Europeans: "The buildings in Shanghai are very much more substantially built than those in Japan. The Native houses are built of wooden framework filled in with bricks and are very easily pulled down in the event of fire."[167] Throughout the sixty years of business in China before World War I, no major fires broke out in Shanghai, and the two great fires in Hong Kong, in 1867 and 1878, did not confirm the rationale for the distinction between Chinese and European.

5 USA/New York[168]

Although the Sun's very first foreign policy was issued in 1836 for one "risk" in New York, the insurance company was one of the last British companies to

165 Ibid.

166 LMA CLC/B/192/019/31522/60, p. 165.

167 LMA CLC/B/192/019/31522/60, p. 160.

168 For a general overview of the American market, see Robin Pearson, "Organizational Forms in Insurance: A Comparison of the USA and Germany During Industrialization,"

establish itself in what had been the largest fire insurance market in the world beginning in the second half of the nineteenth century: from 1836 to 1875, only sixty-one scattered individual policies had been purchased.[169] While other cities were later much more "carelessly" and quickly engaged, in the US it was deemed necessary to wait until it became clear whether the high fire frequency derived from a specific arson culture:

> The truly alarming accounts of Fires which have recently reached England are of a nature which makes it prudent for the Managers to abstain altogether from extending their habilities in the United States, or at least to wait until such intelligence has been received as may tend to remove the existing impression that many of these conflagrations are not the result of any of the common causes of fires, but of arson or incendiarism.[170]

Whenever the fire numbers were particularly high (for example, around 1866), the Sun was asked if it wanted to enter the market, while local insurers raised rates by 25% during these periods (as in February 1867) and were often financially backed against a wall or simply failed.[171] On the one hand, the American market, at least by 1835,[172] was more open than the heavily state-regulated German market;[173] on the other hand, it was obviously more difficult to approach and there was strong resentment against European insurers in America. A satirical one-act play titled "London Assurance" appearing in a New York newspaper in 1871 mocked the effect of the sudden dominance of English insurers in the US market after the Chicago Fire: a customer turned to the London "Whirlpool and Undone Insurance Company" and attempted to

---

in Pearson and Yoneyama, *Corporate Forms*, pp. 114–42 as well as Tebeau, *Eating Smoke*, pp. 78–85 and passim.

169 LMA CLC/B/192/019/31522/275, p. 10f., 17.

170 LMA CLC/B/192/019/31522/275, p. 5.

171 "The Chicago & Boston Fires 1871–72 also tended to send Business to England," ibid., p. 16. Cf. Pearson and Lönnborg, "Regulatory Regimes," p. 62.

172 Earlier, in 1810/1814 in Pennsylvania and New York, foreign companies were barred from managing businesses; however, in 1835, after the major fire in New York, this prohibition had to be abandoned, since the capital power of the British insurers was required; see F. C. Oviatt, 'Historical Study of Fire Insurance in the United States," *Annals of the American Academy of Political and Social Science* 26 (1905), pp. 155–78, here 162.

173 In 1837 Prussia introduced permit requirements for non-Prussian fire insurance enterprises; see Hans Christoph Atzpodien, *Die Entwicklung der preußischen Staatsaufsicht über das private Versicherungswesen im 19. Jahrhundert unter besonderer Berücksichtigung ihres Verhältnisses zum Wirtschaftsliberalismus*, Bonn 1982; Sonja Heiss, *Die Institutionalisierung der deutschen Lebensversicherung*, Berlin 2006, p. 111.

determine its business balances, which seemed to become less credible, smaller, less reliable, and more antiquated with every question.[174] Another, somewhat bitter short story "A British Victory; Or, Time's Revenge" appearing in the *Brooklyn Local Intelligence* of January 13, 1872, addresses the relative financial strength of British fire insurance companies, which allowed them to dominate the American market going back to the wartime period of 1803 to 1814. During this time, some private lenders made good profits on loans to the US state. The American payments on these loans allowed the lenders to establish insurance companies: these companies now threatened domestic insurers, although they had been financed de facto by Americans—a belated British victory, on the financial level at least, following the loss of the old colonies.[175] The Sun observed these developments and made self-critical comments: "Several English Companies Conducted business in a manner unnecessarily offensive to Local Interests—to be regretted that a more just truthful & conciliatory course had not been pursued."[176] In the end, the Sun was not sustainably engaged in the market until 1882 and was then rewarded with high profits, which allowed them to take over the locally operated Watertown Insurance Company with its 156,000 policies.[177] In this regard, it would be useful to further investigate the directly preceding time period, when the Sun tracked the American market very closely and developed the practical skills that allowed them to then relatively securely enter the market on July 31, 1882.

Of the 103 fire insurance companies operating in New York in 1870, most were not older than twenty years (Fig. 82): the oldest firm operating at that time was Knickerbocker Fire Insurance, a mutual society, which was founded in 1787 and lasted until 1890; a total of eleven insurance companies emerged in the initial stage of proliferation between 1821 and 1825. Around 1820, there were seventeen joint-stock insurance companies based in New York, six in Pennsylvania, two in Connecticut, and one each in Rhode Island, New Jersey and Massachusetts;[178] however, many of these early companies fell victim to the major fire of December 1835, resulting in a stagnation in the market and few new companies emerging until about 1850.[179] After this point there was a boom in new businesses, which corresponded with a global trend in the insurance market. In 1853 alone, 19 new insurance companies were established; and 62

174 LMA CLC/B/192/019/31522/275, p. 129.
175 LMA CLC/B/192/019/31522/275, p. 139.
176 Ibid., p. 153.
177 Dickson, *The Sun Insurance Office*, pp. 221–231.
178 Oviatt, "Historical Study," p. 162.
179 Also the judgment in Oviatt, "Historical Study," p. 163.

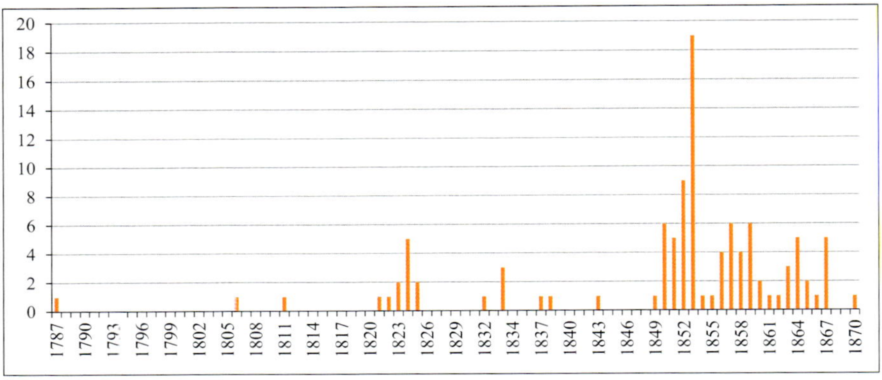

FIGURE 82 Numbers and dates for business origins of fire insurance companies operating in New York in 1870 (LMA CLC/B/192/019/31522/275, p. 102)

of the 103 insurance companies active in 1870 were founded between 1849 and 1859.[180] Most of the 22 foreign insurers who operated in New York in 1882[181] entered the market only after the major fires of 1871/72, which had heavily burdened American insurance companies and, as discussed above, created a point of entry for foreign companies.[182]

What was known about New York, or American cities more generally? Although by the second half of the nineteenth century they often boasted impressive architectural achievements, American cities were certainly associated with great risks, and great fire risk in particular, as demonstrated by the enormous fires of Chicago in 1871, Boston in 1872, and Baltimore in 1904. Christine Meisner Rosen has shown the extent to which these fires indicated significant structural problems in cities that were growing too fast and in turn spurred the adaptation of appropriate structural safety measures.[183] Her study is dedicated to the process of reconstruction within these cities; the earlier fire characteristics of these cities had not previously been examined, which is why these major fires are presented as solitary trigger events for decisions concerning urbanization and the urban development process.[184]

---

180 LMA CLC/B/192/019/31522/275, p. 102 (Summary in table form from *The Spectator*, October 1871, p. 372).

181 LMA CLC/B/192/019/31522/275, p. 184f.

182 Cf. Pearson and Lönnborg, "Regulatory Regimes", pp. 59–86, 63.

183 Meisner Rosen, *The Limits of Power*.

184 Miller, *American Apocalypse*, p. 67 similarly briefly mentions the fact that there were frequent fires in the drought period of 1870/71. Sawislak, *Smoldering City*, pp. 21–67 also

TABLE 7 Statistics on fire numbers and loss distributions determined in 1872 by Marsden (agent of Guardian Insurance) based on data from the New York City fire commissioners, LMA CLC/B/192/019/31522/275, p. 19f

Year	Number of fires	Estimated costs (in $)
1866	746	6,428,000
1867	873	5,711.000
1868	740	4,342,000
1869	850	2,626,000
December 1, 1869-April 1870	354	1,069,000
April 1870–71	1105	7,389,000
1871–72	1370	1,633,253

Total number of fires	**1868**	**1869**	**1870/71**	**1871/72**
	740	**850**	**1105**	**1370**
Confined to one building	698	807	1033	1291
Spreading to other buildings	42	43	49	43
Buildings only slightly damaged	675	784	805	606
Buildings severely damaged	86	89	100	55
Buildings destroyed	36	49	17	37

For example, the Sun had a copy of the report by Mr. Marsden, a Guardian Insurance employee, who in 1872 summarized these informative statistics from the fire commissioners of the city of New York over the previous five years (Table 7).

From these statistics, Marsden concluded: "It will thus be seen that the Commissioners are justified in taking credit in their report that fires are generally confined to the Buildings in which they originate & that while the number of Fires has increased with the growth of the City, the value of property destroyed has steadily diminished."[185] This statement was not true for the prolific fire year

evaluates the fire reports in the context of the narrative of the singularity of the event. On the involvement of insurers in the exceptional case of the earthquake of San Francisco followed by the major fire in 1906, see, in addition to Röder, *Rechtsbildung*, and also Andrea Rees Davies, *Saving San Francisco: Relief and Recovery after the 1906 Disaster*, Philadelphia 2012, pp. 105–42.

185 Ibid., p. 20.

of 1870/71, when the drought that so dramatically affected Chicago was also felt in New York, but all in all, these numbers should have been reassuring enough for the Sun to enter the market. Nevertheless, they continued to wait. It may be that the perception of the fires in Chicago and Boston contributed significantly to this hesitation: these fires, which destroyed street after street of brick-built buildings, hit on a nerve during this "progressive" period from the 1850s on, when older constructions of wooden pillars and beams had been increasingly replaced with iron. It was believed that this type of construction would be fireproof when combined with brick—although as brick was considerably more expensive, its use was not widely implemented.[186] One of the Chicago buildings that resisted the flames the longest, the Tribune Building, had the reputation before the fire of being "one of the most absolutely fire-proof buildings ever erected." Even this building could not be saved in the end, so that contemporary witnesses simply had to conclude with bitter humor: "That is, it was fire-proof up to the date of its destruction."[187] In another report, written in 1876, Sun employee George Manvell initially offers a general impression of the cityscape that was scarcely related to the business of insurance at all. He describes the "celebrated 5th Avenue" as "monotonous" and maintains that Marsden's somewhat disparaging characterization of Broadway as "a mixture of Whitechapel & Tottenham Court Road" is not incorrect, comparing the split use of the walkway to Hamburg and Copenhagen. The "un-English" impression presented by New York was primarily a result of the many advertisements on the buildings. In terms of fire danger, Manvell emphasizes the often-intense heating employed in the basement of the buildings, which dried the wooden beams; however, he notes that the roads are much wider than in Europe, which impedes the spread of fire and facilitates firefighting efforts. He points out the widespread use of bricks in construction but states that wood was also still used—for example, in the form of the ubiquitous wooden window blinds. The so-called "dry goods district" between Worth, Thomas, Church Street, and Broadway comprised the largest concentration of value in New York—its worth was estimated at $25 million.[188]

> The burning of a few blocks in one of these quarters would cause a loss to be reckoned by millions, & would mercilessly reduce to Bankruptery a whole row of Insurance Companies. ... No Insurer can traverse the Dry

186 Sara E. Wermiel, *The Fireproof Building: Technology and Public Safety in the Nineteenth Century American City*, Baltimore 2000.

187 Miller, *American Apocalypse*, p. 67.

188 Augustine E. Costello, *A history of the New York Fire Departments, Volunteer and Paid, Derived from Original Sources and Authorities*, New York 1887, Ch. 50, part 5.

> Goods District, the neighbourhood of the Cotton Stores & Stock Exchanges, or John Street & Maiden Lane without a Shudder. Only a few blocks there need burn down to cause a loss of hundreds of Million Dollars … assuming the simultaneous occurrence in different parts of the City of 2 large fires, a scarcely water supply—as is too often the case & at the same time a strong sea breaze—we have the Elements of a Catastrophe compared with which even that of Chicago would sink into insignificance.[189]

Although some observations appeared pejoratively inflected, it also seems they were reported to London out of dutiful skepticism, since after all, it was impossible for visitors to avoid conceding the strengths of the city: "After all this I must say that I like New York & in some things they are certainly before us."[190] There was no other country in which the Sun operated where such a judgment was reported to the headquarters.

In 1882, when the Sun finally decided to enter the insurance market through its takeover of the Watertown Company, precise financial information was available for each of the twenty-two foreign insurers in New York. A very reliable average value for the ratio between premium revenue and loss payments had been calculated because there were 167 years of data to fall back on, even though each individual company had been operating in New York for only between two and twenty-one years: the "loss ratio" hovered at 60.9% of the premium payments.[191] The danger of fire and associated damages in the city had therefore by no means been underestimated, and compared with the major Indian cities, New York initially seemed to be a much more dangerous and perhaps less profitable market. The ultimately high fire danger, however, obviously also led to a very high demand for the services offered by "fire insurance," resulting in an American market that quickly became profitable for foreign companies.

6 Comparative Analysis

It would seem that the pure numbers of the loss ratio addressed at the beginning of this chapter— i.e., the ratio between premium income and claims payments (nearly 9% for Bombay, 27% and 42% for Shanghai and Hong Kong respectively, 60% for New York, 82% for Hamburg, 87% for Istanbul)—while

189 LMA CLC/B/192/019/31522/276, p. 195.

190 LMA CLC/B/192/019/31522/276, p. 144–46.

191 Cf. tables in LMA CLC/B/192/019/31522/275, p. 184f.

themselves very diverse, conceal urban worlds that are even more different from each other. The fire and fire-danger statistics of these cities emerged as a response to entirely different conditions.

In particular, the files of the Sun Fire Office offer insight into how insurance ultimately employed important practical and simple mechanisms to effect globalization during the second half of the nineteenth century, as well as offering an illustration of the perceptual schemata at work: First, there is clearly a difference between Western European cities and those perceived as non-Western. The Sun's foreign Memorandum Books did not contain maps of Hamburg and even not for New York that meticulously defined insurable and uninsurable zones—though detailed fire insurance maps certainly existed in general for England and the US at that time. The dichotomies of "European" vs. "native" towns, or "Christian"/"European" vs. "Muslim" were also absent in such Western regions at that time. Ethnic segregation, which concerning US cities around 1900 has already been thoroughly thematized in conjunction with disaster history,[192] seems initially not to play a decisive role in the cartographic tradition of fire insurance:[193] the separation between good and bad risks was determined almost solely by differences in building materials. Today, segregation historians do in fact use fire insurance maps because of their detailed rendering of the housing stock around 1880/1900, but they project back Census and city directory data (as for Washington, Nashville and Charleston) to get a clear impression about the way how (and how differently from one city to the other) black/white segregation and coexistence was realized in the fast growing cities of the United States.[194] An initial result of the comparison is thus that the "overvalued" predominance of colonial and "Orientalist" classification strategies really had little to do with systematic economic considerations foremost for Colonial Cities or Treaty Port Cities in a large Asia from Istanbul to China and Japan. In India, it even became apparent that the "native town" was no more vulnerable to fire than the European part of the city and thereby represented an equally good or—due to the relatively low demand for insurance—bad market. The fact that a fire insurance company as important

192 For "racialized views" during the San Francisco disaster regarding interactions with the residents of Chinatown, see Davies, *Saving San Francisco*, passim.

193 On early hazard mapping, particularly in relation to American cities, see Tebeau, *Eating Smoke*, pp. 101–08; van Manen, "Les plans d'assurance incendie de Goad," p. 172; Monmonier, *Cartographies of Danger*, p. 286.

194 J. R. Logan and M. J. Martinez, "The Spatial Scale and Spatial Configuration of Residential Settlement: Measuring Segregation in the Postbellum South," in: *American Journal of Sociology* 123, 4 (2018), pp. 1161–1203; Cornel Zwierlein, Florian Wagner, "Close distance. Social segregation in trading empires and colonies - An introduction", *Journal of Modern European History* 18, 2 (2020), pp. 140-155.

as the Sun (in addition to the other companies already operating in India) apparently took many years to recognize the basic causal relationship associated with the monsoon climate and using tropical timber as a building material is really quite astounding. In Istanbul, the flagship Sixth district burned just as readily as the Muslim Stamboul did a few years earlier. The overzealous inspection narratives, and the "expert" reports by Woods, Atkins, and Manvell seem almost to be nothing but a waste of paper. The whole process functioned according to trial-and-error. It took twenty years for companies to collect the necessary data—namely, their own and others' loss ratios—that constituted perhaps the only reliable criteria for answering the question of whether engaging in insurance practice was advisable. The typically scanty information that could be derived from outside sources about the fire characteristics of the cities in any given location was apparently rarely used as a management tool. In China, the mixing and complex blending of ethnic boundaries within the Treaty Ports further demonstrates how basic patterns of perception concerning the native/European binary, which was de facto used as a business management control, reproduced itself almost ad absurdum. Instead abandoning it in the face of evidence, the Sun encouraged increasingly fine subdifferentiation on an involutional level, at least in terms of the economic process of premium calculation.

Although the material collected did suggest certain conclusions, such conclusions were not really elicited from the data in London; rather, this collection of data was taken as a whole and accompanied by the processes of preliminary probing and the establishment of agencies and branches of business and used to create a qualitative and quantitative memory of the very different relationships between "city" and "fire," or "city" and "fire hazard," for the entire world. As such, in some sense these data exist in a transverse relationship to the predefined schemata: the fire safety of Indian cities did not fit with the apparent affinity between "safety" and "modernity," as envisaged in the European narrative of development, which was consciously or unconsciously acknowledged by the major players in the market; nevertheless, the "native towns" of India were categorized as premodern anyway. It was tempting to reduce the great fire of Hamburg to a consequence of premodern mortgages, as was also done in the case of Istanbul, when the major fires of 1865/70 were resignedly recorded simply as proof of a completely failed Tanzimat and "unmodernizable" unsafe conditions associated with the Oriental city. The fire danger of the rapidly expanding American cities, however, was classified in a different manner. In this case, insurers perceived "progress"—it was even tentatively noted that the Americans were in many ways superior to the old colonizers—but this necessarily brought new threats along with it.

Overall, the way in which the insurance giants of the nineteenth century operated during their global expansion into countries and markets unknown to them is not so different from the early days of fire insurance in London, when Barbon attempted to identify fire statistics and gross damage/loss averages: certainly, Barbon knew London, but his state of knowledge concerning prior and reliable "data" on its fire ecology was quite similar to that of his followers two centuries later in different cities of the world. In the globalization process, pluralization effects are employed both to determine the "reality of the environmental conditions" (do the cities burn or not?) and to influence forms of perception. The observation of both processes at such different times identifies transepochal comparisons—i.e., comparisons of processes concerning the separation of modernity in the developmental diachrony of "old Europe" (seventeenth century) with the processes of cultural diffusion in the synchrony of globalization (nineteenth century)—as a serious task of historical science.

In the first chapter (Ch. 2), the spatial dimension of maritime transport insurance was highlighted and distinguished from later developments. The previous two chapters (Ch. 4, 5), exposed aspects of temporalization (concepts and practices of growth and improvement and the similarity to sustainability thinking) regarding the insurance practices of the Enlightenment. With globalization, space seems to reenter as the dominant perceptual matrix, especially concerning contact with non-Western societies and hybrid colonial societies. While in the context of the "known" Western markets of Germany and the US, individual analyses dominate, the harsh distinction between native and European dominated the perceptions of the offices and agents for the Middle and Far East and for colonial regions so thoroughly that city maps and maps reifying those ethnic segregations became the main orienting tool. Eventually, the idea of value stabilization and, in some cases, the possibility of value growth through insurance established in Europe, also resonated in the "European towns" of the colonial cities; thus, the temporalizing aspect of insurance was also relevant. However, this was more of an initial condition, first implemented in the European-influenced parts of the cities and countries, which were, in a sense, located on one side of the fundamental distinction between native and European like in a small biotope or laboratory. Plotting the contours of the safe, normal world on the maps along the borders of the colonial European ghettos, which reflect the trend toward the tendential self-referentiality of the colonial economy in general, allows "space" to reemerge in a different way as the dominant principle in the global implementation of insurance as a specific tool of security production and as a security regime. Zoning and space boundaries are once again transposed and superimposed onto the cultural tool and technique of security production during its global transfer—despite the epistemic linkage to futurity. Nevertheless, the Enlightenment notion—long since taken for granted on a technical level—of the paradoxically contrafactual expectation of safety and security as the normal situation, of fire and disaster as exceptional events, remains deeply embedded as a driving force within global relationships in Western mental maps. It is as if the colonial agents wanted to delineate the borders of small "normal secure societies" transferred out of Europe in each world region on their (real) maps by distinguishing between native and European zones without fully understanding the complexity of the encounter and of the process itself. The agents' unconscious guiding principle would have been something like the diffusion and unilinear expansion of the Enlightenment's "secure normal society" across the globe, but the reality that was found, produced, and experienced, fractured and pluralized the guiding patterns of expectation in the wake of globalization. This resulted in breaches and also in the unexpected questioning of patterns of perception that, in the long term, had distinct repercussions on a decentered European home society that recognized the plurality of relationships between different types of cities, different forms of security regimes, and different forms of modernity.

CHAPTER 7

Conclusion

How much security is necessary to achieve "modernity," and how much modernity is required for the form of security investigated here? At the outset of this study, I suggested that the sources as well as the current complex historiography makes it inherently impossible to address such questions in any simple form, and that to some extent perhaps the initial suggestion that there must be a link at all between certain measurable forms of fire security, fire safety, and modernity, might prove wrong. It is thus not possible at the end of this study, to situate the equivocation of security and modernity as a result of the historical investigation. The globalization of insurance practices, in particular, provided tangible evidence of the multiplicity of security regimes, as well as of modernity; the climate conditions in India alone clearly led to a completely different framework of expectations concerning security, which European insurers had long failed to understand. Although it is possible to interpret this situation using Shmuel Eisenstadt's "multiple modernities" framework, it should be emphasized that the transfer and accommodation processes concerning the counterfactual security expectation that spread from Europe through the rest of the world are themselves based on the resilient core of the ('single') concept of modernity, which even Eisenstadt could not escape.

Premium-insurance-type security production seems to be an element of European "modernity" and the modern era. The insurance contract itself already existed in the practical world of merchants and, like a troublesome obstruction, burrowed into the world of academic law. In the period 1630–50 the concept of insurance became part of a kind of economic, legal, and technical *Querelle des Anciens et des Modernes*, finally proudly defined by natural law pioneers like Stypmann as an invention of "moderni." This perception of insurance as an achievement of modernity became universal during the Enlightenment period. The normally safe and secure society was a constructed and (consciously or unconsciously) advertised, promoted and supported idea of Enlightened administrators and projectors. As such, it had to confront the "premodern" expectations and conceptions of community and society—or to go even further: it was in that way producing the difference between the premodern and the modern by the yardstick of the existence of insurance-like security production.

I have emphasized that these distinctions and this new way to conceive of urban and later territorial realities and life, in terms of security politics and

 | DOI:10.1163/9789004431225_008

regimes to be planned, shaped, and controlled by the authorities were part of a major change, which also established a buffer between “nature” and “culture” or “civilization”. Nature, its impact and causality became walled off (or should become walled of) from human life. Reality was still very different, and the fire gap spread only slowly; cities still burned down completely throughout the long eighteenth century in Central Europe, as elsewhere. The fire gap as large momentum in the historical fire ecology of Central Europe characterized above mostly for those spectacular very large fires that could consume whole cities is still incompletely representing the every-day threat, which was also present at the micro-level of villages, and included the partial burning of several houses and blocks, frequent but hard to measure in those times. The construction of that image and ideal of a normal secure society was therefore anticipating or surpassing reality to a large extent, and it is worthwhile, and necessary, to reconstruct the parts and patterns of this frame of thought and perception as a historical object in its own right. The level of *policey* norms, technology, administrative communication, surveying and controlling by Cameralist administrators and all those elements of the *real-assecuration* are one part; but the discourses on insurance as an important form of financial post-damage recovery, the development of visual representation as a function of memory, but also of admonishing and advertising, next to all forms of the projects and scientific experiments that were also involved, and which Leibniz coined the *verbal-assecuration* of human societies, are also of great importance. Here, it is the average form that is important—it was the multitude of unknown and small engravers and painters in villages and small cities who were active in modelling that frame of thought and were acquainted with what a burning city and village looks like as an always present threat, more important than a Rafael of Urbino painting one famous picture of Aeneas saving Anchises from burning Troy in the Vatican. I have shown how, even here, the forms used to represent the catastrophic moment were changing decisively—partially causing and stimulating the evolution of fire protection means, partially bearing proof of a relationship between the reaction, correlation or coevolution of these visual (and narrative) forms and the evolution of administrative and fire protection techniques. In this way, the threat to be avoided entered the discourse of normal secure society in its narrative and visual forms, at the same time as it entered as a more or less calculable risk into the Cameralist discourse on fire insurances. Both changes and evolutions are part of what was called above the evolution from Hazardous and (charitable) Solidarity Communities to Risk Communities, while processes of globalization also imply the emergence of global risk communities (due to the global expansion of the large insurance companies) and of global solidarity communities, which arise only briefly in moments of a large catastrophe, perceived as worldwide.

The role of religion and the differences between the Christian denominations in Europe with regard to the spread, evolution and institutionalization of insurance have been addressed: it was a complex relationship since the earliest times, when canon lawyers were reasoning about the permissivity of insurance contracts or its treatment as usury. Fire threat and catastrophes were always occasions for preaching and theological reasoning—but I have stressed that the sometimes apparently strict opposition between theological reasoning about God's providence and the reasoning about secular care, provisioning, calculating risk in forms of insurance, is otherwise not to be found in the German or English sources of the long eighteenth century—both were coexisting forms used to perceive and cope with the hazards and the recovery from damage. Similarly, I stressed that the remarkable elective affinity of certain types used to institutionalize the premium insurance principle, with the Christian denominations prevailing in a given country or region in Europe (the "clover" scheme) cannot be explained in the way that a certain theologoumenon (and a different exegesis concerning that theologoumenon within those denominations) would, by way of entelechial logics, have caused that difference. The conditional or explanatory meaning of it lies is rather in the different form of how state, economics, religion and forms of planning and conceiving of the future are combined in a given country as a product of a prevailing denominational culture ("confessional culture"), The driving force is not the theological discourse as such. Nevertheless, religion and denominations are of great importance, and narrower histories of economy that do not address the cultural and religious contexts of an "ephemeral" idea and accounting trick or *technique* as the premium insurance principle, are bypassing the problem.

Conceiving of "insuring" as simply a cultural technique also cleared the path for using the archives of a global insurer like the *Sun Fire Office* for a different history of globalization and cultural contact, although, decisively and admittedly, only from the point of view of the European observers—and it has to be added that the existing literature until now does not suggest a huge treasure of sources (e.g. from Ottoman, Indian or Chinese archives) that could replace or fundamentally correct the (Orientalist, biased) vision as preserved in the *Sun Fire* records.[1] I have tried to show the very prevalent biases from these records themselves, revealing how the British were equating very different types of cities and how they introduced ethnic and proto-racial forms of perception into the risk calculation itself, while the fire hazards that arose (or, as in Indian

1 The most important parts of the more recent studies on late Ottoman Insurance history or on Malaysia as quoted in chapter 6, note 23 are likewise reconstructed from the London records, not from any other sources that might be thought to contain more 'truth' or 'reality'.

cities: did not arise) showed their forms of expert analysis were economically dysfunctional. Those processes of globalization can, unfortunately, only be analyzed by relying on British sources, as it is only here that large archival records of early global insurers exist, and this is not the case, for instance, for the rare German insurers that were present in Asia or Africa in those times.

In the course of globalization, the colonial "European/non-European" dichotomy therefore also seemed to superimpose itself on the distinction between "modern" and "premodern". Of course, the consciousness about a given set of conditions, a behavior, a practice, or a whole city belonging or not belonging to modernity is implicit, for example, when Istanbul at the end of the nineteenth century is compared to London of the seventeenth century. The already noted "return of space" should be recalled in this context. If, on the one hand, the British ′one and unique′ model of insurance seems to spread across the globe, such observations indicate that perhaps the beginning of a new process of differentiation and dis-integration can be found here as well. As such, the global spread of insurance may serve as both indicator and promoter of a universalization of the secure normal society of the Enlightenment. However, rather than facing "failure" when confronted with unusual or resistant conditions, this approach transforms itself into a multiplicity of security/insecurity relationships. Since even today, insurance coverage in the megacities of the so-called third world is mostly limited to the inner, formerly colonial centers, and huts in the slums burn regularly like the villages and cities of European premodernity, it would seem that the spatialization of the "modern/premodern" distinction—which was recognized by insurance agents in the colonial world around 1850—has remained fairly stable over the period of the last 150 years.

Based on many other situations and observations, this could lead to the idea that this notion of secure modernity should be conceived as locally and temporally diverse, that there was, perhaps, an exception sandwiched between early and late modernity, and that the notion of a normal secure society increasingly seems like a construct of the Enlightenment, which itself embodies a historical and spatial exception: historical, as it was perhaps valid only for that short period of a long eighteenth and a good part of the nineteenth century, and spatial, as it seems best to fit to Europe alone, while the analysis of global developments and realities leads rather to a concussion of those concepts (as of old-European period distinctions in general) also with regard to this specific object, "insurance", as historicized.

On the other hand, the current trends in international politics—namely the expansion of the classic concept of security to human security and beyond—indicate that the universal claim to individual and collective security that initially developed during the European Enlightenment and the driving force associated with this claim have by no means completely disappeared and, are now part of a Western discourse through the United Nations and international relations that also tends to equalize global realities to some extent, through means and perceptions that are often rooted in colonial and pre-colonial times by way of universalization. This concerns a wide range of other disciplines, however, of other fields of security conditions, security politics, and security regimes, which are not part of the thematic focus of the current study. Nevertheless, the first institutions engaged in the serial observation of socio-natural disasters, and thus insecurity, were fire insurance institutions, and if we are used to these today, as consumers of every kind of news in the post-television and internet era, with graphs, diagrams and pictures mapping hazard prone and hazard hit regions of the world today or in the recent past; if we are familiar with a large diversity of representations, photographs and measurements of all kinds of natural hazards in statistical or individual form-then this is all, to some extent, a heritage of the first methods used to institutionalize insurance within company or administration forms: And this lead to a different way of observing the world, to ways of taking hazards and damages *into account*, of representing this world *as* secure or *as* threatened, and to calculate it. And so this is a heritage of the "normal secure society" that emerged around 1700 somewhere in northern Europe, and eventually in the "laboratories" of London and Hamburg.

APPENDIX 1

Academic Legal Treatises and Dissertations on the *Assecuratio* (in Chronological Order)

Note: I give here a list of Northern European treatises exclusively devoted to the question of insurances. As shown and explained above in Ch. 2, there are certainly earlier treatments of the premium insurance contract since the late thirteenth and fourteenth century by Canon and Roman lawyers, Pedro de Santarém being usually recognized to have written the first long treatise at the end of the fifteenth century that appeared later in print. Often, chapters or subchapters of longer treatises on commercial law are devoted to that question, and north of the Alps we can find some mentions and discussions as early as with Conrad Summenhart, then with Rutger Ruland much earlier than in the monograph of 1630 commented upon extensively in Ch. 2, in his *Tractatus De Commissariis et commissionibus* (first edition of 1597), with Johannes Borcholten (*Commentaria in vtilissimum C. Tit. De Nautico Foenore, Helmstedt 1596*), with Hieronymus Pantzschmann, Hermann Vultejus, Anton Hering, Reiner Bachoff von Echt, Paulus Christinaeus and, as mentioned, with Grotius. However, those are either spurious treatments which are just repeating or summarizing the southern discussion and state of knowledge (Straccha, Scaccia …) or (in Grotius' case), we are dealing with a very new embedment within the framework of the emerging universal concept of natural law, as explained above. Reading the dissertations and treatises backwards from the late seventeenth and eighteenth century, Ruland's 1630 treatise and advisory text reveals to be the starting point of a tradition always referred to in the North.

1. Rutger Ruland, *Erledigung einer schweren, in Assecuration Sachen vorgestalter Fragen, ob nemlich, wann einer auff ein Schiff versichern lest, aber kein Praemium bezahlet und hernacher zur See Schaden erfolget, als dann denselben er als Assecuratus von den Assecuratoribus mit Rechte zu fordern vermüge?: alles theils auß d. gemeinen beschriebenen Rechten … abgefasset*, Hamburg 1630
2. Nicolaus v. Schaffshausen, *Dissertatio de assecurationibus, vulgo: von Assecurancien oder Versicherungen* […], Hamburg 1638 [Copy ÖNBW]
3. Johann Adam Schragius [Argentoratensis], *Disputatio inauguralis juridica de assecurationis contractu* [*September 15, 1642*], Straßburg 1642 [Copy ULB Jena]
4. Johann Martin Brandes, *De assecuratione*, resp. Simon Heinrich Sanneman, Rinteln 1664 [Copy HLB Wiesbaden]
5. Michael Friedrich Leder, *Tractatio juridica de jure assecurationum* […], resp. Albertus Fridericus a Syborg [Magdeburgensis], Wittenberg 1667.
6. Reinhold Curicke, "Diatriba de assecurationibus [first 1667]", in: Johann Gottlieb Heineccius (Hg.): *Scriptorum de iure nautico et maritimo fasciculus*, Magdeburg 1740, pp. 827–838.

 | DOI:10.1163/9789004431225_009

7. Bartholomaeus Leonhardus Svendendörfer, *De Assecuratione*, resp. Georgius Conradus Büttner [Dresdensis], Leipzig 1668 [Copy BSB München]
8. Johannes tho Westen [Hamburgensis], *De Assecuratione*, praes. Jacob. Brandmüller, 20.X.1671, in: Brandmüller, *Dispvtationvm jvridicarvm selectiorvm in inclyta Basiliensis Vniversitate disquisitioni publicae ac solenni expositarum* [...] Basel 1675, pp. 1–46 [Copy BSB München]
9. Joh. Wolff. Textor, *Disp. XIX De Contractu Assecurationis, habita Heidelb. m. Majo 1674. Resp. Dn. Joh. Ottone Hamburgensi*, in: Textor, *Disputationes Academicae, duabus partibus distinctae* [...], Frankfurt/M. 1698 [Copy BSB München]
10. Heinrich Coccejus, *De assecuratione vulgo Assurantz*, resp. Henricus ab Aschen Frankfurt/O 1693 [Copy BSB München]
11. Johann Christoph Ochß: De contractu assecurationis, Lugduni Bat. 1699 [UB Frankfurt].
12. Conrad Vegesack [Hamburgensis]: Disputatio juridica inauguralis de Assecuratione. [...] D. Salomonis van Til [...] Leiden 1704.
13. Melchior Dethmarus Grollmann: Dissertatio Juridica, De Jure Assecurationis = Vom Assorantz Recht, resp. Frider. August. de Clettenberg Et Rhoda, Giessen 1708 [Copy ÖNBW/Halle ULB].
14. Johann Nicolaus Hertius, *Paroemia XLIII. Wagen gewinnt/wagen verliert, in: De paroemiis juris germanicis libri tres*, in: *Commentationes Atque Opuscula De Selectis Et Rarioribus Ex Iurisprudentia Universali, Publica, Feudali Et Romana, Nec non Historia Germanica Argumentis*, Frankfurt/M 1716, pp. 455–480.
15. J. A. Cronius, *De jure assecurationum, vom Assecurantz-Recht*, Rostock 1725 [Copy UB München].
16. Jacobus van Ghesel [Amstel. Batavus], *Dissertatio juridica inauguralis theoretico-practica de assecuratione*, Leiden 1727 [Copy UB FU Berlin].
17. Martinus den Beer [Schiedamensis Batavus], *Dissertatio Juridica Inauguralis, De Assecurationibus* [12.9.1729], Utrecht 1729 [Copy UB Rostock].
18. Augustin Balthasar, *Dissertatio inavgvralis De pecvnia assecvrationis in locat. condvctione praediorvm solvta* [...], resp. Martinvs Gvstavvs Fischer, Greifswald 1739 [Copy BSB München].
19. Johannes Werlhofius, *Dissertatio de Instrvmento assecvrationis vvlgo polizza* [...] [resp] Martin Lvcas Schele, Hambvrg. [September 23, 1707], Leipzig 1736 [Copy BSB München].
20. Matthäus Meier [Bremensis], *De assecurationibus mercatorum tractatus*, Bremen 1739 [Copy 'The Making of the Modern World'].
21. Janus Bollard, *Dissertatio* [...] *de assecuratione*, Leiden 1749 [Copy ÖNBW].
22. Franciscus Michael Poppe [Hamburgensis], *Dissertatio inavgvralis ivridica sistens partem commentationis de litivm assecvrationis cavsa orientivm decisione* [...], Göttingen 1752 [Copy BSB München].

23. Hermann Becker, *Dissertatio Inavgvralis, De Indole Contractvs, Locationi Condvctioni Adiecti Assecvrationis, Vvlgo Sic Dicti: (Vom Vorschuß der Pächter)*, resp. Christian Ludwig Johannes Behm [Neostadio-Megapolitanvs] [23.11.1757], Rostock 1757 [Copy ULB Halle].
24. Scato Gockinga, *Dissertatio juridica de assecurationis contractu a romanorum foenore nautico longe diverso*, Groningen 1764 [Copy WLB Stuttgart].
25. Rudolph Gülich [Hamburgensis], *Dissertatio inavgvralis ivridica de termino a qvo, praescriptionis actionis ex contractv assecvrationis secvndvm ordinationem assecvrationis Hambvrgensem. Tit. XVII. Art. I. II.*, Göttingen 1776 [Copy BSB München].
26. Joachim Dehn, *De assecuratione maritima: Commentatio juridica*, Göttingen 1788 [Copy SUB Göttingen].
27. Friedrich Hermann Otto, *Observationes Inavgvrales Ivridicae De Contractv Assecvrationis* [...], Göttingen 1788 [Copy SUB Göttingen].
28. Simon Heinrich Gondela [Bremensis], *Dissertatio inavgvralis ivridica De contractv assecvrationis*, Göttingen 1788 [Copy SUB Göttingen].
29. Johannes Petrus Sieveking, *Dissertatio inauguralis juridica de assecuratione maritima nomine alterius contracta*, Göttingen 1790 [German transl. *Von der Assecuranz für Rechnung eines ungenannten Versicherten*, Göttingen 1791, Copy UB Rostock].
30. *Von Versicherungen und Bodmereien ingleichen von Havereien und Seeschäden nach dem allgemeinen Gesetzbuche für die Preussischen Staaten*, Berlin 1792 [SB Berlin].
31. Johann Matthias Hasse [Hamburgensis], *Dissertatio inavgvralis jvridica De indole atqve effectibvs instrvmenti adsecvrationis, qvod vulgo polizzam vocant, praesertim ex jvre Hambvrgensi spectati*, Göttingen 1796.

APPENDIX 2

Chronological List of Cameralist Fire Insurance Foundations in Germany

After Ebel, *Quellennachweis*, amended and enlarged.

November 30, 1676	Hamburg
May 12, 1685	*Rescript des Kurfürsten betr. Feuer-Cassen-Ordnung an die Bürgermeister und Ratsmannen der Residentzien Berlin, Cölln und Friedrichswerder* (Friedrichsstadt 1698, Berliner Feuersozietät 1718)
1685	Brandkasse Hannover
February 15, 1685	*Ordnung der General-Feuer-Cassa in der Stadt Magdeburg*/ January 20, 1721 *Feuersozietät für das Herzogtum Magdeburg und die Grafschaft Mansfeld*
1700	Private „Große Feuerkasse" in Dresden
January 26, 1701	*Feuer-Ordnung auffm Lande in der Chur- und Mark Brandenburg* (contains paragraphs concerning the fire insurance)
March 18, 1705	Private „Kleine Feuerkasse" in Dresden
October 15, 1705	*Feuer-Cassen-Reglement* for all provinces of Prussia (valid until 1711, then transformation in several particular fire insurances)
1715	*Brandsozietätsordnung mehrerer aneinander stoßender Donah'scher Familiengüter* (Reichertswalde, Schlobitten, Carwinden, Lauch, Schlodien)
1718	Feuerkasse in the Leipzig District
November 4, 1719	*Feuer-Sozietät* for the cities of the Duchy of Hinterpommern (1711 already Fire insurance of Wilhelmsburg-Neuhof)
1719	*Feuersozietät* for the Principality of Halberstadt
November 17, 1720	*Reglement für die Feuer-Societät der Städte des Herzogtums Altpommern* (since that date, several other smaller institutions were founded in Pommern)
May 1, 1722	*Feuersozietät* for the cities of the duchy Cleve and the county of Mark
1723	Ordinance for the fire insurance of the cities of the Principality of Minden and the County of Ravensberg

 | DOI:10.1163/9789004431225_010

April 5, 1729	*Allgemeine Brand-Casse* Saxony
1731	probably foundation of a fire insurance in Bückeburg (Schaumburg-Lippe); *Feuersozietät* of Schaumburg-Lippe 1755
June 6, 1742	*Feuersozietät* of Lower Saxony's cities; 1749/1750 *Brand-Assecurations-Societät* Principalities of Grubenhagen, Calenberg and Göttingen; 1754 Celle
June 6, 1742	*Feuersozietät* for the cities of Lower Silesis; November 24, 1742 *Feuersozietät* for the Lowlands (*platte Land*) of the province Silesia
1748	Ordinance concerning the *Feuersozietät* of the Principality Querfurt
November 3, 1749	*Feuersocietät* for the city of Wrocław/Breslau
April 15, 1751	*Hoch-Fürstlich Anhalt-Bernburgische Feuer-Societäts-Cassen-Ordnung*
February 11, 1752	*Allgemeine Brand- und Assekurations-Societät für die Grafschaft Lippe*
November 20, 1752	Ordinance concerning the *Brand-Assecurations-Societät* in the principality of Lüneburg
December 14, 1752	Lithuanian *Städtesozietät* (*Feuersozietät* of the Gumbinner district)
January 15, 1753	*Brand- und Assecurantz-Casse-Ordnung der Stadt Wismar* (and principality of Wismar)
July 18, 1753	*Landesbrandversicherungsanstalt* Braunschweig-Wolfenbüttel
1754/1768	*Feuersozietät* for the Lowlands of the principality of Moers
April 1, 1754	*Rescriptum Regium* concerning the foundation of the fire insurances in the cities and villages of East Frisia and Harlingen (= later the Landschaftliche Brandkasse, the fire insurance administrated by the estates of the county)
September 7, 1754	*Brandkatasterordnung für Stade*
May 24, 1754	*Brand-Casse* of the duchies of Bremen and (February 1, 1755) Verden (earlier: system of fire contracts)
July 10, 1754	*Brandversicherungsordnung* for the territory of the margraves of Brandenburg-Ansbach
1754/1756/1773	*Herzogliche Württembergische Feuercasse*
September 25, 1758	*Baden-Durlachische Brand-Assecurations-Societät*

December 11, 1758	Reunion of all fire guilds of the land and the cities in Schleswig-Holstein into one *Brandassekuranzkasse* (October 29, 1759)
1760	Private fire insurance association in Krefeld
1760/June 14, 1770	*Brandassekurations-Sozietät* for the principality of Bayreuth
1761	*Eines hochedlen Rates der freien Stadt Lübeck generale Brand-Assecuranz-Casse* (1712 *Renovierte Brandcassa* for the brewery houses in Lübeck, earlier fire guilds)
April 19, 1762	*Brandcasse* for Anhalt-Dessau
July 31, 1764	Ordinance of the fire insurance in Breisgau (Anterior Austria)
September 5, 1764	*General-Brand-Versicherungs-Societät* in the counties of Oldenburg and Delmenhorst
December 12, 1765	Ordinance of the prince-bishop of Hildesheim for the establishment of a *Brandversicherungs-Gesellschaft*
1766	*Markgräflich Baden-Badische Brandversicherungsordnung*
September 12, 1766	*Feuersozietät for the Knights of the Circle of Stargard*
April 21, 1767	*Feuersozietät* for the countries of Nassau-Weilburg
April 27, 1767	Ordinance concerning the establishment of a fire insurance in Hesse-Kassel
January 18, 1768, January 2, 1769 and February 19, 1776	Fire insurance of buildings in the prince-bishopric of Würzburg
April 15, 1768	*Brand-Versicherungs-Gesellschaft* for the cities and the Lowlands (*plattes Land*) of the bishopric and the diocese of Münster
May 14, 1768	*Feuer-Societät* for the Lowlands (*platte Land*) in the kingdom of Prussia (East Prussia)
July 15, 1768	Edict concerning the establishment of a Fire insurance association (*Brand-Versicherungs-Gesellschaft*) in the territories of the elector of Cologne
August 23, 1768	*Brand-Assecuranz- und Indemnisations-Sozietät* [Association for fire insurance and damage compensation) of the duchy of Sachsen-Weimar-Eisenach
1769	*Feuersozietät* in Mecklenburg-Schwerin for the possession of the nobles (the *ritterschaftlichen Güter*)
March 1, 1769	*Brand-Assekuranz-Sozietät* Rügen
March 21, 1769	Edict concerning the establishment of a fire insurance association (Principality of Paderborn)

April 9, 1769	Ordinance concerning the foundation and rules of the fire insurance institution, Duchy of Zweibrücken, *Deux-Ponts*
October 20, 1769	*Verordnung betreffend Brand- und Feuerassecuration* Bamberg
December 11, 1769	Brandassekurations-Sozietät Gotha
June 10, 1770	*Fürstliches Reglement wegen Errichtung einer Feuer-Assecuranz-Societät für Anhalt-Köthen*
1771	Urban Fire Insurance for buildings in Stettin
March 26, 1771	*Brand-Versicherungs-Ordnung eines hochweisen Raths der hl. Reichs-Stadt Frankfurt*
November 26, 1774	*Verordnung betreffend die für die Nassau-Katzenelnbogischen Lande aufgerichtete allgemeine Brand-Assecuranz-Gesellschaft/ Brandkassen-Verordnung* for the countries of Nassau-Orange
1774	*Fürstlich-Hohenlohe-neusteinische Gemeinschaftliche Brand-Assecurations-Gesellschaft*
1774/1779/1792	*Brandversicherungsanstalt* for the Principalities of Öttingen-Spielberg-Wallerstein-Baldern and Sötern-Hohenlohe-Neuenstein
January 8, 1776	Reglement for the *Brand-Assekurations-Institut* of the duchy of Sachsen-Altenburg
August 1, 1777	*Brandassekurations-Ordnung* of the Principality of Hessen-Darmstadt
January 1, 1778	*Fürstenbergische Feuersocietät* (Verhandlungen über Aufnahme Rottweils 1791)
1780	*Feuersozietät* of the Province of Posen
June, July 15, 1780	Fire Insurance association of the Electorate of Mainz
July 15, 1780/January 1, 1788	Fire Insurance Society of the Electorate of Trier
1781/82	Fire Insurance Association of the nobles in Rostock, Fire Insurance Rostock and for the Lowlands (*plattes Land*) March 31, 1782
April 10, 1782	Fire Insurance Society of the Imperial City of Rothenburg
June 22/December 30, 1782	*Brand-Assekurations-Gesellschaft* Nürnberg
May 1, 1783/1784	Fire Insurance of Eichstätt/for the principality of Eichstätt
1784	Fire Society of the Swabian Imperial Knights (*Kanton Donau*)
April 5, 1784	*Ober-Amts-Patent, die Errichtung der Oberlausitzschen Brand-Versicherungs-Societät betreffend*
January 1, 1785	Fire Insurance for the principalities of Waldeck and Pyrmont

1785	Fire damage compensation association for all counties and possessions of the dynasty of Waldenburg
July 30, 1785	Fire insurance association of the Mecklenburg cities
December 17, 1785	*Feuer-Sozietät* for the Lowlands (*platte Land*) in Western Prussia
April 28, 1786	*Brandschadensversicherungsanstalt* for the bishopric of Augsburg
1786/July 1, 1787	*Gesellschaftliche Brandversicherungsanstalt verschiedener schwäbischer Reichsabteien Ochsenhausen, Elchingen, Kaisersheim, Roggenburg, Wittenhausen, Meresheim, Gutenzel, Göfflingen, Isny und des Reichsgotteshauses St. Ulrich in Augsburg*
1787	*Feuer-Assecuranz* of the *Bailiwicks* Alsatia and Burgundy of the Teutonic Order
1791	Fire Insurance Ordinance of the county Friedberg-Scheer
1791	Fire Insurance Association of the Possessions of the Thurn und Taxis on the Holy Roman Empire's ground (*Reichslande*)
June 23, 1792	Fire Damage indemnation association administrated by the estates for the Hospital Biberach
1793	Fire Insurance institution of the Imperial abbey of Ottobeuren
September 30, 1793	*Hochfürstlich Hohenlohe- und Waldenburg-Schillingsfürstische Feuer-Entschädigungsgesellschaft*
1797	*Freywillige Brandversicherungs-Gesellschaft im Fürstentum Schwarzburg-Rudolstadt*
January 17, 1799	Ordinance concerning the Fire Insurances of buildings in Bavaria
September 17, 1799	*Altbaierische Brandversicherungsgesellschaft für die Provinzen Ober- und Niederbayern, Oberpfalz und die Herzogtümer Neuburg und Sulzbach*
November 26, 1801	*Bergische Feuer-Versicherungs-Sozietät*
April 28, 1802	Fire Insurance Association of the Principality of Ellwangen
October 1, 1803	Fire Insurance Institution of the Principality and County of Löwenstein
July 24, 1805	Fire Insurance Association Wied
April 10, 1808	Fire Association of Hohenzollern-Sigmaringen
November 26, 1817	Fire Insurance Institution of the Palatinate

APPENDIX 3

Cameralist Treatises of *Brandkassen* and Insurances

Note: I start that list with Marperger, as we find here the first longer paragraphs devoted to fire insurance within a polemic printed treatise against abuses to be observed with charitable collections that fits the type of 'academic cameralism' though the first cameralist professorships were created only five years later. This does not mean, as shown above in Ch. 4, 1.2, that we would not find earlier texts, advisory reports for princes or other genres that can already be included into the epistemic field of "Cameralism", even long before Leibniz, but surely earlier than Marperger. Next to Beckmann, already Friedrich Jakob Barthold, *Disputatio de incendiariis famosis, von Mordbrennern*, Frankfurt am Main 1690, Ch. IV, num. 54–59 (ed. 1719, p. 49–63) devoted a paragraph to the 1676 Hamburg fire insurance ("Feuer-ordnung"). The Hamburg institution is mentioned also by Franz Ernst Vogt, *De lytro incendiario quod vocant Brand-Schatzung*, 1713, § 23, p. 47, where he distinguishes the damage compensation money to be paid for burning down cities within the context of the laws of war from that new kind of damage compensation in times of peace as it was institutionalized in Hamburg ("Addendum forte fuisset, quomodo lytrum incendiarium differat [... thirdly] a pecunia, quae certo aerario, quod vocatur Feuer-Cassa, infertur"). However, as becomes clear, either those texts had not appeared in print, or they cover only marginally the new institution of fire insurance by way of comparison or a short note wihin different contexts of texts on arson and the laws of war.

1. M., P. J. [Marperger, P. J.,] *Von dem Unfug Des täglich überhand nehmenden Brand-Bettelns* [...] s.l., s.a. [ca. 1715/1722].
2. Anon., "Gedanken von der Einrichtung und dem Nutzen der in den Fürstenthümern Calenberg, Göttingen und Grubenhagen zu errichtenden Brand-Assecurationssocietät, nebst der Untersuchung einiger dagegen etwa zu machenden Einwürfe", *Hannoverische Gelehrte Anzeigen* 1751, 9. Stück, pp. 291–300.
3. Zincke, Georg Heinrich, "Von den nöthigen Policey-Anstalten in Städten, wegen derer Folgen und Wirckungen der Unglücks-Fälle, die man nicht verhüten/ können; insonderheit aber denen Feuer- und Brand-Cassen." = preface to Id., *Leipziger Sammlungen von Wirthschafftlichen, Policey- Cammer- und Finantz-Sachen*, vol. VII, 1751/1752, pp. III-LIV.
4. Magens, Nikolaus, *Versuch über Assecuranzen, Havereyen und Bodmereyen*, Hamburg 1753.
5. Moser, Johann Jakob, *Nachricht von einer freywilligen Feuer-Cassa* ..., s.l. [Stuttgart/Württemberg] 1754.
6. Moser, Johann Jakob, *Erläuterte Nachricht wegen der nun würcklich eröffneten freywilligen Feuer-Cassa* ..., s.l. [Stuttgart/Württemberg] 1754.

 | DOI:10.1163/9789004431225_011

7. Justi, Johann Heinrich Gottlieb von, "Vorschlag von Verbindung der Feuerassecuranzsocietäten mit einer Leihebanco auf die Häuser", in Id., *Neue Wahrheiten zum Vortheil der Naturkunde und des gesellschaftlichen Lebens der Menschen*, vol. 1, Stück 1: *Vor die Monate Jenner und Hornung* 1754, Leipzig 1754, pp. 561–582.
8. Justi, Johann Heinrich Gottlieb von, "Vorschlag, wie durch die Feuer-Assecuranz-Anstalten eine Leihebank errichtet werden könne, um Nahrung und Gewerbe zu befördern," in *Göttingische Policey=Amts Nachrichten auf das Jahr 1755. oder vermischte Abhandlungen zum Vortheil des Nahrungsstandes* [...] Nr. XLVIII, (December 15) 1755, pp. 189–191.
9. Justi, Johann Heinrich Gottlieb von, *Staatswirthschaft, oder systematische Abhandlung aller ökonomischen und Cameral-Wissenschaften*, 2 Vol. 2nd ed., Leipzig 1758 (first 1755), pp. 276–289.
10. "Gedancken von den gemeinschaftlichen Feuers -Cassen mehrerer benachbarter Reichs-Stände," in *Wöchentliche frankfurtische Abhandlungen zur Erweiterung der nothwendigen, brauchbaren und angenehmen Wissenschaften* 1 (1755), Part 1, num. XV, pp. 225–250.
11. J.V.E., "Vorschlag, wie die Versicherung der Gebäude zu Ersezung des Feuerschadens auf eine leichte Art einzurichten 1756", in *Physikalisch-Oekonomische Realzeitung, oder gemeinnüzliche Wochenschrift*, 17. Stück, col. 674–683, 702–716.
12. Justi, Johann Heinrich Gottlieb von, *Grundsätze der Polizeywissenschaft*, Göttingen 1756, 14. Hauptstück "Von denen Anstalten, wieder die Hinternisse eines blühenden Nahrungsstandes", pp. 178–231.
13. "Policeinachricht von der in diesem Jahr in dem Markgraftum Anspach errichteten Brandversicherungsanstalt," in *Physikalisch-Oekonomische Realzeitung, oder gemeinnüzliche Wochenschrift*, 29.-30. Stück 1756, col. 879–892.
14. "Nachricht von der neuen allgemeinen und freiwilligen Brandversicherungsanstalt des Herzogtums Würtemberg," in *Physikalisch-Oekonomische Realzeitung, oder gemeinnüzliche Wochenschrift* 1756, col. 1117–1120.
15. Mengel, Christian Gottlieb, "Von Assurance- oder Versicherungs-Cassen," in *Kopenhagener Magazin* (= *Oeconomische Gedanken zu weiterem Nachdenken eröffnet* 1758 [o. 1759], vol. 1, Teil IV, 8, Cap. III), pp. 47–50.
16. Justi, Johann Heinrich Gottlieb von, *Die Grundfeste zu der Macht und Glückseligkeit der Staaten oder ausführliche Vorstellung der gesamten Polizeiwissenschaft*, Königsberg 1760 (Aalen 1965), vol. 1, VI. Buch, XXIV. Hauptst., II. Abschn. § 714f.; vol. 1, Buch VIII, Hauptstück 32, pp. 763–771: "Von denen Maasreguln wider die Unglücksfälle, und insonderheit von den Assecuranzen".
17. Justi, Johann Heinrich Gottlieb von, *Oekonomische Schriften über die wichtigsten Gegenstände der Stadt- und Landwirthschaft*, Berlin and Leipzig 1760, pp. 473–478.
18. Bielfeld, Jakob Friedrich, *Institutions politiques* 1760, tom. I, chap VII, § 17.
19. Hoffmann, Gottfried Daniel, *Dissertatio politico ivridica de assecvratione aedium*, def. Ferdinand Christoph Harpprecht, Tübingen 1761.

20. "Über die Einrichtung einer Brandversicherungs-Gesellschaft/Von Brand-Assecuranzen," in *Gnädigst privilegirtes Leipziger Intelligenzblatt: in Frag- und Anzeigen, für Stadt- und Land-Wirthe, zum Besten des Nahrungsstandes* 1763 N. 15; 1768, N. 51; 1776, p. 453f.; 1781, p. 382; 1787, p. 206; 1801, pp. 281ff.
21. Schreber, D. Daniel Gottfried, *Sammlung verschiedener Schriften, welche in die öconomischen, Policey- und Cameral- auch andere Wissenschaften einschlagen*, vol. 15 [1765], pp. 196–207.
22. "Von dem Nutzen einer Brandassecurationssocietät," in *Strelitzische Anzeigen: nebst nützliche Beyträge zu den Strelitzischen Anzeigen Neubrandenburg 1765*, Stück 5 u. 6 = *Hannoverisches Magazin* St. 4 (1765).
23. Sonnenfels, Josef v., *Grundsätze der Polizey, Handlungs- und Finanzwissenschaft*, 3 parts, Wien 1765–1787, I, 368–389.
24. J.C.B., "Vorschläge, wie eine Brandgewährleistungs-Gesellschaft, in der Kürze und ohne viele Schwierigkeiten auch ohne baare Anlage, füglich errichtet werden könne," in *Giessische wöchentlich-gemeinnützige Anzeigen und Nachrichten* 1767, pp. 339–343, 363f.
25. Hennings, Johann Nicolaus, *Versuch in politischen Schriften über die Staatswirthschaft, die Handlung und Manufacturen*, 1762–1769, p. 43, 80.
26. Plenciz, Lepold, *Von dem Ursprunge und Wachsthum der Handlung*, praes. v. Sonnenfels, Joseph, Vienna 1769, pp. 185–192.
27. Förster, Johann Christian, *Versuch einer Einleitung in die Cameralpolizey- und Finanzwissenschaften*, Halle 1771.
28. Fredersdorf, Leopold Friedrich, *Anweisung für angehende Justitz-Beamte und Unterrichter*, vol. 1, Lemgo 1772, pp. 144–152.
29. Waser, Joh. Heinrich, [gewesenen Pfarrers zum Creutz, und der physic. Gesellschaft in Zürich ordentlichen Mitglieds,] *Betrachtungen über die Zürcherischen Wohnhäuser, vornemlich in Absicht auf die Brandcassen und Bürger-Protocoll: samt einigen andern dahin einschlagenden öconomisch-politischen Bemerkungen*, Zürich 1778.
30. Pfeiffer, Johann Friedrich, *Natürliche aus dem Endzweck der Gesellschaft entstehende Allgemeine Policeiwissenschaft. Erster Theil* [...], Frankfurt/M 1779, VIII. Hauptstück, Kap. 3: "Wider die Unglücksfälle, sind vernünftige Maßregeln zu nehmen," pp. 332–340.
31. Pfeiffer, Ferdinand Friedrich, *Gedanken über Versicherungs-Anstalten: hauptsächlich zum Vorteil der Landwirtschaft entworffen, und* [...] *wärend den Akademischen öffentlichen Prüfungen verteidigt*, Stuttgart 1780.
32. "Ueber Assecuranz-Sachen", in J.C.Sinapius, *Kaufmännische Hefte*, Stück 7, Altona 1781, pp. 662–674, Stück 9, Altona 1781, pp. 827–847, 1031–1047, Stück 12, Altona 1781, pp. 1143–1156.
33. Pfeiffer, Johann Friedrich, *Grundsätze der Universal-Cameral-Wissenschaft oder deren vier wichtigsten Säulen nämlich der Staats-Regierungskunst, der*

Policey-Wissenschaft, der allgemeinen Staats-Oekonomie, und der Finanz-Wissenschaft, zu akademischen Vorlesungen und zum Unterricht angehender Staatsbedienten gewidmet [...], Frankfurt am Main 1783, Ch. 32: "Von den Verwahrungsmitteln wider die Unglücksfälle", pp. 567–580.

34. Fetzer, [Johann Jakob], *Uiber [sic] Brandschadens-Versicherungsanstalten, ein patriotischer Vorschlag eines Reichsstädters*, Reutlingen 1783.

35. Becker, W.G., "Assekuranz-Projekt eines Juden zu Entschädigung der Landleute in Wetterschlags-Fällen und Viehseuchen; mit Anmerkungen von mehrern Freunden der Ephemeriden," in *Ephemeriden der Menschheit* 1783, vol. 2, pp. 10–34.

36. Anon., "Bemerkungen über die besten Verhütungs-Mittel, daß die gemeinnützliche Einrichtung der Feuerschadens Versicherung der Gesellschafft nicht zum ungebührlichen Nachtheil gereichen möge; bey Gelegenheit des Brandes zu Göppingen 1782," in Johann Georg Meusel, *Historische Litteratur für das Jahr 1784*, Erlangen 1784, pp. 78–94, 169–179.

37. Lamprecht, D.G.F., *Versuch eines vollständigen Systems der Staatslehre mit Jnbegriff ihrer beiden wichtigsten Hauptheile der Polizei- und Kameral- oder Finanzwissenschaft zum Gebrauch academischer Vorlesungen*, vol. 1, Berlin 1784, pp. 408–423.

38. Anon., "Ist es nicht rathsam bey allgemeinen Brandassecuranzen, dass Kirchen und geistliche Gebäude ihre eigene Assecuranz unter sich haben?," in *Wittenberg(i)sches Wochenblatt* 1785, St. 23 und 24.

39. Beckmann, Johann, "Assecuranz," in Id., *Beyträge zur Geschichte der Erfindungen*, vol. 1, Stück 2, Leipzig 1785/86, pp. 204–222.

40. Bergius, Johann Heinrich Ludwig, Art. "Assecuranzanstalten," and Art. "Feuerassecuranzanstalten 1786," in: *Polizey und Cameralmagazin in welchem nach alphabetischer Ordnung die vornehmsten und wichtigsten bey dem Polizey- und Cameralwesen vorkommenden Materien nach richtigen und vernünftigen Grundsätzen practisch abgehandelt und durch landesherrliche Gesetze und hin und wieder wirklich gemachte Einrichtungen erläutert werden*, vol. I (A-B), pp. 60–99; III (F), *Neue, verb. Aufl.*, Vienna 1786, pp. 62–121.

41. Rößig, Carl Gottlob, *Lehrbuch der Polizeywissenschaft*, Jena 1786, p. 235f.

42. *Anweisung für den Bürger und Landmann wie er sich bey der neuen Einrichtung in Ansehung der erlittenen Brandschäden zu verhalten*, Leipzig, 1787.

43. Jung-Stilling, Johann Heinrich, *Lehrbuch der Staats-Polizey-Wissenschaft*, Leipzig 1788.

44. *Von Brand-Assekuranz-Anstalten überhaupt: mit einem besondern Entwurf zu einer Brand-Assekuranz für den Kanton Bern 1789*, S.l.: Hortin, 1789.

45. Wild, Karl, Wyttenbach, Johannes and Gruner, Gottlieb Siegmund, *Ueber die Errichtung einer Brand-Assecuranz-Anstalt in dem Canton Bern*, Bern and Leipzig 1789.

46. "Ueber die Brandsocietäten," in [Richter, Johann Daniel], *Finanz-Materialien nach allgemeinen verbesserten und practischen Grundsätzen*, Berlin 1789, St. 1, pp. 15–144.
47. Frank, Johann Philipp, *System der landwirtschaftlichen Polizey, besonders in Hinsicht auf Teutschland nach den besten Verordnungen, Vorschlägen und Anstalten*, 3 vol., Leipzig 1789–91.
48. Schrader, L. A. G., "Ueber die Anwendung der Versicherungsanstalten in landwirthschaftlichen Geschäften in besonderer Rüksicht [sic] auf die Herzogthümer Schleswig und Holstein," in *Schleswig-Holsteinische Provinzialberichte*, 4, vol. 2, Nr. 6 (1790), pp. 688–705.
49. Brückner, J. A., *Abhandlung über die Errichtung einer Brand-Assekuranz-Kasse für den Canton Bern. Eine gekrönte Preisschrift*, Zürich 1790.
50. Valentiner, Friedrich, *Ueber die Möbel-Gilden in den Herzogthümern Schleswig und Holstein*, [Kiel:] Gedruckt bey Michael Friederich Bartsch Acad. Buchdr., 1791.
51. Schaeffer, J. L., *Nachricht über die Vortheile der kurtrierischen Brand-Versicherungs-Institut*, [Koblenz] [1791].
52. "Plan einer Schwäbischen Mobilien-Assekurations-Gesellschaft," in *Der Anzeiger. Tageblatt zum Behuf der Justiz, der Polizey und aller bürgerlichen Gewerbe*, vol. 2 (1792), pp. 413, 422.
53. Büsch, Johann Georg, *Theoretisch-praktische Darstellung der Handlung in deren mannigfaltigen Geschäften*, 2 parts, Hamburg 1792.
54. Gäng, Philipp, *Von Versicherungsanstalten wider Feuerschäden und ihrem Nutzen im Allgemeinen* [...], Salzburg 1792.
55. "Reichsstädtische Verbindung zu einer Brandassecuration," in *Der Anzeiger. Tageblatt zum Behuf der Justiz, der Polizey und aller bürgerlichen Gewerbe* 1793, pp. 1249–1251.
56. Prestinary, Bertrand L., *Pflicht der Nächstenliebe in Hinsicht auf die Brandversicherungs-Gesellschaft: nebst Bemerkung der Vortheile, die mit dieser so gemeinnützigen Anstalt verbunden sind* [S.l.] 1794.
57. Bieber, Georg Elert, *Plan zur Errichtung einer für Hamburg möglichst vorteilhaften Versicherungs-Compagnie gegen Feuer-Gefahr*, Hamburg 1795.
58. *Bemerkungen über die vorläufigen Puncte einer (zu errichtenden) Association Hamburgischer Einwohner, zur gemeinsamen Versicherung ihres Mobiliar-Vermögens, auf den Fuß der Hamburgischen Feuer-Casse, in Gemäßheit des von G.E. Bieber bekanntgemachten Plans, welcher auch von den bisherigen Beförderern dieser Association zum Bevollmächtigten gewünscht wird. Und der hoffentlich diese Wünsche erfüllen wird*, Hamburg 1795.
59. Bieber, Georg Elert, *Prüfung der Bemerkungen eines Ungenannten über die vorläufigen Punkte einer Association Hamburgischer Einwohner, zur gemeinsamen Versicherung ihrer Mobilien-Güter usw.*, Hamburg 1795.

60. Büsch, Johann Georg, *Allgemeine Übersicht des Assekuranzwesens als Grundlage zur Beurteilung von Biebers Plan zur Errichtung einer Feuerversicherungs-Compagnie*, Hamburg 1795.
61. Anon., "Vorschlag zu einer Brandversicherungs-Anstalt für die Güterbesitzer und Pachtinhaber in Thüringen" with a comment by Rudolf Zacharias Becker," in: *Reichs-Anzeiger* Nr. 79, Mittwochs, 8.4.1795, col. 761–768; by August Ludewig Kastendieck, Nr. 102, 5.5.1795, Sp. 1001–1009; Nr. 112, 18.5.1795, col. 1097f., 1309–1311, 1340f., 1406.
62. "Beyträge zur Kenntniss des Assecuranzwesens, älterer und neuerer Zeiten," in *Handlungs-Zeitung* 1795, Nr. 5 und 6.
63. Anon., "Ueber die Mängel der Assekuranz-Anstalten," in *Deutsches Magazin* vol. 12 (December), 1796, pp. 603–660.
64. Büsch, Johann Georg, *Versuch einer Geschichte der hamburgischen Handlung*, Hamburg 1797.
65. Valentiner, Friedrich, *Über die zweckmässigsten Brandanstalten in grossen Städten*, Hamburg-Kiel 1797.
66. Reimarus, J.A.H., "Ueber das Alter der Assecuranzen," in *Handlungsbibliothek* vol. 3 (1797), pp. 119–122.
67. Taubenrauch, Just Heinrich (1747–1805), *Anleitung zum gerichtlichen Verfahren bey der, in Chursachsen vermittelst Mandats vom 10. November 1784, neu eingerichteten Versorgung der Brandgeschädigten: zum Nutzen für die Obrigkeiten und die Societäts-Mitglieder*, Leipzig 1798.
68. Bensen, Heinrich, *Versuch eines systematischen Grundrisses der reinen und angewandten Staatslehre für Kameralisten*, Erlangen 1798, pp. 122–129.
69. Bensen, Heinrich, *Über die Einrichtng der Brandversicherungsgesellschaften, in: Ders.: Materialien zur Polizei-, Kameral- und Finanzpraxis für angehende praktische Staatsbeamten*, vol. 1, Heft 1, Erlangen 1801, pp. 146–152.
70. Lorsch, Christian Gottfried, *Welchen Nachtheil bringt die Vereinigung der Stadt und der Landschaft in eine Brandassecurationscasse dem Nürnbergischen Bürger?*, s.l., 1799.
71. Freier, Johann Paul M.D. in Havelberg, *Vorschläge wegen einer in den Königlich Preußischen Staaten zu errichtenden Brand-Kasse für Mobilien, Waaren und Landwirthschaftliche Produkte*, Berlin 1800.
72. V. Berg, G. H., *Handbuch des Teutschen Polizeyrechts*, Teil 3, Hannover 1800, Achtes Hauptstück: "Recht der Policey gegen Unglücksfälle. Erste Abtheilung. Recht der Feuerpolicey."

Afterword

It might be helpful to remind the changes in the English version: The discussion of the early insurance contract has been enriched though the main argument of the closed time horizon encapsulated in the practice as well as in the Roman legal theoretical discussion until around 1600 remains supportable and unchallenged as far as I see (chapter 2). The work on the city fire database has not been renewed and enlarged for this version as it would need more hands; what could and should been done in the same manner as I and the research students Anne Meißner, Max Ammareller, Marianne Timpe, and several others had done for the German cities would be the integration of the Bohemian cities now available through the publication of the Czech Urban History Volumes; for other regions, even such a proximate source is missing, and no one could dare to enter into all the hundred small town and city archives for whole countries where such a preparatory work has not been done—the German, Austrian, and Czech *Städtebücher* were a work of now nearly one hundred years of preparation. Urban History must stand here on the shoulders of a multitude of often unknown archivists and collaborators of those large projects. However, climate history became interested in the relationship between mega-draught years like 1540 and a rise in frequency of large city conflagrations, a discussion which had to be integrated (ch. 3, 1.5). The discussion of the theological framework, the reception of Deism, the Reimarus circle, and the early enlightenment debate about lightning, electricity, and the communication links between the European royal and learned societies and local centers of practical enlightenment as Hamburg had to be strongly refined and reworked (ch. 3, 2.1). The chapter 3, 2.3 on *Ground Zeros* has been added; it was first presented on a panel of the American Historical Association, organized by Robert Spaulding and generously sponsored by the Central European History Society in New York 2015: discussion on the pre-circulated paper with Chris Friedrich and the other participants were very helpful; it adds a reflection to the treatment of the visualization of fire catastrophes concerning the special moment of how to visualize and commemorate the complete annihilation of the physical body of towns—which might have served premodern citizens as a *memento mori* and to increase the consciousness among the burghers that, in the end, the city was nothing stable but just consisting of the community of the inhabitants as a collective as such. Despite the celebration in 2018 of the 300 year jubilee of the *Berlin-Brandenburg* fire insurance (*Feuersozietät*), much of the research on its early history could remain as it stood, however in chapter 3, 3.2 and elsewhere, new findings about the importance of enlightenment prize contests and the role of Leibniz during the early struggles of competition between the Prussian *Generaldirektorium* and the newly founded Academy of sciences concerning matters of fire protection and fire insurance had to be integrated—here I was profiting from what I learned myself while supervising the PhD thesis of Rebecca Rose (born Knapp). In chapter 4, my view on Barbon has not changed in its core, however the literature and discussion on mercantilism exploded anew just after the publication of the German book—and likewise, much new research on early political arithmetic and on London's rebuilding after 1666 has appeared. These new publications do focus less or not at all on insurance history, the *Fire office*, and the link between speculative building and mercantilism, whereby the main body of the text needed no change. But the ongoing debate on mercantilism within trading empires and on the question to what extent mercantilists could conceive of 'growth' as (re-)started by Steven Pincus and continued by Carl Wennerlind and others (as discussed partially by myself in *Imperial Unknowns. The French and the British in the Mediterranean, 1650–1750*, Cambridge 2016, ch. 1), sheds a different light also on how Barbon and Newbold and other early insurance pioneers were conceiving of infrastructural invention as a means of 'colonization of the internal resources of the state' (parallel to 'real' colonization abroad) and of 'growth' (chapter 4, 1.1). The part on Leibniz and Hamburg in chapter 4, 1.2 has much profited and is changed to some extent due to a contribution prepared for the German Studies Association at Kansas in 2014 on a panel organized by Mary Lindemann; I profited much from the discussion by Andre Wakefield and others then present. The question, how we can historicize current terms of political language and international relations such as environmental security, sustainability and how this is related to insurance history has been deepened since the German version in several contributions (to a workshop at the German Historical Institute at Washington on historical accounting practices; a 2018 special issue of the European Journal of Security Research), cf. chapter 5, specially 3. I profited much from a vivid discussion in Harvard's Early Sciences Working Group at the Harvard History of Science Department organized by David Kennedy, Paola Savoia, and Allyssa Metzger and from the discussion e.g. by Katherine Park, Florin-Stefan Morar and Alex Csiszar. The discussion was focusing much on the methodological questions of comparison between the early modern times and the global setting covered by the last chapter. At the History Department, it was mostly Louis Gerdelan who was developing his PhD project on natural hazard and early modern climate history at the time of my stay with whom there was most exchange on these topics; I could continue these discussions also later in 2016 and 2018 profiting again from the Widener Library and the electronic resources for short periods (while being back again to normality in Germany) for the work of updating and rewriting—though it was mostly finalized at Berlin with the perhaps not equal but not negligible resources of the Staatsbibliothek. The last chapter on the global diffusion of fire insurances has grown the most and seems to be also the field where most of new research is happening right now. The parts on early German industrialization (Hamburg) and on China, small paragraphs on Japan, new elements in the chapter on Istanbul, and some general reflections on the history of colonial segregation have been added. These parts have much profited from the ongoing discussion most of all with French friends in natural hazard history during events in Paris, Oxford and Versailles: Thomas Le Roux, Jean-Baptiste Fressoz, Frédéric Grabier, Gregory Quenet, and the group gathered in the special issue of *Le Mouvement social* 2014, a conference organized by Gerrit Schenk at Heidelberg, and the earlier conferences on *Urbs incensa* (Max-Planck Institute of Art History) and on *Flammable Cities* (at the German Historical Institute in Washington, organized by Greg Bankoff, Jordan Sand, and Uwe Lübken).

About a third of the English text was existing somehow before, edited and one part translated by Stephen Walsh (†) and others, the full translation of the rest from the reworked German text was done by Monica Birth (b. Hoesch)—a great thanks to her and her patience with a hundreds of emails and questions on early modern wording as well as on postmodern sociological terms—not all seemed really translatable without some slight change of meaning and semantic flavor. After some further changes and additions, another two-and-a-half layers of editing by Miranda Critchley, Emma Taylor, Mallory Martin Hope, and Rebecca van Hove were added. I still remember with gratitude all those mentioned in the *Nachwort* of the German Book, reworked from the Bochum Habilitation thesis, which, however, shall not be repeated here in entirety: Ongoing discussions remained since those times with Christof Mauch, who was—together with Helmuth Trischler and Frank Uekötter –essential for including the German version into the successful series of LMU Munich's Rachel Carson Center; discussions also with the former PhD students and now colleagues Magnus Ressel (on slavery insurance) and Eleonora Rohland (on Hurricane disasters and long-term narratives in environmental history). New intersections of Environmental and Mediterranean History are now at the core of the collaboration with the Alexander-von-Humboldt's foundation Anneliese-Maier-Award winner Alan Mikhail during the procedure of nomination and the travel grant program and other projects run by it: *Premium insurance* is one of Braudel's Mediterranean influences in Northern Europe and the world, thus for me, my fields of historical research remain in a sometimes surprising but always lively harmony.

The book appears now when just moving, at least for some years, maybe for longer, to the Friedrich-Meinecke-Institut of the Freie Universität Berlin, with gratitude for the welcome by the new colleagues. After having taught already the first course on the early modern history of disaster and catastrophe there, in a strange semester itself threatened by a pandemic quarantine situation (COVID), me and my diligent students have recognized again that those themes belong very much to the core of History. For financial support, I must, finally, express my deep gratitude to the Heisenberg Program of the German Research Foundation (GZ Zw 164–8/1 and –9/1).

Berlin, December 22, 2020

Sources and Literature

Unpublished Sources

Aachen AMB Generali, Archive of *Aachen-Münchner Versicherung*
  Fach Bücher
  Kästen, years 1825 to 1850
Aachen, Stadtarchiv
  RA II Allg. Akten 577, 1068
  RA II Edikte II 1
  RA II Beamtenprotokolle vol. 1
Berlin, Bundesarchiv (BA, Berlin-Lichterfelde)
  R 1001/826
Berlin, Geheimes Staatsarchiv Preußischer Kulturbesitz (GStPK)
  HA I, Rep. 9 (Allgemeine Verwaltung), Nr. C6a1; Nr. G1 Kollektenwesen Generalia; Nr. G1 Kollektenwesen Spezialia, G2
  HA II, Abt. 14 (Kurmark)
    tit. CLXIV Stadt Ruppin
    tit. CXC Stadt Zehdenick
    tit. CCXVII, Nr. 7, 8, 11–17, 20–22, 29, 31, 33; sect. a, Nr. 3, Bd. 1; Bd. 2.
Berlin, Landesarchiv (LA)
  A Rep. 005-05 Nr. 8, old shelfmark Stadtarchiv 27423
Cologne
  HStA Köln Ratsprotokolle 103, 129
  HStA Köln Verfassung und Verwaltung (V+V) N12; N17; N21; N23; N336; N860; N862; N863; N864
  HStA Köln Edikte 6
  HStA Köln Bibliothek Ec 127, § 37
Düsseldorf
  Hauptstaatsarchiv, Reg. Aachen, Abt. I Nr. 31
Göttingen
  Archiv der Akademie der Wissenschaften, Scient. 196, vol. 10, Fasz. 29
Hamburg
  Best. 111-1, Cl. VII,
    Lit. Fe Nr. 1, vol. 1; vol. 4;
    Lit. Fe Nr. 2, vol. 1b, Nr. 4; vol. 2, Nr. 7/8; vol. 3; vol. 5; vol. 7; vol. 10; vol. 11; vol. 12a; vol. 12b; vol. 13; vol. 14; vol. 19; vol. 27; vol. 28a, Nr. 27; vol. 36, nr. 1a; vol. 40
    Lit. Fe Nr. 3, vol. 3; vol. 5, Nr. 13b; vol. 2; vol. 9, nr. 9; vol. 13, nr. 1; vol. 14; vol. 16; vol. 20
    Lit. Fe Nr. 15, vol. 8, fasc. 2 (1736–1738)
    Lit. La Ko 1, vol. 1, c3

Best. 333-1/1, Nr. IV a2, vol. 1
Best. Reichskammergericht
H 127, L 35 part 1 and 2
Best. 621-1
Best. 622-1 Repsold 7
Best. Bibliothek A 456/18, A 640–77

London
The National Archives Kew
PRO State Papers (SP) 97/20; 97/24; 105/110; 110/72; 105/118

London, London Metropolitan Archives (LMA),
Journals of the Court of Common Council 49
Archive of the Sun Fire Office
Foreign Letter Books
CLC/B/192/019/18249, vol. 1 to 150
Memorandum Books:
CLC/B/192/019/31522/ vol. 60 to 63, 106, 107, 109, 111, 151 152, 153, 154, 156, 158, 160, 161, 164, 175, 176, 257, 258, 259, 260, 262, 266, 275, 276
Foreign Accounts CLC/B/192/019/38852/1 and 2; 38853/1
CLC/B/192/019/11935K.

Mühlhausen, Stadtarchiv
Th. 61/38
10/W 1–7 Bd. 16

Paris
Archives de l'Académie royale des Sciences
Procès verbaux des séances 13 [1676], 35 [1716], 44 [1725], 51 [1732], 54 [1735]

Potsdam
BLHAP
Rep. 2 D 680, 686, 706, D 721, D 807, D 815, D 842, S 4126, S 5526
Rep. 3 Nr. 10445, 11435, 11855
Rep. 8 Drossen 646, 647, 648, 650, Lenzen 40A, Strasburg 829, 836, 840, Wriezen 630, 631

Roubaix
Archives nationales du monde du travail, 46 AQ 1 until 10 (*Procès-verbaux du conseil d'administration 1819–1846*)

Solingen
Stadtarchiv S-3143

Strasbourg
Archives municipales, série III 93/3

Vatican, Rome
Archivio Apostolico Vaticano, Miscellanea, Armadio II, vol. 90

Published Primary Sources*

A Letter to a Gentleman of the Insurance Office, Concerning the Cities Insuring Houses, London 1681.

A Letter to a member of Parliament, occasioned, by the growing Poverty of the Nation, from the want and Decay of Trade, and wrong Management thereof, Edinburgh 1700.

Ahlwardt, Peter, *Bronto-Theologie, oder Vernünftige und theologische Betrachtungen über den Blitz und Donner* [...], Greifswald 1747.

Althusius, Johannes, *Politica methodice digesta*, Herborn 1614.

An Advertisement from the Insurance-Office for Houses, September 16, s.l., 1681.

Angliae tutamen: or, the Safety of England. Being an Account of the Banks, Lotteries, Mines, Diving, Draining, Lifting, and other Engines, and many pernicious Projects now on foot; tending to the Destruction of Trade and Commerce, and the Impoverishing this Realm [...], London 1695.

Anon., *An die hin und her zerstreueten vormaligen Einwohner der durch Feuer zerstörten Stadt Gera und an ein mitleidiges Publikum*, Cuba bey Gera 1781.

Anon., *Beschreibung der bedenklichen Verwüstung der berühmten Stadt Gera im Voigtlande* [...], Basel 1780.

Anon., *Das Geraische Tabeera oder Beschreibung der erschrecklichen Feuersbrunst welche im Herbst des Jahres 1780 die hochgräflich Reußische Kauf- und Handelstadt Gera ohne alle Rettung völlig in die Asche legte* [...], s.l., 1780.

Anon., *Einige Nachrichten von dem undenkbar grossen Unglück welches die Handelsstadt Gera, am 18. September 1780 betroffen hat*, s.l. 1780.

Anon., *Gedanken über den Brand zu Gera*, s.l. 1780.

Anon., *Gute Gedanken eines Bürgers bey dem höchst traurigen Schicksale seiner geliebten Vaterstadt Gera*, Gera 1781.

Anon., *Klagen eines Ungenannten über Gera am 18.ten des Septembers 1780*, ed. Friedrich Wilhelm Streit, Altenburg 1781.

Arnisaeus, Henning, *De Republica, sev relectionis politicae libri duo* [...], Strasbourg 1636.

Ashburner, Walter (ed.), *Nomos Rodiōn nautikos = The Rhodian Sea-Law*, Oxford 1909, reprint Aalen 1976.

B., *Die Größe Gottes in der Zerstörung von Gera zu zeigen und nähere die unglückliche Stadt betreffende Nachrichten* [...], s.l. 1781.

Bank-Credit or the usefulness and Security of the Bank of Credit Examined; in a Dialogue between a Country Gentleman and a London Merchant, London 1683.

Barbon, Nicholas, *A Letter to A Gentleman in the Country, Giving an Account of the Two Insurance-Offices; the Fire-Office & Friendly Society*, London 1684.

* The legal and cameralist treatises on Insurances are not to be found here. They are listed separately in the appendices 1 and 3.

Barbon, Nicholas, *A Discourse of Trade*, ed. Jacob H. Hollander (1690; Baltimore 1905).

Barbon, Nicholas, *A Letter to A Gentleman in the Country, Giving an Account of the Two Insurance-Offices; the Fire-Office & Friendly Society*, London 1684.

Barbon, Nicholas, *A Proposal for Raising the Publick Credit, By Setting up an Office for Transferring and Discounting Tallies* [1697].

Barbon, Nicholas, *An Apology for the Builder: Or a Discourse Shewing the Cause and Effects of the Increase of Building*, London 1685.

Barbon, Nicholas, *An Enquiry, Whether it be the Interest of the City To Insure Houses from FIRE; And whether the Insured may expect Any Advantage thereby, more than from the Insurance-Office already Setled*, S.l. 1681.

Barbon, Nicholas, *Discourse shewing the great advantages that New-Buildings, And the Enlarging of Towns and Cities Do bring to a Nation*, London 1678.

Beckmann, Johann, "Assecuranz," in *Beyträge zur Geschichte der Erfindungen*, vol. 1.2, Leipzig 1785/86.

Behr, Christian August, *Auf das Gera am 18. September 1780 betroffene traurige Schicksal*, Cuba bei Gera 1781.

Bieber, Georg Ehlert, *Plan zur Errichtung einer für Hamburg möglichst vortheilhaften Versicherungs-Compagnie gegen Feuers-Gefahr entworfen und seinen Mitbürgern gewidmet*, S.l. (Hamburg 1794).

Bieber, Georg Ehlert, *Prüfung der Bemerkungen eines Ungenannten über die Vorläufigen Puncte einer Association Hamburgischer Einwohner, zur gemeinsamen Versicherung ihrer Mobilen-Güter*, S.l. (i.e. Hamburg 1795).

Böckh, Simon, [...] *Eine Christliche Erinnerungs- und Gedächtnus-Predigt deß höchstkläglichesten Abbrands/der/deß Heil. Röm. Reichs-Stadt Giengen*, Ulm 1655.

Böckh, Simon, *Cometarum praesagitio*, Ulm 1681.

Brandes, Johann Martin, *De assecuratione. Resp. Simon Heinrich Sanneman*, Rinteln 1664.

Coccejus, Heinrich, *De assecuratione vulgo Assurantz. Resp. Henricus ab Aschen*, Frankfurt an der Oder 1693.

Copia [...] *Fewr: oder Brand-Ordnung Vigilate* [...], Hamburg 1647.

Copia Dero [...] *Feur- oder Brand-Ordnung Concordia* [...], Hamburg 1643.

Defoe, Daniel, *An Essay on Projects*, London 1697.

Della formazione de' Fulmini [Verona: 1747].

Der Stadt Hamburg Anno 1685. Neu Revidirte Feuer=Ordnung [...], Hamburg 1685.

Der Stadt Hamburg Neu revidirte Feuer= Ordnung 1676.

Der Stadt Hamburg Wacht- und Feuer-Ordnung [...], Hamburg 1643.

Dreyhaupt, Johann Christoph von, *Pagus neletici et nvdzici, oder diplomatisch-historische Beschreibung des Saal-Creyses* [...], part 2, Halle 1773, p. 388f.

Edzard, Jodocus, *Feuer-Predigt über den Evangelischen Spruch: Text. Matth. 22. v. 7. Der Herr zündet ihre Stadt an. Darinn gehandelt wird, was von denen, in dieser guten

Stadt, unlängst leyder oftmahls entstandenen Feuersbrünsten zu halten, und was hinferner zu Abwendung derselben in acht zu nehmen von nöthen sey, gehalten am XX. Sonntage nach Trinit. in der Kirchen St. Michaelis, und auf etlicher Bußeiferigen Begehren in Druck gegeben, Hamburg 1637, reprint Hamburg 1750.

Edzard, Jodocus, *Grundlegungs-Predigt, da der erste Stein zur Neuen Kirchen in der Neuen Stadt Hamburg geleget worden, welche dem grossen Nahmen deß Heil. Michaels, deß Groß-Fürsten und lebendigen Sohns Gottes, zu Ehren soll gebauet und consecriret werden* [original copy 26 April 1649], reprint Hamburg 1750.

Edzard, Jodocus, *Tempel-Predigt HAGGAI, Erkläret und wiederholet, die Christliche Gemeine in der Newen- oder Vor-Stadt, und andere Christliebende Hertzen anzumahnen dem HERRN Salvatori daselbst sein Haus zu bawen* [original date 5 April 1646], reprint Hamburg 1750.

Gedächtniß Christoffs von Wrißberg, des Kriegs Obristen, aus fast lauter alten ungedruckten Urkunden zusammen gezogen, Und zur Verbesserung verschiedener Geschichtschreiber des 16ten Jahrhunderts, insonderheit des Schmalckaldischen Krieges, nunmehro zum Druck befördert, ed. Johann Justus Losius, Hildesheim 1742.

Gericke, Johann Peter, *Der Herr im Feuer wurde der Gemeine zu St. Michaelis, am Palmsonntage 1750. in einer Predigt, nachdem das prächtige Gebäude der Michaelis Kirche durch einen am 10. März erfolgten Wetterstrahl angezündet und bis auf den Grund verbrannt worden, aus Ps. 50, 1. 2. 3. in der Kirche zu St. Marien Magdalenen vorgestellet, und auf Verlangen dem Druck übergeben*, Hamburg 1750.

Glaser, Johann Friedrich, *[A]usführliche Beschreibung der glücklich abgelaufenen großen Feuerprobe, welche mit seinem erfundenen brandabhaltenden Holzanstriche* [...] *öffentlich gemacht worden* [...], Leipzig 1773.

Glaser, Johann Friedrich, "Von Verbesserung der Feuerlöschungsanstalten in den kleinen Städten und auf den Dörfern hiesiger Lande," *Hannoverisches Magazin* 13, no. 44, June 2 (1775), pp. 689–736.

Glaser, Johann Friedrich, *Ausführliche und auf richtige Erfahrung gegründete Abhandlung und Vorschläge, wie thunlicher Weise die meisten Feuerbrünste an und in den Gebäuden wohl verhütet* [...] *werden können*, Dessau 1783.

Glaser, Johann Friedrich, *Beantwortung und Widerlegung verschiedlicher wider seinen erfundenen und in der damit an etlichen kleinen Probehäusern öffentlich angestellten und glücklich abgelaufenen großen Feuerprobe bewährt befundenen brandabhaltenden Holzanstrich gemachten ungegründeten Einwendungen und erregten Zweifel*, Leipzig 1774.

Glaser, Johann Friedrich, *Beweisgründe, daß und warum die in seiner gedruckten Schrift: Abhandlung und Vorschläge, wie die meisten Feuersbrünste zu verhüten, viel wirksamer und gewisser sind, als die bisher gewöhnlichen*, Schleusingen 1784.

Glaser, Johann Friedrich, *Fernere Erörterung und Aufklärung seiner verbesserten Preisschrift, von Verbesserung der Feuerlöschanstalten* [...], Hildburghausen 1779.

Glaser, Johann Friedrich, *Nützliche und durch die Erfahrung bewährte Vorschläge, bey heftigen und geschwinden Feuersbrünsten Häuser und Mobilien sicher zu retten* [...], 4th much expanded and improved edition, Hildburghausen 1772.

Glaser, Johann Friedrich, *Preisschrift, wie das Bauholz in den Gebäuden zu Abhaltung grosser Feuersbrünste, leicht, wohlfeil, bewährt und dauerhaft zuzurichten, daß es nicht leicht Feuer fange oder fortbrenne*, with notes, clarifications, and improvements, Hildburghausen 1762.

Graunt, John, *Natural and Political Observations Mentioned in a Following Index, and Made upon the Bills of Mortality* [...] *With Reference to the Government, Religion, Trade, Growth, Ayre, and Diseases of the Said City*, London 1662.

Hale, Matthew, *The Primitive Origination of Mankind Considered and Examined According to the Light of Nature*, London 1677.

Hederich, Benjamin, *Gründliches mythologisches Lexicon, worinnen so wohl die fabelhafte, als wahrscheinliche und eigentliche Geschichte der alten römischen, grichischen und ägyptischen Götter und Göttinnen* [...] *zusammen getragen* [...] [1770], reprint Darmstadt 1996.

Heinemann, Johann Wilhelm, [G]*ekrönete Abhandlung über die Feuerlöschungsanstalten in kleinen Städten und auf den Dörfern*, Lemgo 1777.

Hertel, Christoph Ernst, *Die erste Predigt nach dem am 18. Sept. 1780. Ausgebrochenen Geraischen Brand Von dem gottgefälligen Verhalten bey den Unglücksfällen unserer Nebenmenschen*, Greiz 1781.

Hertzog Heinrichs von Braunschweig Antwort/Der Keiserlichen Maiestat/auff die vbergebene Supplication/der Mordbrenner halben/gegeben, s.l., 1541 (BSB Munich 4 DHg. 255, Beibd. 9).

Herzberg, Ewald Friedrich, *Vorschläge zur Verbesserung der bisher üblichen Dächer*, Breslau 1774; supplement Breslau 1779.

Heyden, Jan van der, *Beschrijving der nieuwlijks uitgevonden en geoctrojeerde Slang-Brand-Spuiten* [..], Amsterdam 1690.

Heyden, Jan van der, *Fire Engines with Water Hoses and the Method of Fighting Fires now used in Amsterdam*, Canton 1996.

Jung-Stilling, Johann Heinrich, *Lehrbuch der Staats-Polizey-Wissenschaft*, Leipzig 1788, reprint 1970.

Justi, Johann Heinrich Gottlob, "Vorschlag von Verbindung der Feuerassecuranzsocietäten mit einer Leihebanco auf die Häuser," in Johann Heinrich Gottlob Justi, *Neue Wahrheiten zum Vortheil der Naturkunde und des gesellschaftlichen Lebens der Menschen*, vol. 1.1, *Vor die Monate Jenner und* Hornung 1754, Leipzig 1754.

Krafft, Justus Christoph, *Eine Predigt über Luc. 13, 1–5 mit Beziehung auf die Feuersbrunst, welche den 18. Sept. 1780 die Stadt Gera im Voigtlande, in die Asche geleget hat. Nach derselben folgen einige Nachrichten und Briefe, welche diese traurige Begebenheit betreffen*, s.l. 1780.

Lauterbach, Wolfgang Adam, *Q.F.E.F.S. Theseis de Jure in curia mercatorum usitato. Resp. Lucas Langemann.* [*12. Calend. Octobr.*], Tübingen 1655.

Leibniz, Gottfried Wilhelm, *Allgemeiner, politischer und historischer Briefwechsel* (*1676–1679*), Darmstadt 1927.

Leibniz, Gottfried Wilhelm, *Allgemeiner, politischer und historischer Briefwechsel* (*1680–1683*), Leipzig 1938.

Leibniz, Gottfried Wilhelm, *Die Werke. Gemäß seinem handschriftlichen Nachlasse in der Königlichen Bibliothek zu Hannover*, ed. Otto Klopp, series 1, vol. 10, *Historisch-politische und staatswissenschaftliche Schriften*, Hanover 1877.

Leibniz, Gottfried Wilhelm, *Hauptschriften zur Versicherungs- und Finanzmathematik*, ed. Eberhard Knobloch und J.-Matthias Graf von der Schulenburg, Berlin 2000.

Leibniz, Gottfried Wilhelm, *Sämtliche Schriften und Briefe*, Series 4, Politische Schriften, vol. 3 (*1677–1689*), ed. Lotte Knabe and Margot Faak, Berlin 1986.

Lessius, Leonardus, *De iustitia et iure ceterisque virtutibus cardinalibus libri quatuor*, Lyon 1653 [orig.1605].

Lummer, Johann Gottlieb, *Ermunterung an die Geraischen Abgebrannten: Jn ihren Trübsalen Gott und ihren Nebenmenschen zu geben, was sich gebühret. Eine Predigt am 23. Sonnt. nach Trin. 1780 in der Gottesackerkirche zu Gera gehalten*, Gera 1781.

Luther, Martin, "Against Hanswurst, 1541," trans. Eric W. Gritsch, in *Luther's Works*, ed. Helmut T. Lehmann, Church and Ministry III 41, ed. Eric W. Gritsch, Philadelphia 1966, pp. 179–256.

Luther, Martin, "Wider Hans Worst [1541]," WA 51, 461–571.

Machiavelli, Niccolò, *Opere I*, ed. Corrado Vivanti, Turin 1997.

Meier, Matthaeus, *De assecurationibus mercatorum tractatus denuo editus* [*1668*], Bremen 1739.

Molina, Luis, *De iustitia et iure tomi sex*, Antwerp 1615 (orig. 1602/03).

Moser, Johann Jakob, *Nachricht von einer freywilligen Feuer-Cassa, Welche zum Besten der Einwohnere des Herzogthums Würtemberg angeordnet und von Hoch-Fürstlicher Landes-Herrschafft gnädigst bestättiget worden ist*, S.l. 1754.

Müller, Gottlieb, *Einige Predigten über verschiedene Materien aus der Natur- und Geisterlehre an das Licht* [*...gebracht*], Wittenberg 1749.

Mun, Thomas, *England's Treasure by Foreign Trade*, London 1664.

Mylius, Christian Otto (ed.), *Corpus constitutionum Marchicarum oder Königlich preußische und churfürstlich Brandenburgische in der Chur- und Mark Brandenburg, auch incorporirten Landen publicirte und ergangene Ordnungen, edicta, mandata, rescripta, Teil 1–6, nebst Anhang und Continuatio 1–4*, Berlin 1737–55.

Neu-revidirte Feuer-Ordnung der Stadt Hamburg, Hamburg 1750.

Newbold, Augustine *Londons Jmprovement and the Builder's Security asserted, by the Apparent Advantages that will Attend their Easie Charge, in Raising such a Joint-Stock, as may Assure a Re-building of those Houses, which shall hereafter be Destroyed by the Casualties of Fire* (presented January 1, 1679), London 1680.

Obrecht, Georg, *Constitutio Von nothwendiger vnd nützlicher Anstellung eines Aerarij Sancti* […], S.l. 1617.

Observations on the Proposals of the City, to insure Houses in case of fire, London 1681.

Olofsson Rhyzelius, Anders, *Brontologia theologico-historica, thet är enfaldig lära och sanfertig berettelse: om åske-dunder, blixt och skott; bestående af fyra delar* […], Stockholm 1721.

Oraeus, Henricus, *Theatri Europaei Oder Historischer Chronicken Beschreibung Aller Vornembsten vnd Denckwürdigsten Geschichten, so sich hin vnd wieder in Europa, sonderlich im Reich Teutscher Nation von Ao. 1633 biß in Ao. 1638 begeben vnd zugetragen*, part 3, first published by Matthäus Merian, Frankfurt am Main 1670.

Panciroli, Guido *Rerum memorabilium iam olim deperditarum et contra recens atque ingeniose inventarum libri II*, ed. Heinrich Salmuth, 2 vols., Amberg 1599–1602.

Pfeiffer, Johann Friedrich v., *Grundsätze der Universal-Cameral-Wissenschaft oder deren vier wichtigsten Säulen* […], Frankfurt am Main 1783.

Pfeiffer, Johann Friedrich v., *Natürliche aus dem Endzweck der Gesellschaft entstehende Allgemeine Policeiwissenschaft*, 2 vols., Frankfurt am Main 1779, reprint Aalen 1970.

Plahn, Joachim Nicolaus Friedrich, *Das verbrannte Heiligthum Hamburgs. Eine Erzehlung der am 10ten März 1750. durch einen Wetter-Strahl geschehenen Entzündung und gäntzlichen Einäscherung des Thurms und der Hauptkirche zu St. Michaelis* […], Hamburg 1750.

Pomereschius, Johannes, "Oratio in Mortem […] Francisci Stypmanni […]," in *Memoriae jurisconsultorum nostri seculi clarissimorum renovatae decas prima*, ed. Henning Witte, Frankfurt am Main 1676.

Primatt, Stephen, *The City and Countrey Purchaser and Builder* […], London 1668.

Publick Credit, Under consideration, how to be Retreiv'd [1700].

R., L., *To my Honoured Friend Mr. M.T. one of the Committee chosen by the Common Council of London, for the Insuring of Houses from Fire*, s.l. 1682.

Reimarus, Johann Albrecht Heinrich, *Die Ursache des Einschlagens vom Blitze, nebst dessen natürlichen Abwendung von unsern Gebäuden; aus zuverläßiger Erfahrung von Wetterschlägen vor Augen gelegt* [Langensalza: 1769].

Rhetius, Johannes Friedrich, *Q.D.F.E.J. Dissertatio Juridica De Versura* […]. *Resp. Walter Neuhaus.* [*24.9.1666.*], Frankfurt an der Oder 1666.

Rücker, Daniel, *Das durch ein zu Zion vom Herrn angesteckte Feuer zur Buße erweckte Israel wolte zur Erweckung einer hertzlichen Buße, als vorher am 10ten Martii 1750. die herrliche St. Michaelis-Kirche durch einen feurigen Wetter-Strahl entzündet und in die Asche geleget war, an dem ausserordentlichen Buß- und Bet-Tage aus Amos IV. 11, 12. seiner lieben und wehrten Gemeine in der Heiligen Geist-Kirche vorstellen*, Hamburg 1750.

Samlung [sic] der Obrigkeitlichen Verordnungen, Predigten, Gedichte und Kupfer welche bey Gelegenheit der durch einen Wetterstrahl eingeäscherten herrlichen Haupt-Kirche St. Michaelis in Hamburg zum Vorschein gekommen sind [...] (StAHamburg, Bibliothek A 640–77).

Santerna, Petrus, *Tractatus de assecurationibus et sponsionibus* (facsimile), Lisbon 1961.

Schetelig, Johann Andreas Gottfried, *Die von Hamburg empfundenen Rührungen des Zorns und der Güte Gottes, bey Gelegenheit der am 10. Merz 1750. durch einen Wetterstrahl geschehenen Entzündung und Einäscherung der Haupt-Kirche St. Michaelis, in gebundener Rede erwogen*, Hamburg 1750.

Schlosser, Johann Ludwig, *Die Stimme des Herrn im Feuer: am Buß-Tage wegen der durch einen Wetter-Strahl verursachten Einäscherung der S. Michaelis-Kirche in Hamburg zur Erweckung vorgestellt*, Hamburg 1750.

Schmidt, Johann Jakob, *Biblischer Physicus, Oder Einleitung zur Biblischen Natur-Wissenschaft und deren besondern Theilen* [...], Leipzig 1731.

Schragius, Johann Adam, *Disputatio inauguralis juridica de assecurationis contractu* [*15 Sept.1642*], Strasbourg 1642.

Senf, Johann Gottfried, *Eine Predigt von der ernstlichen Entschlüßung eines unter dem Zorngerichte des verzehrenden Feuers liegenden Volks: am 18. Sonntage nach Trinit. 1780. über 2 Sam. 24, 14. nach dem unglücklichen Brande der den 18. September vorher die ganze Stadt Gera in einen Schutthaufen verwandelt hatte*, Cuba bei Gera 1780.

Stein, Walter (ed.), *Akten zur Geschichte der Verfassung und Verwaltung der Stadt Köln im 14. Und 15. Jahrhundert*, 2 vols. Bonn 1893–95.

Straccha, Benvenuto, *Tractatus duo de assecurationibus ex proxenetis atque proxeneticis* [...], Amsterdam 1569/1668.

Summarischer Entwurf wie der zweyte große Fast-Buß- und Bethtag, welcher Donnerstags den 19 Oct. 1780. In den Hochgräflichen Reußischen Herrschaften zu Gera, mit Singen, Lectionen und Predigten allhier gefeyert und gehalten werden soll, Cuba bei Gera 1780.

Supplication/an Keiserliche Maiestat/Der Mortbrenner halben/Auff dem Reychstag zu Regenspurg/ Kayserlicher Myestat vberantwortet, s.l. 1541 (VD16 E 4649, BSB Munich 4 Res 232, Beibd. 2).

Surland, Johann, *Gott ist ein starcker Thurm, und der beste Schirm. Das ist: Eine Christliche Danck-Predigt, aus den Sprüchen Salomonis 18 Cap. Vers. 10. den 9 Martii, Anno Christi 1669. Nach glücklicher Verfertigung des Neuen Thurms an der Kirchen St. Michaelis* [...], reprint Hamburg 1750.

Teller, Johann Friedrich, *Gera in der Asche. Eine Predigt am 18. Sonntage nach Trinitatis 1780*, Zeitz 1780.

The Friendly Society, Or, A Proposal of a New way or Method for Securing Houses from any Considerable Loss by Fire, by way of Subscription, and Mutuall Contribution, s.l. 1684.

The Importance of Publick Credit, London 1699.

Uhrlandt, Carl Friedrich, *Klagelied eines geraischen Stadtkindes bey dem Verlust seiner Mutter,* Cuba bei Gera 1781.

Uhrlandt, Carl Friedrich, *Kurze Beschreibung von dem grossen Brande, der am 18. September 1780. Die hochgräfliche Reußplauische Residenz und Handelsstadt Gera ganz in die Asche geleget hat*, Cuba bei Gera 1780.

Ullmer, James H., "The Macroeconomic Thought of Nicholas Barbon," *Journal of the History of Economic Thought* 29 (2007), pp. 101–16.

Vaget, Bernard, *Wahrhaftiger Bericht von dem gräulichen und erschrecklichen Brande des Thurms St. Nicolai in Hamburg* [*im Jahr 1589*], reprint Hamburg 1750.

Vaz, Alvaro [Valascus], *Decisionum consultationum ac rerum iudicatarum in regno lusitaniae libri duo*, Frankfurt am Main 1608 [orig. 1593].

Veckinchusen, Hildebrand, *Hildebrand Veckinchusen. Briefwechsel eines deutschen Kaufmanns im 15. Jahrhundert,* ed. Wilhelm Stieda, Leipzig 1921.

Verordnung [i.e. regulation, edict] *wegen des, bey Gelegenheit der, am 10ten Märtz, des 1750sten Jahres, durch einen Wetter-Strahl, geschehenen Entzündung des Thurmes zu St. Michaelis* [...] *auf den 19ten dieses Monats, besonders angesetzten Buß-Fast- und Bet-Tages*, Hamburg 1750.

Vier Revidirte Articuli, Der Boots=Leute Schlaven= Gelder Einnahm betreffend geplacidirt den 27 Januarii Anno 1679, Hamburg 1679.

W., "Ueber die Abhaltung des Feuers vom Bauholze," *Hannoverisches Magazin* 11 (1773), p. 1631f.

Wagner, Friedrich, *Der gerechte Feuer-Eifer Gottes über Hamburg. Als die neue St. Michaelis-Kirche den 10 Merz 1750 durch einen Blitz in die Asche geleget war, am ausserordentlichen Buß- Fast- und Bet-Tage den 19 Merz zu St. Marien Magdalenen aus Amos IV. 11. 12 vorgestellet: Nebst einigen Historischen Nachrichten*, Hamburg 1750.

Wahrmund, Theophil [pseud.], *Die Jammer-gedrückte, Hülffleistend erquickte, Und Kronen-beglückte Rhein- und Neckar-Pfalz* [...] *Wie solche, samt andern benachbarten Ländern und angränzenden Flüssen,* [...] *von denen alles-verheerenden Franzosen, so übel zugerichtet, und jämmerlich verwiestet worden*, s.l. 1691.

Wellmann, Johann Christian, *Das von der göttlichen Regierung an denen Mordbrennern, welche in der Nacht zwischen den 19. und 20. May 1723 die Lebusische Vorstadt zu Franckfurt an der Oder* [...] *angesteckt bewiesene Denk-Mahl* [...], Frankfurt an der Oder 1725.

Werner, Heinrich, *Der Stadt Hamburgk Fewer-Ordnung/Wornach Dero Bürgere/Einwohnere/Vnterthanen vnd jedermänniglich daselbst in begebenen Fewersnöthen sich richten vnd verhalten sollen*, Hamburg 1637.

Whiston, J., *The causes of our present calamities in reference to the Trade of the Nation fully discovered* London 1696.

Winkelmann, Johann Justus, *Oldenburgische Friedens- und der benachbarten Oerter Kriegshandlungen*, Oldenburg 1671.

Zimmermann, Joachim Johann Daniel, *Zwo Predigten nach der am 10ten März 1750 durch einen Wetterstrahl verursachten Einäscherung der Hauptkirche zu St. Michaelis in Hamburg*, Hamburg 1750.

Zincke, Georg Heinrich, "Vorrede," in *Leipziger Sammlungen von Wirthschafftlichen, Policey-Cammer- und Finantz-Sachen*, vol. 7, Leipzig 1751, pp. iii-liv.

Published Secondary Sources

"Das schöne nahrhafte Göppingen in Asche"—Die Brandkatastrophe von 1782 und der Wiederaufbau der Stadt, [Stadtarchiv Göppingen], DVD, Göppingen s.d.

100 Jahre Hamburg-Bremer Feuer-Versicherungs-Gesellschaft Hamburg 1854–1954, s.l.

Adam, Barbara, *Time and Social Theory*, Philadelphia 1990.

Adam, Ulrich, *The Political Economy of J.H.G. Justi*, Oxford 2006.

Adams, W.M., "The Origins of Sustainable Development," in Redclift, *Sustainability*, vol. 2, pp. 69–100.

Addobbati, Andrea, *Commercio, rischio, guerra: il mercato delle assicurazioni marittime di Livorno (1694–1795)*, Rome 2007.

Ahlberg, Nils, *Stadsgrundningar och planförändringar: svensk stadsplanering 1521–1721*, Uppsala 2005.

Ahlberg, Nils, *Stadsplaneförändringar i Sverige 1550–1875*, 2 vols., Uppsala 1995.

Akasoy, Anna, "The Man-Made Disaster: Fire in Cities in the Medieval Middle East," *Historical Social Research* 32 (2007), pp. 75–87.

Akbulut, Hüseyin, *Chumhuriyet Dönemine Kadar Osmanli Imparatorluğuńda Sigortacilik*, PhD thesis Istanbul Üniversitesi 2014.

Alff, Kristen, "Levantine Joint-Stock Companies, Trans-Mediterranean Partnerships, and Nineteenth-Century Capitalist Development," *Comparative Studies in Society and History* 60, 1 (2018), pp. 150–177.

Allemeyer, Marie Luise, *"Kein Land ohne Deich …!" Lebenswelten einer Küstengesellschaft in der Frühen Neuzeit*, Göttingen 2006.

Allgemeinen Betrachtungen über die Triebe der Thiere, ed. Jürgen von Kempski, Göttingen 1982.

Alpers, Svetlana, *The Art of Describing: Dutch Art in the Seventeenth Century*, Chicago 1983.

Amend-Traut, Anja, *Wechselverbindlichkeiten vor dem Reichskammergericht*, Cologne 2009.

Andervont Edie, Carolyn, "New Buildings, New Taxes, and Old Interests: An Urban Problem of the 1670s," *The Journal of British Studies* 6, 2 (1967), pp. 35–63.

Andraschek-Holzer, Ralph, "Topographische Ansichten Österreichs (in den Grenzen der Republik) 1500–1800," in Pauser, Josef, Scheutz, Martin and Winkelbauer,

Thomas (eds.), *Quellenkunde der Habsburgermonarchie (16.-18. Jahrhundert). Ein exemplarisches Handbuch*, Vienna 2004, pp. 1048–59.

Appleby, Joyce, "Ideology and Theory: The Tension between Political and Economic Liberalism in Seventeenth-Century England," *The American Historical Review* 81 (1976), pp. 499–515.

Arcy Wood, Gillen d', *Tambora. The eruption that changed the World*, Princeton 2014.

Ardaman, Emel, "Perspective and Istanbul: The Capital of the Ottoman Empire," *Journal of Design History* 20 (2007), pp. 109–30.

Arlinghaus, Franz-Josef, *Zwischen Notiz und Bilanz. Zur Eigendynamik des Schriftgebrauchs in der kaufmännischen Buchführung am Beispiel der Datini/di Berto-Handelsgesellschaft in Avignon (1367–73)*, Frankfurt am Main 2000.

Ashburner, Walter (ed.), *Nomos Rodiōn nautikos = The Rhodian Sea-Law*, Oxford 1909, reprint Aalen 1976.

Aslanian, S.D., *From the Indian Ocean to the Mediterranean. The global trade networks of Armenian merchants from New Julfa*, Berkeley 2011.

Atzpodien, Hans Christoph, *Die Entwicklung der preußischen Staatsaufsicht über das private Versicherungswesen im 19. Jahrhundert unter besonderer Berücksichtigung ihres Verhältnisses zum Wirtschaftsliberalismus*, Bonn 1982.

Baasch, Ernst, *Die Hansestädte und die Barbaresken*, Kassel 1897.

Baasner, Rainer, *Abraham Gotthelf Kästner, Aufklärer (1719–1800)*, Tübingen 1991.

Bächtold-Stäubli (ed.), *Handwörterbuch des Aberglaubens*, 10 vols., Berlin 1929–42.

Baer, Marc David, "The Great Fire of 1660 and the Islamization of Christian and Jewish Space in Istanbul," *International Journal of Middle East Studies* 36 (2004), pp. 159–81.

Baer, William C., "Is Speculative Building Underappreciated?," *Urban History* 34 (2007), pp. 296–316.

Baer, William C., "The Institution of Residential Investment in Seventeenth-Century London," *Business History Review* 76 (2002), pp. 515–51.

Bähr, Andreas, *Furcht und Furchtlosigkeit. Göttliche Gewalt und Selbstkonstitution im 17. Jahrhundert*, Göttingen 2013.

Baier, Christof, *Bürgerhäuser in Templin und Zehdenick. Die Theorie vom bürgerlichen Wohnhaus und die Praxis des provinzialstädtischen Bürgerhausbaus in Brandenburg-Preußen im 18. Jahrhundert*, PhD thesis, Humboldt University Berlin 2006.

Baillet, Lina, "Schwendi, lecteur de Machiavel," *Revue d'Alsace* 112 (1986), pp. 119–97.

Baker, Timothy, *London: Rebuilding the City after the Great Fire*, Chichester 2000.

Balachandran, G., "India: From a Colonial Outpost to a Leading Market," in Peter Borscheid and Niels Viggo Haueter (eds.), *World insurance: the evolution of a global risk network*, Oxford 2012, pp. 444–71.

Bald, Rudolf and Dollberger, A., *700 Jahre Stadtrecht Herbstein. Geschichte und Entwicklung einer Stadt*, Lauterbach/Hessen 1962.

Ballestracci, D., "La lotta contro il fuoco (XIII-XVI secolo)," in *Città e servizi sociali nell'Italia dei secoli XII-XV*, Pistoia 1990, pp. 417–38.

Bankoff, Greg, Lübken, Uwe and Sand, Jordan (eds.), *Flammable Cities: Urban Conflagration and the Making of the Modern World*, Madison 2012.

Banse, Gerhard and Müller, Hans-Peter (eds.), *Johann Beckmann und die Folgen. Erfindungen—Versuch der historischen, theoretischen und empirischen Annäherung an einen vielschichtigen Begriff*, Münster 2007.

Barkow, J.H., Cosmides, L. and Tooby, J. (eds.), *The Adapted Mind*, Oxford 1992.

Barnouw, Jeffrey, "The Pursuit of Happiness in Jefferson and its Background in Bacon and Hobbes" *Interpretation: A Journal of Political Philosophy* 11 (1983), pp. 225–48.

Basas Fernández, M., *El seguro marítimo en Burgos (siglo XVI)*, Bilbao 1963.

Bayerl, Günter and Beckmann, Jürgen (eds.), *Johann Beckmann (1739–1811). Beiträge zu Leben, Werk und Wirkung des Begründers der Allgemeinen Technologie*, Münster 1999.

Bayerl, Günter, "Die Anfänge der Technikgeschichte bei Johann Beckmann und Johann Heinrich Moritz von Poppe," in *Die technikhistorische Forschung in Deutschland von 1800 bis zur Gegenwart*, ed. Wolfgang König, Kassel 2007, pp. 13–32.

Bayerl, Günter, "Prolegomenon der 'Großen Industrie'. Der technisch-ökonomische Blick auf die Natur im 18. Jahrhundert," in Werner Abelshauser (ed.), *Umweltgeschichte. Umweltverträgliches Wirtschaften in historischer Perspektive*, Göttingen 1994, pp. 29–56.

Bayly, Christopher A., *Indian Society and the Making of the British Empire*, Cambridge 1988.

Bec, Christian, *Les marchands écrivains. Affaires et humanisme à Florence 1375–1434*, Paris 1967.

Beck, Ulrich, *Risikogesellschaft. Auf dem Weg in eine andere Moderne*, Frankfurt am Main 1986.

Beck, Ulrich, *Weltrisikogesellschaft*, Frankfurt am Main 2007.

Beck-Saiello, Émilie, *Pierre Jacques Volaire, 1729–1799: dit le Chevalier Volaire*, Paris 2010.

Becker, Ulrich and Hockerts, Hans-Günther and Tenfelde, Klaus (eds.), *Sozialstaat Deutschland. Geschichte und Gegenwart*, Bonn 2010.

Behrens-Abouseif, Doris, "The Fire of 884/1479 at the Umayyad Mosque in Damascus and an Account of Its Restoration," *Mamlūk Studies Review* 8 (2004), pp. 279–97.

Behringer, Wolfgang (ed.), *Kulturelle Konsequenzen der "Kleinen Eiszeit"*, Göttingen 2005.

Behringer, Wolfgang, *Kulturgeschichte des Klimas. Von der Eiszeit bis zur globalen Erwärmung*, Munich 2007.

Beier, Adrian, *Mag. Adrian Beiers Jenaische Annalen (1523–1599)*, ed. Herbert Koch, Jena 1928.

Bellows, Barbara L., *Benevolence Among Slaveholders: Assisting the Poor in Charleston, 1670–1860*, Baton Rouge 1993.

Bennassar, Bartolomé (ed.), *Les catastrophes naturelles dans l'Europe médiévale et moderne*, Toulouse 1996.

Bensa, Enrico, *Il contratto di assicurazione nel Medio Evo*, Genova 1884.

Berg, Jan van den, "English Deism and Germany: The Thomas Morgan controversy," *The Journal of Ecclesiastical History* 59 (2008), pp. 48–61.

Berger, Alfred *Ein Streifzug durch die Nossener Geschichte zur 1000-Jahrfeier der Landschaft, 750-Jahrfeier des Ortes und 100-Jahrfeier der Einführung der Städteordnung. Gedenkschrift zum Heimat- und Schulfest Pfingsten 1936*, Nossen 1936.

Bergère, Marie-Claire, *Histoire de Shanghai*, Paris 2002.

Bergh, Govaert C.J.J. van den, *Die holländische elegante Schule. Ein Beitrag zur Geschichte von Humanismus und Rechtswissenschaft in den* Niederlanden 1500–*1800*, Frankfurt am Main 2002.

Bergmann, Rudolf, *Geschichte des rheinischen Versicherungswesens*, Essen 1928.

Bernal, Antonio-Miguel, "La Aseguración marítima en la carrera de Indias: Seguros, Crédito y Actividad Financeria (Ss. XVI—XVIII)," *Atti e memorie della Società Savonese di Storia patria* 32/33 (1996/97), pp. 83–125.

Bernhardt, Christoph (ed.), special issue "Kleine Städte," *Informationen zur modernen Stadtgeschichte* 2 (1999).

Bertucci, Paola, "Sparking Controversy: Jean Antoine Nollet and Medical Electricity south of the Alps," *Nuncius* 20 (2005), pp. 153–87.

Bhattacharya, Sabyasachi (ed.), *The Cambridge Economic History of India*, New Delhi 2005.

Bibliothèque nationale de France (ed.), *La Révolution de la photographie instantanée 1880–1990*, Paris 1996.

Bickers, Robert and Henriot, Christian (eds.), *New Frontiers: Imperialism's New Communities in East Asia, 1842–1953*, Manchester 2000.

Bieber, Johann Ehlert, *Verzeichniss aller Feuersbrünste in Hamburg und der Umgegend, von 1281 bis 1830*, Hamburg 1830.

Biela, Adam, *Analogy in Science. From a Psychological Perspective*, Frankfurt am Main 1991.

Biéler, André, *Calvin's Economic and Social Thought*, ed. Edward Dommen, trans. James Greig, orig. French, 1961; Geneva 2006.

Binding, Günther, Linscheid-Burdich, Susanne, and Wippermann, Julia, *Planen und Bauen im frühen und hohen Mittelalter nach den Schriftquellen bis 1250*, Darmstadt 2002.

Biondi, Carminella et al. (eds.), *La Quête du bonheur et l'expression de la douleur dans la littérature et la pensée françaises*, Geneva 1995.

Black, Jeremy, "A Military Revolution? A 1660–1792 Perspective," in Rogers, Clifford J. (ed.), *The Military Revolution Debate, Readings on the Military Transformation of Early Modern Europe*, Boulder 1995, pp. 95–114.

Black, Jeremy, *A Military Revolution? Military Change and European Society, 1550–1800*, Basingstoke 1991.

Black, Jeremy, *European Warfare, 1660–1815*, London 1994.

Blackbourn, David, *The Conquest of Nature: Water, Landscape and the Making of Modern Germany*, London 2006.

Blickle, Peter, Hüglin, Thomas O., Wyduckel, Dieter (eds.), *Subsidiarität als rechtliches und politisches Ordnungsprinzip in Kirche, Staat und Gesellschaft. Genese, Geltungsgrundlagen und Perspektiven an der Schwelle des dritten Jahrtausends*, Berlin 2002.

Blum, Edgar, "Les assurances terrestres en France sous l'Ancien Régime," in *Revue d'histoire économique et sociale* 8 (1920), pp. 95–104.

Bodemann, Eduard, "Leibnizens volkswirthschaftliche Ansichten und Denkschriften," *Preußische Jahrbücher* 53 (1884), pp. 378–404, 392–96.

Boehart, William, *Politik und Religion: Studien zum Fragmentenstreit* (*Reimarus, Goeze, Lessing*), Hamburg 1987.

Bohn, Robert, "Von Sklavenkassen und Konvoifahrten. Die arabischen Seeräuber und die deutsche Seefahrt im 17. und 18. Jahrhundert," in Thomas Stamm-Kuhlmann, Jürgen Evert, Birgit Aschmann, and Jens Hohensee (eds.), *Geschichtsbilder. Festschrift für Michael Salewski zum 65. Geburtstag*, Stuttgart 2003, pp. 25–37.

Boiteux, Louis Augustin, *La fortune de mer. Le besoin de sécurité et les débuts de l'assurance maritime*, Paris 1968.

Bolzoni, Lina, *La stanza della memoira. Modelli letterari e iconografici nell'età della stampa*, Torino 1995.

Böning, Holger, Schmitt, Hanno, Siegert, Reinhart (eds.), *Volksaufklärung. Eine praktische Reformbewegung des 18. und 19. Jahrhunderts*, Bremen 2007.

Bonß, Wolfgang, *Vom Risiko. Unsicherheit und Ungewißheit in der Moderne*, Hamburg 1995.

Borsay, Peter and Proudfoot, Lindsay (eds.), *Provincial towns in early modern England and Ireland: change, convergence, and divergence*, Oxford 2002.

Borscheid, Peter and Drees, Annette (eds.), *Versicherungsstatistik Deutschlands 1750–1985*, St. Katharinen 1988.

Borscheid, Peter and Haueter, Niels Viggo, *World Insurance: The Evolution of a Global Risk Network*, Oxford 2012.

Borscheid, Peter and Umbach, Kai, "Systemwettbewerb, Institutionenexport und Homogenisierung. Der Internationalisierungsprozess der Versicherungswirtschaft im 19. Jahrhundert," *Jahrbuch für Wirtschaftsgeschichte* (2008), pp. 207–26.

Bourguet, Marie-Noëlle, *Déchiffrer la France: la statistique départementale à l'époque napoléonienne*, Paris 1988.

Bossert, Gustav, "Die Liebestätigkeit der evangelischen Kirche Württembergs von der Zeit des Herzogs Christoph bis 1650," in *Württembergische Jahrbücher für Statistik und Landeskunde 1905*, Stuttgart 1906, vol. 1, pp. 1–28.

Bowers Isaacson, Lisa Margaret, *The Building Society: Speculative Building in the West End of London, 1660–1760*, PhD thesis, Princeton University 1992.

Bowie Moniz, Amanda, "Cosmopolitanism and Philanthropy in the Early American Republic," *GHI Bulletin Supplement* 5 (2008), pp. 9–22.

Bowie Moniz, Amanda, *From Empire to Humanity: The American Revolution and the Origins of Humanitarianism*, Oxford 2016.

Brakensiek, Stefan, "Projektemacher. Zum Hintergrund ökonomischen Scheiterns in der Frühen Neuzeit, in *Fiasko—Scheitern in der Frühen Neuzeit*, Bielefeld 2015, pp. 39–58.

Brather, Hans-Stephan (ed.), *Leibniz und seine Akademie. Ausgewählte Quellen zur Geschichte der Berliner Sozietät der Wissenschaften, 1697–1716*, Berlin 1993.

Braude, Benjamin and Lewis, Bernard (eds.), *Christians and Jews in the Ottoman Empire: The Functioning of a Plural Society*, 2 vols., New York 1982.

Braudel, Fernand, *Civilisations matérielle, économie et capitalisme XVe-XVIIIe siècle. II. Les jeux de l'échange*, Paris 1979.

Braudel, Fernand, *La Méditerranée et le monde méditerranéen à l'époque de Philippe II* [first ed. 1949], 3 vols., 9th ed. Paris 1990.

Brauner, Christine, "Recommendation und Reklame. Niederrheinische Brandspritzenmacher und Praktiken der Werbung in der Frühen Neuzeit," *Zeitschrift für Historische Forschung* 46, 1 (2019), pp. 1–45.

Braunschweigische Landesbrandversicherungsanstalt (ed.), *Festschrift 225 Jahre LBVA. Braunschweigische Landesbrandversicherungsanstalt 1754–1979*, Braunschweig 1979.

Breen, Timothy H., *The Marketplace of Revolution: How Consumer Politics Shaped American Independence*, New York 2004.

Brett-James, Norman G., "A Speculative London Builder of the 17th Century, Dr. Nicholas Barbon," *Transactions of the London and Middlesex Archaeological Society* 6 (1933), pp. 110–45.

Brett-James, Norman G., *The Growth of Stuart London*, London 1935.

Brewer, John, *The Sinews of Power: War, Money and the English State, 1688–1783*, Cambridge/Mass. 1988.

Briese, Olaf and Günther, Timo, "Katastrophe. Terminologische Vergangenheit, Gegenwart und Zukunft," *Archiv für Begriffsgeschichte* 51 (2009), pp. 155–95.

Briese, Olaf, *Für des Staates Sicherheit. Das Löschwesen im 19. Jahrhundert und die Gründung der ersten Berufsfeuerwehr Deutschlands in Berlin 1851*, Berlin 2018.

Briffa, Keith R., "Annual Climate Variability in the Holocene. Interpreting the Message of Ancient Trees," *Quaternary Science Reviews* 19 (2000), pp. 87–105.

Brint, Stephen, "*Gemeinschaft* Revisited: A Critique and Reconstruction of the Community Concept," *Sociological Theory* 19 (2001), pp. 1–23.

Broc, Numa, *La géographie de la Renaissance (1420–1620)*, Paris 1980.

Broich, Hans-Josef, “Erkelenz. Geschichtliches Portrait einer Stadt,” *Der Niederrhein* 75 (2008), pp. 114–17.

Brundtland, Gro Harlem (ed.), *Our Common Future: The World Commission on Environment and Development*, Oxford 1987.

Buck, August, “Aus der Vorgeschichte der Querelle des Anciens et des Modernes in Mittelalter und Renaissance,” *Bibliothèque d'Humanisme et Renaissance* 20 (1958), pp. 527–41.

Burgdorf, Wolfgang, *Reichskonstitution und Nation. Verfassungsreformprojekte für das Heilige Römische Reich Deutscher Nation im politischen Schrifttum von 1648 bis 1806*, Mainz 1998.

Burkhardt, Johannes, “Die Friedlosigkeit der frühen Neuzeit. Grundlegung einer Theorie der Bellizität Europas,” *Zeitschrift für historische Forschung* 24, 4 (1997), pp. 509–74.

Busch, Werner, “Die Naturwissenschaften als Basis des Erhabenen in der Kunst des 18. und frühen 19. Jahrhunderts” in *Jahrbuch des Historischen Kollegs* (2004), pp. 83–110.

Büttner, Nils, *Die Erfindung der Landschaft. Kosmographie und Landschaftskunst im Zeitalter Bruegels*, Göttingen 2000.

Calabi, Donatella, *Venezia in Fumo. I grandi incendi della città-fenice*, Bergamo 2006.

Caradonna, Jeremy L., *Sustainability. A History*, Oxford 2014.

Caradonna, Jeremy L., *The Enlightenment in Practice. Academic Prize Contests and Intellectual Culture in France, 1670–1794*, Ithaca 2012.

Cardoso, José Luís, “El terremoto de Lisboa de 1755 y la política de regulación económica del Marqués de Pombal”, *Historia y Política* 16 (2006), pp. 209–236.

Carey, Daniel and Finlay, Christopher J. (eds.), *The Empire of Credit: The Financial Revolution in the British Atlantic World, 1688–1815*, Dublin 2011.

Carl, Horst, “ ‘Brüssel 1695’: Kriegszerstörungen und ihre Visualisierung im späten 17. und 18. Jahrhundert,” in Emich, Birgit and Signori, Gabriela (eds.), *Kriegs/Bilder in Mittelalter und Früher Neuzeit*, Berlin 2009, pp. 295–315.

Carl, Horst, *Okkupation und Regionalismus. Die preußischen Westprovinzen im Siebenjährigen Krieg*, Mainz 1993.

Carroll, John M., “Slow Burn in China: Factories, Fear, and Fire in Canton,” in Robert Peckham (ed.), *Empires of Panic. Epidemics and Colonial Anxieties*, Hong Kong 2015, pp. 35–55.

Carroll, Siobhan, “Crusades Against Frost: Frankenstein, Polar Ice, and Climate Change in 1818”, *European Romantic Review* 24, 2 (2013), pp. 211–230.

Carruthers, Bruce G., “Politics, Popery, and Property: A Comment on North and Weingast,” *The Journal of Economic History* 50 (1990) pp. 693–98.

Carruthers, Bruce G., *City of Capital. Politics and Markets in the English Financial Revolution*, Princeton 1996.

Casares, Gabriel Tortella et al. (eds.), *Historia del seguro en España*, Madrid 2014.

Castor, Markus A., "Augenblick und Zeitfluß—Bemerkungen zu einigen Konstanten in der Malerei Jan Vermeers," *Pantheon* 54 (1996), pp. 84–101.

Ceccarelli, Giovanni, "Coping with Unknown Risks in Renaissance Florence: insurers, Friars and Abacus Teachers," in Cornel Zwierlein (ed.), *The Dark Side of Knowledge. Histories of Ignorance, 1400 to 1800*, Leiden/Boston 2016, pp. 117–38.

Ceccarelli, Giovanni, "Risky Business: Theological and Canonical Thought on Insurance from the Thirteenth to the Seventeenth Century," *Journal of Medieval and Early Modern Studies* 31 (2001), pp. 607–58.

Ceccarelli, Giovanni, *Un mercato del rischio. Assicurare e farsi assicurare nella Firenze rinascimentale*, Venice 2012.

Cecchi Aste, Elena (ed.), *Il carteggio di Gaeta nell'Archivio del mercante pratese Francesco di Marco Datini. 1387–1405*, Gaeta 1997.

Çelik, Zeynep, *The Remaking of Istanbul: Portrait of an Ottoman City in the Nineteenth Century*, Berkeley 1993.

Cevolini, Alberto, "Die Einrichtung der Versicherung als soziologisches Problem," *sociologia internationalis* 1 (2010), pp. 65–89.

Chakrabarty, Dipesh, *Provincializing Europe: Postcolonial Thought and Historical Difference*, Princeton 2000.

Chambers, Frank M., "The 'Little Ice Age'", *The Holocene* 26, 3 (2016), pp. 335–337.

Charm of Bombay: An Anthology of Writing in Praise of the First City in India (ed.), R.P. Karkaria, Bombay 1915.

Chassagnette, Axelle, "Mettre en carte: fonctions pratiques, intellectuelles et sociales de la cartographie allemande au XVIe siècle," in Pantin, Isabelle and Péoux, Gérald (eds.), *Mise en forme des savoirs à la Renaissance. À la croisée des idées, des techniques et des publics*, Paris 2013, pp. 79–101.

Chattopadhyay, Swati, *Representing Calcutta. Modernity, nationalism, and the Colonial Uncanny*, London 2005.

Chaudhuri, Binay Bhushan (ed.), *Economic History of India from Eighteenth to Twentieth Century*, New Delhi 2005.

Chevreau, Emmanuelle, "La 'lex rhodia de iactu'—un exemple de la réception d'une institution étrangère dans le droit romain," *The Legal History Review* 73 (2005), pp. 67–80.

Clark, Geoffrey et al. (eds.), *The Appeal of Insurance*, Toronto 2010.

Clark, Geoffrey, "Slave Insurance in late Medieval Catalonia," in *Sicherheit in der Frühen Neuzeit. Norm, Praxis, Repräsentation*, eds. Christoph Kampmann and Ulrich Niggemann, Cologne 2013, pp. 418–29.

Clark, Geoffrey, *Betting on Lives. The Culture of Life Insurance in England, 1695–1775*, Manchester 1999.

Clark, Peter, "Small towns 1700–1840," in Peter Clark (ed.), *The Cambridge Urban History of Britain*, vol. 2, *1540–1840*, Cambridge 2000, pp. 733–74.

Clark, Peter, *Small Towns in Early Modern Europe*, Cambridge 1995.

Cocco, Sean, *Watching Vesuvius: A History of Science and Culture in Early Modern Italy*, Chicago 2012.

Cockerell, Hugh A. L. and Green, Edwin (eds.), *The British Insurance Business. A Guide to its History and Records*, 2nd ed., Sheffield 1994.

Collet, Dominik, "Risk and Resilience", in *Concepts of Urban-Environmental History*, eds. Sebastian Haumann, Martin Knoll, Detlev Mares, Bielefeld 2020, pp. 79–94.

Conklin, Carli N., *The Pursuit of Happiness in the founding Era: An Intellectual History*, Columbia 2019.

Conrad, Anne, "Aufgeklärte Elite und aufzuklärendes Volk," in Conrad, Anne, Herzig, Arno, and Kopitzsch, Franklin (eds.), *Das Volk im Visier der Aufklärung. Studien zur Popularisierung der Aufklärung im späten 18. Jahrhundert*, Hamburg 1998, pp. 1–15.

Conze, Werner, "Sicherheit, Schutz," in *Geschichtliche Grundbegriffe. Historisches Lexikon zur politisch-sozialen Sprache in Deutschland*, vol. 5, eds. Otto Brunner, Werner Conze, Reinhart Koselleck, Stuttgart 1984, pp. 831–62.

Cooper, Michael, *"A More Beautiful City": Robert Hooke and the Rebuilding of London after the Great Fire*, Stroud 2003.

Cooper, Michael, "Robert Hooke's work as surveyor for the City of London in the aftermath of the Great Fire," *Notes and Records of the Royal Society of London* 51 (1997), pp. 161–74; 52 (1998), pp. 25–38, 205–220.

Copenhaver, Brian P., "The Historiography of Discovery in the Renaissance: The sources and Composition of Polidore Vergil's De inventoribus rerum, I-III", *Journal of the Warburg and Courtauld Institutes* 41 (1978), pp. 192–214.

Cordes, Albrecht, "Auf der Suche nach der Rechtswirklichkeit der mittelalterlichen *Lex mercatoria*," *Zeitschrift der Savigny-Stiftung für Rechtsgeschichte, Germanistische Abteilung* 118 (2001), pp. 168–84.

Cordes, Albrecht, *Spätmittelalterlicher Gesellschaftshandel im Hanseraum*, Cologne 1998.

Cortekar, Jörg, *Glückskonzepte des Kameralismus und Utilitarismus: Implikationen für die moderne Umweltökonomik und Umweltpolitik*, Göttingen 2007.

Corwin, Nancy A., *The Fire Landscape: Its sources and its development from Bosch through Jan Brueghel I with special emphasis on the Mid-Sixteenth Century Bosch 'revival,'* PhD diss. University of Washington 1976.

Cosgrove, Denis, "Prospect, Perspective and the Evolution of the Landscape Idea," *Transactions of the Institute of British Geographers* 10 (1985), pp. 45–62.

Cosgrove, Denis, *The Palladian landscape. Geographical Change and its Cultural Representations in Sixteenth Century Italy*, Leicester 1993.

Costello, Augustine E., *A history of the New York Fire Departments, Volunteer and Paid, Derived from Original Sources and Authorities*, New York 1887.

Craciun, Adriana, "Writing the Disaster: Franklin and *Frankenstein*", *Nineteenth-Century Literature* 65, 4 (2011), pp. 433–480.

Crome, Bruno, *Kulturgeschichtliche Miniaturen vom Ausgang des Mittelalters aus einer alten Chronik*, Göttingen 1921.

Cronon, William, *Nature's Metropolis: Chicago and the Great West*, New York 1990.

Crosby, Alfred W., *The Measure of Reality: Quantification and Western Society, 1250–1600*, Cambridge 1997.

Crosland, Maurice, "The Congress on Definitive Metric Standards, 1798–1799: The First International Scientific Conference?", *Isis* 60, 2 (1969), pp. 226–231.

Crouzet-Pavan, Élisabeth, Crouzet, Denis and Desan, Philippe (eds.), *Cités humanistes, cités politiques: 1400–1600*, Paris 2014.

Cunningham, H. and Innes, Joanna (eds.), *Charity, Philanthropy and Social Reform in Europe and North America 1690–1850*, London 1998.

d'Ester, Karl, "Fuggerzeitungen," in Walther Heide (ed.), *Handbuch der Zeitungswissenschaft*, vol. 1, Leipzig 1940, col. 1187–98.

Dahlström, Kristina, Skea, Jim, and Stahel, Walter R., "Innovation, Insurability and Sustainable Development: Sharing Risk Management between Insurers and the State," *The Geneva Papers on Risk and Insurance* 28, 3 (2003), pp. 394–412.

Dalby, Simon, "Firepower: Geopolitical Cultures in the Anthropocene", *Geopolitics* 23, 3 (2018), pp. 718–742.

Danhauser, Konrad, *Topographische Geschichte der Stadt Aichach und ihrer Umgebung*, Munich 1858.

Darnton, Robert, "The Pursuit of Happiness," *Wilson Quarterly* 19 (1995), pp. 42–52.

Daston, Lorraine J., "The Domestication of Risk: Mathematical Probability and Insurance 1650–1830," in Krüger, Lorenz, Daston, Lorraine J., Heidelberger, Michael (eds.), *The Probabilistic Revolution*, vol. 1: *Ideas in History*, Cambridge/Mass. 1987, pp. 237–60.

Daston, Lorraine, *Classical Probability in the Enlightenment*, Princeton 1988.

Daunton, Martin J. (ed.), *Charity, Self-Interest and Welfare in the English Past*, London 1996.

Davies, Andrea Rees, *Saving San Francisco: Relief and Recovery after the 1906 Disaster*, Philadelphia 2012.

Davies, Kenneth G., "Joint-Stock Investment in the Later Seventeenth Century," *The Economic History Review* 4, 3 (1952), pp. 283–301.

Dawkins, C. J., "Measuring the Spatial Pattern of Residential Segregation," in: *Urban Studies* 41, 4 (2004), 833–851.

De Ruysscher, Dave, "How Normative were Merchant Guidebooks? Of Customs, Practices, and … Good Avice (Antwerp, Sixteenth Century)," in: Pihlajamäki, Heikki et al. (ed.), *Understanding the Sources of Early Modern and Modern Commercial Law. Courts, Statues, Contracts, and Legal Scholarship*, Leiden 2018, pp. 144–165.

Deane, Phyllis, *The First Industrial Revolution* [Cambridge 1979].

Decock, Wim, *Theologians and Contract Law: The Moral Transformation of the* Ius commune (*ca. 1500–1650*), Leiden 2013.

Delsalle, Paul, " 'L'orvale de feu' dans les villes et les campagnes comtoises (1493–1634)," in Vion-Delphin and Lassus (eds.), *Les hommes et le feu*, pp. 253–73.

Delumeau, Jean, *La peur en Occident* (*XIVe-XVIIIe siècles*). *Une cité assiégée*, Paris 1978.

Delumeau, Jean, *Le péché et la peur. La culpabilisation en Occident, XIIIe-XVIIIe siècles*, Paris 1983.

Delumeau, Jean, *Rassurer et proteger, Le sentiment de sécurité dans l'Occident d'autrefois*, Paris 1989.

Desai, G. R., *Life Insurance Business in India*, Delhi 1973.

Desan, Philippe, *L'imaginaire économique de la Renaissance*, Fasano 2002.

Devisscher, Hans, "Enkele schilderijen van Enghelbert Ergo," in Balis, Arnout (ed.), *Rubens and His World*, Antwerp 1985, pp. 235–44.

Devisscher, Hans, *Kerstiaen de Keuninck*, Freren 1987 (= *Vlaamse schilders uit de tijd van de grote meesters*, deel 3).

Dickson, Peter G. M., *The Financial Revolution in England. A Study in the Development of Public Credit, 1688–1756*, London 1967.

Dickson, Peter G. M., *The Sun Insurance Office 1710–1960. The History of Two and a Half Centuries of British Insurance*, London 1960.

Dickson, Peter, *The Sun Insurance Office 1710–1960. The History of Two and a Half Centuries of British Insurance*, London 1960.

Die Flämische Landschaft, 1520–1700, ed. Kulturstiftung Ruhr Essen/Kunsthistorischen Museum Wien (Ausstellungskatalog: Essen, Villa Hügel, 23. Aug.—30. Nov. 2003), Lingen 2003.

Dini, Bruno, "Art. Brief, Briefliteratur, Briefsammlungen. F. Handels- und Kreditbrief II. Italien," *LexMA* 2 (1983), p. 681f.

Dohrn-van Rossum, Gerhard, *Die Geschichte der Stunde. Uhren und moderne Zeitordnungen*, Munich 1992.

Doll, Andreas, *Von der vis maior zur höheren Gewalt. Geschichte und Dogmatik eines haftungsentlastenden Begriffs*, Frankfurt am Main 1989.

Donahue Kuretsky, Susan, "Jan van der Heyden and the Origins of Modern Firefighting. Art and Technology in Seventeent-Century Amsterdam," in Bankoff, Lübken and Sand (eds.), *Flammable Cities*, pp. 23–43.

Donahue, Charles, "Medieval and Early Modern *Lex mercatoria*, " *Chicago Journal of International Law* 5, 1 (2004), pp. 21–37.

Dorin, Rowan, *Banishing Usury: The Expulsion of Foreign Moneylenders in Medieval Europe, 1200–1450*, PhD thesis Harvard University, 2015.

Dorwart, Reinhold A., "The Earliest Fire Insurance Company in Berlin and Brandenburg, 1705–1711," *The Business History Review* 32, no. 2 (1958), pp. 192–203.

Dosquet, Emilie, "Die Verwüstung der Pfalz als (Medien-)Ereignis. Von der rheinländischen Kriegshandlung zum europäischen Skandal," in *Krieg und Kriegserfahrung im Westen des Reiches 1568–1714*, ed. Andreas Rutz, Göttingen 2017, pp. 333–370.

Dossal, Mariam, "Knowledge for Power: The Significance of the Bombay Revenue Survey, 1811–1827," in Banga (ed.), *Ports and Their Hinterlands*, pp. 227–43.

Dossal, Mariam, *Imperial Designs and Indian Realities: The Planning of Bombay City, 1845–1875*, Oxford 1991.

Douglas, Mary, "Risk as a Forensic Resource," *Daedalus* 119 (1990), pp. 1–16.

Doumerc, Bernard, " 'Par Dieu écrivez plus souvent!' La lettre d'affaires à Venise à la fin du Moyen Âge," in Société des historiens médiévistes de l'enseignement supérieur (ed.), *La circulation des nouvelles*, pp. 99–109.

Drayton, Richard, *Nature's Government: Science, Imperial Britain, and the "Improvement" of the World*, New Haven 2000.

Dreyer, Thomas, *Die 'Assecuranz- und Haverey-Ordnung' der Freien und Hansestadt Hamburg von 1731*, Frankfurt am Main 1990.

Dreyfus, Michel, Voldman, Danièle, Ruffat, Michèle, and Viet, Vincent (eds.), *Se protéger, être protégé: Une histoire des assurances sociales en France*, Rennes 2015.

Dross, Fritz, "Gottes elektrischer Wille? Zum Düsseldorfer 'Blitzableiter-Aufruhr' 1782/83," in Jörg Engelbrecht (ed.), *Landes- und Reichsgeschichte. Festschrift für Hansgeorg Molitor zum 65. Geburtstag*, Bielefeld 2004, pp. 281–302.

Du Plessis, Paul J., *Letting and Hiring in Roman Legal Thought: 27 BCE-284CE*, Leiden 2012.

Duchhardt, Heinz, "Staatenkonkurrenz und Fürstenrivalität. Krieg und Frieden in Europa 1700–1714," in Marcus Junkelmann (ed.), *Theatrum belli. Die Schlacht von Höchstädt 1704 und die Schlösser von Schleißheim und Blenheim*, Herzberg 2004, pp. 3–11.

Duncan, O. D. and B., "A methodological analysis of segregation indices," in: *American Sociological Review* 20 (1955), 210–217.

Dupuy, Jean-Pierre, *Pour un catastrophisme éclairé*, Paris 2002.

Dym, Warren Alexander, *Divining Science. Treasure Hunting and Earth Science in Early Modern Germany*, Leiden—Boston 2011.

Ebel, Wilhelm, "Über Sklavenversicherung und Sklavereiversicherung," *Zeitschrift für die Versicherungswissenschaften* 52 (1963), pp. 207–30.

Ebel, Wilhelm, *Die Hamburger Feuerkontrakte und die Anfänge des deutschen Feuerversicherungsrechts*, Weimar 1936.

Ebert-Weidenfeller, Andreas, *Hamburgisches Kaufmannsrecht im 17. und 18. Jahrhundert. Die Rechtsprechung des Rates und des Reichskammergerichts*, Frankfurt am Main 1992.

Eisenstadt, Shmuel N., "Multiple Modernities," *Daedalus* 129 (2000), pp. 1–29.

Eisenstadt, Shmuel N., "The Axial Age: The Emergence of Transcendental Visions and the Rise of Clerics," *Comparative Civilizations and Multiple Modernities*, part 1, Leiden 2003, pp. 197–217.

Elliott, J.H., *Empires of the Atlantic World. Britain and Spain in America 1492–1830*, New Haven 2006.

Ellisen, Otto A., "Einbeck im 16. Jahrhundert," *Zeitschrift des Harz-Vereins* 27 (1894), pp. 549–60.

Emigh Dylan Riley, Rebecca Jean and Ahmed, Patricia (eds.), *Antecedents of Censuses from Medieval to Nation States. How Societies and States Count*, Heidelberg 2016.

Engelhardt, Ulrich, "Zum Begriff der Glückseligkeit in der kameralistischen Staatslehre des 18. Jahrhunderts (J. H. G. v. Justi)", *Zeitschrift für Historische Forschung* 8 (1981), pp. 37–79.

Ennen, Leonard, *Geschichte der Stadt Köln: meist aus den Quellen des Kölner Stadt-Archivs*, 5 vols., Cologne 1863–80.

Ericson, Richard V., Doyle, Aaron and Barry, Dean, *Insurance as Governance*, Toronto 2003.

Erlin, Matt, *Necessary Luxuries. Books, Literature, and the Culture of Consumption in Germany, 1770–1815*, Ithaca/NY 2014.

Ernst, Wolfgang, "Wandlungen des 'vis maior'-Begriffes in der Entwicklung der römischen Rechtswissenschaft," *Index. Quaderni camerti di studi romanistici* 22 (1994), pp. 293–321.

Ertman, Thomas, *Birth of the Leviathan: Building states and regimes in medieval and early modern Europe*, Cambridge 1997.

Escobar, Arturo, *Encountering Development: The Making and Unmaking of the Third World*, Princeton 1995.

Evers, Adalbert and Nowotny, Helga: *Über den Umgang mit Unsicherheit. Die Entdeckung der Gestaltbarkeit von Gesellschaft*, Frankfurt am Main 1987.

Ewald, François, *Der Vorsorgestaat*, epilogue Ulrich Beck, Frankfurt am Main 1991.

Ewald, François, *L'Etat providence*, Paris 1986.

Ewen, Cecil L., *Lotteries and Sweepstakes: An Historical, Legal, and Ethical Survey of Their Introduction, Suppressing and Re-establishment in the British Isles*, New York 1932.

Eyll, Klara van, *"... genannt Colonia." 150 Jahre Kölnische Feuer-Versicherungs-Gesellschaft AG 1839–1989*, Cologne 1989.

Eyll, Klara van, "Wirtschaftsgeschichte Kölns vom Beginn der Preussischen Zeit bis zur Reichsgründung," in Kellenbenz (ed.), *Zwei Jahrtausende*, vol. 2, pp. 163–266.

Face, Richard D., "Techniques of Business in the Trade between the Fairs of Champagne and the South of Europe in the Twelfth and Thirteenth Centuries," *Economic History Review* 10, 2 (1957), pp. 427–38.

Fairbank, John King, *Trade and Diplomacy on the China Coast: The Opening of the Treaty Ports, 1842–1854*, Cambridge, Mass. 1964.

Faure, David and Köll, Elisabeth, "China: The Indigenization of Insurance," in Borscheid and Haueter, *World Insurance*, pp. 472–494.

Favier, René and Remacle, Claudine (eds.), *Gestion sociale des risques naturels/Gestione sociale dei rischi naturali*, Valle d'Aosta 2007.

Favier, René (ed.), *Les pouvoirs publics face aux risques naturels dans l'histoire*, Grenoble 2002.

Favier, René and Granet-Abisset, Anne-Marie (eds.), *Histoire et mémoire des risques naturels*, Grenoble 2000.

Favier, René and Pfister, Christian (eds.), *Solidarité et assurance. Les sociétés européennes face aux catastrophes (17e -21e siècles)*, Grenoble 2008.

Febvre, Lucien, "Pour l'histoire d'un sentiment: le besoin de sécurité," *Annales* 11 (1956), pp. 244–47.

Fehl, Gerhard, "Perspektivischer Stadtraum. Modellmäßiger Hausbau. Zum planmäßigen Stadt- und Wohnungsbau zwischen 15. und 18. Jahrhundert," in Harlander, Tilman (ed.), *Stadtwohnen. Geschichte, Städtebau, Perspektiven*, Munich 2007, pp. 19–45.

Ferdinand, Christine Y., "Selling it to the Provinces: News and commerce Round Eighteenth-Century Salisbury", in John Brewer and Roy Porter (eds.), Consumption and the World of Goods, London 1993, pp. 393–411.

Festschrift 225 Jahre Braunschweigische Landes-Brandversicherungsanstalt 1754–1979, s.l., s.d. [1979].

Feuersozietät (Berlin, West) (ed.), *260 Jahre Feuersozietät Berlin*, Berlin 1978.

Field, Jacob F., "Charitable Giving and its Distribution to Londoners after the Great Fire, 1666–1676," *Urban History* 38, 1 (2011), pp. 3–23.

Field, Jacob F., *London, Londoners and the Great Fire of 1666. Disaster and Recovery*, Abingdon, New York 2018.

Finkelstein, Andrea, "Nicholas Barbon and the Quality of Infinity," *History of Political Economy* 32 (2000), pp. 83–102.

Flachenecker, Helmut and Kiessling, Rolf (eds.), *Städtelandschaften in Altbayern, Franken und Schwaben. Studien zum Phänomen der Kleinstädte während des Spätmittelalters und der Frühen Neuzeit*, Munich 1999.

Folin, Marco and Preti, Monica (eds.), *Wounded Cities: The Representation of Urban Disasters in European Art (14th-20th Centuries)*, Leiden 2015.

Forstmoser, Peter and Herger, Nikodemus, "Managing Reputational Risk: A Reinsurer's View," *The Geneva Papers on Risk & Insurance* 31 (2006), pp. 409–24.

Foster, Nicholas H. D., "Foundation Myth as Legal Formant. The Medieval Law Merchant and the New *Lex Mercatoria*," *forum historiae iuris* (18 Mar 2005, http://www.forhistiur.de/zitat/0503foster.htm).

Fouquet, Gerhard, "Für eine Kulturgeschichte der Naturkatastrophen. Erdbeben in Basel 1356 und Großfeuer in Frankenberg 1476," in Andreas Ranft and Stephan Selzer (eds.), *Städte aus Trümmern. Katastrophenbewältigung zwischen Antike und Moderne*, Göttingen 2004, pp. 101–31.

Fouquet, Gerhard, *Bauen für die Stadt. Finanzen, Organisation und Arbeit in kommunalen Baubetrieben des Spätmittelalters. Eine vergleichende Studie vornehmlich zwischen den Städten Basel und Marburg*, Cologne 1999.

Françon, Marcel, "Le Langage mathématique de J.-J. Rousseau," *Isis* 40 (1949), pp. 341–44.

Françon, Marcel, "Sur le langage algébrique de Rousseau," *Annales de la Société Jean-Jacques Rousseau* 33 (1953–55), pp. 243–46.

Frangioni, Luciana, *Milano fine Trecento. Il carteggio milanese dell'Archivio Datini di Prato*, 2 vols., Florence 1994.

Frängsmyr, Tore et al. (eds.), *The Quantifying Spirit in the Eighteenth Century*, Berkeley 1990.

Frank, Andre Gunder, *ReOrient: Global Economy in the Asian Age*, Berkeley 1998.

Freist, Dagmar, *Absolutismus*, Darmstadt 2008.

Frentz, Eva-Christine, *Das Hamburgische Admiralitätsgericht (1623–1811). Prozeß und Rechtsprechung*, Frankfurt am Main 1985.

Friedman, Lawrence J. and McGarvie, Mark D. (eds.), *Charity, Philanthropy and Civility in American History*, Cambridge 2003.

Frierson, Cathy, *All Russia is burning! A Cultural History of Fire and Arson in Late Imperial Russia*, Seattle 2002.

Fronmüller, Georg Tobias Christoph, *Chronik der Stadt Fürth*, 2nd ed., Fürth 1887.

Frost, Lionel E. and Jones, Eric L., "The Fire Gap and the Greater Durability of Nineteenth Century Cities," *Planning Perspectives* 4 (1989), pp. 333–47.

Frost, Lionel, "Coping in Their Own Way: Asian Cities and the Problem of Fires," *Urban history* 24, 1 (1997), pp. 5–16.

Fumaroli, Marc and Lecoq, Anne-Marie, *La querelle des Anciens et des Modernes: XVIIe—XVIIIe siècles*, Paris 2001.

Füssel, Marian, *Der Siebenjährige Krieg. Ein Weltkrieg im 18. Jahrhundert*, Munich 2010.

Garcia i Sanz, Arcadi and Ferrer i Mallol, Maria-Teresa: *Assegurances i canvis marítims medievals a Barcelona*, 2 vols., Barcelona 1983.

Garland, David, "'Governmentality' and the Problem of Crime: Foucault, Criminology, Sociology," in *Theoretical Criminology* 1 (1997), pp. 173–214.

Garrioch, David, "Towards a fire history of European cities (late Middle Ages to late nineteenth century)," *Urban History* 46, 2 (2019), pp. 202–224.

Gebhardt, Oskar, *Abriss der Geschichte und Topographie von Markt-Redwitz* [...], Markt-Redwitz 1906.

Geertz, Clifford, *Local Knowledge: Further Essays in Interpretative Anthropology*, New York 1983.

Gerdelan, Louis D., "The Royal Society, Port Royal and the Great Trans-Atlantic Earthquake of 1692," *Studi Storici*, 60, 4 (2019), pp. 845–874.

Gerkens, Jean-François, " 'vis maior' and 'vis cui resisti non potest,' " in Rena van den Bergh (ed.), *Ex iusta causa traditum: Essays in honour of Eric H. Pool*, Unisa 2005, pp. 109–20.

Gerwen, Jacques van and Leeuwen, Marco H.D. van (eds.), *Studies over zekerheidsarrangementen. Risico's, risicobestrijding en verzekeringen in Nederland vanaf de Middeleeuwen*, Amsterdam 1998.

Ghosh, Peter, *Max Weber and the Protestant Ethic: Twin Histories*, Oxford 2014.

Giddens, Anthony, *Modernity and Self-Identity: Self and Society in the Late Modern Age*, Stanford 1991.

Giddens, Anthony, *The Politics of Climate Change*, Cambridge 2009.

Gilbert, Adrian, *The New Jerusalem: Rebuilding London: Christopher Wren, the Great Fire and the Royal Society*, London 2002.

Gisler, Monika and Hürlimann, Katja and Nienhaus, Agnes (eds.), *Naturkatastrophen/Catastrophes naturelles*, Zürich 2003.

Gläser, Jochen, *Neue Begriffe, alte Schwächen. Virtuelle Gemeinschaft*, in Jäckel, Michael and Mai, Manfred (eds.), *Online-Vergesellschaftung? Mediensoziologische Perspektiven auf neue Kommunikationstechnologien*, Wiesbaden 2005, pp. 51–72.

Gläser, Manfred (ed.), *Lübecker Kolloquium zur Stadtarchäologie im Hanseraum*, vol. 1, *Stand, Aufgaben und Perspektiven*, Lübeck 1997.

Glaser, Rüdiger, *Klimageschichte Mitteleuropas. 1200 Jahre Wetter, Klima, Katastrophen*, Darmstadt 2008.

Goedde, Lawrence O., "Naturalism as Convention. Subject, Style, and Artistic Self-Consciousness in Dutch Landscape," in Wayne Franits (ed.), *Looking at Seventeenth-Century Dutch Art*, Cambridge 1997, pp. 129–43.

Goldenbaum, Ursula, "The public discourse of Hermann Samuel Reimarus and Johann Lorenz Schmidt in the *Hamburgische Berichte von gelehrten Sachen* in 1736," in Mulsow, Martin (ed.), *Between Philology and Radical Enlightenment. hermann Samuel Reimarus (1694–1768)*, Leiden 2011, pp. 75–101.

Goose, Nigel, "The Rise and Decline of Philanthropy in Early Modern Colchester: The Unacceptable Face of Mercantilism?" *Social History* 31 (2006), pp. 469–487.

Gorski, Philip S., "Calvinism and Revolution: The Walzer Thesis Reconsidered," in Madsen, Richard et al. (eds.), *Meaning and Modernity*, Berkeley 2001, pp. 78–104.

Gorski, Philip S., "The Little Divergence: The Protestant Reformation and Economic Hegemony in Early Modern Europe," in Swatos and Kaelber, *The Protestant Ethic* Turns 100, pp. 165–90.

Gorski, Philip S., *The Disciplinary Revolution: Calvinism and the Rise of the State in Early Modern Europe*, Chicago 2003.

Gorski, Philip S., *The Protestant Ethic Revisited*, Philadelphia 2011.

Gosden, Peter H.J.H., *Self-Help: Voluntary Associations in Nineteenth-Century Britain*, New York 1974.

Goudsblom, Johan, *Vuur en beschaving* (Amsterdam 1992), i.e. *Fire and civilization*, Harmondsworth 1992 [qtd. German transl. 1995].

Goudsblom, Johann, *Die Entdeckung des Feuers*, Frankfurt am Main 1995.

Graber, Frédéric, "Pour une histoire des formes projet. Du faiseur de projet au projet régulier dans les Travaux Publics (18e-19e siècles)," *Revue d'histoire moderne et contemporaine* 58, no. 3 (2011), pp. 7–33.

Graf, Friedrich Wilhelm, "The German Theological Sources and Protestant Church Politics," in Lehmann and Roth (eds.), *Weber's Protestant Ethic*, pp. 27–49.

Gräf, Holger Th. and Keller, Katrin (eds.), *Städtelandschaft—Réseau urbain—Urban network. Städte im regionalen Kontext in Spätmittelalter und Früher Neuzeit*, Cologne 2004.

Gräf, Holger Th. (ed.), *Kleine Städte im neuzeitlichen Europa*, Berlin 1997.

Gray, Peter and Kendrick, Oliver (eds.), *The Memory of Catastrophe*, Manchester 2004.

Greene, Molly, *A Shared World: Christians and Muslims in the Early Modern Mediterranean*, Princeton 2000.

Grell, Ole Peter and Cunningham, Andrew (eds.), *Health Care and Poor Relief in Protestant Europe 1500–1700*, London 1997.

Grell, Ole Peter, Cunningham, Andrew and Jütte, Robert (eds.), *Health Care and Poor Relief in Eighteenth and Nineteenth Century Protestant Europe*, Aldershot 2002.

Gressel, Hans, *Die Stadt Kyritz. Entwicklung, Verfassung und Wirtschaft bis zur Städteordnung 1808/09*, Berlin 1939.

Groetsch, Ulrich, "The Devil in the Details: The Case of Hermann Samuel Reimarus (1694–1768)," in Martin Mulsow and Asaph Ben-Tov (eds.), *Knowledge and Profanation. Transgressing the Boundaries of Religion in Premodern Scholarship*, Leiden 2019, pp. 273–296.

Groetsch, Ulrich, *Hermann Samuel Reimarus (1694–1768): Classicist, Hebraist, Enlightenment Radical in Disguise*, Leiden 2015.

Grotius, Hugo, *De iure belli ac pacis libri tres* [*orig. 1625*], eds. B. J. A. De Kanter-van Hettinga Tromp, Robert Feenstra, and C.E. Persenaire, Aalen 1993.

Grove, Jean M., Switsur, Roy, "Glacial Geological Evidence for the Medieval Warm Period," *Climatic Change* 26 (1994), pp. 143–69.

Villalba, Ricardo, "Tree-Ring and Glacial Evidence for the Medieval Warm Epoch and the Little Ice Age in Southern South America," *Climatic Change* 26 (1994), 183–197.

Grove, Jean M., *The Little Ice Age*, London 1988.

Grove-Stephensen, F. S., "Slavekassen og de slesvigske søfartsforhold," *Sønderjysk Månedsskrift* 53 (1977), pp. 410–17.

Gründler, Hana, " 'La più orrenda pittura'—Zu einer Ästhetik des Erhabenen in Vasaris *Vite*, in Koppenleitner, Rößler, and Timmann (eds.), *Urbs incensa*, pp. 61–79.

Grunert, Frank, "Die Objektivität des Glücks. Aspekte der Eudämonismusdiskussion in der deutschen Aufklärung." in Frank Grunert and Friedrich Vollhardt (eds.), *Aufklärung als praktische Philosophie*, Tübingen 1998, pp. 351–68.

Guasti, Cesare and Mazzei, Ser Lapo, *Lettere di un notaro a un mercante del secolo XIV*, vol. 2, Florence 1880.

Günther, Johann Arnold, *Ueber die anwendbarsten Vorschläge zur Verhütung von Feuers-Gefahr in Fabriken, und über die Gefahren von Selbst-Entzündungen*, Hamburg 1797.

Gunthert, André, "Esthétique de l'occasion. La naissance de la photographie instantanée comme genre," *Ètudes Photographiques* 9 (2001), pp. 64–87.

Haag, Norbert, *Predigt und Gesellschaft. Die lutherische Orthodoxie in Ulm 1640–1740*, Mainz 1992.

Haar, Ingo, "Ostforschung im Nationalsozialismus. Die Genesis der Endlösung aus dem Geiste der Wissenschaften," in Rainer Mackensen (ed.), *Bevölkerungslehre und Bevölkerungspolitik im "Dritten Reich,"* Opladen 2004, pp. 219–40.

Haar, Ingo, *Historiker im Nationalsozialismus. Die deutsche Geschichtswissenschaft und der Volkstumskampf im Osten*, Göttingen 2000.

Hacking, Ian, *The Emergence of Probability: A Philosophical Study of Early Ideas about Probability, Induction and Statistical Inference*, Cambridge 1976.

Hackmann, Jörg, "Der Kampf um die Weichsel. Die deutsche Ostforschung in Danzig 1918–1945," *Zapiski Historyczne* 58 (1993), pp. 37–57.

Hagena, Wilhelm, *Die Ansichten der deutschen Kameralisten des 18. Jahrhunderts über das Versicherungswesen*, PhD thesis University of Erlangen, 1910.

Hägerstrand, Torsten, *Time and Culture*, Berlin 1985.

Hahn, Andreas, *"…dat zy de aanschouwers schynen te willen aansprecken"—Untersuchungen zur Rolle des Betrachters in der Niederländischen Malerei des 17. Jahrhunderts*, Munich 1996.

Hald, Anders, *A History of Probability and Statistics and Their Applications before 1750*, New York 1990.

Hale, Matthew, *The Primitive Origination of Mankind Considered and Examined According to the Light of Nature*, London 1677.

Halpérin, Jean, "La Notion de sécurité dans l'histoire économique et sociale," in *Revue d'Histoire Économique et Sociale* 30 (1952), pp. 7–25.

Halpérin, Jean, *Les Assurances en Suisse et dans le monde. Leur rôle dans l'évolution économique et social*, Neuchâtel 1946.

Hamilton, John T., *Security—Politics, Humanity and the Philology of Care*, Princeton 2013.

Hamilton, Walter, "A Geographical, Statistical, and Historical Description of Hindostan and The Adjacent Countries [1820]," in *Calcutta in the 19th Century (Company's Days)*, Calcutta 1989, pp. 220–44.

Hanselaer, Johan, "Aeneas' vlucht uit het brandende Troje. Virtuositeit en Allegorie in de zuidnederlandse schilderkunst van de 16de en van de eerste helft van de 17de eeuw," *Gentse bijdragen tot de kunstgeschiedenis en oudheidkunde* 28 (1989), pp. 105–18.

Harms, Wolfgang (ed.), *Deutsche illustrierte Flugblätter des 16. und 17. Jahrhunderts*, 7 vols., Stuttgart 1980–2005.

Harris, John R., "Movements of Technology between Britain and Europe in the Eighteenth Century," in David J. Jeremy (ed.), *International Technology Transfer: Europe, Japan, and the USA, 1700–1914*, Aldershot 1991, pp. 9–30.

Harris, Mark, " 'Inky Blots and Rotten Parchment Bonds': London, Charity Briefs and the Guildhall Library," *Historical Research* 66 (1993), pp. 98–110.

Härter, Karl (ed.), *Repertorium der Policeyordnungen der Frühen Neuzeit,* vol. 6, *Reichsstädte 2: Köln*, 2 vols., ed. Klaus Militzer, Frankfurt am Main 2005.

Hart, Emma, "A British Atlantic World of Advertising? Colonial American 'For Sale' Notices in Comparative Context, *American Periodicals* 24, 2 (2014), pp. 120–127.

Hayez, Jerôme (ed.), "Le carteggio Datini et les correspondances pratiques des XIVe-XVIe siècles ", special issue of *Mélanges de l'École française de Rome. Moyen Âge* 117, 1 (2005).

Hayez, Jerôme, "La gestion d'une relation épistolaire dans les milieux d'affaires toscans à la fin du Moyen Âge," in *La circulation des nouvelles au Moyen Âge*, ed. Société des historiens médiévistes de l'enseignement supérieur, Paris/Rome 1994, pp. 63–83.

Heege, Andreas, *Einbeck 1540. Brandstiftung! Der Einbecker Stadbrand vom 26. Juli 1540. Archäologischer Befund und politische Hintergründe*, Einbeck 2005.

Heering, Peter and Hochadel, Oliver (eds.), *Playing with Fire. Histories of the Lightning Rod*, Philadelphia 2009.

Heikamp-Wagner, Helga, *Jan van der Heyden*, Amsterdam 1972.

Heilbron, John L., *Electricity in the 17th and 18th Centuries. A Study of Early Modern Physics*, Mineola 1999.

Heilbron, John L., *Weighing Imponderables and Other Quantitative Science around 1800*, Berkeley 1993.

Heiss, Sonja, *Die Institutionalisierung der deutschen Lebensversicherung*, Berlin 2006.

Helama, Samuli, Meriläinen, Jouko, and Tuomenvirta, Heikki, "Multicentennial Megadrought in Northern Europe Coincided with a Global El Niño-Southern Oscillation Drought Pattern during the Medieval Climate Anomaly," *Geology* 37, 2 (2009), pp. 175–78.

Helmer, Georg, *Die Gegenseitigkeitsversicherung. Eine Schöpfung germanischer Völker*, Vienna 1937.

Helmer, Georg, *Die Geschichte der privaten Feuerversicherung in den Herzogtümern Schleswig und Holstein, insbesondere die Geschichte der Entstehung und Entwicklung der Brandgenossenschaften oder "Brandgilden" in Schleswig-Holstein*, Berlin 1925.

Hennerdal, P. and Meinild Nielsen, M., "A Multiscalar Approach for Identifying Clusters and Segregation Patterns That Avoids the Modifiable Areal Unit Problem," in: *Annals of the American Association of Geographers* 107, 3 (2017), 555–574.

Hennock, E. Peter, "Social Policy in the Bismarck Era: A Progress Report," *German History* 21 (2003), pp. 229–38.

Hennock, E. Peter, *The Origin of the Welfare State in England and Germany, 1850–1914*, Cambridge 2007.

Henriot, Christian, " 'Little Japan' in Shanghai: an insulated community, 1875–1945," in Bickers and Henriot, *New Frontiers*, p. 148.

Hersche, Peter, "Max Weber, Italien und der Katholizismus," *Quellen und Forschungen aus Italienischen Archiven und Bibliotheken* 76 (1996), pp. 362–82.

Hildebrand, Bernhard, "Aalen im Jahre 1634. Der große Stadtbrand und seine Ursachen," *Aalener Jahrbuch* (1982), pp. 67–81.

Hildebrand, Bernhard, "Die Ereignisse um 1634. Im allgemeinen und unter besonderer Berücksichtigung der Schlacht bei Nördlingen und des Schicksals der Stadt Aalen," *Nordschwaben. Zeitschrift für Landschaft, Geschichte, Kultur und Zeitgeschehen* 12, 3 (1984), pp. 123–31.

Himmelfarb, Gertrude, *The Idea of Poverty: England in the Early Industrial Age*, London 1985.

Hing, Lee Kam, *A Matter of Risk. Insurance in Malaysia, 1826–1990*, Singapore 2012.

Hintze, Otto, *Der Commissarius und seine Bedeutung in der allgemeinen Verwaltungsgeschichte* (1910), reprinted in Otto Hintze, *Staat und Verfassung. Gesammelte Abhandlungen zur allgemeinen Verfassungsgeschichte*, ed. Gerhard Oestreich, Göttingen 1970, pp. 242–74.

Hippel, Wolfgang von, "Von der Mühsal der Modernisierung—die Einführung des metrischen Systems in der linksrheinischen Pfalz", Aus südwestdeutscher Geschichte: FS Hans-Martin Maurer, ed. Wolfgang Schmierer et al., Stuttgart 1994, pp. 580–594.

Hirsch, Eike Christian, *Der berühmte Herr Leibniz. Eine Biographie*, Munich 2007.

Hochadel, Oliver, " 'Hier haben Wetterableiter unter den Augsburger Gelehrten eine kleine Revolution gemacht': Die Debatte um die Einführung der Blitzableiter in Augsburg, 1783–91," *Zeitschrift des Historischen Vereins für Schwaben* 92 (1999/2000), pp. 139–64.

Hoffmann, Alfred (ed.), *Österreichisches Städtebuch*, Vienna 1968–1999: vol. 1, *Die Städte Oberösterreichs*, ed. Herbert Knittler (1968); vol. 2, *Die Städte des Burgenlandes*, ed. Renõ Deák, 2nd revised ed. (1996); vol. 3, *Die Städte Vorarlbergs*, ed. Franz Baltzarek and Johannes Pradel (1973); vol. 4, *Die Städte Niederösterreichs*, 3 subvols., ed. Friederike Goldmann (1976–1988); vol. 5, *Die Städte Tirols*, 5.1, *Bundesland Tirol*, 5.2 *Südtirol*, ed. Franz-Heinz Hye (1980/2001); vol. 6, *Die Städte der Steiermark*, 2 subvols., ed. Friederike Goldmann (1990/1995); vol. 7, *Die Stadt Wien*, ed. Peter Csendes and Ferdinand Opll (1999) still to be published are: vol. 6.1, *Graz*, vol. 6.2, *Die Städte der Steiermark: A-H*; vol. 8, *Kärnten*.

Hoffmann, Friedrich Hermann, *Die Kunstdenkmäler von Oberpfalz und Regensburg*, Bezirksamt Parsberg 4, Munich 1906.

Hohrath, Daniel, "Spätbarocke Kriegspraxis und aufgeklärte Kriegswissenschaften. Neue Forschungen und Perspektiven zu Krieg und Militär im 'Zeitalter der Aufklärung,'" *Aufklärung* 12 (2000), pp. 5–47.

Holenstein, André P., *"Gute Policey" und lokale Gesellschaft im Staat des Ancien Régime. Das Fallbeispiel der Markgrafschaft Baden(-Durlach)*, Tübingen 2003.

Holenstein, André, " 'Gute Policey' und lokale Gesellschaft. Erfahrung als Kategorie im Verwaltungshandeln des 18. Jahrhunderts," in *"Erfahrung" als Kategorie der Frühneuzeitgeschichte*, Munich 2001, pp. 433–50.

Holtz, Sabine, *Theologie und Alltag. Lehre und Leben in den Predigten der Tübinger Theologen 1550–1750*, Tübingen 1993.

Home, Robert, "British colonial civic improvement in the early twentieth century: E. P. Richards in Madras, Calcutta, and Singapure," in: *Planning Perspectives* 31, 4 (2016), pp. 635–644.

Home, Robert, *Of Planting and Planning: The Making of British Colonial Cities*, London 1997.

Hook, Brian, *Shanghai and the Yangtze Delta: A City Reborn*, Oxford 1998.

Horsefield, John Keith, *British Monetary Experiments 1650–1710*, London 1960.

Houssaye Michienzi, Ingrid, *Datini, Majorque et le Maghreb (14e-15e siècles). Réseaux, espaces méditerranéens et stratégies marchandes*, Leiden 2013.

Hrdlička, Josef, "Beten oder löschen? Feuer in den böhmischen Städten der Frühen Neuzeit," in Lorenzetti, Luigi and Giannò, Vanessa (eds.), *Al fuoco! Usi, rischi e rappresentazioni dell'incendio dal Medioevo al XX secolo*. Dibattiti & documenti 13/Studi di storia alpina/Studies on Alpine History 6., Milan and Lugano 2010, pp. 91–108.

Hudson, Wayne, *Enlightenment and Modernity. The English Deists and Reform*, London 2009.

Hufton, Owen, *The Poor of Eighteenth-Century France 1750–1789*, Oxford 1974.

Huggins, William, "Sketches in India, Treating on Subjects connected with the Government; Civil and Military Establishments; Characters of the European, and Customs of the Native Inhabitants [1824]," in *Calcutta in the 19th Century*, pp. 411–26.

Hughes, Malcolm K. and Diaz, Henry F., "Was There a 'Medieval Warm Period,' and If So, Where and When?," *Climatic Change* 26 (1994), pp. 109–42.

Hughes, Thomas P., *Rescuing Prometheus*, New York 2000.

Humpert, Klaus and Schenk, Martin, *Entdeckung der mittelalterlichen Stadtplanung. Das Ende vom Mythos der "gewachsenen Stadt"*, Stuttgart 2001.

Hunt, B. G., "The Medieval Warm Period, the Little Ice Age and Simulated Climatic Variability," *Climate Dynamics* 27 (2006), pp. 677–94.

Iacob, Dan Dumitru, "Incendiul din Iaşi, din 25 Iunie 1833", *Historia Urbana* 26 (2018), pp. 67–96.

İnalcik, H., "Istanbul," *The Encyclopaedia of Islam* 4 (1978), pp. 224–48.

Index deutschsprachiger Zeitschriften 1750–1815, ed. Klaus Schmidt, Hildesheim 1997.

Innes, Joanna, "The State and the Poor: Eighteenth-Century England in European Perspectives," in Brewer, John and Hellmuth, Eckhart (eds.), *Rethinking Leviathan: The Eighteenth-Century State in Britain and Germany*, London 1999, pp. 225–281.

Irwin, Julia F., "Raging Rivers and Propaganda Weevils: Transnational Disaster Relief, Cold War Politics, and the 1954 Danube and Elbe Floods," *Diplomatic History* 40, 5 (2016), pp. 893–921.

Isenmann, Moritz (ed.), *Merkantilismus. Wiederaufnahme einer Debatte*, Stuttgart 2014.

Itkonen, Esa, *Analogy as Structure and Process: Approaches in Linguistics, Cognitive Psychology and Philosophy of Science*, Amsterdam 2005.

Jaeckel, Ralph, '*… daß so wohl der Regularität nach, und zum Zierath der Strassen, als vor Feuers-Gefahr sicher gebauet werde …'—Die Erneuerung der Städte in Brandenburg-Preußen im Zeitalter des Absolutismus zwischen 1640 und 1806/11. Untersuchungen zu den städtebaulichen Retablissements und Etablissements in Städten der südlichen und östlichen Kurmark sowie der Neumark Brandenburg*, PhD thesis, Humboldt University Berlin 1999.

Jaeggli-Leepin, Alvin, *Burgen und Schlösser in der Schweiz. Nach alten Vorlagen*, Frankfurt am Main 1962.

Jakob, Andreas, " 'Der Ort stieg aus seiner Asche viel schöner empor'—Der Brand der Altstadt Erlangen am 14. August 1706 und ihr Wiederaufbau als moderne Planstadt bis 1712," *Erlanger Bausteine zur fränkischen Heimatforschung* 51 (2006), pp. 9–47.

Jarrige, François and Reynaud, Bénedicte, "Les usines en feu. L'industrialisation au risque des incendies dans le textile (France 1830–1870)," *Le mouvement social* 249 (2014), pp. 141–62.

Jaspers, Karl, *Vom Ursprung und Ziel der Geschichte*, Munich 1949.

Jitschin, Adrian, *Über die Versicherung der Inder. Indische Lebensversicherung 1874 bis 1988*, PhD thesis, University of Marburg 2011.

Johanek, Peter and Post, Franz-Joseph (ed.), *Vielerlei Städte. Der Stadtbegriff*. Cologne 2004.

Johanek, Peter, "Landesherrliche Städte—kleine Städte. Umrisse eines europäischen Phänomens," in Jürgen Treffeisen (ed.), *Landesherrliche Städte in Südwestdeutschland*, Sigmaringen 1994, pp. 9–25.

Johnson, Linda Cooke, *Shanghai: From Market Town to Treaty Port*, 1074–1858, Stanford 1995.

Jonas, Hans, *Das Prinzip Verantwortung. Versuch einer Ethik für die technologische Zivilisation*, Frankfurt am Main 1979.

Jones, Colin, "The Military Revolution and the Professionalisation of the French Army under the Ancien Régime," in Rogers, *The Military Revolution Debate*, pp. 149–67.

Jones, Eric L., Porter, Stephen and Turner, Michael, *A Gazetteer of English Urban Fire Disasters, 1500–1900*, Norwich 1984.

Jones, James F., "Prolegomena to a History of Happiness in the Eighteenth Century," *The French American Review* 6 (1982), pp. 283–95.

Jongh, Eddy de, "Realism and Seeming Realism in Seventeenth-Century Dutch Painting [1971]," in Franits, *Looking at Seventeenth-Century Dutch Art*, pp. 21–56.

Jongh, Eddy de, *Questions of Meaning: Theme and Motif in Dutch Seventeenth-Century Painting,* trans. and ed. Michael Hoyle, Leiden 2000.

Jordan, W. K., *Philanthropy in England, 1480–1660: A Study of the Changing Patterns of English Social Aspirations*, London 1959.

Jordan, W. K., *The Charities of London, 1480–1660*, London 1961.

Jung-Stilling, Johann Heinrich, *Lehrbuch der Staats-Polizey-Wissenschaft*, Leipzig 1788, reprint 1970.

Jutikkala, Eino, "Great Fires of Finnish Cities," in Körner, Martin (ed.), *Stadtzerstörung und Wiederaufbau*, vol 1, *Zerstörung durch Erdbeben, Feuer und Wasser*, Bern 1999, pp. 239–54.

Kadens, Emily, "The Medieval Law Merchant: The Tyranny of a Construct," *Journal of Legal Analysis* 7, 2 (2015), pp. 251–89.

Kadens, Emily, "The Myth of the Customary Law Merchant," *Texas Law Review* 90 (2012), pp. 1153–1206.

Kaelber, Lutz, "Weber's lacuna: Medieval Religion and the roots of rationalization," *Journal of the History of Ideas* 57, 3 (1996), pp. 465–85.

Kapse, S. R., *Police Administration in Bombay, 1600–1865*, Bombay/Delhi/Nagpur 1987.

Kaufmann, Franz-Xaver, *Sicherheit als soziologisches und sozialpolitisches Problem. Untersuchung zu einer Wertidee hochdifferenzierter Gesellschaften*, 2nd ed., Stuttgart 1973.

Kaufman-Osborn, Timothy V., *Creatures of Prometheus: Gender and the Politics of Technology*, Lanham 1997.

Kauß, Dieter, "Der Göppinger Stadtbrand vom 25. August 1782 und seine Folgen," *Hohenstaufen* 13 (1986), pp. 135–55.

Kaviraj, Sudipta, "Modernity and Politics in India," *Daedalus* 129 (2000), pp. 137–62.

Keene, Derek, "Fire in London: Destruction and Reconstruction, A.D. 982–1676," in Körner, *Wiederaufbau*, pp. 187–211.

Keller, Katrin, "Armut und Tod im alten Handwerk: Formen sozialer Sicherung im sächsischen Zunftwesen des 17. und 18. Jahrhunderts," in Johanek, Peter (ed.), *Städtisches Gesundheits- und Fürsorgewesen vor 1800*, Cologne 2000, pp. 199–223.

Keller, Katrin, *Kleinstädte in Kursachsen. Wandlungen einer Städtelandschaft zwischen Dreissigjährigem Krieg und Industrialisierung*, Vienna 2001.

Keller, Vera, *Knowledge and the Public Interest, 1575–1725*, Cambridge 2015.

Kempe, Michael and Rohr, Christian (eds.), *Coping with the Unexpected. Natural Disasters and Their Perception*, Strond 2003.

Keneley, Monica and Verhoef, Grietjie, "Establishing Insurance Markets in Settler Economies: A Comparison of Australian and South Africa Insurance Markets, 1820–1910," *African Historical Review* 47, 1 (2015), pp. 76–105.

Keneley, Monica, "Business Strategies and Conditions of Uncertainty: The Rise of Mutual Life Insurers in Colonial Australia," in Pearson and Yoneyama (eds.), *Corporate Forms*, pp. 169–92.

Keneley, Monica, "The Origins of Formal Collusion in Australian Fire Insurance, 1870–1920," *Australian Economic History* 42 (2002), pp. 54–76.

Keuscher, Jacob, *Die Feuerbrände der Stadt Bingen in den Jahren 1403, 1490, 1540, 1689 und 1850. Ein Beitrag zur Geschichte dieser Stadt*, Dannstadt 1853.

Keussen, Hermann, "Hilfe der Stadt Köln nach dem Aachener Stadtbrande 1656," *Zeitschrift des Aachener Geschichtsvereins* 51 (1929), p. 392.

Keynes, Geoffrey, *A Bibliography of Sir William Petty, F.R.S., and of Observations on the Bills of Mortality by John Graunt, F.R.S.*, Oxford 1971.

Keyser, Erich (ed.), *Deutsches Städtebuch. Handbuch städtischer Geschichte*, 5 vols., Stuttgart 1939–1974: vol. 1, *Nordostdeutschland* (1939); vol. 2, *Mitteldeutschland* (1942); vol. 3, *Nordwest-Deutschland*, vol. 3.1, *Niedersachsen/Bremen. Niedersächsisches Städtebuch* (1952); vol. 3.2, *Westfalen. Westfälisches Städtebuch* (1954); vol 3.3, *Landschaftsverband Rheinland. Rheinisches Städtebuch* (1956); vol. 4, *Südwest-Deutschland* vol. 4.1, *Land Hessen. Hessisches Städtebuch* (1957); vol. 4.2.1, *Badisches Städtebuch* (1959); vol. 4.2.2, *Württembergisches Städtebuch* (1961); vol. 4.3, *Land Rheinland-Pfalz und Saarland. Städtebuch Rheinland-Pfalz und Saarland* (1964); vol. 5, *Bayern. Bayerisches Städtebuch*, vol. 5.1, *Unter-, Mittel- und Oberfranken*, (1971); vol 5.2 *Ober-, Niederbayern, Oberpfalz und Schwaben* (1974).

Keyser, Erich, "Neue deutsche Städteforschung," in StB vol. 1, p. 1–10.

Keyser, Erich, "Neue Forschungen zur Geschichte der deutschen Städte," *Blätter für deutsche Landesgeschichte* 83 (1937), pp. 46–53.

Kidambi, Prashant, *The Making of an Indian Metropolis. Colonial Governance and Public Culture in Bombay, 1890–1920*, Aldershot 2007.

Kiesselbach, Arnold G., *Die wirtschafts- und rechtsgeschichtliche Entwickelung der Seeversicherung in Hamburg*, Hamburg 1901.

Kinsbruner, Jay, *The Colonial Spanish-American City: Urban Life in the Age of Atlantic Capitalism*, Austin 2005.

Kissling, Peter, "Reichsstädtische und territoriale Policeyen. Ein Vergleich der Reichsstadt und der Grafschaft Kempten im 16. Jahrhundert," in Peter Blickle, Peter Kissling, and Heinrich Richard Schmidt (eds.), *Gute Policey als Politik im 16. Jahrhundert. Die Entstehung des öffentlichen Raumes in Oberdeutschland*, Frankfurt am Main 2003, pp. 229–83.

Kittsteiner, Heinz D., *Die Entstehung des modernen Gewissens*, Frankfurt am Main 1991.

Klein, Dietrich, *Hermann Samuel Reimarus (1694–1768). Das theologische Werk*, Tübingen 2009.

Klueting, Harm, *Die Lehre von der Macht der Staaten. Das außenpolitische Machtproblem in der "politischen Wissenschaft" und in der praktischen Politik im 18. Jahrhundert*, Berlin 1986.

Knapp, Rebecca, *Eine Wissensgeschichte der Feuersicherheit. Kommunikation über Brandbekämpfung zwischen 1600 und 1800*, PhD Diss., Ruhr University Bochum 2013.

Knight, Frank H., *Risk, Uncertainty and Profit*, Boston 1921.

Koch, Jürgen W., *Der Hamburger Spritzenmeister und Mechaniker Johann Georg Repsold (1770–1830), ein Beispiel für die Feinmechanik im norddeutschen Raum zu Beginn des 19. Jahrhunderts*, Hamburg 2001.

Koch, Peter "Versicherungswesen," in Adalbert Erler (ed.), *Handwörterbuch zur deutschen Rechtsgeschichte*, vol. 5, Berlin 1998, pp. 815–26.

Koch, Peter, "Berlins Bedeutung für die Versicherungswirtschaft," *Versicherungswirtschaft* 46 (1991), pp. 236–50.

Koch, Peter, "Der Einfluss der Reformation auf die Entwicklung des Versicherungswesens [1983]," in *Beiträge zur Geschichte des deutschen Versicherungswesens*, vol. 1, Karlsruhe 1995, pp. 171–81.

Koch, Peter, *Geschichte der Versicherungswissenschaft in Deutschland*, Karlsruhe 1998.

Koch, Peter, *Pioniere des Versicherungsgedankens. 300 Jahre Versicherungsgeschichte in Lebensbildern, 1550–1850*, Wiesbaden 1968.

Koenen, Ulrike, "Die brennende Stadt. Sodom in spätantiken Darstellungen," *Actes du XIe Congrès international d'archéologie chrétienne* [1986], *Collection de l'Ecole Française de Rome* 123, 3, Rome 1989, pp. 1355–70.

Konrad, Franz, *Der Kirchenkasten zu Wallau an der Lahn. Bericht über die Jahresrechnungen der Kirchenkästen von Wallau/Lahn und Weifenbach von 1588 bis 1832*, Wallau/Lahn 2004.

Konvitz, Josef W., *Cartography in France, 1660–1848: science, engineering, and statecraft*, Chicago 1987.

Kopitzsch, Franklin, *Grundzüge einer Sozialgeschichte der Aufklärung in Hamburg und Altona*, 2 vol.,Hamburg: 1982.

Koppenleitner, Fionie, Rößler, Hole and Thimann, Michael (eds.), *Urbs incensa. Ästhetische Transformationen der brennenden Stadt in der Frühen Neuzeit*, Munich 2011.

Krämer, Margaretha, "Der Brand von Troja und von Rom—Sieben Versionen von Pieter Schoubroeck," in Hürkey, Edgar J. (ed.), *Kunst, Kommerz, Glaubenskampf. Frankenthal um 1600*, Worms 1995, pp. 97–102.

Krapivin, Vladimir F. and Varotsos, Costas A., *Globalization and Sustainable Development. Environmental Agendas*, Berlin 2007.

Kraus, Thomas R., "Der Aachener Stadtbrand vom 2. Mai 1656 und seine Folgen," *Zeitschrift des Aachener Geschichtsvereins* 109 (2007), pp. 35–99.

Kreager, Philip, "Death and Method: The Rhetorical Space of 17th-Century Vital Measurement," in Eileen Magnello and Anne Hardy (eds.), *The Road to Medical Statistics*, Amsterdam 2002, pp. 1–36.

Kreager, Philip, "New Light on Graunt," in *Population Studies* 42 (1988), pp. 129–40.

Kretzschmar, Johannes, "Zur Geschichte Herfords im 30jährigen Kriege. Mit einem Plane der Stadt von 1638," *Westfälische Zeitschrift* 58 (1900), pp. 5–29.

Krieger, Karl-Friedrich, "Die Entwicklung des Seerechts im Mittelmeerraum von der Antike bis zum Consolat de Mar," *Jahrbuch für internationales Recht* 16 (1973), pp. 179–208.

Kruft, Hanno-Walter, *Städte in Utopia. Die Idealstadt vom 15. bis zum 18. Jahrhundert zwischen Staatsutopie und Wirklichkeit*, Munich 1989.

Krüger, Kersten, "Die Idealstadt der frühen Neuzeit, insbesondere in Nordeuropa," in Frank Braun (ed.), *Städtesystem und Urbanisierung im Ostseeraum in der frühen Neuzeit. Wirtschaft, Baukultur und historische Informationssysteme*, Münster 2004, pp. 11–47.

Krüger, Kersten, *Finanzstaat Hessen 1500–1567. Staatsbildung im Übergang vom Domänenstaat zum Steuerstaat*, Marburg 1980.

Kuča, Karel, *Města a městečka v Čechách, na Moravě a ve Slezsku* [*Cities and small towns in Bohemia, Moravia and Silesia*], vol. 1–8, Prague 1996–2011.

Kunitz, Claudia, *Aachen nach dem Stadtbrand. Wege und Maßnahmen zum Wiederaufbau 1656–1661*, MA thesis, RWTH Aachen University 1994.

Kunreuther, Howard, "The Failure of EIL Coverage: Gridlock in Environmental Insurance," *Environment* 29, 1 (1987), pp. 18–35.

Kuter, Mikhail, Gurskaya, Maria, Papakchian, Armina, "The Early Practice of Maritime Insurance Accounting in the First Proprietorship of Francesco di Marco Datini in Pisa in 1382–1406", in *Digital Science 2019, Advances in Intelligent Systems and Computing 1114*, eds. *Antipova Tatiana* and Alvaro Rocha, Heidelberg 2020, pp. 299–313.

Kuter, Mikhail, Gurskaya, Marina, Aleinikov, Dmitriy, "The earlier synthetic balance sheet of Datini's Company in Avignon (1410): The combined accounting system", *De Computis. Revista Española de Historia de la Contabilidad* 17, 1 (2020), pp. 7–35.

La Compagnie d'Assurances générales contre l'incendie 1819–1919. Cent ans d'Assurance et de Sagesse Financière, Paris 1920.

La Torre, Antonio, *L'assicurazione nella storia delle idee. La risposta giuridica al bisogno di sicurezza economica. Ieri e oggi*, 2nd ed., Milan 2000.

Ladwig-Winters, Simone, *300 Jahre Feuersozietät Berlin-Brandenburg*, Berlin 2018.

Lamarre, Christine, "Les incendies de villes dans les géographies du XVIIIe siècle," in Lassus and Vion-Delphin, *Les hommes*, pp. 283–92.

Lamb, Hubert H., "The Early Medieval Warm Epoch and Its Sequel," *Palaeogeography, Palaeoclimatology, Palaeoecology* 1 (1965), pp. 13–37.

Lammert, Gottfried, *Geschichte der Seuchen, Hungers- und Kriegsnoth zur Zeit des Dreissigjährigen Krieges*, Wiesbaden 1890.

Lampart, Fabian et al. (eds.), *Der Zweite Dreißigjährige Krieg. Deutungskämpfe in der Literatur der Moderne*, Baden-Baden 2019.

Lancilotti, S. and Tassoni, Alessandro, "Pensieri diversi [1620], lib. X: Ingegni antichi e moderni," in *Prose politiche e morali*, by S. Lancilotti and Alessandro Tassoni, ed. Giorgio Rossi, vol. 2, Bari 1930, pp. 298–340.

Landes, David, *The Unbound Prometheus*, 2nd ed., Cambridge 2005.

Landwehr, Achim, *Policey im Alltag. Die Implementation frühneuzeitlicher Policeyordnungen in Leonberg*, Frankfurt am Main 2001.

Landwehr, Götz, *Das Seerecht der Hanse (1365–1614). Vom Schiffordnungsrecht zum Seehandelsrecht*, Hamburg 2003.

Landwehr, Götz, *Die Haverei in den mittelalterlichen deutschen Seerechtsquellen*, Göttingen 1985.

Götz Landwehr, "Zur Begriffsgeschichte der Haverei vom 16. bis 18. Jahrhundert," in Jayme, Erik et al. (eds.), *Festschrift für Hubert Niederländer zum 70. Geburtstag*, Heidelberg 1991, pp. 57–69.

Langenhuyzen, Ton, "Zekerheid en Brand vóór 1800," in Gerwen, van Leeuwen (eds.), *Studies*, pp. 203–222.

Langer, Herbert, "Handel und Bildung in pommerschen Seestädten im 16. bis 17. Jahrhundert," in *Stadt und Literatur im deutschen Sprachraum der Frühen Neuzeit*, eds. Klaus Garber et al., vol. 1, Tübingen 1998, pp. 729–736.

Larguier, Gilbert, "La solidarité face aux risques naturels en Languedoc (XVIIe-XVIIIe siècles)," in Favier, René and Pfister, Christian (eds.), *Solidarité et assurance*, pp. 37–56.

Laslett, Peter, "Natural and Political Observations on the Population of Late Seventeenth-Century England: on the Work of Gregory King and John Graunt," in Kevin Schurer and Tom Arkell (eds.), *Surveying the People: The Interpretation and Use of Document Sources for the Study of Population in the Later Seventeenth Century*, Oxford 1992, pp. 6–30.

Lassus, François and Vion-Delphin, François (eds.), *Les hommes et le feu de l'Antiquité à nos jours. Du feu mythique et bienfaiteur au feu dévastateur*, Besançon 2007.

Lauring, Kåre, *Byen brænder. Københavns brand 1728*, Copenhagen 2003.

Le Bras, Hervé, *Naissance de la Mortalité: l'origine politique de la statistique et de la démographie*, Paris 2000.

Le Goff, Jacques, *Pour un autre Moyen Âge. Temps, travail et culture en Occident*, Paris 1978.

Le Roy Ladurie, Emmanuel, *Histoire humaine et comparée du climat*, vol. 1, *Canicules et glaciers, XIIIe-XVIIIe siècles*, Paris 2006.

Leeuwen, Marco H. D. van, *De rijke Republiek: gilden, assuradeurs en armenzorg 1500–1800* (= *Zoeken naar zekerheid—Risico's, preventie, verzekeringen en andere zekerheidsregelingen in Nederland 1500–2000*, eds. Jacques van Gerwen and Marco H. D. van Leeuwen, part 1), The Hague 2000.

Lehmann, Hartmut (ed.), *Max Webers Religionssoziologie in interkultureller Perspektive*, Göttingen 2003.

Lehmann, Hartmut and Roth, Guenther (eds.), *Weber's Protestant Ethic: Origins, Evidence, Contexts*, Cambridge 1993.

Lehmann, Hartmut, *Max Webers "Protestantische Ethik." Beiträge aus der Sicht eines Historikers*, Göttingen 1996.

Lehmann, Jörn, *Die Stadt Liebenwalde. Ein Streifzug durch die Geschichte der Ortsteile Freienhagen, Hammer, Kreuzbruch, Liebenthal, Liebenwalde und Neuholland*, Karwe 2004.

Lehmann, Mathias, *Der Dreißigjährige Krieg im Musiktheater während der NS-Zeit* [...], Hamburg 2004.

Lehmann, Rudolf, *Geschichte der Niederlausitz*, Berlin 1963.

Lenci, Marco, "Riscatti di Schiavi Cristiani dal Maghreb. La Compagnia della SS. Pietà di Lucca (Secoli XVII-XIX)," *Società e Storia* 31 (1986), pp. 53–80.

Leonard, Adrian B., "Underwriting British Trade to India and China, 1780–1835," *The Historical Journal* 55, 4 (2012), pp. 983–1006.

Leonard, Adrian B. (ed.), *Marine Insurance. Origins and Institutions, 1300–1850*, New York 2016.

Leonardi, Monica, "Der Vesuv und die phlegräischne Felder. Feuer, Wasser, schwefel und Gestein", in *'Kennst Du das Land'. Italienbilder der Goethezeit*, eds. Frank Büttner, Herbert W. Rott, Cologne 2005, pp. 255–271.

Lessius, Leonardus, *De iustitia et iure ceterisque virtutibus cardinalibus libri quatuor*, Lyon 1653 [orig.1605].

Levesque, Catherine, *Journey through Landscape in Seventeenth-Century Holland*, Pennsylvania 1994.

Leybourn, William, *A Platform for Purchasers, Guide for Builders, Mate for Measurers*, London 1668.

Lichtblau, Klaus, "The Protestant Ethic versus the 'New Ethic,'" in Lehmann and Roth (eds.), *Weber's Protestant Ethic*, pp. 179–93.

Licoppe, Christian, *La formation de la pratique scientifique. Le discours de l'expérience en France et en Angleterre (1630–1820)*, Paris 1996.

Lieshout, Carry van, "'The Most Valuable Menas of Extinguishing the Destroying Fires': Fire-fighting and the London Water Companies since the Long Eighteenth Century", *The London Journal* 42, 1 (2017), pp. 53–69.

Lignereux, Yann, *Lyon et le roi. De la ,bonne ville' à l'absolutisme municipal (1594–1654)*, Seyssel 2003.

Lilja, Sven, "Wooden Towns on Fire. Fire Destruction and Human Reconstruction of Swedish Towns Prior to 1800," in Körner, *Stadtzerstörung*, pp. 255–75.

Lilja, Sven, *Städernas folkmängd och tillväxt. Sverige (med Finland) ca 1570-tal till 1810-tal (Historisk tätortsstatistik 2.)*, Stockholm 1996.

Limmer, Karl, *Entwurf einer urkundlich-pragmatischen Geschichte der Lausitzen*, Ronneburg 1839.

Lindemann, Margot, *Nachrichtenübermittlung durch Kaufmannsbriefe. Brief-Zeitungen in der Korrespondenz Hildebrand Veckinchusens (1398–1428)*, Munich 1978.

Lindemann, Mary, *Patriots and Paupers: Hamburg 1712–1830*, Oxford 1990.

Linden, Marcel van der, *Social Security Mutualism: The Comparative History of Mutual Benefit Societies*, Bern 1996.

Lingelbach, Gabriele, *Spenden und Sammeln: Der westdeutsche Spendenmarkt bis in die 1980er Jahre*, Göttingen 2009.

Linnemann, Kai Arne, *Das Erbe der Ostforschung. Zur Rolle Göttingens in der Geschichtswissenschaft der Nachkriegszeit*, Paderborn 2002.

Löffler, Ulrich, " 'Gott wollte verschonen, nun tödteten wir' Das Erdbeben von Lissabon und der Siebenjährige Krieg im Spiegel protestantischer Predigt," in Lauer and Unger, *Erdbeben*, pp. 437–58.

Löffler, Ulrich, *Lissabons Fall—Europas Schrecken. Die Deutung des Erdbebens von Lissabon im deutschsprachigen Protestantismus des 18. Jahrhunderts*, Berlin 1999.

Logan, J. R. and Martinez, M. J., "The Spatial Scale and Spatial Configuration of Residential Settlement: Measuring Segregation in the Postbellum South," in: *American Journal of Sociology* 123, 4 (2018), pp. 1161–1203.

Long, James and Stocqueler, J.H., *British Social Life in Ancient Calcutta (1750 to 1850)*, ed. with notes P. Thankappan Nair, Calcutta 1983.

Looz-Corswarem, Clemens Graf von, *Das Finanzwesen der Stadt Köln im 18. Jahrhundert*, Cologne 1978.

Lorenz, Chris, *Wozu noch Theorie der Geschichte?. über das ambivalente Verhältnis zwischen Gesellschaftsgeschichte und Modernisierungstheorie*, Stuttgart 2004.

Lorenz, Martina, "Physik im Hamburgischen Magazin (1747–1767)—Publizistische Utopie und Wirklichkeit," *Zeitschrift des Vereins für Hamburgische Geschichte* 80 (1994), pp. 13–46.

Loso, Michael G., "Summer Temperatures during the Medieval Warm Period and Little Ice Age Inferred from Varved Proglacial Lake Sediments in Southern Alaska," *Journal of Palaeolimnology* 41 (2009), pp. 117–28.

Lowood, Henry E., "The Calculating Forester: Quantification, Cameral Science, and the Emergence of Scientific Forestry Management in Germany," in Frängsmyr, Tore et al. (eds.), *The Quantifying Spirit*, pp. 315–342.

Luh, Jürgen, *Kriegskunst in Europa 1650–1800*, Cologne 2004.

Luhmann, Niklas, *Il rischio dell'assicurazione contro i pericoli*, trans. and introd. Alberto Cevolini, Rome 2013.

Lummer, Johann Gottlieb, *Ermunterung an die Geraischen Abgebrannten: Jn ihren Trübsalen Gott und ihren Nebenmenschen zu geben, was sich gebühret. Eine Predigt am 23. Sonnt. nach Trin. 1780 in der Gottesackerkirche zu Gera gehalten*, Gera 1781.

Lürmann, Paul, *Der große Brand der Stadt Greußen in der Nacht vom 16. zum 17. Oktober 1834*, Greußen/Thuringia 1934.

Luserke-Jacqui, Matthias, *Schiller-Studien. Der ganze Mensch und die Ästhetik der Freiheit*, Tübingen 2018.

Lyth, Peter and Trischler, Helmuth (eds.), *Wiring Prometheus: Globalisation, History and Technology*, Aarhus 2004.

Maass, Ludwig, *Die Brandgilden insbesondere in Schleswig-Holstein*, Stuttgart 1910.

MacKinnon, Malcolm H., "Calvinism and the Infallible Assurance of Grace: The Weber Thesis reconsidered," *British Journal of Sociology* 39 (1988), pp. 143–77.

MacKinnon, Malcolm H., "The Longevity of the Thesis: A Critique of the Critics," in Lehmann and Roth (eds.), *Weber's Protestant Ethic*, pp. 211–243.

Magnan, Serge, "L'assurance incendie. Du risque industriel et la concurrence de 1820 à nos jours," *Risques* 18 (1994), pp. 133–49.

Magnusson, Lars, *Mercantilism: The Shaping of an Economic Language*, London 1994.

Manen, Niels van, "Les plans d'assurance incendie de Goad: Cartographie des risques d'incendie et normalisation des risques industriels (1885–1903)," *Le Mouvement Social* 249 (2014), pp. 163–85.

Marquard, Johann, *Tractatus Politico-Juridicus De Iure Mercatorum et Commercium singulari*, Frankfurt 1662.

Martuccelli, Danilo, *Sociologies de la modernité. L'itinéraire du XXe siècle*, Paris 1999.

März, Olaf, *Fließende Übergänge zwischen Stadt und Land. Ein sozial- und wirtschaftsräumlicher Vergleich in Braunschweig-Wolfenbüttel im 18. Jahrhundert*, Hannover 2014.

Marzagalli, Silvia, *Les boulevards de la fraude. Le négoce maritime et le blocus continental, 1806–1813*, Villeneuve d'Ascq 1999.

Masselos, Jim, "Changing Definitions of Bombay: City State to Capital City," in Indu Banga (ed.), *Ports and Their Hinterlands in India (1700–1950)*, New Delhi 1992, pp. 273–316.

Mattern, Michael, "Die Stadtbefestigung Buxtehudes im 17. Jahrhundert und ihr Bild in den schwedischen Festungsplänen," *Heimatliches Buxtehude* 5 (1997), pp. 115–29.

Matterson, Clarence H., *English Trade in the Levant, 1693–1753*, PhD diss. Harvard University 1936 (Harvard University Archives HU 90.3052).

Maunowy, Jean-Louis, *La genèse des innovations. La création technique dans l'activité de la firme*, Paris 1968.

Mauzi, Robert, *L'idée du bonheur dans la littérature et la pensée françaises au XVIIIe siècle*, Paris 1960 (1994).

Mayet, Paul, *Landwirthschaftliche Versicherung* [...] *Vorschläge zur Besserung der Lage des Japanischen Landmanns, im Auftrag des Kais. Jap. Ministeriums des Innern*, Tokyo 1888.

Mazauric, Simone, *Fontenelle et l'invention de l'histoire des sciences à l'aube des Lumières*, Paris 2007.

McCarthy, Kathleen D., *American Creed: Philanthropy and the Rise of Civil Society, 1700–1865*, Chicago 2003.

McCormick, Ted, *William Petty And the Ambitions of Political Arithmetic*, Oxford 2009.

McKellar, Elizabeth, *The Birth of Modern London. The Development and Design of the City 1660–1720*, Manchester 1999.

Meisner Rosen, Christine, *The Limits of Power: Great Fires and the Process of City Growth in America*, Cambridge 1986.

Melis, Federigo, "Intensità e regolarità nella diffusione dell'informazione economica generale nel Mediterraneo e in Occidente alla fine del Medioevo," *in: Mélanges en l'honneur de Fernand Braudel. Bd. 1: Histoire économique du monde méditerranéen 1450–1650*, Toulouse 1973, pp. 389–424/b.

Melis, Federigo, "Movimento di popoli e motivi economici nel Giubileo del 1400," in *Miscellanea Gilles Gérard Meersseman*, ed. Gilles Gérard Meersseman, 2 vols., Padua 1970, vol. 1, pp. 343–67.

Melis, Federigo, *Aspetti della vita economica medievale. Studi nell'Archivio Datini di Prato*, Siena 1962.

Melis, Federigo, *Documenti per la storia economica dei secoli XIII-XVI*, Florence 1972.

Melis, Federigo, *I trasporti e le comunicazioni nel medioevo, a cura di Luciana Frangioni*, Florence 1984.

Melis, Federigo, *Origini e sviluppi delle assicurazioni in Italia (secoli XIV-XVI)*, vol. 1, *Le fonti*, intro. Bruno Dini, Rome 1975.

Melis, Federigo, *Sulle fonti della storia economica. Appunti raccolti alle lezioni del Prof. Federigo Melis*, ed. von Bruno Dini, Milan 1985.

Mellas, Akylas (ed.), *Pera: The Crossroads of Constantinople*, Athens s.d. (ca. 2003).

Mellmann, Katja, *Emotionalisierung. Von der Nebenstundenpoesie zum Buch als Freund. Eine emotionspsychologische Analyse der Literatur der Aufklärungsepoche*, Paderborn 2006.

Melosi, Martin V., "The place of the city in environmental history," *Environmental History Review* 17 (1993), pp. 1–23.

Mercier-Faivre, Anne-Marie and Thomas, Chantal (eds.), *L'invention de la catastrophe au XVIIIe siècle. Du châtiment divin au désastre naturel*, Geneva 2008.

Meyer, Annette, *Von der Wahrheit zur Wahrscheinlichkeit. Die Wissenschaft vom Menschen in der schottischen und deutschen Aufklärung*, Tübingen 2008.

Meyer, Torsten, "Sicherheit durch Erfindungen. Zur sozialen Bedeutung der Technologie im ausgehenden 18. Jahrhundert," in Gerhard Banse and Hans-Peter Müller (eds.), *Johann Beckmann und die Folgen. Erfindungen—Versuch der historischen, theoretischen und empirischen Annäherung an einen vielschichtigen Begriff*, Münster 2001, pp. 191–200.

Michaels, Ralf, "Legal Medievalism in *Lex Mercatoria* Scholarship," *Texas Law Review* 90 (2012), pp. 259–68.

Milanesi, Marica, *Tolomeo sostituito. Studi di storia delle conoscenze geografiche nel XVI secolo*, Milan 1984.

Miller, Ross. *American Apocalypse: The Great Fire and the Myth of Chicago*, Chicago 1990.

Mintsker, Yair, *The defortification of the German city, 1689–1866*. Cambridge 2012.

Mitzman, Arthur, *The Iron Cage: A Historical Interpretation of Max Weber*, New York 1970.

Modéer, Kjell A, "Aktenversendung und Greifswalder Juristenfakultät," in *Justitia in Pommern*, ed. Dirk Alvermann and Jürgen Regge, Münster 2004.

Mohnhaupt, Heinz, "'Jura mercatorum' durch Privilegien. Zur Entwicklung des Handelsrechts bei Johann Marquard (1961–1668)," in Gerhard Köbler (ed.), *Wege europäischer Rechtsgeschichte, Festschrift für Karl Kroeschell*, Frankfurt am Main 1987, pp. 308–23.

Moll-Murata, Christine, "Merchant and Craft Guilds", in: Ead., *State and Crafts in the Qing Dynasty (1644–1911)*, Amsterdam 2018, pp. 321–347.

Monmonier. Mark, *Cartographies of Danger: Mapping Hazards in America*, Chicago 1997.

Moore, Seán, "'Vested' Interests and Debt Bondage: Credit as Confessional Coercion in Colonial Ireland," in Carey and Finlay, *Empire of Credit*, pp. 209–28.

Morgan, John, "The representation and experience of English urban fire disasters, c. 1580–1640", *Historical Research* 89, 244 (2016), pp. 268–293.

Mori, Massimo, "Glück und Autonomie. Die deutsche Debatte über den Eudämonismus zwischen Aufklärung und Idealismus" *Studia Leibnitiana* 25 (1993), pp. 27–42.

Morich, Heinrich, "Die früheren Brandkatastrophen in Clausthal und Zellerfeld," in *450 Jahre Clausthal*-Zellerfeld *1532–1982*, Clausthal-Zellerfeld 1982, pp. 74–76.

Mulcahy, Matthew, "Urban Catastrophes and Imperial Relief in the Eighteenth-Century British Atlantic World: Three Case Studies," in Massard-Guilbaud, Geneviève, Platt, Harold L., Schott, Dieter (eds.), *Cities and Catastrophes. Villes et catastrophes*, Frankfurt am Main 2002, pp. 105–22.

Müller, Kurt and Krönert, Gisela, *Leben und Werk von Gottfried Wilhelm Leibniz. Eine Chronik*, Frankfurt am Main 1969.

Müller, Laurenz, *Diktatur und Revolution. Reformation und Bauernkrieg in der Geschichtsschreibung des "Dritten Reiches" und der DDR*, Stuttgart 2004.

Mulsow, Martin, "Qu'est-ce qu'une constellation philosophique? Propositions pour une analyse des réseaux intellectuels," *Annales HSS* 64, no. 1 (2009), pp. 81–109.

Mulsow, Martin, *Konstellationsforschung*, Frankfurt am Main 2005.

Munch-Petersen, Thomas, *Defying Napoleon. How Britain bombarded Copenhagen and seized the Danish Fleet in 1807*, Stroud 2007.

Murphy, Anne L., *The Origins of English Financial Markets: Investment and Speculation before the South Sea Bubble*, Cambridge 2009.

Müssener, Alexander, *Die Entwicklung der Aachener Feuer-Versicherungs-Gesellschaft im 19. Jahrhundert unter besonderer Berücksichtigung ihrer Allgemeinen Versicherungsbedingungen*, Hamburg 2008.

Mylius, Christian Otto (ed.), *Corpus constitutionum Marchicarum oder Königlich preußische und churfürstlich Brandenburgische in der Chur- und Mark Brandenburg, auch incorporirten Landen publicirte und ergangene Ordnungen, edicta, mandata, rescripta, Teil 1–6, nebst Anhang und Continuatio 1–4*, Berlin 1737–55.

Nanz, Klaus-Peter, *Die Entstehung des allgemeinen Vertragsbegriffs im 16. bis 18. Jahrhundert*, Munich 1985.

Nardi, Loris de, "El caso fortuito: fundamentos culturales y religiosos de una categoría jurídica y de una cosmovisión", *Derecho PUCP* 84 (2020), pp. 337–354.

Nassehi, Armin, *Der soziologische Diskurs der Moderne*, Frankfurt am Main 2006.

"Natural Disasters, Big Fires and epidemics in the Cities", special issue *Historia urbana* [Editura Academiei Române] 26 (2018).

Nehlsen-Von Stryk, Karin, *Die venezianische Seeversicherung im 15. Jahrhundert*, Ebelsbach 1986.

Neukom, Raphael et al. (eds.), "No evidence for globally coherent warm and cold periods over the preindustrial Common Era", *Nature* 571 (2019), pp. 550–554.

Neumann, Christoph K., "Modernitäten im Konflikt. Der Sechste Munizipal-Bezirk von Istanbul, 1857–1912," in *Istanbul: vom imperialen Herrschersitz zur Megapolis. Historiographische Betrachtungen zu Gesellschaft, Institutionen und Räumen*, Munich 2006, pp. 351–75.

Nicklas, Thomas, *Um Macht und Einheit des Reiches. Konzeption und Wirklichkeit der Politik bei Lazarus von Schwendi (1522–1583)*, Husum 1995.

Niekerk, Johan Petrus van, *The Development of the Principles of Insurance Law in the Netherlands from 1500 to 1800*, 2 vols., Cape Town 1998.

Nightingale, Paul, "Tacit knowledge and Engineering Design," in Anthonie Meijers (ed.), *Philosophy of Technology and Engineering Sciences*, Amsterdam 2009, pp. 351–79.

Nigro, Giampiero, *Mercanti in Maiorca: il carteggio datiniano dall'Isola (1387–1396)*, Florence 2003.

Nipperdey, Justus, *Die Erfindung der Bevölkerungspolitik. Staat, politische Theorie und Population in der Frühen Neuzeit*, Göttingen 2012.

Nippert, Klaus, *Nachbarschaft der Obrigkeiten. Zur Bedeutung frühneuzeitlicher Herrschaftsvielfalt am Beispiel des Hannoverschen Wendlands im 16. Und 17. Jahrhundert*, Hanover 2000.

Norgaard, Richard B., "Sustainable Development: A Co-Evolutionary View," in Redclift, *Sustainabilty*, vol. 2, pp. 131–48.

North, Douglas C. and Weingast, Barry R., "Constitutions and Commitment: The Evolution of Institutional Governing Public Choice in Seventeenth-Century England," *The Journal of Economic History* 49 (1989), pp. 803–32.

Nouet, Noel, *The Shogun's City. A History of Tokyo*, Sandgate 1990.

Novak, Maximilian E. (ed.), *The Age of Projects*, Toronto 2008.

O'Brian, Anne, *Philanthropy and Settler Colonialism*, New York 2015.

Oberkrome, Willi, *Volksgeschichte. Methodische Innovation und völkische Ideologisierung in der deutschen Geschichtswissenschaft*, Göttingen 1993.

Oliver-Smith, Anthony and Hoffmann Susanna M. (eds.), *Catastrophe and Culture*, Sante Fe 2002.

Origo, Iris, *The Merchant of Prato, Fran di Marco Datini 1335–1410*, New York 1954.

Ormrod, David, *The Rise of Commercial Empires: England and the Netherlands in the Age of Mercantilism, 1650–1770*, Cambridge 2003.

Otro Serrera, Juan Miguel, " 'Incendio de Troya' de Christian de Keuninck," *Archivo Español de Arte* 59, 236 (1986), pp. 420–23.

Oviatt, F. C., "Historical Study of Fire Insurance in the United States," *Annals of the American Academy of Political and Social Science* 26 (1905), pp. 155–78.

Özkaynak, Begüm, Devine, Pat and Rigby, Dan, "Operationalising Strong Sustainability: Definitions, Methodologies and Outcomes," in Redclift, *Sustainability*, vol. 2, pp. 369–394.

Pahlow, Louis, "Ein 'Geschöpf der ganzen civilisirten handeltreibenden Welt.' Der Versicherungsvertrag im Deutschland des 19. Jahrhunderts," *Zeitschrift für neuere Rechtsgeschichte* 29 (2007), pp. 18–46.

Papenbrock, Martin, " 'Das Brennende Troja' in den Bildern von Pieter Schoubroeck. Zur Revision des Vaterlanndsbegriffs in Kunst und Philosophie um 1600," in Koppenleitner, Rößler, and Tiemann (eds.), *Urbs incensa*, pp. 45–60.

Pearce, David W. and Warford, Jeremy J., *World without End: Economics, Environment, and Sustainable Development*, Oxford 1993.

Pearson, Robin and Lönnborg, Mikael, "Regulatory Regimes and Multinational Insurers before 1914," *Business History Review* 82 (2008), pp. 59–86.

Pearson, Robin and Yoneyama, Takau (eds.), *Corporate Forms and Organizational Choice in International Insurance*, Oxford 2015.

Pearson, Robin, "Ein Wachstumsrätsel: Feuerversicherung und die wirtschaftliche Entwicklung Großbritanniens 1700–1850," *Zeitschrift für Unternehmensgeschichte* 44 (1999), pp. 218–34.

Pearson, Robin, *Insuring the Industrial Revolution: Fire Insurance in Great Britain, 1700–1850*, Aldershot 2004.

Peeters, Herwig, "Sustainable Development and the Role of the Financial World," *Environment, Development and Sustainability* 5 (2003), pp. 197–230.

Peeters, Jan, "De stad als landschap de stad als stilleven," in *Kunstschrift/Openbaar Kunstbezit* 36 (1992), pp. 20–24.

Pelc, Ortwin, "Hamburg 1842. Eine Katastrophe verändert die Stadt," in Claudia Horbas and Ortwin Pelc (eds.), *Es brannte an allen Ecken zugleich. Hamburg 1842. Ausstellungskatalog des Museums für Hamburgische Geschichte*, Heyde 2002, pp. 9–15.

Pene Vidari, Gian Savino, "Sulla classificazione del contratto d'assicurazione nell'età del diritto comune," in *Rivista di storia del diritto italiano* 71 (1998), pp. 113–37.

Percy, Sholto], *The Percy Anecdotes*, Revised Edition, 2 vols., New York 1845.

Perdikas, Panayotis, "Die Entstehung der Versicherung im Mittelalter. Geschichtliche Grundlagen im Verhältnis zu Seedarlehen, Überseekauf, commenda und Bodmerei," *Zeitschrift für die gesamte Versicherungswissenschaft* 55 (1966), pp. 425–509.

Perdikas, Panayotis, "Die Palermo-Verträge und die Entstehung der Versicherung," *Zeitschrift für die gesamte Versicherungswissenschaft* 59 (1970), pp. 151–61.

Pereira, Alvaro S., "The Opportunity of a Disaster: the Economic Impact of the 1755 Lisbon Earthquake," *The Journal of Economic History* 69, 2 (2009), pp. 466–99.

Petersson, Astrid, *Zuckersiedergewerbe und Zuckerhandel in Hamburg im Zeitraum von 1814 bis 1834. Entwicklung und Struktur zweier wichtiger Hamburger Wirtschaftszweige des vorindustriellen Zeitalters*, Stuttgart 1998.

Petit, Maxime, *Les grands incendies*, Paris 1882, pp. 273–92.

Petkov, Kiril, *The anxieties of a Citizen Class: The Miracles of the True Cross of San Giovanni Evangelista, Venice 1370–1480*, Leiden 2014.

Pfister, Christian, "The 'Black Swan' of 1540: Aspects of a European Megadrought," in Claus Leggewie and Franz Mauelshagen (eds.), *Climate and Culture, Climate Change and Cultural Transition in Europe*, Leiden 2018, pp. 156–194.

Pfister, Christian, *Wetternachhersage. 500 Jahre Klimavariationen und Naturkatastrophen* (1496–1995), Bern 1999.

Piergiovanni, Vito (ed.), *From Lex Mercatoria to Commercial Law*, Berlin 2005.

Pincus, Steven, "Rethinking Mercantilism: Political Economy, the British Empire, and the Atlantic World in the Seventeenth and Eighteenth Centuries," *The William and Mary Quarterly* 69, 1 (2012).

Pinwinkler, Alexander, "Volk, Bevölkerung, Rasse and Raum: Erich Keyser's Ambiguous Concept of a German History of Population, ca. 1918–1955," in Ingo Haar and Michael Fahlbusch (ed.), *German Scholars and Ethnic Cleansing*, New York 2005, pp. 86–99.

Plamper, J., "The History of Emotions: An Interview with William Reddy, Barbara Rosenwein, and Peter Stearns", *History and Theory* 49 (2010), pp. 237–65.

Poblotzki, Siegfried, *Geschichte der Herrschaft, der Stadt und der Pfarrei Pleystein*, Pleystein 1980.

Pohl, Hans. "Wirtschaftsgeschichte Kölns im 18. und beginnenden 19. Jahrhundert," in *Zwei Jahrtausende Kölner Wirtschaft*, ed. Hermann Kellenbenz, vol. 2: *Vom 18. Jahrhundert bis zur Gegenwart*, Cologne 1975, pp. 9–162.

Poirier, Jean-Paul, *Le tremblement de terre de Lisbonne 1755*, Paris 2005.

Pomeranz, Kenneth, *The Great Divergence. Europe, China, and the Making of the Modern World Economy*, Princeton 2000.

Popplow, Marcus, *Neu, nützlich und erfindungsreich: die Idealisierung von Technik in der frühen Neuzeit*, Münster 1998.

Porter, Stephen, *The Great Fire of London*, Phoenix Mill 1996.

Porter, Theodore M., *Trust in Numbers: The Pursuit of Objectivity in Science and Public Life*, Princeton 1995.

Portney, Kent E., *Taking sustainable Cities seriously. Economic Development, the Environment, and Quality of Life in American Cities*, Cambridge/Mass. 2003.

Pouilloux, Didier, *Mémoires d'assurances. Recueil de sources françaises sur l'histoire des assurances du XVIème au XIXème siècle*, Mayenne 2011.

Pribyl, Kathleen, "A survey of the impact of summer droughts in southern and eastern England, 1200–1700", *Climate of the Past* 16 (2020), pp. 1027–1041.

Priesdorff, Wilhelm Werner Kurt von, *Offizier-Stammliste des Grenadier-Regiments König Friedrich Wilhelm IV. (1. Pommerschen)*, no. 2, Berlin 1906.

Pröve, Ralf, *Stehendes Heer und städtische Gesellschaft im 18. Jahrhundert. Göttingen und seine Militärbevölkerung 1713–1756*, Munich 1995.

Ptaszyński, Maciej, *Beruf und Berufung. Die evangelische Geistlichkeit und die Konfessionsbildung in den Herzogtümern Pommern, 1560–1618*, Göttingen 2017.

Pyne, Stephan, *Vestal Fire: An Environmental History, Told through Fire, of Europe and Europe's Encounter with the World*, Seattle 1997.

Pyne, Stephen *Fire in America: A Cultural History of Wildland and Rural Fire*, Seattle 1997.

Pyne, Stephen, *Vestal Fire: An Environmental History, Told through Fire, of Europe and Europe's Encounter with the World*, Seattle 1997.

Quenet, Grégory, *Les tremblements de terre aux XVIIe et XVIIIe siècles. La naissance d'un risque*, Paris 2005.

Quirke, Brendan J., "Leasing the Environment," *Sustainable Development* 4 (1996), pp. 98–102.

Radkau, Joachim, "'Nachhaltigkeit' als Wort der Macht: Reflexionen zum methodischen Wert eines umweltpolitischen Schlüsselbegriffs," in Duceppe-Lamarre, François and Engels, Jens Ivo (eds.), *Umwelt und Herrschaft in der Geschichte*, Munich 2008, pp. 131–36.

Radkau, Joachim, *Max Weber. Die Leidenschaft des Denkens*, Munich 2005.

Rasmus-Brandt, J., "Pity and Fear. A note of Raphael's 'Incendio del Borgo,'" *Acta ad Archaeologiam et Artium Historiam Pertinentia. Institutum Romanum Norvegiae Roma* N.S. 1 (1981), pp. 259–74.

Rassem, Mohammed, Stagl, Justin, and Rose, Wolfgang (eds.), *Geschichte der Staatsbeschreibung: ausgewählte Quellentexte 1456–1813*, Berlin 1994.

Rassem, Mohammed, *Statistik und Staatsbeschreibung in der Neuzeit. vornehmlich im 16.-18. Jahrhundert*, Paderborn 1980.

Raymond, André, "Islamic City, Arab City: Orientalist Myth and Recent Views," *British Journal of Middle Eastern Studies* 21 (1994), pp. 3–18.

Raynes, Harold E., *A History of British Insurance*, 2nd ed., London 1964.

Redclift, Michael (ed.), *Sustainability: Critical Concepts in the Social Sciences*, 4 vols., London 2005.

Reddaway, Thomas F., *The Rebuilding of London after the Great Fire*, London 1940.

Reddy, William M., "Against Constructionism. The Historical Ethnography of Emotions," in *Current Anthropology* 38, no. 3 (1997), pp. 327–51.

Reddy, William M., "The Logic of Action. Indeterminacy, Emotion, and Historical Narrative," in *History and Theory* 40 (2001), pp. 10–33.

Reinhold, Werner, *Chronik der Stadt Dahme und der Umgegend*, Dahme 1845.

Reinisch, Ulrich, *Der Wiederaufbau der Stadt Neuruppin nach dem großen Brand von 1787*, Worms 2001.

Reitemeier, Arnd, *Außenpolitik der Spätmittelalter. Die diplomatischen Beziehungen zwischen dem Reich und England (1377–1422)*, Paderborn 1999.

Reith, Gerda, "Uncertain Times: The Notion of 'Risk' and the Development of Modernity," *Time & Society* 13 (2004), pp. 383–402.

Renouard, Yves, "Comment les papes d'Avignon expédiaient leur courrier," *Revue historique* 180 (1937), pp. 1–29.

Renouard, Yves, *Les relations des papes d'Avignon et des compagnies commerciales et bancaires de 1316 à 1378*, Paris 1941.

Ressel, Magnus and Zwierlein, Cornel, "The Ransoming of North European Captives from Northern Africa. A Comparison of Dutch, Hanseatic and English Institutionalization of Redemption from 1610 to 1645," in *Seeraub im Mittelmeerraum*, eds. Nikolas Jaspert and Sebastian Kolditz, Munich 2013, pp. 377–406.

Ressel, Magnus and Zwierlein, Cornel, "The Ransoming of North European Captives from Northern Africa. A Comparison of Dutch, Hanseatic and English Institutionalization of Redemption from 1610–1645," in Jaspert, Nikolas and Kolditz, Sebastian (eds.), *Seeraub im Mittelmeerraum. Piraterie, Korsarentum und maritime Gewalt von der Antike bis zur Neuzeit*, Paderborn 2013, pp. 377–406.

Ressel, Magnus, *Zwischen Sklavenkassen und Türkenpässen. Nordeuropa und die Barbaresken in der Frühen Neuzeit*, Berlin 2012.

Rieu, Josiane, "La temporalisation de l'espace" in *Le Paysage à la Renaissance*, ed. Yves Giraud (Fribourg: 1988), pp. 297–310.

Rimbach, Karl Ernst, *Feuersozietät Berlin 1718–1968*, Berlin 1968.

Ringrose, David R., *Spain, Europe, and the 'Spanish Miracle,' 1700–1900*, Cambridge 1996.

Röder, Tilman J., *Rechtsbildung im wirtschaftlichen "Weltverkehr." Das Erdbeben von San Francisco und die internationale Standardisierung von Vertragsbedingungen (1871–1914)*, Frankfurt am Main 2006.

Roeder, Gustaf, *Württemberg. Vom Neckar zur Donau. Landschaft, Geschichte, Kultur, Kunst*, Nuremberg 1972.

Rogowski, Hellmut, *Verfassung und Verwaltung der Herrschaft und Stadt Jever von den Anfängen bis zum Jahre 1807*, Göttingen 1967.

Rohland, Eleonora, *Hurricanes in New Orleans, 1718–1965: A History of Adaptation*, PhD diss., University of Bochum, 2014.

Romani, Roberto, *National Character and Public Spirit in Britain and France, 1750–1914*, Cambridge 2002.

Roo, Gert de and Miller, Donald (eds.), *Compact Cities and Sustainable Urban Development. A Critical Assessment of Policies and Plans from an International Perspective*, Aldershot 2000.

Rosa, Hartmut, *Beschleunigung. Die Veränderung der Zeitstrukturen in der Moderne*, Frankfurt am Main 2005.

Rosental, Claude, "Toward a Sociology of Public Demonstrations," *Sociological Theory* 31, 4 (2013), pp. 343–65, 343–46.

Rosseaux, Ulrich, *Städte in der Frühen Neuzeit*, Darmstadt 2006.

Rossi, Guido, *Insurance in Elizabethan England: The London Code*, Cambridge 2016.

Roth, Guenther, "Weber the Would-Be Englishman: Anglophilia and Family History," in Lehmann and Roth (eds.), *Weber's Protestant Ethic*, pp. 83–121.

Rothermund, Dietmar, *Indiens wirtschaftliche Entwicklung. Von der Kolonialherrschaft bis zur Gegenwart*, Paderborn 1985.

Rousseau, Jean-Jacques, *Émile*, in *Œuvres completes*, vol. 4, Paris 1969, 245–868.

Rowe, Michael, *From Reich to State: The Rhineland in the Revolutionary Age*, Cambridge 2003.

Rowe, William T., *China's Last Empire: The Great Qing*, Cambridge 2009.

Roy, Tirthankar, *The Economic History of India 1857–1947*, 2nd ed., Oxford 2006.

Rudert, Thomas and Zückert, Hartmut (eds.), *Gemeindeleben. Dörfer und kleine Städte im östlichen Deutschland (16.-18. Jahrhundert)*, Cologne 2001.

Ruffat, Michèle, "French Insurance from the *ancien régime* to 1946: Shifting Frontiers between State and Market," *Financial History Review* 10 (2003), pp. 185–200.

Ryan, R. J., *A History of the Norwich Union Fire and Life Insurance Societies from 1797 to 1914*, PhD thesis, University of East Anglia 1983.

Sachs, Stephen E., "From St. Ives to Cyberspace: The Modern Distortion of the Medieval 'Law Merchant,'" *American University International Law Review* 21 (2006), pp. 685–812.

Said, Edward, *Orientalism*, London 2003.

Salgaro, Silvino (ed.), *Cristoforo Sorte e il suo tempo*, Bologna 2012.

Salgaro, Silvino, " 'Conoscere i luoghi senza leggere le lettere de' loro nomi.' Cristoforo Sorte cartografo," in Salgaro, *Cristoforo Sorte*, pp. 305–52.

Sand, Jordan and Wills, Steven, "Governance, Arson, and Fire Fighting in Edo, 1600–1868," in Greg Bankoff, Uwe Lübken, and Jordan Sand (eds.), *Flammable Cities: Urban Fire and the Making of the Modern World,* Madison, Wisconsin 2011, pp. 44–62.

Sandl, Marcus, Ökonomie des Raumes. Der kameralwissenschaftliche Entwurf der Staatswirtschaft im 18. Jahrhundert, Cologne 1999.

Sapori, Armando, *Studi di storia economica (secoli XIII, XIV, XV)* [first ed. 1940], Florence 1982.

Sarkar, Benoy Kurmar, *Social insurance Legislation and Statistics*, Calcutta 1937.

Sawislak, Karen, *Smoldering City: Chicagoans and the Great Fire, 1871–1874*, Chicago 1995.

Schäfer, Wilhelm, *Urkundliche Beiträge und Forschungen zur Geschichte der Feuerversicherung in Deutschland*, 2 vol., Hannover 1911.

Schaffer, Simon and Shapin, Steven, *Leviathan and the Air Pump. Hobbes, Boyle, and the Experimental Life*, 2nd ed., Princeton 2011.

Schaps, Regina, *Hysterie und Weiblichkeit: Wissenschaftsmythen über die Frau*, Frankfurt am Main 1982.

Scheel, Günter, "Leibniz' Wirken für Kaiser und Reich im Jahre 1688 in Wien nach bisher unbekannten Quellen," in *Leibniz und Europa. VI. Internationaler Leibniz-Kongreß*, part 1: presentations, Hanover 1994, pp. 697–704, 699.

Scheller, Benjamin, "(Un-)sichere Häfen: Häfen als Hotspots maritimer Risiken und Risikokommunikation im Mittelmeerraum des 15. Jahrhunderts", *Historische Anthropologie* 26, 1 (2018), pp. 43–65.

Scheller, Benjamin, "Die Geburt des Risikos", *Historische Zeitschrift* 304, 2 (2017), pp 305–331.

Schellong, Dieter, "Calvinismus und Kapitalismus. Anmerkungen zur Prädestinationslehre Calvins," in Hans Scholl (ed.), *Karl Barth und Johannes Calvin*, Neukirchen-Vluyn 1995, 74–106.

Schellong, Dieter, "Der 'Geist' des Kapitalismus und der Protestantismus: Eine Max-Weber-Kritik," in Faber, Richard (ed.), *Der Protestantismus: Ideologie, Konfession oder Kultur?,* Würzburg 2003, pp. 231–53.

Schenk, Gerrit Jasper (ed.), *Historical Disaster Experience*, Heidelberg 2017.

Schenk, Gerrit Jasper and Engels, Jens Ivo (eds.), *Historical Disaster Research: Concepts, Methods and Case Studies*, Cologne 2007.

Scherner, Karl Otto, "Allgemeine Rechtsgrundsätze und Rechtsvergleichung im europäischen Handelsrecht des 17. und 18. Jahrhunderts," *Ius Commune* 7 (1978), pp. 118–35.

Scherner, Karl Otto, “Die Wissenschaft des Handelsrechts,” in *Handbuch der Quellen und Literatur der neueren europäischen Privatrechtsgeschichte*, vol. 2.1: *Wissenschaft (1500–1800)*, ed. Helmut Coing, Munich 1977.

Scherner, Karl Otto, “Zum Sonderrecht der Kaufleute bei Johann Marquard,” in Hans-Peter Haferkamp (ed.), *Usus modernus pandectarum: Römisches Recht, Deutsches Recht und Naturrecht in der Frühen Neuzeit*, Cologne 2007, 199–217.

Scherner, Karl-Otto, “Die Wissenschaft des Handelsrechts,” in Helmut Coing (ed.), *Handbuch der Quellen und Literatur der neueren europäischen Privatrechtsgeschichte*, vol. 2.1, *Wissenschaft (1500–1800)*, Munich 1977, pp. 797–997.

Schetelig, Johann Andreas Gottfried, *Die von Hamburg empfundenen Rührungen des Zorns und der Güte Gottes, bey Gelegenheit der am 10. Merz 1750. durch einen Wetterstrahl geschehenen Entzündung und Einäscherung der Haupt-Kirche St. Michaelis, in gebundener Rede erwogen*, Hamburg 1750.

Scheuner, Ulrich, “Die Staatszwecke und die Entwicklung der Verwaltung im deutschen Staat des 18. Jahrhunderts,” in Kleinheyer, Gerd and Mikat, Paul (eds.), *Beiträge zur Rechtsgeschichte. Gedächtnisschrift für Hermann Conrad*, Paderborn 1979, pp. 467–89.

Schewe, Dieter, *Geschichte der sozialen und privaten Versicherungen im Mittelalter in den Gilden Europas*, Berlin 2000.

Schilling, Heinz, *Die Stadt in der Frühen Neuzeit*, 2nd ed., Munich 2004.

Schilling, Lothar (ed.), *Der Absolutismus—ein unersetzliches Forschungskonzept?*, Munich 2008.

Schilling, Ruth, *Johann Friedrich Glaser (1707–1789). Scharfrichtersohn und Stadtphysikus in Suhl*, Vienna 2015.

Schliemann, Friedrich, “Zur Geschichte der Stadt Neukalen,” *Archiv für Landeskunde in den Großherzogthümern Mecklenburg und Revüe der Landwirthschaft* 12 (1862), pp. 511–57, 641–58.

Schlögl, Rudolf, *Glaube und Religion in der Säkularisierung. Religiosität in der katholischen Stadt, Köln, Aachen, Münster 1700–1840*, Munich 1995.

Schlosser, Johann Ludwig, *Die Stimme des Herrn im Feuer: am Buß-Tage wegen der durch einen Wetter-Strahl verursachten Einäscherung der S. Michaelis-Kirche in Hamburg zur Erweckung vorgestellt*, Hamburg 1750.

Schluchter, Wolfgang and Graf, Friedrich Wilhelm (eds.), *Asketischer Protestantismus und der “Geist” des modernen Kapitalismus. Max Weber und Ernst Troeltsch*, Tübingen 2005.

Schmidt, Burghart, *Hamburg im Zeitalter der Französischen Revolution und Napoleons 1789–1813*, 2 vols., Hamburg 1998.

Schmidt, Gunnar, “Splashes & Flashes. Über High-Speed-Visualisierungen und Wissensformen,” in Inge Hinterwaldner and Markus Buschhaus (eds.), *The Picture's Image. Wissenschaftliche Visualisierung als Komposit*, Munich 2006, pp. 180–95.

Schmidt, Johann Jakob, *Biblischer Physicus, Oder Einleitung zur Biblischen Natur-Wissenschaft und deren besondern Theilen* [...], Leipzig 1731.

Schmitt-Lermann, Hans, *Der Versicherungsgedanke im deutschen Geistesleben des Barock und der Aufklärung*, Munich 1954.

Schneckenburger, Matthias, *Vergleichende Darstellung des lutherischen und reformierten Lehrbegriffs*, Stuttgart 1855.

Schneider, Alfons Maria, "Brände in Konstantinopel," *Byzantinische Zeitschrift* 41 (1941), pp. 382–403.

Schneider, Ivo, "Geschichtlicher Hintergrund und wissenschaftliches Umfeld der Schriften," in Leibniz, *Hauptschriften*, pp. 591–623.

Schneider, Ivo, "Why Do We Find the Origin of the Calculus of Probabilities in the Seventeenth Century," in J. Hintikka et al. (eds.), *Probabilistic thinking*, Dordrecht 1981, pp. 3–24.

Schneider, Jane (ed.), *Italy's "Southern Question": Orientalism in One Country*, Oxford 1998.

Schrader, F.R., "Aus den Tagen des großen Brandes," ed. Theodor Schrader, *Mittheilungen des Vereins für Hamburgische Geschichte* 15 (1892), pp. 378–83.

Schröder, Asta, "Die Stadt Olpe," in Josef Wermert (ed.), *Olpe. Geschichte von Stadt und Land.* vol. 1, *Von den Anfängen bis zum Ende des Ersten Weltkrieges*, Olpe 2002, pp. 219–62.

Schröder, Hans and Kellinghusen, Anton H., *Lexikon der hamburgischen Schriftsteller bis zur Gegenwart VII*, Hamburg 1875–79.

Schulz, Jürgen, "Cristoforo Sorte and the Ducal Palace in Venice," *Mitteilungen des Kunsthistorischen Instituts in Florenz* 3 (1962), pp. 193–208.

Schulz, Jürgen, "New Maps and Landscape Drawings by Cristoforo Sorte," *Mitteilungen des Kunsthistorischen Instituts in Florenz* 20 (1976), p. 107.

Schulz, Jürgen, *La cartografia tra scienza e arte: carte e cartografi nel Rinascimento italiano*, Modena 2006.

Schulze, Winfried, *Deutsche Geschichtswissenschaft nach 1945*, Munich 1989.

Schurer, Kevin, *Surveying the People. The Interpretation and Use of Document Sources for the Study of Population in the Later Seventeenth Century*, Oxford 1992.

Schwaiger, Clemens, *Das Problem des Glücks im Denken Christian Wolffs. Eine quellen-, begriffs- und entwicklungsgeschichtliche Studie zu Schlüsselbegriffen seiner Ethik*, Stuttgart-Bad Cannstatt 1995.

Schwarz, P., "Der große Reutlinger Stadtbrand im September 1726," in *Reutlinger Geschichtsblätter* 14 (1976), pp. 7–43.

Scott, James C., *Seeing Like a State: How Certain Schemes to Improve the Human Condition Have Failed*, New Haven 1998.

Seifert, Arno, *Der Rückzug der biblischen Prophetie von der Neueren Geschichte. Studien zur Geschichte der Reichstheologie des frühneuzeitlichen deutschen Protestantismus*, Cologne 1990.

Seng, Eva-Maria, *Stadt—Idee und Planung. Neue Ansätze im Städtebau des 16. und 17. Jahrhunderts*, Munich 2003.

Serafy, Salah El, "The Environment as Capital," in Redclift, *Sustainability*, vol. 3, pp. 49–57.

Serrão, Vitor, "O mito do herói redentor: a representação de Eneias na pintura do Portugal restaurado," *Quintana* 1 (2002), pp. 71–81.

Setzler, Wilfried, Schönhagen, Benigna, and Binder, Hans-Otto, *Kleine Tübinger Stadtgeschichte*, Tübingen 2006.

Sewall, Hannah Robie, *The Theory of Value before Adam Smith*, New York 1901.

Sieferle, Rolf-Peter, *Der unterirdische Wald: Energiekrise und industrielle Revolution*, Munich 1982.

Siekmann, Mechthild, "Die Brandversicherung im Hochstift Münster 1768–1805. Entstehung, Arbeitsweise, Quellen," *Westfälische Forschungen* 31 (1982), pp. 154–68.

Simon, Oskar, "Die Entwicklung des Immobiliar-Feuerversicherungswesens in Preußen bis zum Jahre 1866," *Annalen des Deutschen Reiches* 21 (1888), pp. 62–95.

Slack, Paul, "The Politics of Consumption and Englands's Happiness in the Later Seventeenth Century," *English Historical Review* 122 (2007), pp. 609–31.

Slack, Paul, *From Reformation to Improvement: Public Welfare in Early Modern England*, New York 1999.

Slack, Paul, *The Invention of Improvement. Information and Material Progress in Seventeenth-Century England*, Oxford 2015.

Smetshurst, Richard J., *Agricultural Development and Tenancy Disputes in Japan, 1870–1940*, Princeton 1986.

Smiley, S. L., "Mental maps, segregation, and everyday life in Dar es Salaam, Tanzania," in: *Journal of Cultural Geography* 30, 2 (2013), 215–244.

Smith, Justin H., "Leibniz on Natural History and National History," in *History of Science* 50, 4 (2012), pp. 377–401.

Soave, M., "Malta e gli schiavi ebrei," *Corriere israelitico* 17 (1878–79), pp. 54–57; 81–82; 101–03; 125–27; 147–49; 172–73.

Sofonea, Traiano, "Leibniz und sein Projekt zur Errichtung staatlicher Versicherungsanstalten," *Schweizerische Versicherungszeitschrift* 25 (1957), pp. 144–48.

Sombart, Nicolaus, *Nachdenken über Deutschland. Vom Historismus zur Psychoanalyse*, Munich 1987.

Sorte, Cristoforo, *Osservazioni nella pittura, in: Paola Barocchi* (ed.), *Trattati d'arte del cinquecento fra manierismo e controriforma*, vol. 1, Bari 1960, pp. 271–301.

Spalding, Almut, "Der Fragmenten-Streit und seine Nachlese im Hamburger Reimarus-Kreis," *Aufklärung* 24 (2012), pp. 11–28.

Spaulding, Robert Mark, "Changing Patterns of Rhine Commerce in the Era of French Hegemony, 1793–1813", *Vierteljahrschrift für Sozial- und Wirtschaftsgeschichte* 100, 4 (2013), pp. 413–431.

Speck-Nagel, Kurt Gustav, *Die vorderösterreichischen Landstände im 15. und 16. Jahrhundert*, 2 vols., Karlsruhe 1991.

Spies, Martina, *Feuerversicherung, Waisen-und Kreditkassen bei ostschwäbischen Reichsklöstern vor der Säkularisierung und ihre Auflösung*, Munich 2007.

Spohn, Thomas, "'Sollen recht ordentlich bauen sonder Resoniren. F W'—Hausbau und Stadtplanung unter preußischem Einfluß, dargelegt am Wiederaufbau der märkischen Städte und Flecken im 18. Jahrhundert," *Der Märker* 39 (1990), pp. 191–206, 250–61.

Spohn, Thomas, "Der Stadtbrand als Chance. Wiederaufbauplanungen für westfälische Städte vom 17. bis ins 19. Jahrhundert," in *Feuer! Stadtbrand in Westfalen*, Münster 1990/1991, pp. 49–72.

Spohn, Thomas, "Sollen recht ordentlich bauen," *Der Märker* 39 (1990), pp. 191–206, 250–61.

Spohn, Thomas, "Zur Einleitung," in Thomas Spohn (ed.), *Bauen nach Vorschrift? Obrigkeitliche Einflussnahme auf das Bauen und Wohnen in Nordwestdeutschland, 14.-20. Jh.*, Münster 2002.

Spooner, Frank, *Risk at Sea: Amsterdam Insurance and Maritime Europe, 1766–1786*, Cambridge 1983.

Sprandel, Rolf, "Die Interferenz von Gesellschaften und Genossenschaften im hansischen Handel," in Nils Jörn, Detlef Kattinger, and Horst Wernicke (eds.), *Genossenschaftliche Strukturen in der Hanse*, Cologne 1999, pp. 79–100.

Städtebuch Brandenburg und Berlin, ed. Evamaria Engel, Stuttgart 2000.

Stahel, Walter R., "The Role of Insurability and Insurance," *The Geneva Papers on Risk and Insurance* 28, 3 (2003), pp. 374–81.

Stahl, Karl Joseph, *Hadamar Stadt und Schloß. Eine Heimatgeschichte anläßlich der 650-Jahrfeier der Stadtrechteverleihung an die Stadt Hadamar 1974*, Hadamar 1974.

Statut des im Jahre 1818 gestifteten Vereins zur Rettung bei Feuers-Gefahr, Danzig 1819.

Steenweg, Helge, "Einbeck im Zeitalter der reformatorischen Bewegung," in *Geschichte der Stadt Einbeck*, ed. Horst Hülse on behalf of the Einbecker Geschichtsverein e.v., vol. 1, *Von den Anfängen bis zum Ende des 18. Jahrhunderts*, Einbeck 1991, pp. 135–54.

Stefani, Giovanni, *L'assicurazione a Venezia dalle origini alla fine della Serenissima*, Trieste 1956.

Steiger, Johann Anselm, "Ist es denn ein Wunder? Die aufgeklärte Wunderkritik. Oder: Von Spinoza zu Reimarus," in *500 Jahre Theologie in Hamburg*, ed. Johann Anselm Steiger, Berlin 2005.

Stein, Erwin, "Theoria cum praxi: Leibniz als technischer Erfinder," in Reydon, Thomas A.C., Heit, Helmut, Hoyningen-Huene, Paul (eds.), *Der universale Leibniz. Denker, Forscher, Erfinder*, Stuttgart 2009, pp. 155–183, 166–179.

Stein, Walter (ed.), *Akten zur Geschichte der Verfassung und Verwaltung der Stadt Köln im 14. Und 15. Jahrhundert*, 2 vols. Bonn 1893–95.

Steiner, Andreas, *Das nervöse Zeitalter. Der Begriff der Nervosität bei Laien und Ärzten in Deutschland und Österreich um 1900*, Zurich 1964.

Stern, P.J. and Wennerlind, C. (eds.), *Mercantilism reimagined. Political economy in early modern Britain and its empire*, Oxford 2014.

Stichweh, Rudolf, "Ungleichzeitigkeit in der Weltgesellschaft. Zur Unterscheidung von Tradition und Moderne," in *Die Weltgesellschaft. Soziologische Analysen*, Frankfurt am Main 2000, pp. 207–19.

Stine, Scott, "On the Medieval Climatic Anomaly," *Current Anthropology* 41, 4 (2000), pp. 627f.

Stobart, Jon, *Sugar & Spice. Grocers and Groceries in Provincial England, 1650–1830*, Oxford 2016.

Stöcker, Heinrich, *Krieg, Pest und Brand im Großenhainer Land*, Großenhain 1998.

Stoichita, Victor I., " 'Lochi di foco'—La città ardente nella pittura del Cinquecento," in Pfisterer, Ulrich and Seidel, Max (eds.), *Visuelle Topoi. Erfindung und tradiertes Wissen in den Künsten der italienischen Renaissance*, Munich 2003, pp. 439–51.

Stoichita, Victor I., " 'Lochi di foco.' La ville en flammes dans la peinture des XVIe et XVIIe siècles," in Dimiter Daphinoff and Edgar Marsch (eds.), *Das Feuer. Beiträge zu einem interdisziplinären Gespräch,* Freiburg in Switzerland 1998, pp. 91–116.

Stoob, Heinz and Johanek, Peter (eds.), *Deutsches Städtebuch* [Neufassung], vol. 1, *Schlesisches Städtebuch*, ed. Waldemar Grosch (1995); vol. 2, *Städtebuch Brandenburg und Berlin*, eds. Evamaria Engel, Lieselotte Enders, Gerd Heinrich and Winfried Schich, with Harald Engler (2000); vol. 3.2, *Städtebuch Hinterpommern*, eds. Peter Johanek and Franz-Joseph Post, with Thomas Tippach and Roland Lesniak, 2003.

Stoob, Heinz, "Grundrißbild und Entwicklung der Altstadt Hamm bis 1830," in Herbert Zink (ed.), *750 Jahre Stadt Hamm*, Hamm 1971.

Ströbele, Werner, "Der Stadtbrand von 1726—ein Ereignis schafft Bilder," in *Stadt—Bild—Geschichte. Reutlingen in Ansichten aus fünf Jahrhunderten. Katalog und Ausstellung*, Reutlingen 1990, pp. 49–56.

Stroup, Alice, *A Company of Scientists. Botany, Patronage, and Community at the Seventeenth-Century Parisian Royal Academy of Sciences*, Berkeley 1990.

Stuber, Martin, *Wälder für Generationen. Konzeptionen der Nachhaltigkeit im Kanton Bern (1850–1880)*, Vienna 2008.

Sturdy, David J., *Science and Social Status: The Members of the Academie des Sciences, 1666–1750*, Woodbridge 1995.

Suerbaum, Werner, *Vom antiken zum frühmittelalterlichen Staatsbegriff. Über Verwendung und Bedeutung von res publica, regnum, imperium und status von Cicero bis Jordanis*, Münster 1976.

Supple, Brian, *The Royal Exchange Assurance: A History of British Insurance, 1720–1970*, Cambridge 1970.

Sutherland, Ian, "John Graunt: A Tercentenary Tribute," *Journal of the Royal Statistical Society Series A* 126, 4 (1963), pp. 537–56.

Sutton, Peter C. (ed.), *Jan van der Heyden (1637–1712)*, New Haven 2006.

Swatos, William H. and Kaelber, Lutz (eds.), *The Protestant Ethic Turns 100: Essays on the Centenary of the Weber Thesis*, Boulder 2005.

Sydow, Jürgen, *Geschichte der Stadt Tübingen*, vol. 1, *Von den Anfängen bis zum Übergang an Württemberg 1342*, Tübingen 1974.

Tacke, Martin et al. (eds.), *"Der Mahler Ordnung und Gebräuch in Nürnberg": Die Nürnberger Maler(zunft)bücher ergänzt durch weitere Quellen, Genealogien und Viten des 16., 17. und 18. Jahrhunderts*, Munich 2001.

Tardif, Robert et al., "Last Millennium Reanalysis with an expanded proxy database and seasonal proxy modeling", *Climate of the Past* 15 (2019), pp. 1251–1273.

Tebeau, Mark, *Eating Smoke. Fire in Urban America, 1800–1950*, Baltimore 2003.

Temin, Peter and Voth, Hans-Joachim (eds.), *Prometheus Shackled: Goldsmith Banks and England's Financial Revolution after 1700*, Oxford 2013.

Temin, Peter and Voth, Hans-Joachim, "Private Borrowing during the Financial Revolution: Hoare's Bank and Its Customers, 1702–24," *Economic History Review* 61 (2008), pp. 541–64.

Tenenti, Alberto and Tucci, Ugo, *Storia di Venezia. Temi. Il mare*, Rome 1991, pp. 663–86.

Tenenti, Alberto, "I tassi assicurativi sulla piazza di Venezia: secc. XVI-XVII," *Studi Veneziani* 10 (1985), pp. 15–55.

Tenenti, Alberto, "Profitti assicurativi mediterranei nella seconda metà del cinquecento," *Atti e memorie della Società Savonese di Storia patria* 32/33 (1996/97), pp. 127–48.

Tenenti, Alberto, *Il prezo del rischio. L'assicurazione mediterranea vista da Ragusa (1563–1591)*, Jouvence 1985.

Tenenti, Alberto, *Naufrages, corsaires et assurances maritimes à Venise (1592–1609)*, Paris 1959.

Tennstedt, Florian, "Risikoabsicherung und Solidarität. Bismarck, Lohmann und die Konflikte um die gesetzliche Krankenversicherung in ihrer Entstehungsphase," in Herbert Obinger and Elmar Rieger (eds.), *Wohlfahrtsstaatlichkeit in entwickelten Demokratien. Herausforderungen, Reformen und Perspektiven*, Frankfurt am Main 2009, pp. 65–94.

The Friendly Society, Or, A Proposal of a New way or Method for Securing Houses from any Considerable Loss by Fire, by way of Subscription, and Mutuall Contribution, s.l. 1684.

The Gazetteer of Bombay City and Island, 3 vols., Bombay 1909.

Theurot, Jacky, " 'Au feu!' Les bourgs et les ville du comté de Bourgogne face à l'incendie (XIVe et XVe siècles)," in Lassus and Vion-Delphin (eds.), *Les hommes et le feu*, pp. 233–251.

Tigges, Michael, *Geschichte und Entwicklung der Versicherungsaufsicht*, Karlsruhe 1985.

Tinniswood, Adrian, *By Permission of Heaven. The story of the Great Fire of London*, London 2004.

Tinniswood, Adrian, *By Permission of Heaven: The Story of the Great Fire of London*, London 2003.

Tönnies, Ferdinand, *Gemeinschaft und Gesellschaft. Abhandlung des Communismus und des Socialismus als empirischer Culturformen* (orig. 1887), Darmstadt 2005.

Trebilcock, Clive, *The Phoenix Assurance and the Development of British Insurance*, vol. 1, *1782–1870*, Cambridge 1985.

Trempler, Jörg, *"Katastrophen." Ihre Entstehung aus dem Bild*, Berlin 2013.

Trinkaus, Charles "Antiquitas versus Modernitas: An Italian Humanist Polemic and its Resonance," *Journal of the History of Ideas* 48 (1987), pp. 11–21.

Trivellato, Francesca, *The Familiarity of Strangers: The Sephardic Diaspora, Livorno, and Cross-Cultural Trade in the Early Modern Period*, New Haven 2009.

Troitzsch, Ulrich, "Übersetzung, Bearbeitung, Plagiat, kritische Übertragung. Johann Beckmanns Beyträge zur Geschichte der Erfindungen in englischen und japanischen Ausgaben," in *Technik, Arbeit und Umwelt in der Geschichte. Günter Bayerl zum 60. Geburtstag*, ed. Torsten Meyer, Münster 2006, pp. 95–109.

Ugawa, Kaoru, "The Great Fire of Edo (Tokyo) in 1657", in Körner, *Stadtzerstörung*, vol. 1, pp. 213–238.

Ullmer, James H., "The Macroeconomic Thought of Nicholas Barbon," *Journal of the History of Economic Thought* 29 (2007), pp. 101–16.

Ullmer, James H., "William Petty, Nicholas Barbon, Multiples, and a New Hypothesis of Inevitable Discovery," *Journal of the History of Economic Thought* 31, 3 (2009), pp. 341–59.

Underdown, David, *Fire from Heaven. Life in an English Town in the Seventeenth Century*, New Haven 1992.

Ursinus, Michael, "Zur Diskussion um 'millet' im Osmanischen Reich," *Südost-Forschungen* 48 (1989), pp. 195–207.

Van Boom, Willem H., "Insuring vs. investing in litigation: a comparative legal history of litigation insurance and claim investment", *Comparative Legal History* 8, 1 (2020), pp. 2–26.

Varanini, Gian Maria, "Governo del territorio e raffigurazioni cartografiche. La Terraferma veneta tra Quattrocento e tardo Cinquecento," in Salgaro, *Sorte*, pp. 87–107.

Veckinchusen, Hildebrand, *Hildebrand Veckinchusen. Briefwechsel eines deutschen Kaufmanns im 15. Jahrhundert*, ed. Wilhelm Stieda, Leipzig 1921.

Ventzke, M., "Fürsten als Feuerbekämpfer: Handlungsmotive einer sich wandelnden Hofgesellschaft am Ende des 18. Jahrhunderts," in Id. (ed.) *Hofkultur und aufklärerische Reformen in Thüringen. Die Bedeutung des Hofes im späten 18. Jahrhundert*, Cologne 2002, pp. 223–35.

Verhandlungen und Schriften der Hamburgischen Gesellschaft zur Beförderung der Künste und nützlichen Gewerbe, 7 vols., Hamburg 1792–1807.

Vérin, Hélène, *La gloire des ingénieurs. L'intelligence technique du XVIe au XVIIIe siècle*, Paris 1993.
Vetter, Klaus, *Zwischen Dorf und Stadt. Die Mediatstädte des kurmärkischen Kreises Lebus*, Weimar 1996.
Vetter, Roland, *"Die ganze Stadt ist abgebrannt." Heidelbergs zweite Zerstörung im Pfälzischen Erbfolgekrieg 1693*, Karlsruhe 2009.
Victor, Peter A., *Indicators of Sustainable Development: Some Lessons from Capital Theory*, in Redclift, *Sustainability*, vol. 3, pp. 58–79.
Vier Revidirte Articuli, Der Boots=Leute Schlaven= Gelder Einnahm betreffend geplacidirt den 27 Januarii Anno 1679, Hamburg 1679.
Viner, Jacob, *International Trade and Economic Theory*, Glencoe 1952.
Virol, Michèle, *Vauban. De la gloire du roi au service de l'état*, Seyssel 2003.
Voigt, Christopher, *Der englische Deismus in Deutschland. Eine Studie zur Rezeption englisch-deistischer Literatur in deutschen Zeitschriften und Kompendien des 18. Jahrhunderts*, Tübingen 2003.
Völkel, Markus, "'Romanität'/'Germanität,'" in Wolfgang Schmale (ed.), *Kulturtransfer. Kulturelle Praxis im 16. Jahrhundert*, Innsbruck 2003, pp. 247–63.
Von der Ackerbürgerstadt zum Oberzentrum. 90 Jahre Haus- und Grundbesitzerverein Weiden und Umgebung, Weiden 1998.
Vovelle, Michel, Piété *baroque et déchristianisation en Provence au XVIIIe siècle. Les attitudes devant la mort d'après les clauses des testaments*, Paris 1973.
Vries, Jan de, *European Urbanization 1500–1800*, London 1984.
Wagner, Adolph, "Versicherungswesen," in Gustav Schönberg (ed.), *Handbuch der Politischen Ökonomie*, 3rd ed, vol. 2, Tübingen 1891, pp. 939–1026.
Wagner, Peter, *Modernity as Experience and Interpretation: A New Sociology of Modernity*, Cambridge 2008.
Wagner, Peter, *Theorizing Modernity: Inescapability and Attainability in Social Theory*, London 2001, pp. 128–33.
Wakefield, Andre, "Leibniz and the Wind Machines," *Osiris* 25 (2010), pp. 171–188.
Waligore, Joseph, "The Christian Deist Writings of Benjamin Franklin," *The Pennsylvania Magazine of History and Biography* 140, 1 (2016), pp. 7–29.
Walker, Matthew F., "The Limits of collaboration: Robert Hooke, Christopher Wren and the Designing of the Monument to the Great Fire of London," *Notes & Records of the Royal Society* 65 (2011), pp. 121–43.
Walker, Matthew, *Architects and Intellectual Culture in Post-Restoration England*, Oxford 2017.
Walter, Mack, *German home towns: community, state, and general estate, 1648–1871*, 2nd ed., intro. James J. Sheehan, Ithaca 1998.
Walther, François, *Katastrophen. Eine Kulturgeschichte vom 16. bis ins 21. Jahrhundert*, Stuttgart 2010.

Ward Fay, Peter, *The Opium War, 1840–1842*, 2nd ed., Chapel Hill 1997.

Warde, Paul, "*Sustainability*, Resources and the Destiny of States in German Cameralist Thought," in Forrester, Katrina and Smith, Sophie (eds.), *Nature, Action and the Future. Political Thought and the Environment*, Cambridge 2018, pp. 43–69.

Warnke, Martin, "Das Erdbeben von Lissabon 1755—eine Bewährung der Aufklärung," *Zeitschrift des Vereins für Hamburgische Geschichte* 95 (2009), pp. 1–22.

Weber, Josef, *Zur Geschichte der Stadt Elzach*, Elzach 1978.

Weber, Max,*Gesamtausgabe*, ed. Horst Baier, part 1: *Schriften und Reden*, vol. 10: *Zur Russischen Revolution von 1905*, Tübingen 1989.

Weber, Max, *Die protestantische Ethik I. Eine Aufsatzsammlung*, ed. Johannes Winckelmann, Gütersloh 1991.

Weber, Max, *Die protestantische Ethik II. Kritiken und Antikritiken*, ed. Johannes Winckelmann, 5th ed., Gütersloh 1987.

Weber, Max, *Gesamtausgabe*, part 1: *Schriften und Reden 1892–1899*, vol. 4.1, *Landarbeiterfrage, Nationalstaat und Volkswirtschaftspolitik*, Tübingen 1993.

Wehner, Christoph, *Die Versicherung der Atomgefahr: Risikopolitik, Sicherheitsproduktion und Expertise in der Bundesrepublik Deutschland und den USA 1945–1986*, Göttingen 2017.

Weinstein, Barbara, "Developing Inequality," *American Historical Review* 113 (2008), pp. 1–18.

Wellmann, Johann Christian, *Das von der göttlichen Regierung an denen Mordbrennern, welche in der Nacht zwischen den 19. und 20. May 1723 die Lebusische Vorstadt zu Franckfurt an der Oder* [...] *angesteckt bewiesene Denk-Mahl* [...], Frankfurt an der Oder 1725.

Wengenroth, Ulrich, "Innovationspolitik und Innovationsforschung," in Gerd Graßhoff and Rainer C. Schwinges (eds.), *Innovationskultur. Von der Wissenschaft zum Produkt*, Zürich 2008, pp. 61–77.

Wengenroth, Ulrich, "Vom Innovationssystem zur Innovationskultur. Perspektivwechsel in der Innovationsforschung," in Johannes Abele et al. (eds.), *Innovationskulturen und Fortschrittserwartungen im geteilten Deutschland*, Cologne 2001, pp. 23–32.

Wennerlind, Carl, *Casualties of Credit. The English Financial Revolution, 1620–1720*, Cambridge/Mass. 2011.

Wermiel, Sara E., *The Fireproof Building: Technology and Public Safety in the Nineteenth Century American City*, Baltimore 2000.

Werner, Heinrich, *Der Stadt Hamburgk Fewer-Ordnung/Wornach Dero Bürgere/Einwohnere/Vnterthanen vnd jedermänniglich daselbst in begebenen Fewersnöthen sich richten vnd verhalten sollen*, Hamburg 1637.

Werner, Theodor-Gustav, "Das kaufmännische Nachrichtenwesen im späten Mittelalter und in der frühen Neuzeit und sein Einfluss auf die Entstehung der handschriftlichen Zeitung." *Scripta Mercaturae* 2 (1975), pp. 3–51.

Wetter, Oliver and Pfister, Christian, "An underestimated record breaking event—why summer 1540 was likely warmer than 2003," *Climate of the Past* 9 (2013), pp. 41–56.

Wetter, Oliver et al., "The year-long unprecedented European heat and drought of 1540—a worst case," *Climatic change* 125 (2014), pp. 349–63.

Wheelock, Arthur K., "Accidents and Disasters," in Susan Donahue Kuretsky (ed.), *Time and Transformation in Seventeenth-Century Dutch Art*, New York 2005, pp. 73–82.

Whiston, J., *The causes of our present calamities in reference to the Trade of the Nation fully discovered* London 1696.

White, Sam, *The Climate of Rebellion in the Early Modern Ottoman Empire*, Cambridge 2011.

Whitmer, Kelly J., "Projects and Pedagogical Expectations: Inside P.J. Marperger's 'Golden Clover Leaf' (*Trifolium*), 1700–1730, in: Notes and Records of the Royal Society 72, 2 (2018), pp. 139–157.

Wigelsworth, Jeffrey R., " 'God always acts suitable to his character, as a wise and good being': Thomas Chubb and Thomas Morgan on Miracles and Providence," in Hudson, Wayne, Lucci, Diego and Wigelsworth, Jeffrey R. (eds.), *Atheism and Deism Revalued. Heterodox Religious Identities in Britain, 1650–1800*, Aldershot 2014, pp. 157–72.

Wigelsworth, Jeffrey R., *Deism in Enlightenment England: The Politics, and Newtonian Public Science*, Manchester 2009.

Wildeboer, Hans, "De ontwikkeling van de brandspuit in de zeventiende en achttiende eeuw," in *Jaarboek voor de Geschiedenis van Bedrijf en Techniek* 10 (1993), pp. 9–41.

Wilke, Jürgen, "Von der Lotterie zur patriotischen Opfergabe. Vorstufen und Erscheinungsformen von Spendenkampagnen in der älteren Presse," in *Massenmedien und Spendenkampagnen. Vom 17. Jahrhundert bis in die Gegenwart*, Cologne 2008, pp. 15–72.

Williams, Elgin, "Nicholas Barbon, An Early Economic Realist," *Southern Economic Journal* 11 (1944), pp. 45–55.

Wilson, Peter H., "Social Militarization in Eighteenth-Century Germany," *German History* 18 (2000), pp. 1–39.

Wolff, Eberhard, "Der 'willkommene Würgeengel': Verstehende Innenperspektive und 'genaue' Quelleninterpretation—am Beispiel des erwünschten Kindertods in den Anfängen der Pockenschutzimpfung," in Martin Dinges (ed.), *Neue Wege in der Seuchengeschichte*, Stuttgart 1995, pp. 105–41.

Wolff, Eberhard, *Einschneidende Maßnahmen: Pockenschutzimpfung und traditionale Gesellschaft im Württemberg des frühen 19. Jahrhunderts*, Stuttgart 1998.

Wolff, Larry, *Inventing Eastern Europe: The Map of Civilization on the Mind of the Enlightenment*, Stanford 1994.

Wolkan, Rudolf, "Studien zur Reformationsgeschichte Nordböhmens," *Jahrbuch der Gesellschaft für die Geschichte des Protestantismus in Österreich* 4, 2 (1883), pp. 67–95.

Wright, Conrad Edick, *The Transformation of Charity in Postrevolutionary New England*, Boston 1992.

Yamamoto, Zipporah Lax, *After 9/11: Transformations of Memory into History*, Diss. UCLA 2011.

Yeh, Catherine Vance, "Representing the City: Shanghai and Its Maps," in David Faure and Tao Tao Liu (eds.), *Town and Country in China: Identity and Perception*, Basingstoke 2002, pp. 166–202.

Yoneyama, Takau, "Tsuneta Yano, Founder of the First Mutual Company in Japan," in Robin Pearson and Takau Yoneyama (eds.), *Corporate Forms and Organizational Choice in International Insurance*, Oxford 2015, pp. 29–46.

Zachlod, Christian M., "Katastrophenhilfe, Kreditwesen und Konjunktur. Zur Wirtschafts- und Sozialgeschichte der staatlichen Feuerversicherung in der Mitte des 18. Jahrhunderts am Beispiel des Hochstifts Hildesheim," in Gerhard Lauer and Thorsten Unger (eds.), *Das Erdbeben von Lissabon und der Katastrophendiskurs im 18. Jahrhundert*, Göttingen 2008, pp. 553–74.

Zappavigna, M., *Tacit Knowledge and Spoken Discourse*, London 2013.

Zelle, Carsten, *"Angenehmes Grauen." Literaturhistorische Beiträge zur Ästhetik des Schrecklichen im Achtzehnten Jahrhundert*, Hamburg 1987.

Ziegler, Hannes, "Im Schlaf der Sicherheit. Die Endlichkeit des Religionsfriedens und die Zukunft des Reiches (1606–1618)," in: Christoph Kampmann, Angela Marciniak, Wencke Meteling (eds.), *'Security turns its eye exclusively to the future'. Zum Verhältnis von Sicherheit und Zukunft in der Geschichte*, Baden-Baden 2018, pp. 189–206.

Zillhardt, Gerd, *Der Dreißigjährige Krieg in zeitgenössischer Darstellung. Hans Heberles "Zeytregister" (1618–1672). Aufzeichnungen aus dem Ulmer Territorium*, Ulm 1975.

Zimmermann, Walther, Berger, Hugo, Klocke, Friedrich von, and Bauermann, Johannes (eds.), *Handbuch der historischen Stätten Deutschlands*, vol. 3 *Nordrhein-Westfalen*, Stuttgart 1963.

Zwierlein, Cornel "Sicherheit durch Versicherung. Ein frühneuzeitliches Erfolgsmodell," in: Christoph Kampmannn and Ulrich Niggemann (eds.), *Sicherheit in der Frühen Neuzeit. Norm—Praxis—Repräsentation*, Cologne 2013, pp. 381–99.

Zwierlein, Cornel, "Consociatio," in Francesco Ingravalle and Corrado Malandrino (eds.), *Il lessico della "Politica" di Johannes Althusius. L'arte della simbiosi santa, giusta, vantaggiosa e felice*, Florence 2005, pp. 277–93.

Zwierlein, Cornel, "Das Glück des Bürgers. Eine begriffsgeschichtliche Skizze zur optimistischen Grundmentalität der Aufklärung," in Hans-Edwin Friedrich, Fotis Jannidis, and Marianne Willems (eds.), *Bürgerlichkeit im 18. Jahrhundert*, Tübingen 2006, pp. 71–113.

Zwierlein, Cornel, "Die Auswirkungen von spatial turn und Kulturtransferheuristiken auf das Epochenkonzept 'Frühe Neuzeit' (exemplifiziert anhand der Transfers des Versicherungsprinzips)," in Michael North (ed.), *Kultureller Austausch. Bilanzen und Perspektiven der Frühneuzeitforschung*, Vienna 2009, pp. 43–67.

Zwierlein, Cornel, "Durée, stabilité et grandeur urbaine: de la cité humaniste à la métropole moderne," in Élisabeth Crouzet-Pavan, Denis Crouzet, and Philippe Desan (eds.), *Cités humanistes, cités politiques* (1400–1600,), Paris 2014, pp. 237–51.

Zwierlein, Cornel, "Frühe Neuzeit, *multiple modernities*, globale Sattelzeit," in *Frühe Neue Zeiten. Zeitwissen zwischen Reformation und Revolution*, ed. Achim Landwehr, Bielefeld 2012, pp. 389–405.

Zwierlein, Cornel, "Gegenwartshorizonte im Mittelalter: Der Nachrichtenbrief vom Pergament- zum Papierzeitalter," *Jahrbuch für Kommunikationsgeschichte* 12 (2010), pp. 1–60.

Zwierlein, Cornel, "Grenzen der Versicherbarkeit als Epochenindikatoren? Von der europäischen Sattelzeit zur Globalisierung des 19. Jahrhunderts," *Geschichte & Gesellschaft* 38 (2012), pp. 423–52.

Zwierlein, Cornel, "Insurances as Part of *Human Security*, their Timescapes, and Spatiality," *Historical Social Research* 35, 4 (2010), pp. 253–74.

Zwierlein, Cornel, "Katastrophe und Prävention—Leibniz, Brandgefahr und Versicherung," in Friedrich Beiderbeck et al. (eds.), *Umwelt und Weltgestaltung. Leibniz' politisches Denken in seiner Zeit*, Göttingen 2015, pp. 433–62.

Zwierlein, Cornel, "Klimageschichte und Kulturgeschichte der Frühen Neuzeit. Zum Problem des interdisziplinären Dialogs," in Philip Ajouri, Katja Mellmann, and Christoph Rauen (eds.), *Empirie in der Literaturwissenschaft*, Münster 2013, pp. 331–51.

Zwierlein, Cornel, "Perceiving Urban Fire Regimes in Europe and China (1830 to 1870): British Fire Insurance Businesses and the Sudden Challenge of Globalisation," in Gerrit J. Schenk (ed.), *Historical Disaster Experiences*, Heidelberg 2017, pp. 327–52.

Zwierlein, Cornel, "Pluralisierung und Autorität. Tentative Überlegungen zur Herkunft des Ansatzes und zum Vergleich mit gängigen Großerzählungen," in *Pluralisierungen. Konzepte zur Erfassung der Frühen Neuzeit*, eds. Jan-Dirk Müller, Wulf Oesterreicher and Friedrich Vollhardt, Berlin 2010, pp. 3–30.

Zwierlein, Cornel, "Reformation als Rechtsreform. Bucers Hermeneutik der lex Dei und sein humanistischer Zugriff auf das römische Recht," in Christoph Strohm (ed.), *Martin Bucer und das Recht. Beiträge zum internationalen Symposium vom 1. bis 3. März 2001 in der Johannes a Lasco Bibliothek Emden*, Geneva 2002, pp. 29–82.

Zwierlein, Cornel, "Return to Premodern Times?—Contemporary Security Studies, the Early Modern Holy Roman Empire and Coping with Achronies," *German Studies Review* 38, 2 (2015), pp. 373–92.

Zwierlein, Cornel, "*Se-curare, sine cura, se-curitas, assecuratio*: Innovationen der Sicherheitsproduktion in der Renaissance," in Melville, Gert, Vogt-Spira, Gregor, and Breitenstein, Mirko (eds.), *Sorge*, Cologne 2015, pp. 109–35.

Zwierlein, Cornel, "Security, Nature and Mercantilism in the Early British Empire," *European Journal of Security Research* 3 (2018), pp. 15–34.

Zwierlein, Cornel, "Sicherheitsgeschichte. Ein neues Feld der Geschichtswissenschaften," *Geschichte & Gesellschaft* 38 (2012), pp. 365–86.

Zwierlein, Cornel, "Tod und Leben. Das Reichskammergericht und die Lebensversicherung im 18. Jahrhundert," in *Kulturen des Risikos im Mittelalter und in der Frühen Neuzeit*, ed. Benjamin Scheller, Berlin 2019, pp. 147–173.

Zwierlein, Cornel, "Vom Londoner Brand 1666 bis 9/11. Feuergefahr und Feuerexperten seit dem 17. Jahrhundert," *Geschichte in Wissenschaft und Unterricht* 67, 11/12 (2016), pp. 711–730.

Zwierlein, Cornel, *Discorso und Lex Dei. Die Entstehung neuer Denkrahmen im 16. Jahrhundert und die Wahrnehmung der französischen Religionskriege in Italien und Deutschland*, Göttingen 2006.

Zwierlein, Cornel, "Orient contra China: Eusèbe Renaudot's Vision of World History (c. 1700)," *Journal of the History of Ideas* 81, 1 (2020), pp. 23–44.

Zwierlein, Cornel, "Storia della sicurezza in Europa/Storia europa della sicurezza", *De Europa* 3, 1 (2020), pp. 19–35 (DOI: http://dx.doi.org/10.13135/2611-853X/4435).

Zwierlein, Cornel, Graf, Rüdiger and Ressel, Magnus (eds.), *The Production of Human Security in Premodern and Contemporary History*, Cologne 2010.

Zwierlein, Cornel, *Imperial Unknowns. The French and British in the Mediterranean, 1650–1750*, Cambridge 2016.

Index Locorum

Index Nominum

Index Rerum